A CLASS-BOOK OF BOTANY

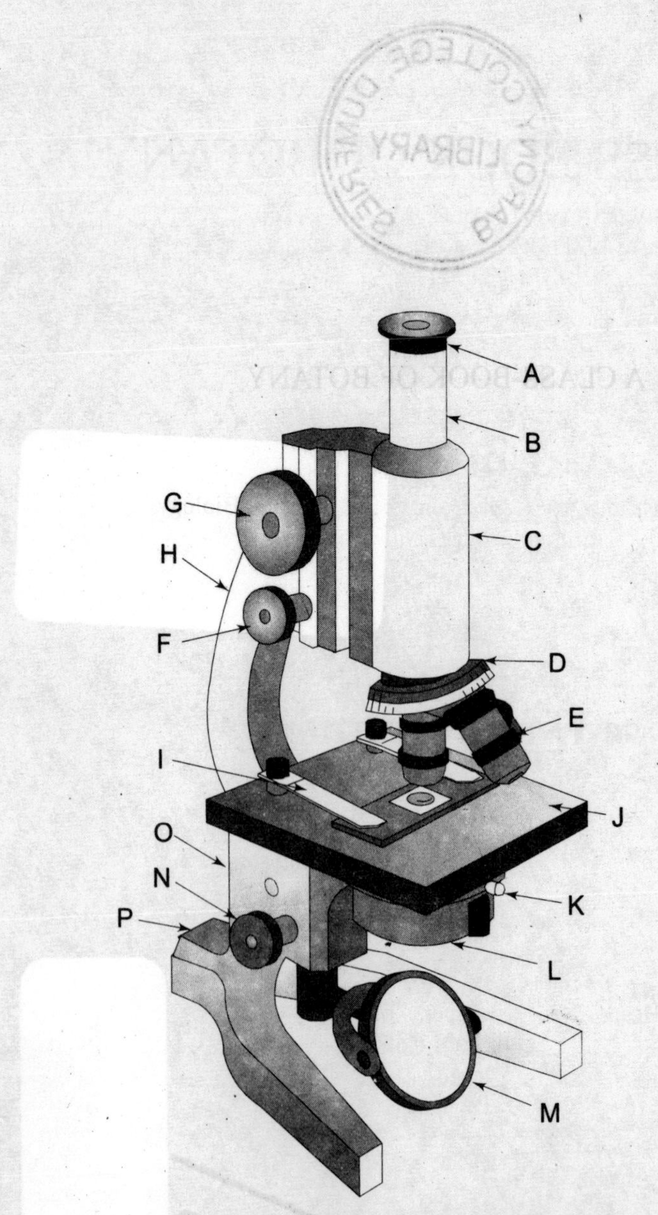

COMPOUND MICROSCOPE

A, eyepiece; *B*, draw tube; *C*, body tube; *D*, nosepiece (revolving); *E*, objective; *F*, fine adjustment; *G*, coarse adjustment; *H*, arm; *I*, clip; *J*, stage; *K*, condenser; *L*, iris-diaphragm; *M*, mirror; *N*, inclination joint; *O*, pillar; and *P*, foot (horseshoe-shaped). Of these, *A*, *E*, *K* and *M* constitute the *optical parts* and the rest constitute the *mechanical parts*.

A CLASS-BOOK OF BOTANY

FOR PRE-UNIVERSITY, INTERMEDIATE, PRE-MEDICAL, HIGHER SECONDARY, AND ALL-INDIA SCHOOL CERTIFICATE STUDENTS

A. C. Dutta, M.Sc.
Formerly Head of the Departments of Botany and Biology,
Cotton College, Guwahati.

SEVENTEENTH EDITION

Revised by
T. C. Dutta
Formerly Head of the Departments of Botany and Biology,
St. Anthony's College,
Satri Bari Road, Guwahati.

YMCA Library Building, Jai Singh Road, New Delhi 110001

Oxford University Press is a department of the University of Oxford.
It furthers the University's objective of excellence in research, scholarship
and education by publishing worldwide in

Oxford New York

Auckland Bangkok Buenos Aires Cape Town Chennai
Dar es Salaam Delhi Hong Kong Istanbul Karachi Kolkata
Kuala Lumpur Madrid Melbourne Mexico City Mumbai Nairobi
São Paulo Shanghai Taipei Tokyo Toronto

Published in India
by Oxford University Press

First published 1980
Revised edition 2000
Sixth impression 2004

ISBN 0-19-565307-6

Printed in India by Ram Printograph, Delhi 110051
and published by Manzar Khan, Oxford University Press
YMCA Library Building, Jai Singh Road, New Delhi 110001

Preface to the Seventeenth Edition

The seventeenth edition of *A Class-book of Botany* has been brought out with a thorough revision of the whole text, including addition of several new topics and rewriting many portions of the text on the basis of recent researches. All the topics have been discussed in an easily understandable way and with proper illustrations. In the introductory chapter, along with the topic on Use of Plants, the importance of forests has also been included and the basic idea about social forestry given, and the topics like origin of life and sources and use of energy, etc. have been elaborated. In this book morphology has been dealt with, for the convenience of the students, including some necessary relevant portions of embryology, instead of dealing embryology as a separate branch. Further, on the basis of recent researches in biochemistry, the mechanisms of photosynthesis, protein synthesis and that of respiration have been rewritten and topics like biofertilizers and others have been included in physiology. Other new additions in the text are a chapter on ecosystem in 'Ecology', and the different techniques of plant breeding in 'Economic Botany'.

In Cryptogams, the structure and reproduction of Saccharomyces (Yeast) fungus has been dealt anew on the basis of the studies made under the electron microscope. Some old diagrams have also been changed and some new diagrams added. Besides these, some other new additions are differences between Bacteria and Fungi, Bryophyta and Pteridophyta, Gymnosperms and Angiosperms, etc. Above all a completely new branch (Part IX) has been added to the text, which includes bio-technology, tissue culture, genetic engineering, etc., considering the recent advances made in life-sciences.

Likewise the readers will also notice many necessary changes and new additions in this book. Thus this thoroughly revised, enlarged and improved, seventeenth edition of *A Class-book of Botany* will fully meet the needs of students of different disciplines, thereby benefiting them. Any suggestions for the improvement of this book will be highly appreciated.

Satri Bari Road
Guwahti

T. C. D

Preface to the First Edition

This book, though intended primarily for the use of Intermediate and Medical students of Calcutta University and of the Dacca Board, covers somewhat wider grounds, and students of other universities, following the same or a slightly higher standard in the curricula, will find the book useful and instructive. Although the generally accepted methods of treatment have been followed, attention may be drawn to certain special features:

1. The text has been illustrated with numerous simple figures and explanatory diagrams drawn by the author himself, in most cases directly from objects which are typical and easily available. The figures and diagrams have been drawn with a view to provide correct and easy appreciation.

2. An attempt has been made to familiarize the students with the meanings of Latin and Greek prefixes and suffixes, and to trace the technical and scientific terms to their respective Latin and Greek roots. This will enable the students to master the subject of terminology more easily.

3. In many cases, more than one example has been given to illustrate a particular form or feature. A large number of English and vernacular names have been introduced to suit the convenience of students. Latin names have been followed by vernacular or English equivalents, or often by both.

The author takes this opportunity of thanking Dr D. Thomson, M.A., B.Sc., Ph.D., I.E.S., Principal, Cotton College, Guwahati, for the encouragement received from him during the preparation of this book. For some of the drawings the author expresses his thanks to his pupils, Madhab Chandra Das and Gour Mohan Das.

Cotton College
Guwahati, Assam
June, 1929

A. C. D

Contents

PART VI GYMNOSPERMS

PART VII ANGIOSPERMS

PART VIII EVOLUTION AND GENETICS

PART IX RECENT ADVANCES IN BOTANY

PART X ECONOMIC BOTANY

Introduction

Biology: botany and zoology. The science that deals with the study of living objects goes by the general name of **biology** (*bios,* life; *logos,* discourse or study). Since both animals and plants are living, biology includes a study of both. Biology is, therefore, divided into two branches: **botany** (*botane,* plant) which deals with plants, and **zoology** (*zoon,* animal) which deals with animals.

Origin and Continuity of Life. We do not know what life really is. It is something mysterious, and we are not in a position to define it. The origin of life is equally mysterious. There is no doubt, however, that life came into existence many millions of years ago in water (sea) as a droplet of protoplasm (*protos,* first; *plasma,* form), i.e. first-formed living substances. But how it did so has remained a mystery. Evidently, divergent views have been expressed on this point from the time of Aristotle (384–322 B.C.), the great Greek philosopher, to the present day. The main views in this respect are as follows: (a) Life was created on the earth by Divine Power. This view, strong though for several centuries, had a background in all religions but had no experimental basis. (b) Spontaneous generation of life from non-living or inorganic materials as a result of certain chemical and physical changes in them under special circumstances. This is called *abiogenesis* (life from non-life). (c) The modern view, based on advanced knowledge of chemistry and physics, is that a soup of organic molecules of amino-acids, nucleic acids and other organic compounds (now known to be constituents of protoplasm) appeared first under special physico-chemical conditions. It may be reasonably assumed that the primitive earth was surrounded by a gaseous atmosphere of ammonia (NH_3), methane (CH_4), water vapour (H_2O) and hydrogen (H_2). Under the influence of ultra-violet rays and violent electric discharges as a source of energy, they reacted with each other in a co-ordinated manner, leading to the synthesis of proteins (combination of amino-acids), nucleic acids (complex phosphorous compounds), fats, sugars and other essential compounds, and acquired for the first time the properties of life, possibly in the form of certain aquatic bacteria. Soviet biochemist, Alexander Ivanovich Oparin (1894) speculated that life probably originated on this earth in this manner.

Professor Harold C. Urey, an Amercian chemist also became interested in Oparin's speculation and started investigating the possibility of origin of life in the way Oparin had thought of. Later, his student Stanley Lloyd Miller (1953), another American chemist, conducted some experiments in the laboratory to verify Oparin's hypothesis. The results they obtained strongly supported Oparin's hypothesis. In his experiments, Stanley Miller used an apparatus, commonly called Urey–Miller apparatus in which he allowed ammonia, methane, hydrogen gas and water vapour to pass near an electric spark. The electric discharge (like the lightning of the ancient earth) brought about the chemical reactions between the gaseous substances and the latter condensed into a hot thin soup of simple amino acids. These experiments, therefore, give strong support to Oparin's hypothesis. This, however, does not explain how life was infused into the soup of organic molecules. The next step is evolution, i.e. once life (protoplasm) came into existence it has gradually given rise to newer and more complex forms of plants and animals in successive stages through many millions of years. Life is thus one continuous flow from the beginning to the present day in divergent lines. This is called *biogenesis* (life from life). Although forms of life have changed, protoplasm has remained the same in both plants and animals. Protoplasm is not formed afresh in nature nor can it be created *in vitro,* and therefore, no new life comes into being.

Scope of Botany. The subject of Botany deals with the study of plants from many points of view. This science investigates the internal and external structures of all kinds of plants from the simplest to the most complex forms; their diverse functions, specially with regard to the manufacture of food required for both plants and animals including human beings, mode of respiration, various kinds of movements exhibited by them; their modes of reproduction; conduction of water and food through the plant body, etc.; their adaptations to the diverse conditions of the environment; their distribution in space and time;

their life-history, relationship with other plants and classification into natural groups; their evolution from the lower and simpler forms to the higher and more complex ones; the laws of heredity (inheritance of characters by the offspring); the varied uses that plants may be put to and the different methods that can be adapted to improve plants in the direction of better quality and higher yield, and even to produce newer types of plants for better uses by mankind.

Importance of Green Plants. Green plants purify the atmosphere by absorbing carbon dioxide gas from it and releasing from their bodies (by the breaking-down of water) an almost equal volume of pure oxygen into it; and they prepare food such as starch, the chief constituent of rice, wheat, potato, etc., from carbon dioxide obtained from the air, and water and inorganic salts obtained from the soil. Both these functions are the monopoly of green plants, and are performed by certain minute green bodies or plastids (see Fig. 1) called chloroplasts (*chloros*, green) of the leaf during the daytime, sunlight being the source of *energy*. Animals, being devoid of them, have no such power. It is evident, therefore, that animals including human beings are deeply indebted to plants for these basic needs, viz., oxygen for respiration and food for nutrition. It may thus be rightly asserted that *the existence of man would be unthinkable and as a matter of fact impossible without plants*.

Uses of Plants. *Importance of Forests:* Forests are an important renewable natural resource. In India, about 22.7% of the total land area is now occupied by forests. Forests play an important role in making the survival of man and animal possible on this earth. Further, the economy of a country also depends on its forest wealth. In general, forests play the following important role:

1. Forests provide natural habitat and food for wild life.
2. Forests provide wood which is used as fuel and also as raw material for use in various industries such as pulp and paper, etc. Commonly, wood is used in making furniture, packing boxes, houses, boats, matchsticks, doors, windows, various house-hold goods, etc. Minor products like bamboo, cane, gums, resins, tannis, fibres, medicines, etc. are also obtained from forests.
3. Forests purify the atmosphere, supplying oxygen gas during photosynthesis.
4. Forest tree-roots bind the soil to check soil erosion and also help in retaining water in the soil.

5. Forests cool the atmosphere giving off water vapour during transpiration.
6. Forests reduce earth's temperature by absorbing carbon dioxide from the atmosphere.
7. Forests influence the local rain fall.
8. Forests are connected with our culture and civilization and provide a sanctuary for the modern city.
9. Forests provide food such as various kinds of roots, leaves, fruits, etc. and medicare for tribal people as well as poor villagers who depend on forest wealth for their livelihood.
10. Forests also serve as gene reserve of important and rare species.

Social Forestry. Social forestry, which started with National Commission on Agriculture (NCA, 1976), aims at ensuring ecological, economic and social security particularly to the people of rural areas. Social forestry involves tree plantation on community lands and waste lands for the mixed production of wood, fodder, grasses, bamboos, cane, fruits, etc. for cottage industry and self-utilisation thereby making the people of rural areas self dependent. At the same time this programme also involves tree plantation around agricultural areas, road sides, sides of the canals, etc. for controlling soil erosion.

Plants have an almost endless variety of uses for human beings. The primary needs of man are of course threefold: food, clothing and shelter. All these are extensively furnished by the plant kingdom. The most essential needs of man is, no doubt, food. This food primarily comes from plants in the form of cereals (rice, wheat, maize, barley, etc.) and millets (smaller grains) supplemented by pulses, vegetables, fruits, vegetable oils, etc. Dessert fruits and green leafy vegetables are rich sources of natural vitamins and certain essential minerals required for proper nutrition of the body. Sugar as a universal sweetener, and various spices as flavouring and seasoning materials are also products of plants. For clothing again plants are indispensable sources of fibres, coarse and fine, used for manufacture of various types of garments. Then again shelter from sun and rain, and protection against enemies have been sought after from time immemorial. In this respect the value of house-building materials such as timber, wood, bamboo, cane, reed, thatch grass, etc., is inestimable. With advance in knowledge man has tried to tap a variety of other plants for his comforts and many uses. Mention in this connection may be made of medicinal (drug) plants, antibiotics (products of certain fungi and bacteria), beverages (tea, coffee and cocoa) as universal drinks, industrial fibres and oils, wood for furniture, boat-building, bridge construction, etc., cork, rubber, resin, turpentine, coal, firewood and a host of other products. The aesthetic value of ornamental plants used for decorating houses and gardens, and avenue trees grown for shade and beauty cannot be underestimated. As a matter of fact plants have formed an indispensable item for human existence and welfare.

Sources and Uses of Energy. Energy (i.e. ability to work) is essential for various vital activities of the cell. Energy exists in different forms such as potential energy, kinetic energy, thermal energy, chemical energy, radiant energy, etc. According to the first law of thermodynamics one form of energy can be transformed into another form but it cannot be created or destroyed. Basically energy changes accompany different physical and chemical processes, taking place within the biological system. The study of the energy transformation in the living system is called bioenergetic. The second law of thermodynamics states that during the change of energy from one form to another, a part of it gets lost (released) in the form of heat energy. In this way, after transformation the capacity of energy to perform work is gradually decreased. The chemical energy which is bound in various organic substances (food) and is available for different life processes or functions is known as bioenergy. In fact life is correlated with *energy-receiving* processes, as in the synthesis of various organic products such as carbohydrates, proteins, fats, nucleic acids, ATP, etc., and *energy-releasing* processes, as in the breakdown of these organic compounds, more particularly glucose, by oxidation. The reactions depend on the transfer of chemical energy from one compound to another. These two processes, viz. receiving energy and releasing energy, are simultaneously going on in the living cells, and are responsible for sustaining life. We cannot thus think of life without energy. The main sources of energy for the living world are as follows:

(a) **Light energy** of the sun is the ultimate source of energy, but it is only the green pigment (chlorophyll) of the green cells of plants that entraps light energy of the sun and converts it to chemical energy. The chemical energy, now available to the chloroplasts, initiates a very important food-manufacturing process called photosynthesis (*photos*, light; *synthesis*, building up) in the presence of light, and drives a series of chemical reactions, leading finally to the synthesis of glucose, starch and other carbohydrates, evidently now laden with potential chemical energy. At the initial stage of the process a part of the chemical energy is transferred to ADP which then becomes energy-rich ATP (see below). (b) **Glucose** is a energy-rich compound. It is the main source of energy supply for all metabolic work in both plant cells and animal cells. Glucose is also the main source of carbon. By oxidation of glucose—a normal and constant feature in all living cells—the potential energy stored in it is released in the form of active energy. The released energy is not, however, *directly* utilized for any metabolic work but most of it is transferred from ADP to ATP for *direct action*. A part of this energy is, however, lost in the form of heat during the transfer. (c) **ATP** (adenosine triphosphate), an active energy-rich phosphate compound, present in all living cells, directly provides the necessary chemical energy for all activities in them, including all biochemical reactions. All this, as stated before, requires the transfer of an energy-rich molecule of one compound to a molecule of another compound. This energy-transfer is directly linked with the transfer of an active phosphate bond of ATP, and ATP becomes ADP (adenosine diphosphate). Actually ADP receives an inorganic phosphate bond + energy (the latter mainly from oxidation of glucose), and ADP becomes ATP. This inter-conversion goes on continually in the living cell. [ADP + nP $\rightleftarrows$ ATP] Of course, each reaction is controlled by a specific enzyme. It may be noted that one molecule of glucose on complete oxidation yields 38 molecules of ATP. ATP is mostly formed in the mitochondria and only partly in the cytoplasm. (d) **Food** supplies necessary materials for growth and is an important source of energy. By oxidation of various food substances such as carbohydrates (particularly glucose), proteins and fats the potential chemical energy stored in them is released to ADP which becomes ATP to act in the usual way. Food is initially formed by green plants, and animals including human beings make use of it. Thus there is a flow of energy from sunlight to green plants to animals, and ultimately it is released from their bodies on death by the action of saprophytic bacteria and fungi.

Cells. Cells are the structural units of which the body of the plant or animal is composed. When the cell was first discovered by Robert Hooke

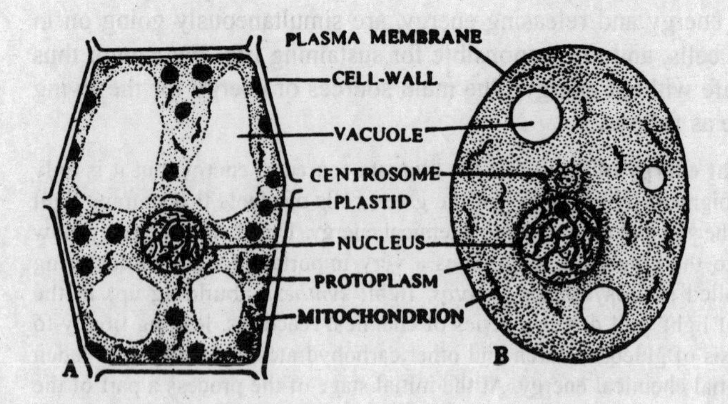

Figure I. A plant cell and an animal cell.

in 1665 in a thin slice of bottle cork, it was regarded as a mere microscopic chamber bounded by a distinct wall—cell-wall. Much later, however, about the year 1838–39, Schleiden—a German botanist—and Schwann—a German zoologist—discovered for the first time that a living substance, i.e. **protoplasm**, filled up the cell. A tiny spherical body, i.e. the **nucleus**, was also found lying embedded in the protoplasm. With rapid improvement of the microscope, attention was focused on these two bodies and their functions soon came to be known. It was soon recognized that the protoplasm and the nucleus were the most important parts of the cell, and the cell-wall a mere by-product of the protoplasm and a structure of secondary importance. On the above basis *the cell is defined as a unit or independent mass of protoplasm with a nucleus in it, enveloped by a distinct cell-wall in the case of a plant but only a thin membrane (plasma membrane) in the case of* an animal. The whole body of the plant or the animal is made of such cells. Cells, when young, are commonly spherical or oval in shape, but as they grow they assume different shapes and perform different functions.

Tissues. In pursuance of a particular function cells *similar in shape and size and having the same origin combine into a bigger unit or group called the tissue.* In the simpler organisms all the functions are performed by a single cell. But in complex forms of plants and animals with differentiated organs there is always a division of labour, i.e. distribution of functions, one group of cells performing one function and another group another function. Each such group of cells is a tissue. There are different kinds of tissues in higher plants and higher animals performing different and distinct functions. Tissues again combine and give rise to tissue systems.

Protoplasm. Protoplasm is the first-formed living substance and is very delicate and complicated. It is the only substance that is endowed with life and is the same in both animals and plants. All vital functions such as nutrition, growth, respiration, reproduction, etc., are performed by it. As the protoplasm dies the cell ceases to perform any of these functions. It is thus described as the *physical basis of life.*

Physical Nature of Protoplasm. Protoplasm is a transparent, foamy or granular, slimy, semi-fluid substance, somewhat like the white of an egg. It is never homogeneous but contains granules of varying shapes and sizes, and can be seen under the microscope. Although often semi-fluid it may be fluid or viscous. It completely fills up the cavity of the young cell, but in a mature cell one or more cavities, called **vacuoles**, appear, filled with water (see Fig. I). In its *active* state protoplasm remains saturated with water, containing 75–90% of it. With decreasing water content its vital activity diminishes and gradually comes to a standstill, as in dry seeds. Protoplasm coagulates on heating, and when killed it loses its transparency.

Protoplasm responds to the action of external stimuli such as the prick of a needle or pin, an electric shock, application of particular chemicals, sudden variation of temperature or of light, etc. On stimulation the protoplasm contracts but expands again when the stimulating agent is removed. This response to stimuli is an inherent power of protoplasm. Protoplasm is semi-permeable in nature, i.e. it allows only certain substances and not all to enter its body. This property is, however, lost when the protoplasm is killed. Under normal conditions the protoplasm of a living cell is in a state of motion which can be seen under a microscope.

Chemical Nature of Protoplasm. Chemically, protoplasm is a highly complex mixture of a variety of chemical substances of which proteins are the chief. All plant proteins are composed of carbon, hydrogen, oxygen, nitrogen, and sulphur, and sometimes also phosphorus. A complex type of protein called *nucleoprotein* is constant in the cytoplasm and the nucleus. Nucleoprotein is composed of phosphorus-containing nucleic acids (see p. 163) and certain specific types of proteins. The exact chemical composition of the living protoplasm, however, cannot be determined because any attempt to analyse it kills it outright with some unknown changes in it. Besides, it undergoes continual changes and its composition is not, therefore, constant. Further, it is not possible to get the protoplasm in a pure state free from foreign bodies. Analysis of the dead protoplasm reveals a long list of elements present in it. Of these oxygen (O), carbon (C), hydrogen (H) and nitrogen (N) are most abundant. Other elements present in smaller quantities are chlorine (Cl), sulphur (S), phosphorus (P), silicon (Si), calcium (Ca), magnesium (Mg), potassium (K), iron (Fe) and sodium (Na); still others are present in mere traces and these are called 'trace' elements. Active protoplasm contains a high percentage of water—usually varying from 75% to 90%. Leaving out this water, the solid matter of the protoplasm contains the following: proteins—40–60%; fats—12–14%; carbo-hydrates—12–14%; and inorganic salts—5–7%.

Characteristics of Living Objects. Life is something mysterious and we are not in a position to define it. All living objects have, however, certain characteristics by which they can be distinguished from the non-living. These are as follows:

(1) **Life-cycle.** All living objects follow a definite life-cycle of birth, growth, reproduction, old age and death. The animal or the plant is born, and gradually it grows into its characteristic form and size. In due course it reproduces to maintain the continuity of the species and also to multiply in number. Ultimately the organism attains old age and dies.

(2) **Protoplasm.** Life cannot exist without protoplasm. It is the actual living substance in both plants and animals, and it is, as Huxley defined it, the physical basis of life. It performs all the vital functions; it shows various kinds of movement and is sensitive to all kinds of stimuli such as light, temperature, chemical substances, electric shock, etc. Protoplasm is a very delicate and complicated substance.

(3) **Cellular Structure.** The whole body of the plant or the animal is composed of cells. A cell is a unit mass of protoplasm, with a nucleus in it, surrounded by a distinct cell-wall in the case of a plant and only a thin delicate membrane (plasma membrane) in the case of an animal (see Fig. I). The cellular structure, as described above, is the characteristic feature of every living organism.

(4) **Respiration.** Respiration is a sign of life. All living beings—plants and animals—respire continuously day and night, and for this process they take in *oxygen gas* from the atmosphere and give out an almost equal volume of *carbon dioxide gas*. By this process the energy stored up in the food and other materials is released and made use of by the protoplasm for its activities.

(5) **Reproduction.** Living beings—animals and plants—possess the power of reproduction, i.e. of giving rise to new young ones like themselves. Non-living objects have no such power. They may mechanically break down into a number of irregular parts; but living objects reproduce according to certain principles.

(6) **Metabolism.** Metabolism is a phenomenon of life. It includes both constructive and destructive changes that are constantly going on in the living body. Constructive changes lead to the formation of food substances and the construction of protoplasm while destructive changes result in their breakdown, ending in the formation of a variety of chemical substances.

(7) **Nutrition.** A living organism requires to be supplied with food. Food furnishes the necessary materials for nutrition and growth, and is a source of energy. Food materials nourishing the plant body or the animal body are much the same in both.

(8) **Growth.** All living objects grow. Some non-living bodies may also grow, as does a crystal. But there is difference in the mode of growth

between the two. The growth of the non-living objects is external, i.e. new particles are deposited on the external surface of their body from outside and as a result they grow; while in living objects, the growth is internal, i.e. it proceeds from within, new particles being secreted by the protoplasm in the interior of their body. Further, in living bodies the growth is the result of a series of complicated processes, both constructive and destructive.

(9) **Movements.** Movements are commonly regarded as a sign of life. Movements in most plants are, however, restricted, as they are fixed to the ground; while most animals move freely. Moving plants and fixed animals are not, however, uncommon among the lower organisms. Movements in plants and animals may be *spontaneous* or *induced.*

(a) **Spontaneous movement** is the movement of an organism or of an organ of a plant or an animal *of its own accord,* i.e. without any external influence. This kind of movement is regarded as a characteristic sign of life. Spontaneous movement is evident in animals with the development of organs of locomotion; while in plants it is exhibited by many unicellular and filamentous algae. Among the 'flowering' plants the best example of spontaneous movement is the Indian Telegraph plant (*Desmodium gyrans;* see Fig. III/30). Besides, the movements of protoplasm (see Fig. II/4) are distinctly visible under the microscope.

(b) **Induced movement** or **irritability**, on the other hand, is the movement of living organisms or of their organs in response to external stimuli. Protoplasm is sensitive to a variety of external stimuli, and when a particular stimulus is applied the reaction is usually in the form of a movement. Thus, when an animal burns itself it immediately moves away from the source of heat. A pinprick or electric shock produces a similar effect. Seedlings grown in a closed box with an open window on one side (see Fig. III/31) grow and bend towards the window, i.e. towards the source of light. Leaflets of sensitive plant (*Mimosa pudica;* see Fig. III/36) and sensitive wood-sorrel (*Biophytum sensitivum;* see Fig. III/35) close up when touched. The tentacles of sundew (see Fig. III/18), a carnivorous plant, bend over the insect from all sides and entrap it, when it falls on the leaf. Leaves of many plants show 'sleep movement'—closing up in the evening and opening again in the morning. No such effect is produced in the case of non-living objects like a log of wood or a bar of metal. Irritability is, however, more pronounced in animals than in plants.

Differences between the Living and the Non-living. It is very difficult to trace the absolute differences between the living and the non-living. Certain points may, however, be cited by way of general differences between the two.

Living	Non-living
(1) **Protoplasm.** All living objects contain protoplasm which is the physical basis of life and performs all the vital functions.	(1) Non-living objects are conspicuous by its absence and, therefore, no vital activity is possible in them.
(2) **Life-cycle.** All living objects follow a definite life-cycle comprising birth, growth, reproduction, old age and death.	(2) Non-living objects have no life-cycle. They may exist in their original state, may disintegrate or may change chemically or otherwise.
(3) **Cellular structure.** A living body is composed of regular cells and is well organized in form and size, both externally and internally.	(3) A non-living body is not cellular nor is it organized (except crystals). It is only made of a mass of particles of one or more kinds.
(4) **Respiration.** This is a very complex vital process, resulting in the breakdown of food with the release of CO_2 and a considerable amount of energy at all times and at all temperatures.	(4) Non-living objects do not respire. But, burning of coal or fire-wood at a high temperature releases CO_2 and some energy. The chemical mechanism, however, is altogether different.
(5) **Metabolism.** Metabolic changes constructive and destructive) are a haracteristic sign of life.	(5) No such changes are found in non-living objects. Chemical change, if any, in them is uncertain and irregular.
) **Nutrition.** Nutrition through d is a regular feature in all living uiganisms.	(6) Non-living objects require no food. Natural wear and tear cannot be made good by the supply of food.
(7) **Reproduction.** All living objects reproduce periodically by one or more methods for continuation of the species.	(7) Non-living objects cannot reproduce their own kind. They may, however, break down mechanically into some irregular pieces.
(8) **Growth.** Growth in living objects is internal, i.e. it proceeds from within the cell as a result of metabolic changes.	(8) Growth, if any, in non-living objects is external, i.e. it proceeds on the external surface only; no metabolic change is involved in this process.
(9) **Movement.** Spontaneous movement is a characteristic sign of life—either movement of protoplasm or of an entire organism or of some of the organs.	(9) Such a movement is never exhibited by non-living objects. The latter, however, can be made to move by some external forces, natural or mechanical.

Distinctions between Plants and Animals. Higher plants and higher animals are readily distinguished from one another by their possession of distinctive organs for the discharge of definite functions; but it is very difficult to make a distinction between unicellular plants and animals. The distinguishing features in general are, however, as follows:

(1) **Cell-wall and Cellulose.** While both plants and animals are cellular in composition, a plant cell is surrounded by a distinct cell-wall made of cellulose or any modification of it. Pure cellulose is not, however, found in fungi. The cell-wall and cellulose are always absent in an animal cell. The latter is surrounded by a thin cytoplasmic membrane called the *plasma membrane*.

(2) **Chlorophyll.** Chlorophyll, the green colouring matter of leaves and tender shoots, is highly characteristic of plants with the exception of fungi and total parasites. Chlorophyll is contained in special protoplasmic bodies, called plastids (see Fig. I), which often occur in large numbers in a cell. Chlorophyll and plastids are conspicuous by their absence in animal cells. Some animals may, however, turn green in colour by feeding upon green parts of plants.

(3) **Utilization of Carbondioxide.** Plants possess the power of utilizing the carbon dioxide of the atmosphere. It is only the green cells that have got this power. Thus during the daytime the green cells of the leaf absorb carbon dioxide from the surrounding air, manufacture sugar, starch, etc., and give out an almost equal volume of oxygen (by the breakdown of water in the process). Animals do not possess this power of utilizing carbon dioxide or of manufacturing food.

(4) **Food.** Green plants absorb raw food materials from outside—water and inorganic salts from the soil and carbon dioxide from the air—and prepare organic food substances out of them, primarily in the leaf, with the help of chlorophyll, in the presence of sunlight. Animals, being devoid of chlorophyll have no power of manufacturing their own food. They have to depend, directly or indirectly, on plants for this primary need. It is also to be noted that plants take in food in solution only, whereas animals can ingest solid food particles.

(5) **Growth.** The regions of growth are localized in plants, lying primarily at the extremities—root-apex and stem-apex—and also in the interior, i.e. growth is both apical, and intercalary; while in animals growth is not localized to any definite region, i.e. all parts grow simultaneously. Moreover, in plants growth proceeds until death; while in animals growth ceases long before death.

(6) **Movements.** Plants grow fixed to the ground or attached to some support, and as such they cannot move from one place to another, except some lower types of plants; while animals move freely in search of food and shelter, and also when attacked; some animals, of course, grow attached to some object, and thus cannot move freely.

(7) **Vacuole and Centrosome.** The vacuole is a common feature of a mature plant cell and is often so much enlarged as to occupy a major portion of it (see Fig. I). In the animal cell the vacuole is somewhat rare and, if present, is small in size. The centrosome, a protoplasmic body, lying close to the nucleus, is a regular feature of the animal cell and is associated with the division of the nucleus. It is very rare in the plant cell, occurring only occasionally in some lower plants.

Binomial Nomenclature. In classifying plants and animals Linnaeus, a Swedish naturalist, first introduced (1735) a system of designating each and every species of plant or animal with a binomial consisting of two parts—the first refers to the genus and the second to the species. A **species** is defined as a group of individuals—plants or animals—resembling one another in almost all respects, differing only in minor details. A **genus** is a group of closely allied species. This system of naming a plant or an animal with two parts (genus and species) is called binomial nomenclature. Since the popular name of a species varies from country to country this system of naming plants and animals has been universally accepted as the correct scientific system. Thus mango is designated as *Mangifera indica,* pea as *Pisum sativum,* onion as *Allium cepa,* garlic as *Allium sativum,* etc.

Main Groups of Plants. There are two main divisions of the plant kingdom, viz., **cryptogams** and **phanerogams**. Cryptogams are lower plants which never bear flowers or seeds; while phanerogams are higher plants which always bear flowers and seeds. So cryptogams are regarded as 'flowerless' or 'seedless' plants, and phanerogams as 'flowering' or 'seed-bearing' plants.

A. **Cryptogams.** The main groups of cryptogams from the lower types to the higher are the following:

(1) **Thallophyta.** Thallophyta are lower cryptogams in which the plant body is not differentiated into the root, stem and leaf. Such an undifferentiated plant body is called a thallus and the thallus-bearing plants are called Thallophyta: (a) **Algae** (sing. alga) are commonly green Thallophyta containing chlorophyll, although this colour may be masked by other colouring matters. They mostly grow in water and are of various

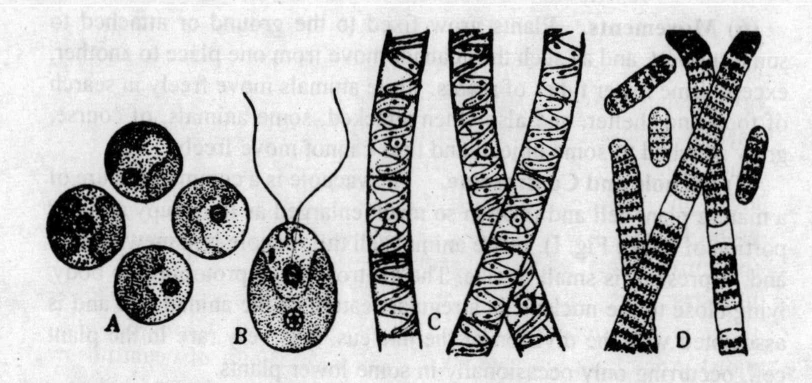

Figure II. Forms of Algae. *A, Protococcus*—unicellular and green; *B, Chlamydomonas*—unicellular, green, ciliate and motile; *C, Spirogyra*—filamentous and green; and *D, Oscillatoria*—filamentous and blue-green.

forms (Fig. II). (b) **Bacteria** (sing. bacterium) are the smallest known organisms, not visible to the naked eye. They are unicellular, non-green, usually spherical or rod-like (Fig. IIIA). They occur almost everywhere—in water, air, and soil, and also in living bodies of plants and animals.

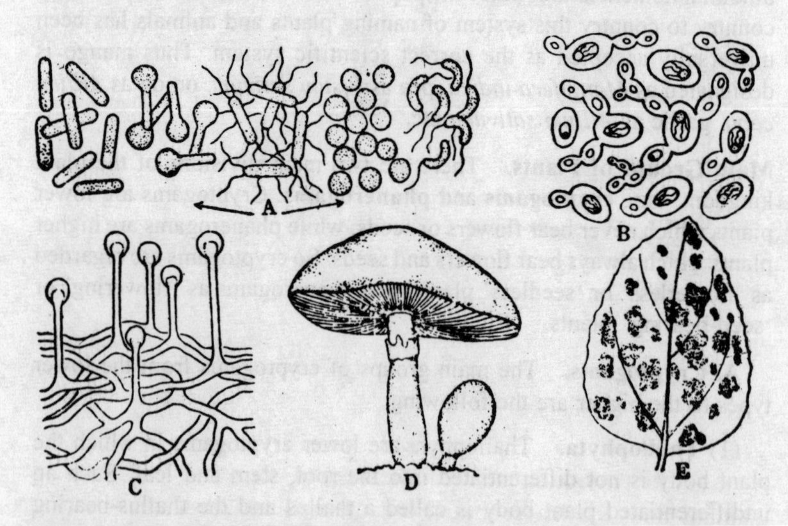

Figure III. Forms of Bacteria and Fungi. *A,* bacteria—five common types; *B,* yeast—a unicellular fungus (mostly budding); *C,* mould—a filamentous fungus; *D,* mushroom—a fleshy fungus; and *E,* a parasitic fungus on a leaf.

(c) **Fungi** (sing. fungus) are non-green Thallophyta containing no chlorophyll. They grow mostly on land either as saprophytes. (see p. 10) or in living bodies as parasites (see p. 9). Like algae they may be of various forms. Common examples of fungi are mould, mushroom, toadstool, puffball, etc. (Fig. IIIB–E). (d) **Lichens** are associations of algae and fungi, e.g. old man's beard (*Usnea*). (2) **Bryophyta** are a group of higher cryptogams in which the plant body may be thalloid (primitive forms) or leafy (advanced forms). They develop some root-like structures, called rhizoids, but no true roots, and the conducting tissue is very simple and primitive. They grow on old damp walls, on moist ground and on bark of trees forming a sort of beautiful green carpet, and are more complicated and more advanced than the Thallophyta. The examples of Bryophyta are *Riccia* (Fig. 52), *Marchantia* (Fig. 57), etc. and mosses (Fig. 65).

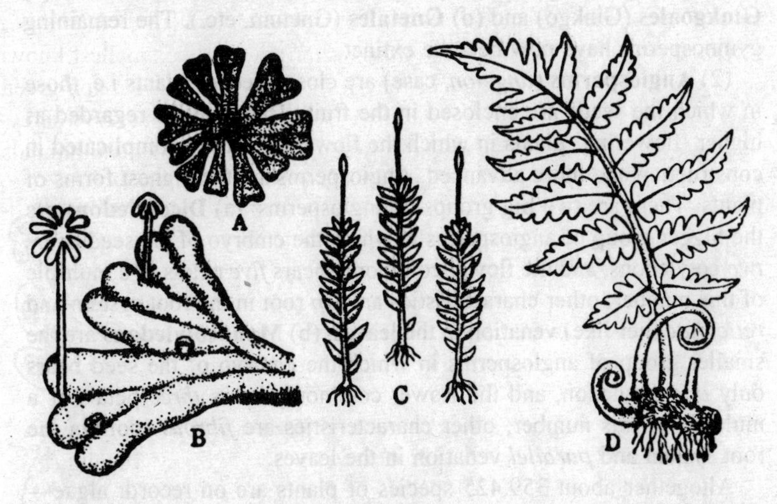

Figure IV. Forms of Bryophyta and Pteridophyta. *A, Riccia* and *B, Marchantia*—two thalloid bryophytes; *C,* moss—a leafy bryophyte; *D, fern*—a pteridophyte.

(3) **Pteridophyta** are the highest group of cryptogams in which the plant body is differentiated into an underground horizontal stem (rhizome) or an erect stem, well-developed green leaves and true roots. The plant body is more complicated with the development of conducting and mechanical tissues. Pteridophyta are more advanced than Bryophyta.

They bear spores on their leaves by which they reproduce and multiply. Ferns (Fig. IVD) and relatives are the common groups.

B. **Phanerogams** or **Spermatophytes.** These are 'flowering' or 'seed-bearing' plants. Their two main characteristics are: (1) formation of pollen-tube for facility of fertilization, and (2) production of seeds for reproduction, multiplication, and efficient protection of the embryo. Phanerogams are the most advanced types of plants with the reproductive shoot modified into a flower (simple or complex), and are divided into two main groups: gymnosperms and angiosperms.

(1) **Gymnosperms** (*gymnos,* naked; *sperma,* seed) are naked-seeded plants, i.e. those in which the seeds are not enclosed in the fruit. They may be regarded as lower 'flowering' plants in which the flowers are unisexual (either male or female), simple in construction and primitive in nature. There are four living orders of gymnosperms: (a) **Cycadales** (Cycad or Cycas, etc.) (b) **Coniferales** (Conifers like Pinus, etc.) (c) **Ginkgoales** (Ginkgo) and (d) **Gnetales** (Gnetum, etc.). The remaining gymnosperms have now become extinct.

(2) **Angiosperms** (*angeion,* case) are closed-seeded plants i.e. those in which the seeds are enclosed in the fruit. They may be regarded as higher 'flowering' plants in which the flowers are more complicated in construction and more advanced. Angiosperms are the highest forms of plants. There are two big groups of angiosperms: (a) **Dicotyledons** are the bigger group of angiosperms in which the embryo of the seed bears *two* cotyledons, and the flower commonly bears *five* petals or a multiple of this number; other characteristics are *tap* root in the root system and *reticulate* (net-like) venation in the leaves; (b) **Monocotyledons** are the smaller group of angiosperms in which the embryo of the seed bears only *one* cotyledon, and the flower commonly bears *three* petals or a multiple of this number; other characteristics are *fibrous* roots in the root system and *parallel* venation in the leaves.

Altogether about 359,425 species of plants are on record: algae—20,000; fungi—90,000; bacteria—2,000; lichens—15,000; liverworts and mosses—23,725; ferns and relatives—9,000; gymnosperms—700; and angiosperms—199,000 (dicotyledons—159,000 and monocotyledons—40,000).

Branches of Botany. Botany, like every other science, may be studied from two aspects—the *pure* and the *applied* or *economic.* Pure botany deals with the study of plants as they form a part of nature, and applied botany as it is applied to the well-being of mankind. The following are the main branches:

(1) **Morphology.** (*morphe,* form; *logos,* discourse or study). This deals with the study of forms and features of different plant organs such as roots, stems, leaves, flowers, seeds and fruits.

(2) **Histology** (*histos,* tissue). The study of detailed structure of tissues making up the different organs of plants, as revealed by the microscope, is called **histology**. The study of gross internal structure of plant organs by the technique of section-cutting is called **anatomy** (*ana,* as under; *temnein,* to cut). **Cytology** (*kytos,* cell) dealing with the cell-structure with special reference to the behaviour of the nucleus, particularly of the chromosomes (see pp. 188), is a newly established branch of histology.

(3) **Physiology** (*physis,* nature of life). This deals with the various functions that the plants perform. Functions may be *vital* or *mechanical;* vital functions are performed by the living matter, i.e. the protoplasm, and the mechanical functions by certain dead tissues without the intervention of the protoplasm; as, for example, bark and cork protect the plant body, and certain hard tissues strengthen it. It is to be noted that structure and function are correlated, i.e. a particular structure develops in response to a particular function.

(4) **Ecology** (*oikos,* home). This deals with the interrelationship between the living organisms (plants and animals) and their environment which includes all the conditions surrounding them.

(5) **Taxonomy** or **Systematic Botany** (*Taxon* = a group, nomos = to name). This deals with the description and identification of plants, and their classification into various natural groups according to the resemblances and differences in their morphological characteristics.

(6) **Organic Evolution.** This deals with the sequence of descent of more complex, more recent and more advanced types of plants and animals from the simpler, earlier and more primitive types through successive stages in different periods of the earth.

(7) **Genetics.** This deals with the facts and laws of inheritance (variation and heredity) of characters from one generation to another.

(8) **Palaeontology.** This deals with the remains of ancient plants and animals preserved in rocks in the form of fossils (*fossilis,* dug out). Fossils give us an idea about the forms of plants and animals that existed in different geological periods of the earth, and also help us to trace the line of evolution from the lower and earlier forms of life to the higher and later forms (see also pp. 530). Palaeobotany is the branch which deals with the study of only fossil plants.

(9) **Economic Botany.** This deals with the various uses of plants and their products, and includes methods for their improvement for better utilization by mankind. It has also several branches: (a) **agronomy** dealing with the cultivation of field crops for food and industry; (b) **horticulture** dealing with the cultivation of garden plants for flowers and fruits; (c) **plant pathology** dealing with the diagnosis, cure and prevention of plant diseases, mainly in field crops and other useful plants, commonly caused by bacteria and fungi, and also deficiency diseases; (d) **pharmacognosy** dealing with the study of medicinal plants with special reference to preparation and preservation of drugs; (e) **forestry** dealing with the study and utilization of forest plants for timber and other forest products; and (f) **plant breeding** dealing with the cross-breeding of plants evolving newer and more improved types with desired characteristics, e.g. higher yield and better quality.

16. **Some Remarkable 'Flowering' Plants.** Certain species of plants show some special features in their habit such as gigantic height or enormous thickness of the tree trunk or great longevity or some other features, as noted below.

Tallest Trees. In Australia *Eucalyptus amygdalina* and *E. regnans* attain a towering height of 114 metres. In America such giants are the two species of *Sequoia,* viz. the mammoth tree (*S. gigantea*) of California and the giant redwood tree (*S. sempervirens*) of Coast Range—112 m.; douglas fir (*Pseudotsuga taxifolia*) of West North America—94 m. In India some of the Himalayan conifers are the tallest trees, e.g. deodar or Himalayan cedar (*Cedrus deodara*), Himalayan spruce (*Picea smithiana = P. morinda*), Himalayan silver fir (*Abies webbiana*)—60 m.; some of the common timber trees are also fairly tall—46 m., e.g. Khasi pine (*Pinus khasya*), Himalayan popular (*Populus ciliata*), *Chickrassia tabularis* (Chikrass), Dipterocarpus *macrocarpus* (Hollong), *D. turbinatus* (Garjan), *Terminalia myriocarpa* (Hollock), *Tetrameles nudiflora* (wood soft), *Ster*-canes (*Calamus zeylanicus* and a few other species), though not tall and erect, are 150–180 m. in length; while some of the lianes, e.g. wood-rose (*Ipomoea tuberosa*) in upper Assam and *Mucuna gigantea* in the Sundarbans, spread extensively and quickly over several tall trees and may be longer than the Rattans.

Thickest Trees (diameter of the trunk). Swamp cypress (*Taxodium mucronatum*) of South Mexico—17 m.; sweet chestnut (*Castanea sativa*) of Italy—17 m.; big cypress tree (*Cupressus macrocarpa*) of California—15 m.; dragon's blood (*Dracaena draco*) of Canary Islands—14 m.; mammoth tree (*Sequoia gigantea*) of California—11 m.; giant redwood tree (*S. sempervirens*) of Coast Range—9 m.; and baobab tree (*Adansonia digitats*) of Africa and South India—9 m.

Oldest Trees. World's *oldest living specimen* is a tall cycad-like tree of *Macrosamia* in the Tambourine Mountain of Australia—over 10,000 years;

big cypress tree of California—6,000 years; dragon's blood—6,000 years, baobab tree—over 5,000 years; bristlecone pine (*Pinus longaeva*) in East California—4,600 years or more; mammoth tree—3,500 years or more; giant redwood tree—somewhat shorter-lived; some species of *Cupresus, Juniperus, Taxus* and *Taxodium*—2–3,000 years; Peepul (*Ficus religiosa*)—2–3,000 years; Himalayan cedar or deodar 1,000 years; *Eucalyptus*—300 years, etc. It may be of special interest to note that maiden hair tree (*Ginkgo biloba*), a big tree first found in East China (later widely cultivated in China and Japan; still later in Europe and America; also planted in some gardens in India) is the *oldest living species* in the world. It originated in the Permian (280 million years ago, possibly earlier), spread widely in the Jurassic (180 million years ago) and has survived till today—a living fossil, indeed.

Largest Cacti. Giant cactus (*Cereus giganteus*) in North America—21 m. high. *C. peruvianus* in South America—12 m. high, *C. grandiflorus* and *C. trangularis* in West Indies (cultivated and wild in India) climb tall trees and reach their tops.

Largest Orchids. Giant orchid (*Grammatophyllum speciosum*) in Malaysia—stem 2–3 m. high and raceme 1.5–2 m. high (specimens up to a total height of 7.6 m. have, however, been recorded); *Galeola falconeri* in Arunachal—the stout inflorescence axis (growing from the stout rhizome)—3 m. high.

Largest Leaves. Certain feather palms like *Raphia ruffia* in Madagascar and *R. taedigera* in South America bear the largest leaves known—20 m. in length; giant aroid (*Amorphophallus titanum*) in Sumatra bears annually a single large pedate leaf—3–4 m. across on a stout stalk—3 m. high.

Giant Water Lily (*Victoria amazonica = V. regia*) is a very handsome aquatic plant of the Amazon river in Brazil, with its huge circular tray-like floating leaves—2 m. in diameter; the margin of each leaf is turned up, while the undersurface of the blade is strongly veined and armed with rigid spines for strength and self-defence. The plant is cultivated at the Indian Botanic Garden near Calcutta.

Pitcher Plant (*Nepenthes khasiana*) is an interesting carnivorous plant of the Garo Hills and the Jaintia Hills of Meghalaya (N.E. India). It is a climbing shrub. In it the leaf-blade takes the curious form of a pitcher with a lid—10–20 cm. or even more in height, meant to capture insects and small animals and digest them for protein food. Malayasia (mainly Borneo) is, however, the natural abode of *Nepenthes* (numbering 67 species), with the extension of certain species (mostly 1 in each case) to N.E. India, Madagascar, Ceylon, South China, Australia and also elsewhere. There are other types of pitcher plants also which are e.g. *Cephalotus* in West Australia, *Darlingtonia* in California and *Sarracenia* in North America.

Largest Flowers. *Rafflesia arnoldi,* root-parasite in Malayasia, bears the largest flowers in the world—50 cm in diameter and 8 kg. in weight. *Sapria himalayana,* a relative of *Rafflesia* and a similar root-parasite in Arunachal and Nagaland—15–30 cm in diameter; and *Magnolia griffithu* in Assam—15 cm in diameter. It may be noted that talipotpalm (*Corypha umbraculiferu*), a tall fan-palm of South India and Ceylon, produces a huge terminal inflorescence known to be the largest—6–8 m. high. The tree is monocarpic like *Agave* and some bamboos. i.e. it dies after flowering once and fruiting. Flowers are bisexual.

Largest Seed (and Fruits). The massive bilobed fruit of double coconut (*Lodoicea maldivica*), a tall dioecious fan-palm of the Seychelles—1 m. or so in length and 18–20 kg. in weight, bears a single bilobed seed known to be the largest in the world (see p. 152). The fruits take a long period of about 10 years to ripen, and they were seen floating in the Indian Ocean long before the tree was discovered in its native place. A giant water melon (*Citrullus lanatus*) grown in New Jersey, weighing 62 kg. is a world record.

Heaviest Wood. It is, the black iron-wood (*Olea laurifolia;* family *Oleaceae*) of South Africa, which is the heaviest wood in the world, weighing 1448 kg. per cu. m.

Largest Families and Genera. Families: orchid family (*Orchidaceae*)—20,000 sp.; sunflower family (*Compositae*)—14,000 sp.; legume family—12,000 sp.; grass family (*Gramineae*)—10,000 sp.; spurge family (*Euphorbiaceae*)—7,000 sp.; and madder family (*Rubiaceae*)—6,000 sp. **Genera:** *Senecio* (*Compositae*)—2,500 sp.; *Astragalus* (*Papilionaceae*) and *Euphorbia* (*Euphorbiaceae*)—about 2,000 sp. each; *Solanum* (*Solanaceae*)—1,500 sp.; *Eupatorium* (*Compositae*), mostly American—1,200 sp.; *Vernonia* (*Compositae*), *Eugenia* (*Myrtaceae*)—mostly American and *Pleurothallis* (*Orchidaceae*)—South American—1,000 sp. each. On the other hand there are a few monotypic families with 1 genus and 1 species, e.g. *Trichopodaceae* (*Trichopus*)—a rigid herb in South India; *Nandinaceae* (*Nandina*)—a shrub in China and Japan (planted in Shillong); *Cephalotaceae* (*Cephalotus*)—a small pitcher plant in West Australia; *Pteridophyllaceae* (Pteridophyllum)—a stemless rhizomatous herb in Japan, etc. In India, the largest families are *Orchidaceae* (1,700 sp.); *Gramineae* (1,133 sp.); *Leguminosae* (951 sp.; *Papilionaceae* dominating); *Compositae* (674 sp.); *Rubiaceae* (489 sp.); *Acanthaccae* (409 sp.); and largest genera are Impatiens (241 sp.); *Primula* (162 sp.); *Strobilanthes* (152 sp.) and *Rhododendron* (125 sp.).

Smallest Plants. *Wolffia arrhiza* and *W. microscopica* of the duckweed family (*Lemnaceae*) are the smallest known 'flowering' plants; they are rootless floating plants as small as pin-heads, evidently bearing minutest florets.

Smallest Seeds. Orchids bear the smallest seeds in the vegetable kingdom; the tiny seeds, each with a microscopic embryo, are like particles of dust borne in millions in each capsule. When the capsule bursts the seeds puff off like powder.

Lightest Wood. The wood of balsa or corkwood (*Ochroma pyramidal* = *O. lagopus;* family *Bombacaceae*), a tall tree of Central America, is known to be the lightest, weighing only 119 kg. per cu. m., as against the common heaviest wood of ironwood tree (*Mesua ferrea;* family *Guttiferae*) of India and ebony (*Diospyros ebenum;* family *Ebenaceae*) of Ceylon, weighing 1190–1246 kg. per cu. m. Balsa wood is used for rafts and floats, and for making models of ships and boats.

Vegetable Milk. *Brosimum galactodendron* (family Moraceae) is the milk-tree or cow-tree of Venezuela; on incision of the stem there is a profuse flow of milky latex used as a substitute for cow's milk; it tastes like milk and is considered wholesome. *Mimusops elata* (family *Sapotaceae*) is the milk-tree of Brazil; its milky latex has the same use.

Bread Fruit. *Artocarpus incisa* (family Moraceae) is the bread fruit tree; fruit is sliced, roasted and eaten like bread.

Part I

MORPHOLOGY

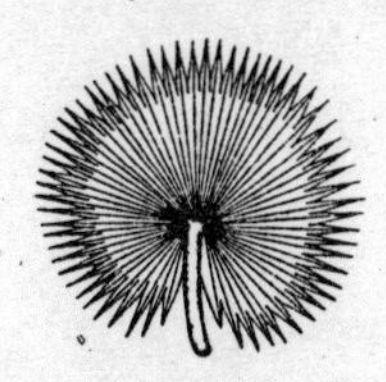

Chapter 1

Diversity of Plant Life

There are not only immense numbers of plants but they also show diversities in various directions—habitat, habit, forms and types, duration of life, mode of nutrition, etc. Many of them have also developed special (modified) organs for the discharge of special functions. Diversity is a special feature of the biological kingdom.

Habitat. The habitat is the natural home of a plant. Each habitat has its own factors, viz. a particular type of climate (rainfall, heat, wind, and light) and a particular type of soil (soil-water, its physical and chemical nature), and it has it's own characteristic flora. Thus certain plants grow in the fresh water of ponds, lakes and rivers, forming what is called the aquatic flora, e.g. water lily, lotus, bladderwort, duckweed, water lettuce, *Vallisneria, Hydrilla,* etc. Close to a pool of water a group of moisture-loving plants are seen to grow, e.g. many grasses, aroids, mosses and ferns. In the saline water of salt lakes and seas an altogether different type of vegetation is seen. Life, however, is harder on land with varying climatic conditions and with different types of soils. In very dry regions or deserts cacti and similar plants form the dominant flora. In dry fields in winter certain weeds make their appearance, while in the same fields during the rains another set of weeds appear. In places with heavy rainfall, evergreen forests grow up, while in places with moderate or low rainfall deciduous forests are seen. Then again at high altitudes certain other types of plants are found, e.g. oak, birch, pine, deodar, fir, etc. Still higher up, as in the Himalayas, certain types of stunted shrubs and small herbs only grow. It is thus evident that particular habitats suit particular types of plants.

Habits of Plants. The nature of the stem, the height of the plant, its duration and mode of life determine the habit of a plant. In habit plants show considerable diversities. Commonly the following terms are used to indicate the general habits of plants.

(1) **Herbs** are small plants with a soft stem. They may vary from a few millimetres to metre or so in height, e.g. duck weed, mustard, radish, sunflower, ginger, *Canna,* etc. (2) **Shrubs** are medium-sized plants, with a hard and woody stem, often much-branched and bushy, e.g. China rose, garden croton, night jasmine, *duranta,* etc. (3) **Trees** are tall plants with a clear, hard and woody stem, e.g. mango, jack, *Casuarina* (B. & H. JHAU), etc. While most shrubs and trees are profusely branched, most palms, although very tall and erect, sometimes 46 m. in height, are unbranched. Some trees take a conical or pyramidal shape, e.g. *Casuarina;* others become dome-shaped, e.g. banyan. Timber trees generally attain a considerable height, e.g. sal tree (*Shorea*) and teak (*Tectona*) 30 m., GARJAN or wood-oil tree (*Dipterocarpus*) 46 m. But the real giants of the plant world are *Eucalyptus* in Australia 98–114 m. and *Sequoia* (the mammoth tree and the giant redwood tree) in America 100–112 m. [See p. xxx]. Some tree trunks are remarkably thick, e.g. baobab tree 9 m. in diameter, mammoth tree 11 m. in diameter, and dragon's blood 14 m. in diameter. [See p. xxx]. On the other hand, plants with a soft stem cannot stand erect. Some of them (the **creepers**) only creep on the ground, e.g. wood-sorrel; some (the **climbers**) climb neighbouring objects by means of special devices (see p. 5). Others (the **twiners**) bodily twine round some support, e.g. country bean, railway creeper, Rangoon creeper, *Clitoria* (B. APARAJITA; H. GOKARNA), etc. Still others (the **lianes**; see p. 8) climb large trees, reach their tops, and often spread over neighbouring trees.

Forms and Types of Plants. There is a considerable diversity of plants ranging from the simplest to the most complicated and gigantic ones. Some plants are tall, or very tall, some medium-sized, some small and some so small that they are invisible to the naked eye. Among those known to us bacteria are the smallest; they are unicellular and only imperfectly seen even under a powerful microscope (see Fig. IIIA). Viruses are still smaller than bacteria but they have no definite cell-structure. Among algae (which include pond-scum and sea-weeds) there are gradations of forms; some are unicellular, e.g. *Protococcus* (see Fig. IIA), while the majority are multicellular (many-celled); the latter may be filamentous, e.g. *Spirogyra* and *Oscillatoria* (see Fig. IIIC–D) or massive, e.g. many sea-weeds. Similarly fungi may be unicellular, e.g. yeast (see Fig. IIIB) or multicellular; the latter may be filamentous, e.g. mould or *Mucor* (see Fig. IIIC), or massive, e.g. mushroom or *Agaricus* (see Fig. IIID). Some plants are thalloid lying flat on the ground. e.g. *Riccia* and *Marchantia* (see Fig. IVA–B). Mosses are short erect plants, with small green leaves, a short axis, and some slender rhizoids (see Fig. IVC). Ferns have already become complex in structure with well-developed leaves, a stem and clusters of roots (see Fig. IVD). Flowering plants, however, show the highest degree of complexity in structure and forms varying from about 2 millimetres to a hundred metres or more. The latter are really gigantic trees, as stated above.

Duration of Life. The life of an individual plant is always limited in duration. Herbs have a short span of life. Those herbs that live for a few months or at most a year are said to be (1) **annuals**, e.g. rice, wheat, maize, mustard, potato, pea, etc. They grow and produce flowers and fruits within this period and then die off. Some herbs may live for two years; such plants are said to be (2) **biennials.** They attain their full vegetative growth in the first year and produce flowers and fruits only in the second year after which they die off, e.g. cabbage, beet, carrot, turnip, etc. (in a tropical climate these vegetables, however, behave like annuals). Some herbs continue to grow from year to year with a new lease of active life for a few months only; the aerial parts of such plants die every year after flowering or in winter, and fresh life begins after a few showers of rain when the underground stem puts forth new leaves. Such plants are said to be (3) **perennials**, e.g. *Canna,* ginger, KACHU (taro), onion, tuberose, etc. Shrubs generally live for a few years. Trees, however, have the greatest longevity. Thus sal tree, teak, wood-oil tree, pines and certain palms (e.g. talipot palm) live for 100–150 years; *Eucalyptus* for 300 years; some conifers (pine-like trees) live for 2–3,000 years or even more; baobab tree and dragon's blood are specially remarkable for their longevity; their age in some cases has been estimated to be 5–6,000 years.

Climbers. Climbers usually develop special organs of attachment by which they cling to neighbouring objects for the support of their body and for aid in climbing. Several climbers, e.g. the twiners, also climb by bodily twining round some support. Climbers may be of the following kinds:

(1) **Rootlet Climbers.** These are plants which climb by means of small adventitious roots given off from their inner side or from their nodes as they come in contact with a supporting plant or any suitable object. Such roots either form small adhesive discs or claws which act as *holdfasts,* or they secrete a sticky juice which dries up, fixing the climbers to their support. Examples may be seen in Indian ivy (*Ficus*

pumila; Fig. 1), ivy (*Hedera helix*), *Piper* (e.g. betel, long pepper, etc.), *Pothos, Hoya, etc.*

(2) **Hook Climbers.** The flower-stalk of *Artabotrys* (B. & H. KANTALI-CHAMPA) produces a curved hook (Fig. 2C) which helps the plant to climb. Often **prickles** and sometimes **thorns** are curved and hooked in certain plants. Thus in cane (*Calamus;* Fig. 3A) a long, slender axis beset with numerous sharp and curved hooks is produced from the leaf-sheath in addition to numerous prickles on it. Climbing rose (Fig. 3B) is provided with numerous curved prickles for the purpose of climbing (and also for self-defence). Glory of the garden (*Bougainvillea;* Fig. 2A) and *Uncaria* (Fig. 2B), both large climbing shrubs, produce curved hooks (thorns) which are used as organs of support for facility of climbing.

(3) **Tendril Climbers.** These are plants which produce slender, leafless, spirally-coiled structures known as **tendrils**, and climb objects with the help of them; tendril twine themselves round some neighbouring object, and help the plants to support their weight and climb easily. Tendrils may be modifications of the stem, as in passion-flower (*Passiflora;* Fig. 4A), vine (*Vitis*) etc., or of leaves, as in pea (*Pisum;* Fig. 4B), wild pea (*Lathyrus;* Fig. 4C), etc.

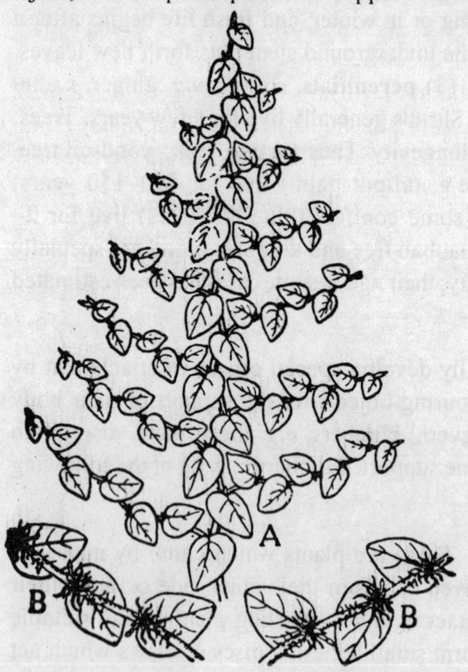

Figure 1. Indian ivy—a rootlet climber; *A*, upper side; *B*, lower side.

(4) **Leaf Climbers.** In some plants a part of the leaf is sensitive to contact with a foreign body. It acts like a tendril and helps the plant to climb. Thus the slender petiole of *Clematis* (Fig. 5A) twines like a tendril round a support and helps the plant to climb. In glory lily (*Gloriosa;* Fig. 5B) the prolonged leaf-apex coils round a

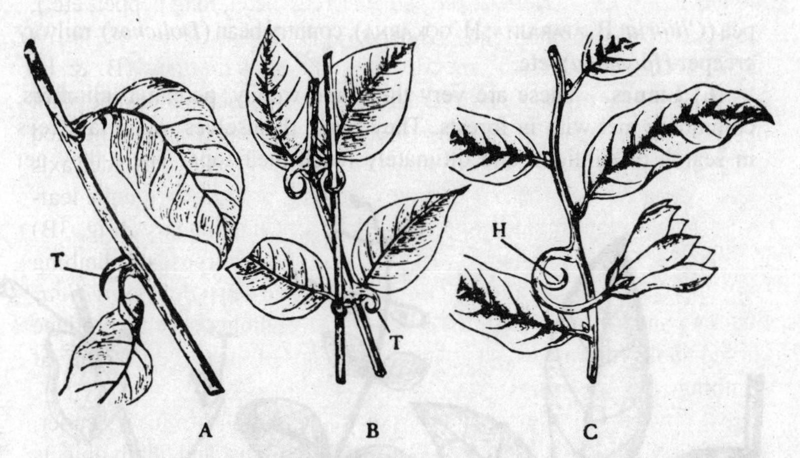

Figure 2. Hook and Thorn Climbers. *A*, glory of the garden; *T*, thorn; *B*, *Uncaria; T*, hooked thorn; C, *Artabotrys; H*, hook.

support like a tendril. The long stiff stalk of the pitcher of pitcher plant (Fig. 5C) also acts as a tendril supporting the pitcher.

(5) **Twiners** or **Stem Climbers.** They are long or short climbers but they have no special organ of attachment like the latter. They climb by bodily twining round some support. Common examples are butterfly

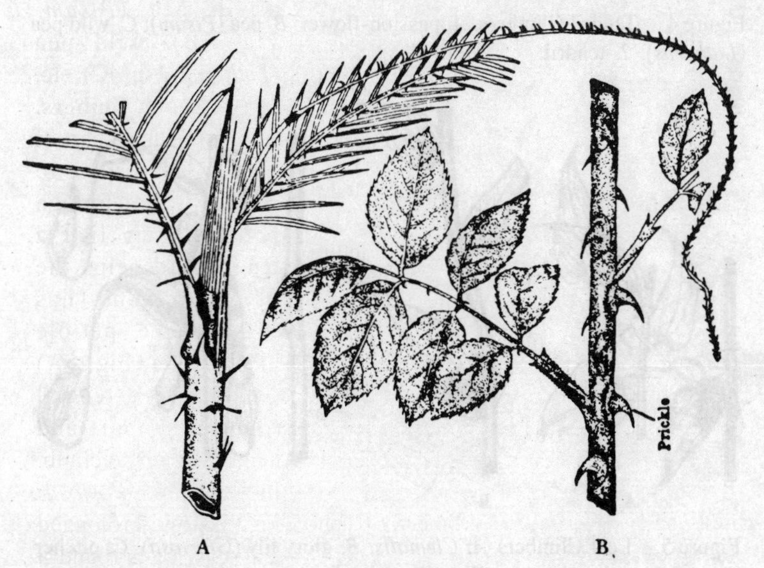

Figure 3. Prickle Climbers. *A*, cane; *B*, rose.

pea (*Clitoria;* B. APARAJITA; H. GOKARNA), country bean (*Dolichos*), railway creeper (*Ipomoea*), etc.

(6) **Lianes.** These are very thick and woody, perennial climbers, commonly met with in forests. They twine themselves round tall trees in search of sunlight, and ultimately reach their tops. There they get

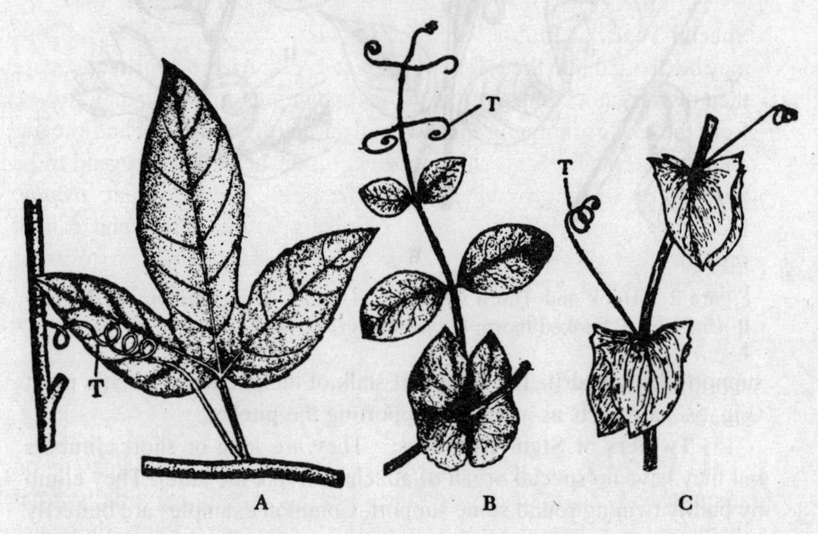

Figure 4. Tendril Climbers. *A,* passion-flower; *B,* pea (*Pisum*); *C,* wild pea (*Lathyrus*); *T,* tendril.

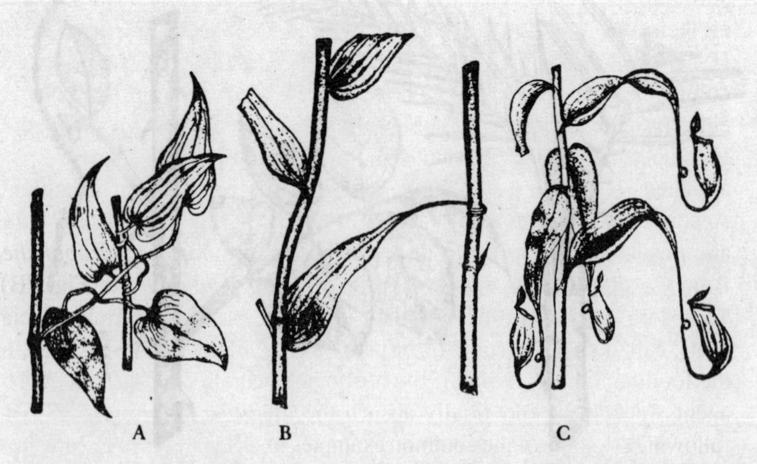

Figure 5. Leaf Climbers. *A, Clematis; B,* glory lily (*Gloriosa*); *C,* pitcher plant (*Nepenthes;* see also Fig. 82).

plenty of sunlight and produce a canopy of foliage, e.g. *Hiptage* (B. MADHABILATA; H. MADHABILATA), camel's foot climber (*Bauhinia vahlii;* B. LATAKANCHAN; H. CHAMBULI)—see Fig. 160), snake climber (*B. anguina;* B. & H. NAGPAT)—Stem and branches wavy, nicker bean (*Entada gigas;* B. & H. GILA), etc.

Special Types of Plants. Depending on the *mode of nutrition* plants may be divided into the following special types. All green plants prepare their own organic food, particularly carbohydrates, mostly in their leaves, from the raw or inorganic materials absorbed from the soil and the air, and nourish themselves with their own food; such plants are said to be **autophytes** or autotrophic plants (*autos,* self; *phyton,* plant; *trophe,* food) or self-nourishing. Non-green plants on the other hand cannot prepare their own carbohydrate food and thus they draw it (together with other kinds of food) from different sources, i.e. their modes of nutrition are different; such plants are said to be **heterophytes** or heterotrophic plants (*heteros,* different) and they are either parasites or saprophytes. There are other types of plants whose mode of nutrition is somewhat peculiar. All such types of plants are as follows.

(1) **Parasites.** Plants that grow upon other living plants (or on animals) and absorb necessary food materials, wholly or partially, from them are called parasites. Among the 'flowering' plants there are different degrees of parasitism. Some are total parasites and others are partial parasites. Total parasites are never green in colour as they absorb all their food from the host plant, e.g. dodder (*Cuscuta;* B. SWARNALATA; H. AKASHBEL—Fig. 6). Partial parasites on the other hand are green in colour and manufacture their food, partially at least; so they do not entirely depend on the host plant, e.g. mistletoe (*Viscum;* B. BANDA; H. BHANGRA—Fig. 8). It is further to be noted that parasites may be attached to the stem and branches or to the root of the host plant. Accordingly they are said to be stem-parasites, e.g. dodder (*Cuscuta*) and mistletoe (*Viscum*) or root-parasites, e.g. broom-rape (*Orobanche;* B. BANIA-BAU; H. SARSON-BANDA—Fig. 7A) and *Balanophora* (Fig. 7B). To absorb food from the host plant a parasite produces certain special roots, called **sucking roots** or **haustoria** (Fig. 6B) which penetrate into the food-containing tissue of the host plant, secrete necessary digestive agents (enzymes) and finally absorb the digested food products. The following are some of the common examples of different types of parasites:

(1) **Stem-parasites:** (a) total—dodder (*Cuscuta*); (b) partial—mistletoe (*Viscum*), *Cassytha* and *Loranthus*.

(2) **Root-parasites:** (a) total—broomrape (*Orobanche*), *Balanophora, Rafflesia,* etc. Broomrape, common in Bihar and Uttar Pradesh, is parasitic on the roots of some field crops like mustard. potato. tobacco. brinjal. etc., *Balanophora* on

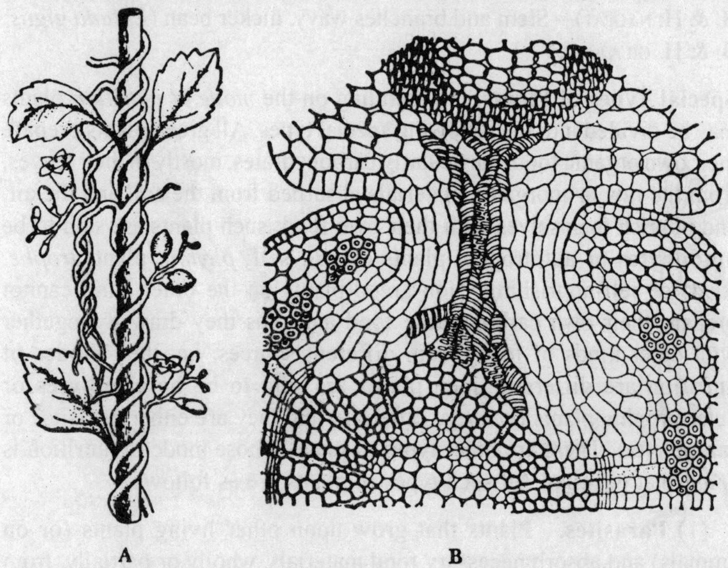

Figure 6. *A,* dodder (*Cuscuta*)—a total stem parasite; *B,* a section through dodder (and the host plant) showing the sucking root (haustorium).

the roots of some forest trees in the Khasi Hills, and *Rafflesia* on the roots of *Vitis,* figs, etc., in Sumatra and Java; (b) partial—sandalwood tree in Mysore, and *Striga* on the roots of some field crops in Mysore and Maharashtra.

It is interesting to note that *Rafflesia arnoldi* (Fig. 9) bears the biggest flower in the world. It measures 50 cm in diameter and weighs 8 kg. Another point of interest is that its stem and root are reduced to a network of slender threads which penetrate into the root of the host plant, ramify through it and draw its nutriment from it. Here and there, the thread-like stem bears giant flowers above the ground flowers. The flowers are unisexual. Another interesting total parasite is *Sapria* found in the hills of Arunachal and Nagaland. It is allied to *Rafflesia* and bears the biggest flower in India measuring 15–30 cm. in diameter.

(2) **Saprophytes** (*sapros,* rotten; *phyta,* plants). These are plants that grow in soils rich in decaying organic substances of vegetable or animal origin, and derive their nutriment from them. They are non-green in colour. Among the 'flowering' plants Indian pipe (*Monotropa;* Fig. 10) and some orchids afford good examples of saprophytes, *Monotropa* grows

in the Khasi Hills at an altitude of 1,800–2,500 metres. Total saprophytes are colourless; and the partial ones are green in colour. Their roots become associated with a filamentous mass of a fungus which takes the

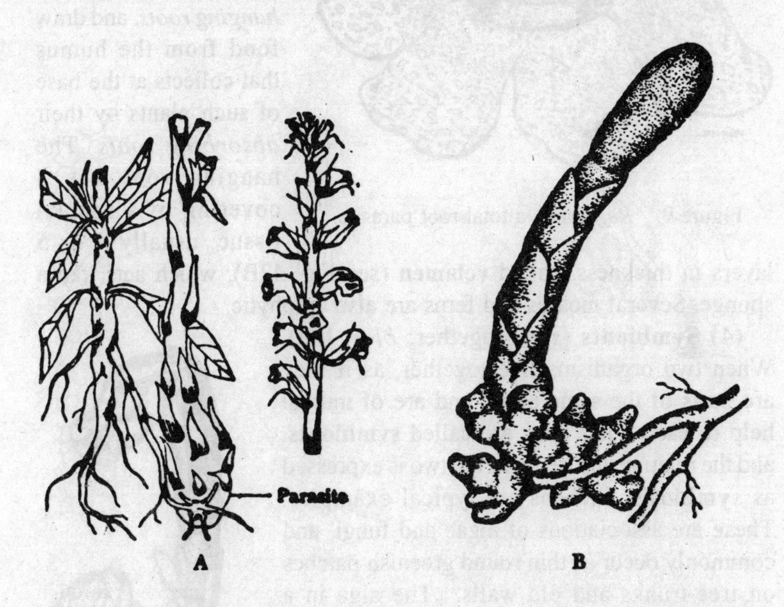

Figure 7. Total root parasites. *A,* broomrape (*Orobanche*); *B, Balanophora.*

place of and acts as the root-hairs, absorbing food material from the decomposed organic substances present in the soil. The association of a fungus with the root of a higher plant is otherwise known as **mycorrhiza**.

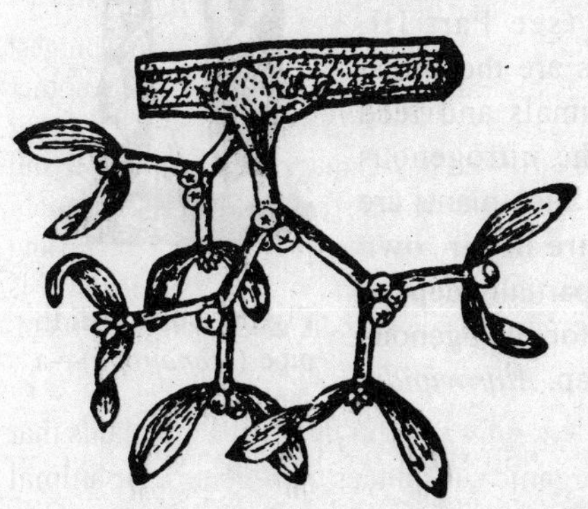

Figure 8. Mistletoe—a partial stem-parasite.

(3) **Epiphytes** (*epi,* upon; *phyta,* plants). These are plants that grow attached to the stem and branches of other plants but do not suck them, i.e. do not absorb food from them unlike the parasites. They are green in colour. Many orchids, e.g. *Vanda* (B. & H. RASNA—see Fig. 42A) are epiphytes. They

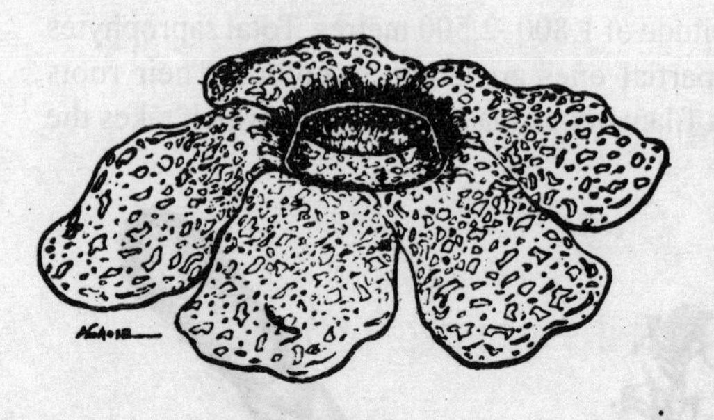

Figure 9. *Rafflesia*—a total root parasite.

absorb moisture from the air and from trickling rain-water by their *hanging roots,* and draw food from the humus that collects at the base of such plants by their *absorbing roots.* The hanging root has a covering of a special tissue, usually 4 or 5 layers in thickness, called **velamen** (see Fig. 42B), which acts like a sponge. Several mosses and ferns are also epiphytic.

(4) **Symbionts** (*sym,* together; *bios,* life). When two organisms live together, as if they are parts of the same plant, and are of mutual help to each other, they are called **symbionts**, and the relationship between the two is expressed as **symbiosis**. Lichens are typical examples. These are associations of algae and fungi, and commonly occur as thin round greenish patches on tree-trunks and old walls. The alga in a lichen being green prepares food and shares it with the fungus, while the latter absorbs water and mineral salts from the surrounding medium, and also affords protection to the alga.

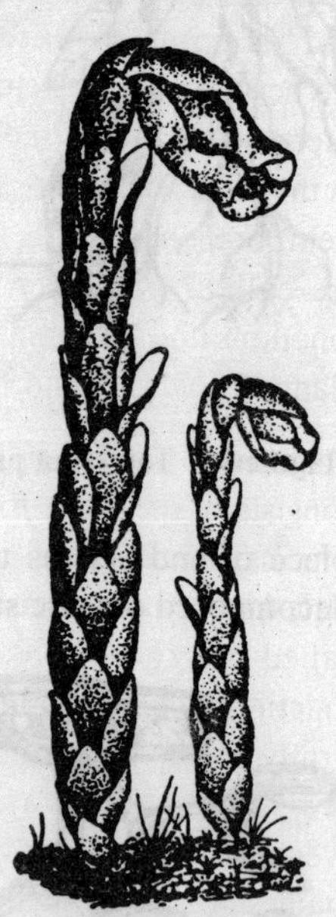

Figure 10. Indian pipe (*Monotropa*)—a saprophyte.

(5) **Carnivorous Plants** (see Part III, Chapter 7). Carnivorous plants are those that capture insects and small animals and feed upon them, absorbing only the nitrogenous compounds from their bodies. Such plants are green in colour and prepare their own carbonaceous food, while they partially depend on insects and other animals for nitrogenous food, e.g. sundew, Venus' fly-trap, *Aldrovanda,* pitcher plant and bladderwort.

Chapter 2

Parts of a 'Flowering' Plant

In response to division of labour (i.e. distribution of work) the plant body is primarily differentiated into the underground root system and the aerial shoot system. The former consists of the main root and the lateral roots, while the latter is differentiated into distinct organs such as the **stem, branches, leaves** and **flowers** (Fig. 11). Of these the roots, stem, branches and leaves are called *vegetative parts,* and the flowers are called *reproductive parts*. These organs have their respective functions and thus contribute to the life, existence and well-being of the plant as a whole, and to the continuation of the race.

Vegetative Parts. The **root system** normally lies underground and consists of the main root and the lateral roots. Each root is tipped by a cap, called the **root-cap**, which protects the tender growing apex; a little higher up, the root bears a cluster of very fine and delicate hairs, called the **root-hairs**. The root system as a whole has two primary functions: *fixation* and *absorption.* The main-root and the lateral roots firmly fix the plant to the ground; while the root-hairs absorb water and raw food materials (mineral salts) from the soil. The **shoot system** (vegetative) on the other hand is normally aerial and consists of the main stem, its branches, and leaves. The main stem and its branches have two chief functions: *support* and *conduction*. These organs give support to the leaves and the flowers and spread them out on all sides, and they conduct water and food through the plant body. These organs, but not the roots, are provided with **nodes** and **internodes**. The leaf appears at the node and is provided with a stalk, called the **petiole**, and a flat green expanded portion, the **leaf-blade** (or **-lamina**) which is interspersed by numerous **veins**, of which the median strong one is called the **mid-rib**. The green leaf-blade manufactures food and is regarded as

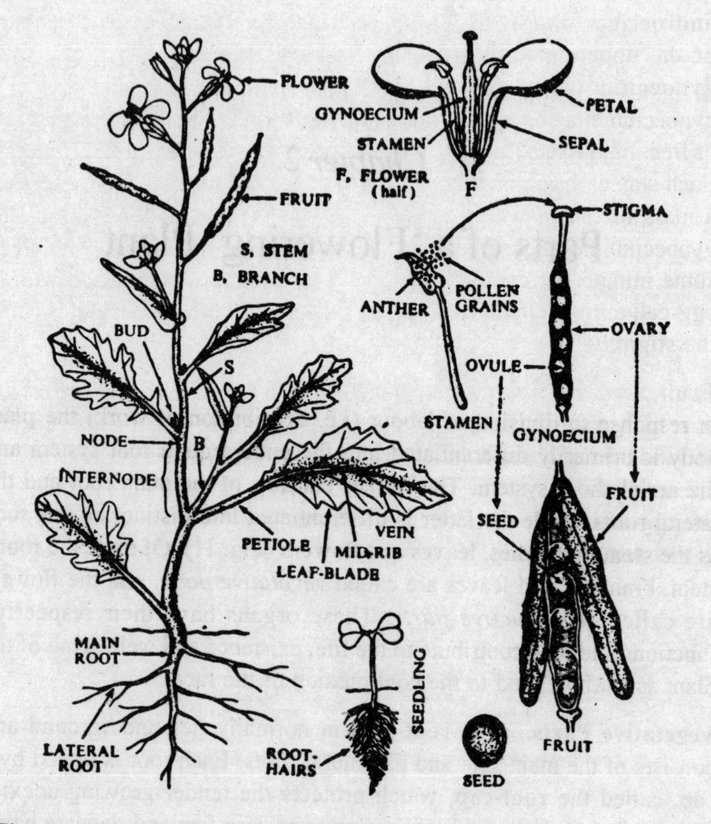

Figure 11. Parts of a 'flowering' plant (mustard plant).

a very important vegetative organ. A **bud** appears in the axil of a leaf, and as it grows and elongates, it gives rise to a branch. There is also a bud at the apex of the stem or the branch, and it is responsible for elongation of that organ by its continued growth.

Reproductive Parts. The **flower** is a highly specialized reproductive shoot. Each typical flower consists of *four* distinct types of members arranged in *four* separate whorls or circles, one above the other, on the top of a long or short stalk. The first or the lowest whorl, often green in colour, is called the **calyx**, and each member of it a **sepal**. The second whorl, often brightly coloured, is called the **corolla**, and each member of it a **petal**. The corolla attracts insects from a distance by its bright colour. The third whorl of the flower is the male whorl, called the

androecium (*andros,* male), and each member of it a **stamen**. The fourth or the uppermost whorl of the flower is the female whorl, called **gynoecium** (*gyne,* female), and each member of it a **carpel**. The gynoecium may be made of one or more carpels, frequently two, united or free. In mustard flower (Fig. 11) there are two carpels united together. Each stamen bears on its top a case, called the **anther**, which contains a mass of fine powdery or dust-like grains—the **pollen grains**. The gynoecium has a chamber at its base, called the **ovary**, which encloses some minute but complex egg-like bodies—the **ovules**, each with an **egg-cell** or ovum in it (see Fig. 131). The top of the gynoecium is called the **stigma**.

Fruit, Seed, and Embryo. Some time after the pollen grains are carried over to the stigma, commonly by insects or wind (*see* pollination, chapter 10) and the fertilization has taken place, the following important changes are noted: the ovary develops into the fruit, the ovule into the seed, and the egg-cell into the embryo. Later as the seed germinates, the embryo grows into a **seedling**.

Chapter 3

The Seed

Seeds soaked in water for a few hours or overnight should be studied by proper dissections under a simple microscope.

PARTS OF A GRAM SEED (Fig. 12)

1. **Seed-coat.** The seed is covered by a brownish coat known as the **seed-coat**. It is made up of two layers or integuments, the outer one being called the **testa** and the inner one the **tegmen**. The testa is brownish in colour and is comparatively thick; while the tegmen is whitish, thin and membranous; it is fused with the testa. The seed-coat affords necessary protection to the embryo which lies within. On one side of the seed, lying above its projected end, a small oval depression may be seen; this is known as the **hilum**. The hilum represents the point of

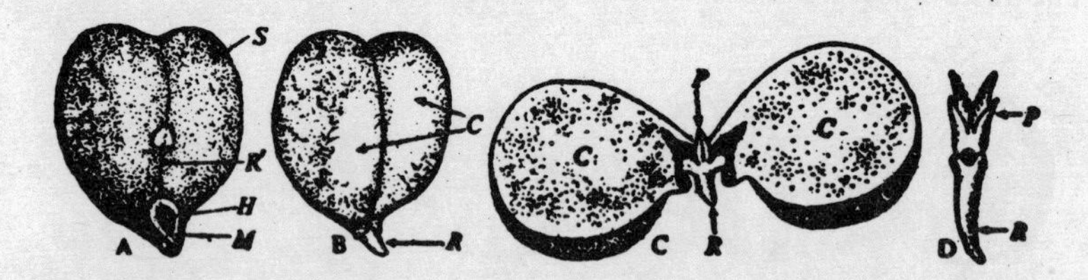

Figure 12. Gram seed: *A*, entire seed; *B*, embryo (after removal of the seed-coat); *C*, embryo with the cotyledons unfolded; and *D*, axis of embryo. *S*, seed-coat; *R'*, raphe; *H*, hilum; *M*, micropyle; *C*, cotyledons; *R*, radicle; and *P*, plumule.

attachment of the seed to its stalk. Just below the hilum a very minute slit may be seen; this minute slit or opening is known as the **micropyle** (*mikros*, small; *pyle*, a gate). When the soaked seed is gently pressed, water and minute air-bubbles are seen to escape through it. Above the hilum the stalk is continuous, with the seed-coat forming a sort of ridge; this ridge which is fused with the testa is called the **raphe**. Through the raphe food is supplied to the embryo.

2. **Embryo.** The entire fleshy body, as seen after removing the seed-coat, is the **embryo** or the baby plant. As the seed germinates it gives rise to a seedling which gradually develops into the gram plant. The embryo consists of *two* white fleshy bodies known as (a) the **cotyledons** or seed-leaves, and (b) a short **axis** to which the cotyledons are attached. The part of the axis lying towards the pointed end of the seed is called (i) the **radicle** (a little root), while the other end lying in between the two cotyledons is known as (ii) the **plumule** (*plumula,* a small feather). The plumule is surrounded at the apex by a number of minute leaves, and as such it looks more or less like a small feather. As the seed germinates the radicle gives rise to the root and the plumule to the shoot. Cotyledons store up food material.

Gram Seed
- seed-coat with testa, hilum, micropyle, raphe and tegmen
- embryo
  - axis with radicle and plumule
  - cotyledons—2, fleshy, laden with food

PARTS OF A PEA SEED (Fig. 13)

1. **Seed-coats.** The seed is somewhat roundish in shape, and is covered by two distinct **seed-coats**. Of the two coats the outer whitish one is called the **testa**; it comes off easily when the seed is soaked in water. The testa encloses another coat which is loose, thin, hyaline and

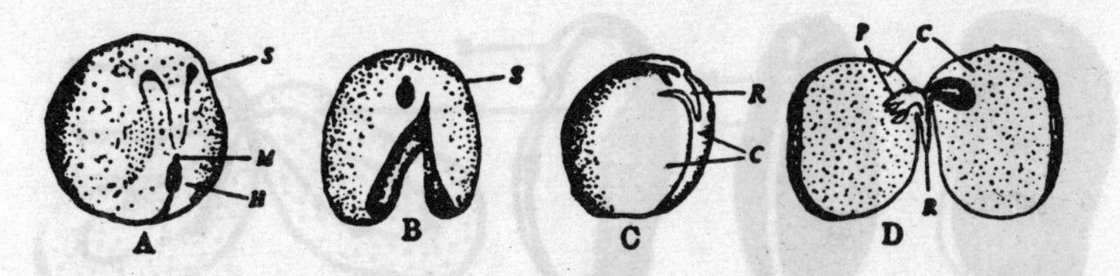

Figure 13. Pea seed. *A,* entire seed; *B,* seed-coat with hilum and micropyle; *C,* embryo (after removal of the seed-coat); *D,* embryo with the cotyledons unfolded. *S,* seed-coat—testa (It encloses a thin membranous tegmen); *M,* micropyle; *H,* hilum; *R,* radicle; *C,* cotyledons; *P,* plumule.

membranous; this inner coat is called the **tegmen**. The seed-coats give necessary protection to the embryo which lies within. On one side of the testa a narrow, elongated scar representing the point of attachment of the seed to its stalk is distinctly visible; this is the **hilum**. Close to the hilum situated at one end of it there is a minute hole; this is the **micropyle**. On germination of the seed the radicle comes out through it. Continuous with the hilum there is a sort of ridge in the testa; this is the **raphe**.

2. **Embryo.** The whole whitish body, as seen after removing the seed-coats, is the **embryo**. It consists of (a) *two* fleshy **cotyledons** or seed-leaves and (b) a short curved **axis** to which the cotyledons remain attached. The portion of the **axis** lying outside the cotyledons, bent inwards and directed towards the micropyle, is (i) the **radicle**, while, the other portion of the axis lying in between the two cotyledons is (ii) the **plumule**. The plumule is crowned by some minute young leaves. The radicle gives rise to the root, the plumule, to the shoot, and the cotyledons store up food material.

Pea Seed
- seed-coats with testa, hilum, micropyle, raphe and tegmen
- embryo
  - axis with radicle and plumule
  - cotyledons—2, fleshy, laden with food

PARTS OF A COUNTRY BEAN SEED (Fig. 14)

1. **Seed-coat.** The country bean seed (*Dolichos lablab;* B. SHIM; H. SEM) is more or less oval, and is covered by a blackish or reddish, hard **seed-coat**. The seed-coat consists of two layers fused together, the outer one being known as the **testa** and the inner one the **tegmen**. At one edge of the seed-coat there is a whitish, elongated ridge; this ridge is called the **raphe**. At the basal portion of the raphe there is a distinct broad scar; this is the **hilum**. At the other end of the raphe away from

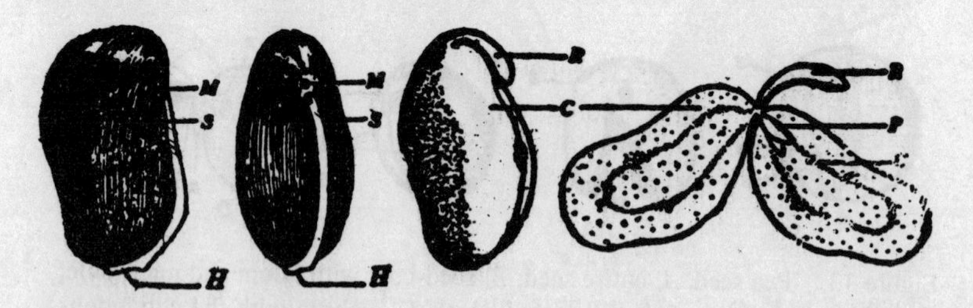

Figure 14. Country bean seed. *M,* micropyle; *S,* seed-coat; *H,* hilum; *R,* radicle; *C,* cotyledons; *P,* plumule.

the hilum there is a minute but distinct hole; this is the **micropyle**. If the soaked seed be gently pressed, water and minute air-bubbles are seen to ooze out through it.

2. **Embryo.** On peeling off the seed-coat a distinct, white, fleshy body is seen occupying the whole space within the seed-coat; this is the **embryo**. It consists of (a) two fleshy **cotyledons** and (b) a curved **axis** to which the cotyledons remain attached. The portion of the axis lying externally with its apex directed towards the micropyle is (i) the **radicle**, and the other portion of the axis lying in between the two cotyledons and composed of minute, young leaves is (ii) the **plumule**.

PARTS OF A CASTOR SEED (Fig. 15)

1. **Seed-coat.** The outer hard and blackish shell with markings on it is the seed-coat or **testa**. At one end of the seed coat there is a white body, an outgrowth formed at the micropyle, called the **caruncle**. Nearly

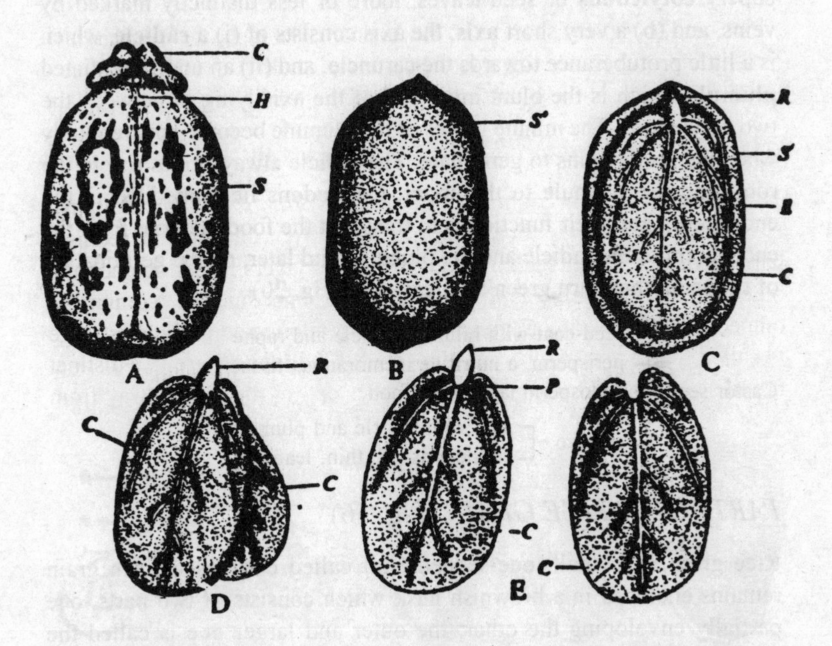

Figure 15. Castor seed. *A,* an entire seed; *B,* endosperm surrounded by perisperm; *C,* the same split open lengthwise; *D,* embryo separated from the endosperm; *E,* cotyledons separated. *Ca,* caruncle; *H,* hilum; *S,* seed-coat; *S',* perisperm, *R,* radicle; *E,* endosperm; *C,* cotyledons; *P,* plumule

hidden by the caruncle a small scar may be seen on the seed-coat, representing the point of attachment of the seed to its stalk; this is the

hilum. Running down from the hilum a ridge may be seen on the seed-coat; this ridge has been formed by the fusion of the stalk with the testa, and is known as the **raphe**.

2. **Perisperm.** On breaking open the seed-coat a thin white papery membrane may distinctly be seen surrounding the endosperm. This is a remnant of the nucellus and is called the **perisperm**. It is a nutritive tissue.

3. **Endosperm** (*endo,* inner or within; *sperm,* seed). On removing the two coats, a white fleshy mass is seen lying inside them—the **endosperm**. It is the food storage tissue of the seed, particularly rich in oil and encloses the embryo.

4. **Embryo.** This lies embedded in the endosperm. Split open the endosperm and observe that the embryo consists of (a) *two* thin, flat and papery **cotyledons** or seed-leaves, more or less distinctly marked by veins, and (b) a very short **axis**; the axis consists of (i) a **radicle**, which is a little protuberance towards the caruncle, and (ii) an undifferentiated **plumule** which is the blunt inner end of the axis lying in between the two cotyledons. The minute leaves of the plumule become apparent only when the seed begins to germinate. The radicle always gives rise to the root, and the plumule to the shoot. Cotyledons lie embedded in the endosperm, and their function is to transport the food material from the endosperm to the radicle and the plumule, and later, on the germination of the seed, they turn green and leafy (see Fig. 20).

Castor seed
- seed-coat with hilum, caruncle and raphe
- perisperm, a nutritive membranous tissue
- endosperm laden with food
- embryo
  - axis with radicle and plumule
  - cotyledons—2, thin, leaf-like

PARTS OF A RICE GRAIN (Fig. 16)

Rice grain is a small, one-seeded fruit called caryopsis. Each grain remains enclosed in a brownish husk which consists of two parts, one partially enveloping the other; the outer and larger one is called the *flowering glume,* while the inner and smaller one is called the *palea*. At the base of the grain are two minute white scales called *empty glumes*. The rice grain and the husk are together known as the paddy grain.

1. **Seed-coat.** On removing the husk a brownish membranous layer is seen adherent to the grain. This layer is made up of the seed-coat and the wall of the fruit fused together.

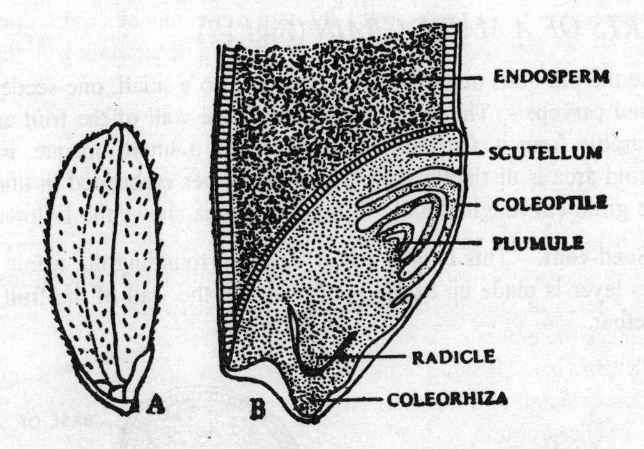

Figure 16. Rice grain. *A,* the grain enclosed in husk; *B,* the grain in longitudinal section (a portion).

2. **Endosperm.** This forms the main bulk of the grain and is its food storage tissue, being laden with reserve food material, particularly starch. In a longitudinal section of the grain it is seen to be distinctly separated from the embryo by a definite layer known as the *epithelium*.

3. **Embryo.** This is very small and lies in a groove at one end of the endosperm. It consists of only (a) *one* shield-shaped cotyledon which is known as the **scutellum**, and (b) a short **axis** which has an upper portion (i) the **plumule**, and a lower portion (ii) the **radicle**. The plumule is surrounded by minute leaves, and the radicle is protected by a cap known as the **root-cap**. The plumule as a whole (growing point and foliage leaves) is surrounded and protected by a plumule-sheath, called **coleoptile**; similarly the radicle is surrounded, by a root-sheath, called **coleorhiza**. The surface layer of the scutellum lying in contact with the endosperm is the *epithelium;* its function is to digest and absorb food material stored in the endosperm.

- Rice grain (enclosed in a husk)
  - Seed-coat with the fruit-wall fused with it.
  - Endosperm ladden with food material and separated from embryo by epithelium.
  - Embryo
    - Axis with radicle and plumule surrounded by coleorhiza and coleoptile respectively.
    - Cotyledon (scutellum)—1, absorbing tissue.

PARTS OF A MAIZE GRAIN (Fig. 17)

Like the previous one the maize grain is also a small, one-seeded fruit called caryopsis. The seed is adherent to the wall of the fruit and not separable from it. On one side of the grain a small, opaque, whitish, deltoid area is distinctly seen. The embryo lies embedded in this area. The grain cut longitudinally through this area shows the following:

1. **Seed-coat.** This is only a thin, layer surrounding the whole grain. This layer is made up of the seed-coat and the wall of the fruit fused together.

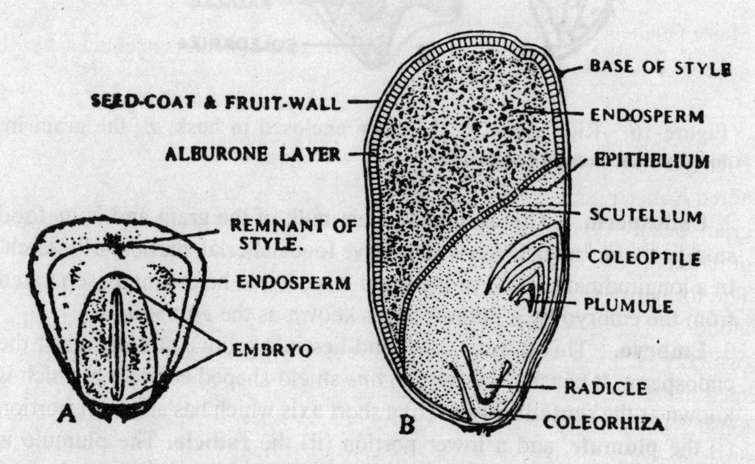

Figure 17. Maize grain. *A,* the entire grain; *B,* the grain in longitudinal section.

2. **Endosperm.** The grain is divide into two unequal portions by a definite layer known as the *epithelium*. The bigger portion is the endosperm, and the smaller portion is the embryo. The endosperm is the food storage tissue of the grain, particularly rich in starch. If a little iodine solution be dropped on the cut surface of the grain the whole of the endosperm becomes black indicating the presence of starch; the embryo takes on a yellowish tinge. Thus the two portions become clearly marked.

3. **Embryo.** This consists of (a) *one* shield-shaped cotyledon known as the **scutellum**, as in the rice grain, and (b) an **axis**. The upper portion of the axis, with minute leaves arching over it is (i) the **plumule**, and the lower portion, provided with the **root-cap**, (ii) the **radicle**. The plumule is surrounded by a plumule-sheath or **coleoptile**, and the radicle

is surrounded by a root-sheath or **coleorhiza**. Coleoptile and coleorhiza are protective sheaths of the plumule and the radicle respectively and are characteristic of the grass family and the palm family. The surface layer of the scutellum lying in contact with the endosperm is the *epithelium;* its function is to digest and absorb food material stored in the endosperm.

Note. In cereals (e.g. rice, wheat, maize, barley and oat), millets and other plants of the grass family (Poaceae) the cotyledon is known as the **scutellum**. It supplies the growing embryo with food material absorbed from the endosperm with the help of the *epithelium.*

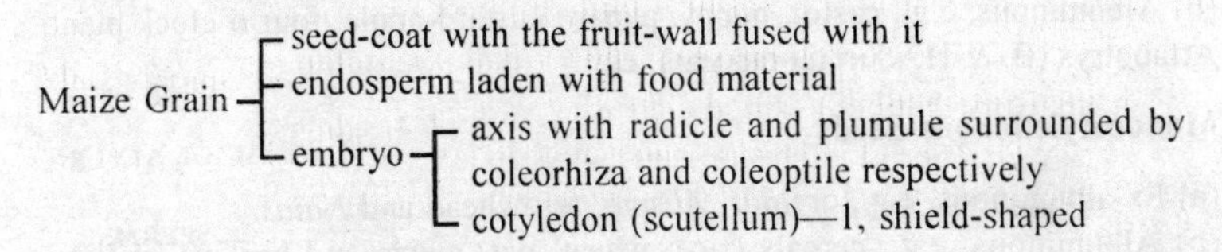

Dicotyledonous and Monocotyledonous Seeds. It must have been noted from studies of the above mentioned seeds that some of them, e.g. gram, bean, pea, castor, etc., bear *two cotyledons* in their embryo; while others, e.g. rice, maize, etc., bear only *one cotyledon* in their embryo. The former types of seeds are said to be dicotyledonous, and the latter mono-cotyledonous. In a dicotyledonous seed the axis is terminal and two cotyledons are lateral in position. Whereas in a monocotyledonous seed the single cotyledon is terminal and the axis is lateral in position. On the basis of the number of cotyledons present in a seed and other characters the 'flowering' plants have been divided into two big classes: **dicotyledons** (with two cotyledons) and **monocotyledons** (with one cotyledon). Dicotyledons far outnumber monocotyledons.

Albuminous and Ex-albuminous Seeds. (1) Seeds that possess a special food storage tissue, called the endosperm, are said to be **albuminous** or endospermic, and those that possess no such special tissue for food storage are said to be **ex-albuminous** or non-endospermic. Mono-cotyledonous seeds are mostly albuminous; while among dicotyledons both are common.

(2) In all seeds the food accumulates in the endosperm tissue at an early stage of seed development. But in albuminous seeds the endosperm continues to store food and to enlarge rapidly. Ultimately in the mature seed it acts as the food storage tissue. In ex-albuminous seeds on the other hand, the food that accumulates in the endosperm tissue at an

early stage of seed-development is utilized by the developing embryo so that the endosperm becomes exhausted.

(3) In albuminous seeds, food being stored in the endosperm, the cotyledon(s) are small and thin; while in ex-albuminous seeds the cotyledons(s) store up food and become thick and fleshy. The food, whether stored in the endosperm or in the cotyledons, is always utilized by the embryo when germination of the seed takes place.[1]

Dicotyledonous Seeds

(a) Ex-albuminous, e.g., gram, pea, bean, gourd, tamarind, mustard, mango, cotton, orange, pulses, sunflower, guava, jack, etc.
(b) Albuminous, e.g., castor, poppy, papaw, custard-apple, four o'clock plant, Artabotrys (B. & H. KANTALI-CHAMPA), etc.

Monocotyledonous Seeds

(a) Ex-albuminous, e.g., orchids, *Alisma,* arrowhead and *Naias.*
(b) Albuminous, e.g., cereals (rice, wheat, oat, maize and barley), millets, grasses (including sugarcane and bamboo), palms, lilies, aroids, etc.

GERMINATION

To study the stages in the germination of seeds they may be sown in moist sand or sawdust. Pre-soaking in water for a few to several hours helps quicker germination.

The embryo lies dormant in the dry seed, but when the latter is supplied with moisture the embryo becomes active and tends to grow and develop into a small seedling. *The process by which the dormant embryo wakes up and begins to grow is known as* **germination**. The first sign of germination is the protrusion of the radicle, often through the micropyle. The radicle then quickly elongates, grows downwards, sometimes forming a loop, and gives rise to the primary root. Its rate of growth is much faster than that of the plumule. The seed-coat bursts and the cotyledons partially or completely separate from each other.

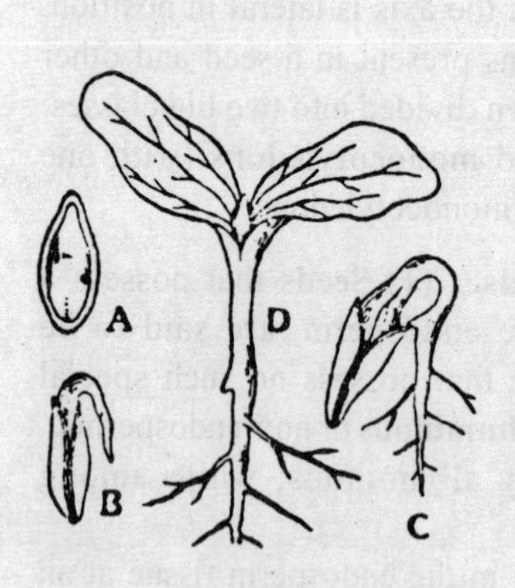

Figure 18. Epigeal Germination, Gourd seed (exalbuminous).

[1]For 'food stored in the seed' see end of Chapter 8, Part III.

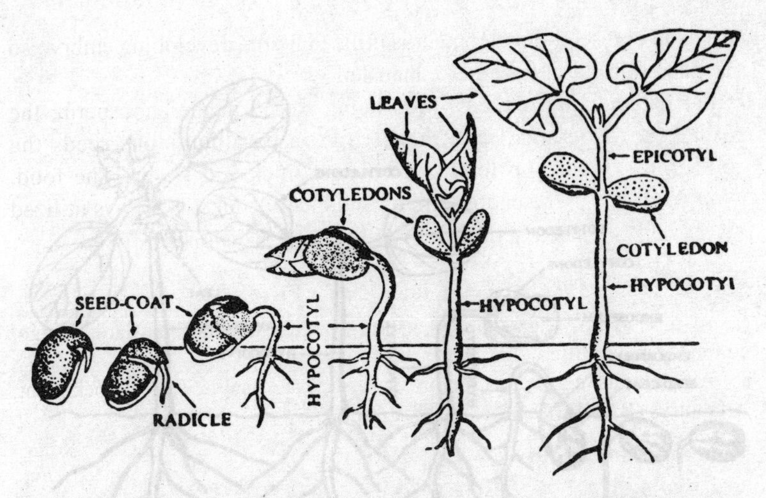

Figure 19. Epigeal Germination. Country bean seed (exalbuminous).

The plumule comes out, grows upwards and gives rise to the shoot. The cotyledons may turn green and become leaf-like in appearance or they may shrivel up and drop. Two kinds of germination are noticed: epigeal and hypogeal.

1. **Epigeal germination** (Figs. 18–20). In some seeds such as bean, gourd, sunflower, cotton, castor, etc., the cotyledons are seen to be pushed upwards by the rapid elongation of the **hypocotyl** (*hypo,* below), i.e. the portion of the axis (radicle) lying immediately below the cotyledonary zone or node (where the cotyledons remain attached to the axis). Germination of this kind is said to be **epigeal** (*epi,* upon; *ge,* earth).

Epigeal Germination of Country Bean Seed (Fig. 19). The seed sown in moist soil absorbs moisture and swells. The seed-coat bursts, and the radicle comes out through the micropyle and grows downwards, forming the primary root. Soon it produces lateral roots. The hypocotyl elongates and lifts the seed out of the ground. The seed-coat drops, and the plumule emerges between the two thick cotyledons. The hypocotyl straightens and two leaves unfold at first. They rapidly grow in size and turn green in colour. During germination the food stored in the thick cotyledons is utilized. Now the green leaves manufacture food for the seedling. As the food is removed from the cotyledons they shrivel up and soon fall off.

Epigeal Germination of Castor Seed (Fig. 20). As germination begins the caruncle drops, and the radicle comes out and grows downwards. It elongates and forms the primary root which soon produces many lateral roots. The

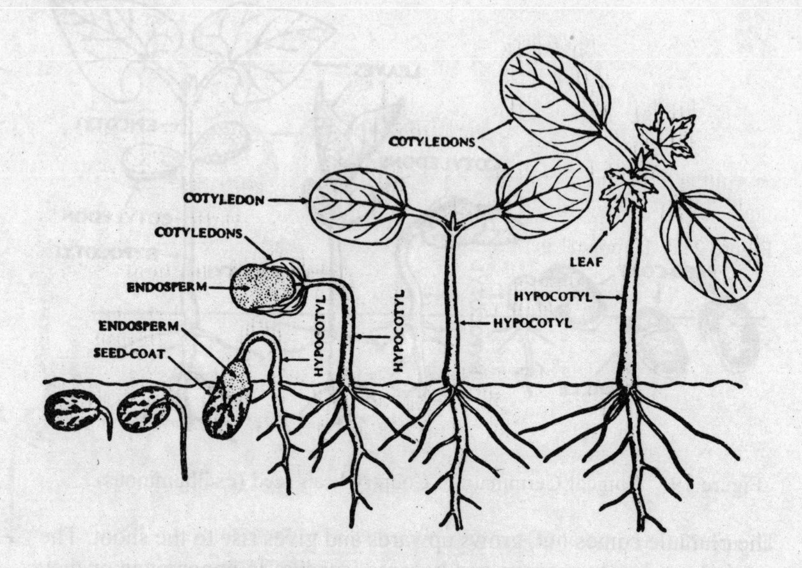

Figure 20. Epigeal Germination. Castor seed (albuminous).

hypocotyl elongates and lifts the seed out of the ground. In the meantime the seed-coat bursts and drops, the perisperm shrivels up, and the food stored in the endosperm gradually moves to the growing parts of the embryo. Evidently the endosperm gets thinner and soon it drops. The hypocotyl straightens up, and the two cotyledons expand and turn green in colour. The minute plumule, hidden between the two cotyledons, now begins to grow. It at first produces two leaves, which quickly grow and turn green in colour. The green leaves and the cotyledons now prepare all the food for the seedling.

2. **Hypogeal Germination** (Figs. 21–22). In other seeds such as gram, pea, mango, litchi, jack, groundnut, etc., the cotyledons are seen to remain in the soil or just on its surface. In such cases the **epicotyl**, i.e. the portion of the axis (plumule) lying immediately above the cotyledonary node or zone, elongates and pushes the plumule upwards. The cotyledons do not turn green, but gradually dry up and fall off. Germination of this kind is said to be **hypogeal** (*hypo,* below; *ge,* earth).

Hypogeal Germination of Pea (Fig. 22). As the seed swells by absorbing moisture the radicle comes out through the micropyle and quickly grows downwards, forming the primary root. The seed-coats burst, and the epicotyl slightly elongates and takes the form of a loop. The plumule, hidden between the two cotyledons tends to emerge out of them. The epicotyl further elongates and cranes its neck. It soon straightens up, and leaves appear from the plumule

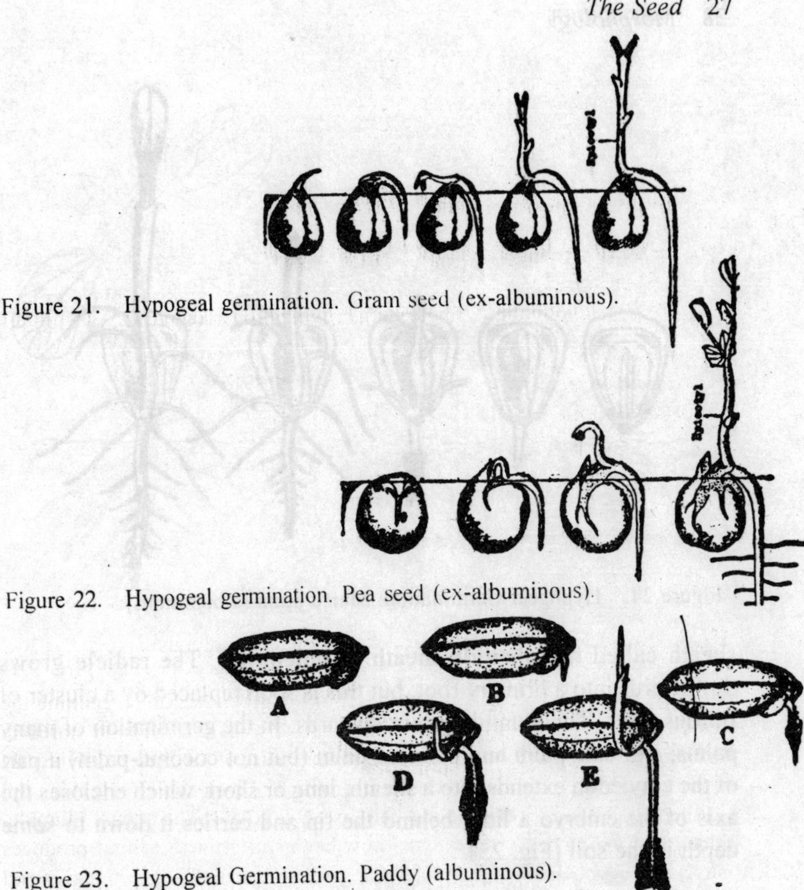

Figure 21. Hypogeal germination. Gram seed (ex-albuminous).

Figure 22. Hypogeal germination. Pea seed (ex-albuminous).

Figure 23. Hypogeal Germination. Paddy (albuminous).

and turn green in colour. In this seed the food is stored in the two fleshy cotyledons, and this food is utilized during germination. Evidently the cotyledons get thinner. All the time, however, they remain in the soil together with the seed-coats. In the meantime the root further elongates, and produces lateral roots. The leaves also enlarge and turn deeper green, and the seedling becomes established in the soil.

Hypogeal Germination of Monocotyledonous Seeds (Figs. 23–25). Monocotyledonous seeds are mostly albuminous and in their germination the cotyledon and the endosperm remain buried in the soil; germination is, therefore, hypogeal (except in the case of onion). In the germination of monocotyledonous seeds like paddy and maize (Figs. 23–24), the radicle makes its way through the lower short, collar-like end of the sheath called the root-sheath or **coleorhiza**; while the plumule breaks through the upper distinct, cylindrical portion of the

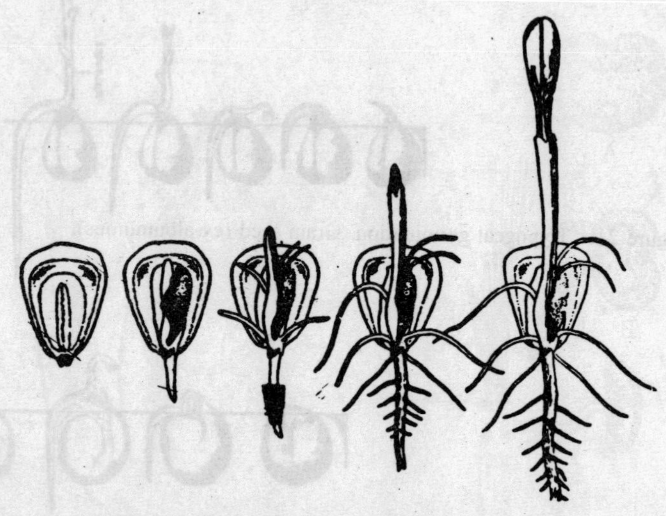

Figure 24. Hypogeal Germination. Maize grain (albuminous).

sheath called the plumule-sheath or **coleoptile**. The radicle grows downwards into a primary root, but this is soon replaced by a cluster of fibrous roots. The plumule grows upwards. In the germination of many palms, e.g. date-palm and palmyra-palm (but not coconut-palm) a part of the cotyledon extends into a sheath, long or short, which encloses the axis of the embryo a little behind the tip and carries it down to some depth in the soil (Fig. 25).

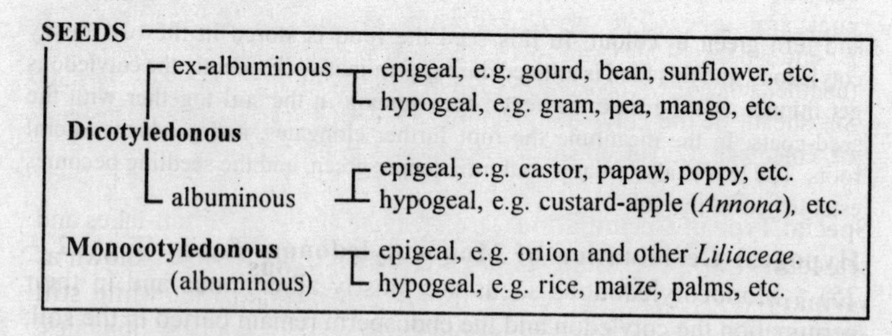

SEEDS

Dicotyledonous	ex-albuminous	epigeal, e.g. gourd, bean, sunflower, etc.
		hypogeal, e.g. gram, pea, mango, etc.
	albuminous	epigeal, e.g. castor, papaw, poppy, etc.
		hypogeal, e.g. custard-apple (*Annona*), etc.
Monocotyledonous	(albuminous)	epigeal, e.g. onion and other *Liliaceae*.
		hypogeal, e.g. rice, maize, palms, etc.

Hypogeal Germination of Maize Grain (Fig. 24). The grain absorbs moisture and swells. The shield covering the embryo at the deltoid area splits and the radicle pierces the collar-like root-sheath or **coleorhiza** and grows downwards, giving rise to the primary root. A few (adventitious) roots also come out. The plumule-sheath or **coleoptile** enclosing the plumule elongates and grows upwards. Soon the plumule breaks through the upper end of the cylindrical coleoptile and gives rise to the first leaf. The seed remains embedded in the soil. From the primary root lateral roots are quickly produced. The food stored in the endosperm is utilized during germination. By the time this food becomes exhausted the leaf, has already turned green and is in a position to manufacture food for the seedling which has now become established in the soil.

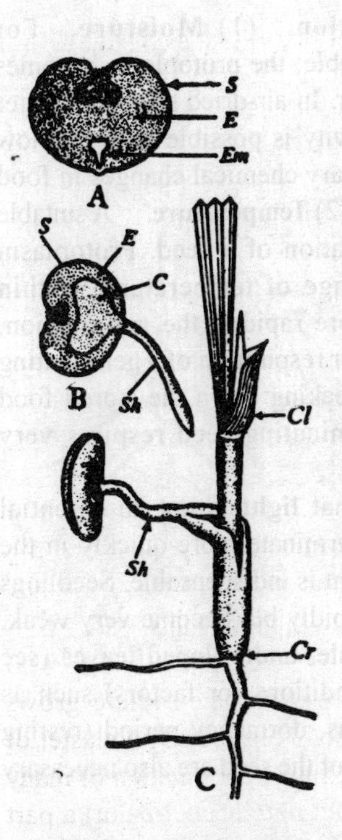

Figure 25. Date-palm seed and its germination. *A*, seed in section; *B*, germinating seed in section; *C*, seedling; *S*, seed-coat and inner fruit-wall; *E*, endosperm; *Em*, embryo (undifferentiated); *C*, cotyledon; *Sh*, sheath of the cotyledon; *Cl*, coleoptile; *Cr*, coleorhiza.

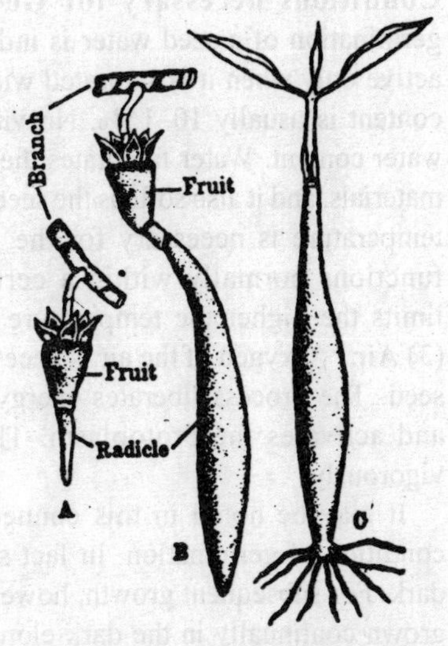

Figure 26. Viviparous germination. *A–B*, stages in germination; *C*, seedling.

Special Type of Germination. Many plants growing in salt-lakes and sea-coasts show a special type of germination of their seeds, known as **vivipary** (Fig. 26). The seed germinates inside the fruit while still attached to the parent tree and is nourished by it. The radicle elongates, swells in the lowest part and gets stouter and heavier. Soon the seedling separates from the parent tree and drops vertically down. The radicle presses into the soft mud, and quickly lateral roots are formed for proper anchorage. Examples are seen in *Rhizosphora* (B. KHAMO), *Sonneratia* (B. KEORA), *Heritiera* (B. SUNDRI), etc.

Conditions necessary for Germination. (1) **Moisture.** For germination of a seed water is indispensable; the protoplasm becomes active only when it is saturated with water. In air-dried seeds the water content is usually 10–15%. No vital activity is possible with this low water content. Water facilitates the necessary chemical changes in food materials, and it also softens the seed-coat. (2) **Temperature.** A suitable temperature is necessary for the germination of a seed. Protoplasm functions normally within a certain range of temperature. Within limits the higher the temperature the more rapid is the germination. (3) **Air.** Oxygen of the air is necessary for respiration of a germinating seed. The process liberates energy by breaking down the stored food and activates the protoplasm. The germinating seed respires very vigorously.

It may be noted in this connection that **light** is not an essential condition of germination. In fact seeds germinate more quickly in the dark. For subsequent growth, however, light is indispensable. Seedlings grown continually in the dark elongate rapidly but become very weak, develop no chlorophyll and bear only pale, undeveloped leaves (see Fig. III/27). Besides, certain internal conditions (or factors) such as presence of auxin, food supply to the axis, dormancy period (resting period) and viability (ability to germinate) of the seed are also necessary for germination.

Three Bean Experiment (Fig. 27). That all the conditions mentioned above are essential for germination can be shown by a simple experiment, known as the **three bean experiment**. Three air-dried seeds are attached to a piece of wood, one at each end and one in the middle. This is then placed in a beaker, and water is poured into it until the middle seed is half immersed in it. The beaker is then left in a warm place for a few days. From time to time water is added to maintain the original level. It is seen that the bean germinates normally because it has sufficient moisture, oxygen and heat. The bottom bean has sufficient moisture and heat but no oxygen. It may be seen to put out the radicle only but further development is checked for want of oxygen. The top bean having only sufficient oxygen and heat, but no moisture, does not show any sign of germination.

This experiment evidently shows that moisture and oxygen are indispensable for germination; the effect of temperature is only indirectly proved. It can, however, be directly proved in the following way. Other

Figure 27. Three bean experiment.

conditions remaining the same, if the temperature be considerably lowered or increased by placing the beaker with the seeds in a freezing mixture or in a bath with constant high temperature. It will be seen that none of the beans will germinate. Thus suitable temperature is also an essential condition for germination.

FUNCTIONS OF COTYLEDONS

(1) In ex-albuminous seeds, as in gram, pea, gourd, tamarind, etc., the cotyledons act as food storage organs and in consequence they become thick and fleshy. The food stored in them is utilized by the embryo when the seed germinates.

(2) In albuminous seeds, as in castor, poppy, four o'clock plant, etc., the cotyledons act as absorbing organs, and they are thin, flat or small. When the seeds germinate, they absorb food from the endosperm and supply it to the radicle and the plumule.

(3) In many seeds showing epigeal germination (i.e. lifting the cotyledons above the ground) the cotyledons may act as food-manufacturing organs. When they are pushed above the ground they generally turn green in colour being exposed to light and then function like ordinary leaves, i.e. they manufacture food in the presence of sunlight.

(4) The cotyledons act as protective organs. They enclose the plumule at the seed stage, and during the early germination period they give it adequate protection.

(5) In monocotyledonous seeds, at the time of germination the cotyledon absorbs food from the endosperm, and at length extends as a sort of sheath, long or short, pushing the radicle and the plumule out of the seed. In many palms as in palmyra-palm and date-palm (but not coconut-palm) a fairly long sheath is produced (see Fig. 25).

Chapter 4

The Root

The **root** is the descending organ of the plant, and is originally the direct prolongation of the radicle of the embryo. It grows downwards, fixes the seedling and later the plant as a whole to the ground, and absorbs raw food materials (water and inorganic salts) from the soil. It is non-green in colour, without nodes and internodes, with no leaves or buds, and is covered at its tip, i.e. the growing point, by a sort of cap known as the **root-cap** (Fig. 33).

Normal Roots. All roots that develop from the radicle, either directly from it or its branches, are called normal roots. The direct prolongation of the radicle forms the **primary root**. If it persists and continues to grow, as in dicotyledons, giving rise to the main root of the plant, it is called the **tap root** (Fig. 28). The tap root normally grows vertically

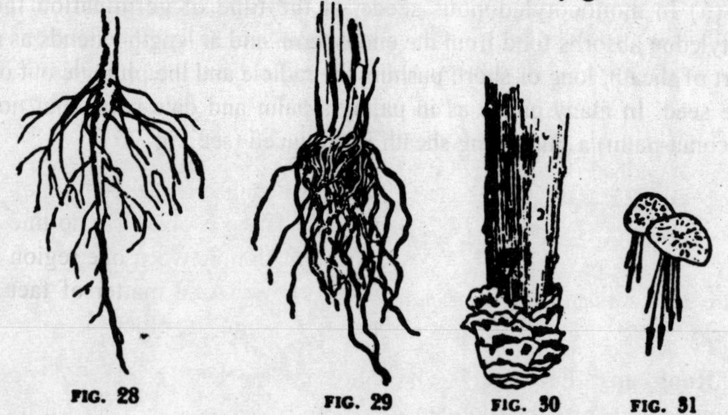

Figure 28. Tap and lateral roots in a dicotyledon. Figure 29. Fibrous roots in a monocotyledon. Figure 30. Multiple root-cap in screwpine (*Pandanus*). Figure 31. Root-pockets in duckweed (*Lemna;* see p. 31).

downwards to a shorter or longer depth. As it grows it produces lateral branches known as the *secondary roots,* and these in turn produce the *tertiary roots*. All these roots together form the **tap root system** of the plant. The lateral roots are produced in *acropetal* succession, i.e. the older and longer roots away from the tip, and, the younger and shorter ones towards it.

Adventitious Roots. Roots that grow from any part of the plant body other than the radicle are called adventitious roots. (1) In monocotyleons where the primary root does not persist, a cluster of slender roots is seen to grow from the base of the stem; such roots arc called **fibrous roots** (Fig. 29). (2) Adventitious roots, solitary or in clusters, also grow from nodes and even internodes, as in many grasses, betel, wood-sorrel, sugarcane, maize, bamboo, etc. (3) They also often grow from stem-cuttings (Fig. 32), as in *Coleus,* rose, garden croton, tapioca, etc. (4) Adventitious roots called **foliar roots** (see Fig. III/28) may also be induced to grow from the petiole or vein of a leaf by the application of certain chemicals, called *hormones,* which are growth-promoting substances.

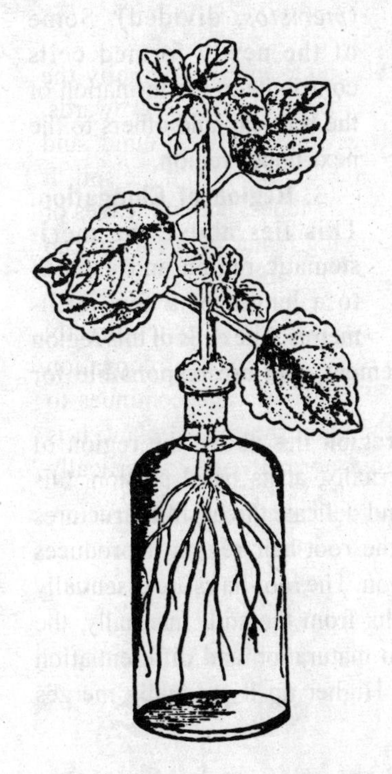

Figure 32. Adventitious roots in *Colecus.*

Regions of the Root (Fig. 33). The following regions may be distinguished in a root from the apex upwards. There is of course no line of demarcation between one region and the other. As a matter of fact one merges into the other.

1. **Root-cap.** Each root is covered over at the apex by a sort of cap or thimble known as the root-cap which protects the tender apex of the root as it makes its way through the soil. The root-cap, if worn out, may be renewed by the underlying growing tissue. It is usually absent in aquatic plants.

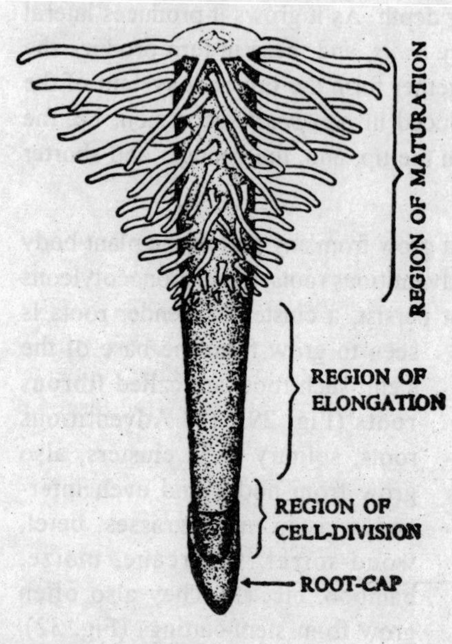

Figure 33. Regions of the root.

2. **Region of Cell Division.** This is the growing apex of the root lying within and a little beyond the root-cap and extends to a length of a few millimetres. The cells of this region undergo repeated divisions, and hence this region is otherwise called the **meristematic region** (*meristos,* divided). Some of the newly formed cells contribute to the formation of the root-cap and others to the next upper region.

3. **Region of Elongation.** This lies above the meristematic region and extends to a length of a few millimetres. The cells of this region undergo rapid elongation and enlargement, and are responsible for growth in length of the root.

4. **Region of Maturation.** This region lies above the region of elongation and extends upwards. Externally, at its basal portion, this region produces a cluster of very fine and delicate thread-like structures known as the **root-hairs**, and above the root-hair region it produces **lateral roots**, both in *acropetal* succession. The root-hairs are essentially meant to absorb water and mineral salts from the soil. Internally, the cells of this region are seen to undergo maturation and differentiation into various kinds of primary tissues. Higher up it gradually merges into the region of secondary tissues.

Characteristics of the Root. There are certain distinctive characteristics of the root by which it can be distinguished from the stem. These are as follows:

(1) The root is the descending organ of the plant and the primary root is the direct prolongation of the radicle; whereas the stem is the ascending organ of the plant and the direct prolongation of the plumule. Roots grow downwards away from light (i.e. positively geotropic);

whereas the stem grows upwards towards light (i.e. positively phototropic). Roots are not normally green in colour; whereas the young stem is normally so.

(2) The root does not normally bear **buds**; while the stem normally bears both vegetative and floral buds for vegetative growth and reproduction. There are, however, cases where the roots are seen to bear vegetative buds (but not floral buds) for vegetative propagation, e.g. sweet potato (*Batatas* = *Ipomoea*), wood-apple (*Aegle*), *Trichosanthes* (B. PATAL; H. PARWAL), Indian redwood (*Dalbergia;* B. SISSOO; H. SHISHAM), and ipecac (*Psychotria*). Such plants are sometimes propagated by root-cuttings, e.g. ipecac (a medicinal plant).

(3) The root ends in, and is protected by a cap or thimble known as the **root-cap** (Fig. 34A); while the stem ends in a bud—the terminal

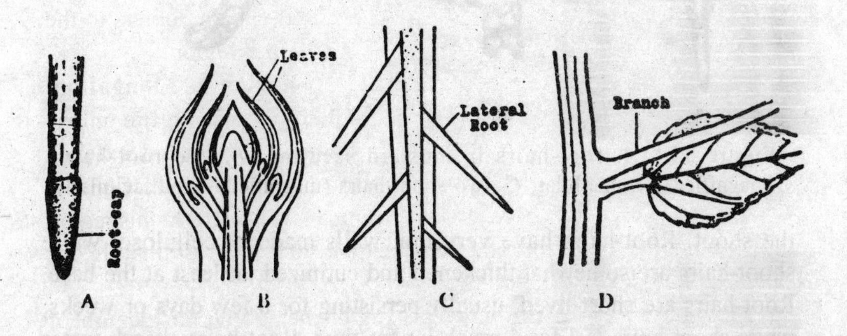

Figure 34. *A,* root-tip; *B,* Stem-apex; *C,* lateral root (endogenous; see also Fig. II/52); *D,* branch (exogenous).

bud (see Fig. 44). A distinct, multiple root-cap is seen in the aerial root of screwpine (*Pandanus;* B. KETUCKY; H. KEORA—Fig. 30).

In water plants like duckweed (*Lemna*), water lettuse (*Pistia*), water hyacinth (*Eichhornia*), etc., a loose sheath which comes off easily is distinctly seen at the apex of each root. This is an anomalous root-cap, called the **root-pocket** (see Fig. 31).

(4) The root bears **unicellular hairs** (Fig. 35A–B); while the stem or the shoot bears mostly **multicellular hairs** (Fig. 35C). Root-hairs occur in a cluster all over the tender part of the young root a little behind the root-cap. But as the root grows, older root-hairs die off and newer ones are always formed close behind the apex. Shoot-hairs, on the other hand, are of various kinds and they remain scattered over the surface of

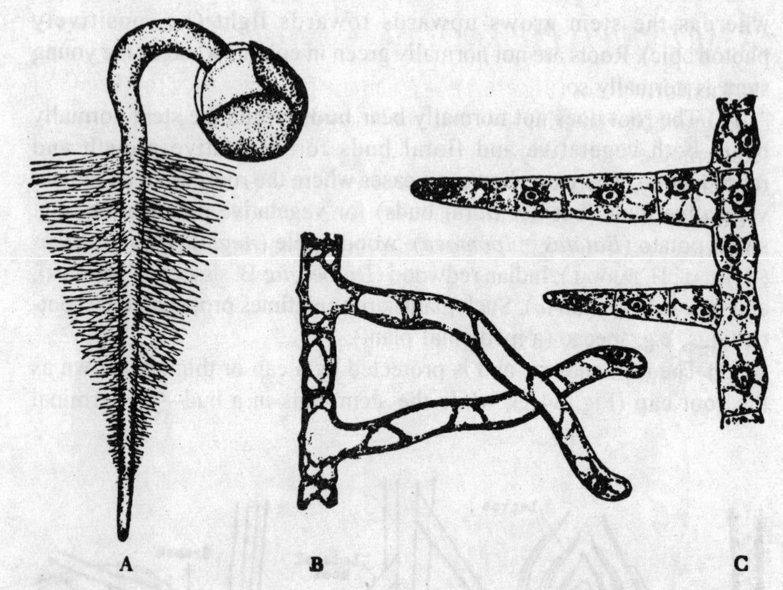

Figure 35. *A*, root-hairs in mustard seedling; *B*, two root-hairs (magnified)—unicellular; *C*, two shoot-hairs (magnified)—multicellular.

the shoot. Root-hairs have very thin walls made of cellulose; while shoot-hairs are somewhat thickened and cutinized, at least at the base. Root-hairs are short-lived, usually persisting for a few days or weeks; while shoot-hairs last for a much longer time. Root-hairs absorb water and mineral salts from the soil, and shoot-hairs prevent evaporation of water from the surface of the plant body and afford protection.

(5) Lateral roots always develop from an inner layer (Fig. 34C); so they are said to be **endogenous** (*endo,* inner; *gen,* producing). Branches, on the other hand, develop from a few outer layers (Fig. 34D); so they are said to be **exogenous** (*exo,* outer).

(6) **Nodes** and **internodes** are always present in the stem, although they may not often be quite distinct; but in the root they are absent.

MODIFIED ROOTS

Specialized functions of varied nature are performed by the modified roots which adapt themselves according to the particular need of the plant. For these purposes the tap root, branch root and adventitious roots may undergo modifications. The following are a few examples of such types:

A. *MODIFIED TAP ROOT* (for storage of food)

1. **Fusiform Root** (Fig. 36A). When the root is swollen in the middle and gradually tapering towards the apex and the base, being more or less spindle-shaped in appearance, it is said to be fusiform, e.g. radish.
2. **Napiform Root** (Fig. 36B). When the root is considerably swollen at the upper part becoming almost spherical, and sharply tapering at the lower part, it is said to be napiform, e.g. turnip and beet.

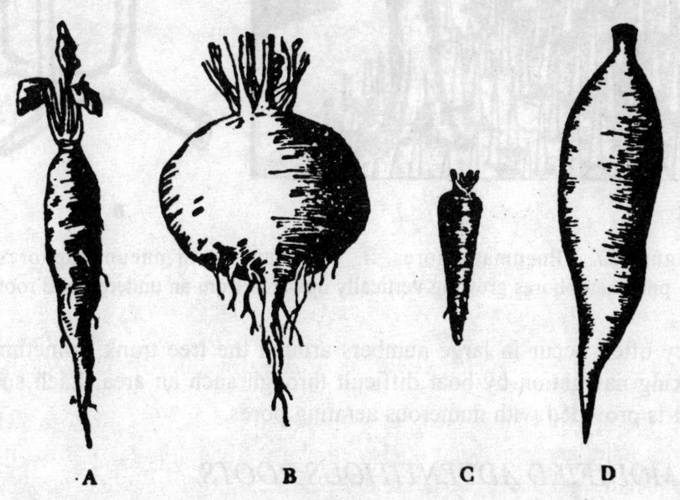

Figure 36. Modified Roots. *A,* fusiform root of radish; *B,* napiform root of turnip; *C,* conical root of carrot; *D,* tuberous root of *Mirabilis.*

3. **Conical Root** (Fig. 36C). When the root is broad at the base and it gradually tapers towards the apex like a cone, it is said to be conical, e.g. carrot.
4. **Tuberous** or **Tubercular Root** (Fig. 36D). When the root is thick and fleshy but does not maintain any particular shape, it is said to be tuberous or tubercular, as in four o'clock plant.

B. *MODIFIED BRANCH ROOT* (for respiration)

Pneumatophores. Many plants growing in estuaries and salt lakes, occasionally inundated by tides, develop special kinds of roots called respiratory roots or **pneumatophores** (Fig. 37) for the purpose of respiration. Such roots grow from the underground roots of the plant but rise vertically upwards and come out of the water like conical spikes.

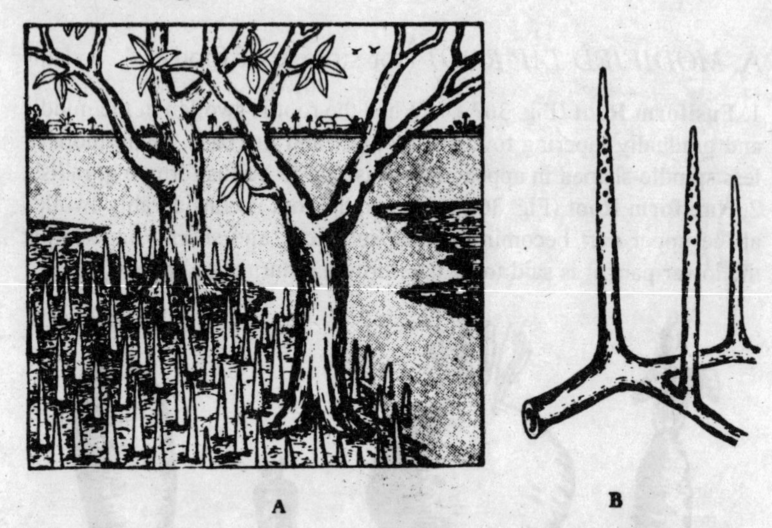

Figure 37. Pneumatophores. *A,* two plants with pneumatophores; *B,* pneumatophores growing vertically upwards from an underground root.

They often occur in large numbers around the tree trunk, sometimes making navigation by boat difficult through such an area. Each such root is provided with numerous aerating pores.

C. *MODIFIED ADVENTITIOUS ROOTS*

(a) *for storage of food*

1. **Tuberous** or **Tubercular Root** (Fig. 38A). This is a swollen root without any definite shape, as in sweet potato. Tuberous roots, whether tap or adventitious, are produced singly and not in clusters.
2. **Fasciculated Roots** (Fig. 38B). When several tubercular roots occur in a cluster or fascicle at the base of the stem, they are said to be fasciculated, as in *Dahlia, Ruellia* and *Asparagus*.
3. **Nodulose Root** (Fig. 38C). When the slender root becomes suddenly swollen at or near the apex, it is said to be nodulose, as in mango ginger (*Curcuma amada;* B. AMADA; H. AMHALDI), turmeric (*Curcuma domestica;* B. HALOOD; H. HALDI) and arrowroot *(Maranta)*.
4. **Moniliform** or **Beaded Root** (Fig. 39A). When there are some swellings in the root at frequent intervals, it is said to be moniliform or beaded, as in Indian spinach (*Basella;* B. PUI; H. POI), *Momordica* (B. KAKROL; H. CHATTHAI), wild vine (*Vitis trifolia;* B. AMAL-LATA; H. AMALBEL) and some grasses.

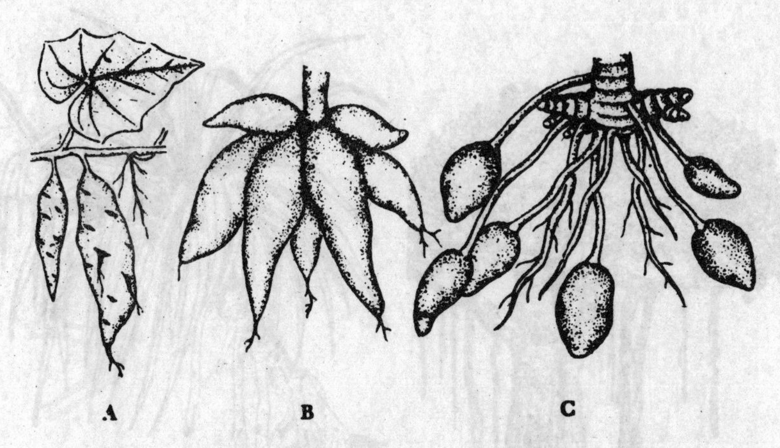

Figure 38. *A,* tuberous roots of sweet potato; *B,* fasciculated roots of Dahlia; *C,* nodulose roots of mango ginger.

5. **Annulated Root** (Fig. 39B). When the root has a series of ring-like swellings on its body, it is said to be annulated, as in ipecac (*Psychotria*).

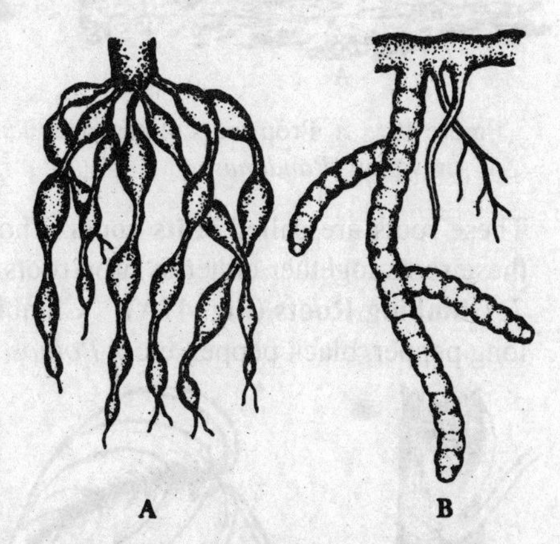

Figure 39. *A,* moniliform roots of Momordica; *B,* annulated roots of ipecac

(b) *for mechanical support*

6. **Prop Roots** and **Stilt Roots** (Fig. 40). In plants like banyan, etc., a number of roots are produced from the heavy, aerial, horizontal branches. These roots grow vertically or obliquely downwards and penetrate into the soil. Gradually they get stouter and act as pillars supporting the branches or the plant as a whole. Such roots are known as **prop roots**. The big banyan tree of the Indian Botanic Garden near Calcutta has produced about 1,600 such roots from its branches. Its age is estimated to be over 200 years, and the circumference of the crown over 400 metres. In July 1982, however, the tree was badly damaged by storm. Whereas in Screwpine, *Rhizophora,* etc. a number of roots come out mainly from the stem. These roots grow downwards and penetrate into soil. Finally growing like pillars they support the main stem or the plant as a whole.

Figure 40. *A,* Prop roots of banyan (*Ficus bengalensis*); *B,* Stilt roots of screwpine (*Pandanus*).

These roots are called **stilts roots**. Though some authors consider all these roots together either as prop roots or stilt roots.

7. **Climbing Roots** (Fig. 41A). Climbing plants like *Piper* (e.g. betel long pepper, black pepper, etc.), *Pothos, Hoya,* etc., produce roots from

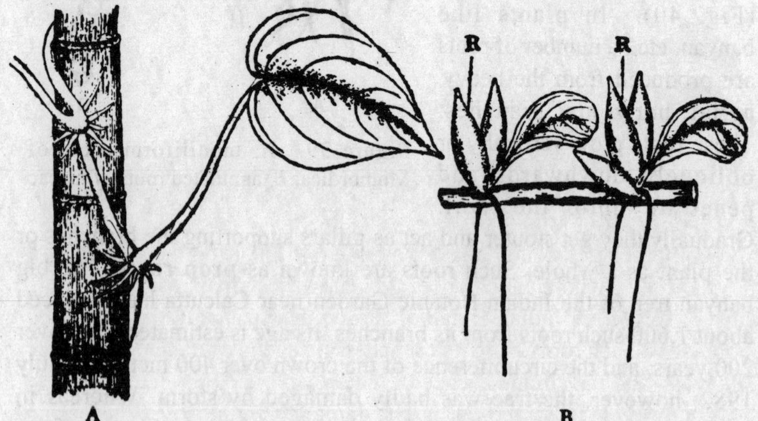

Figure 41. *A,* climbing roots of betel (*Piper betle*); *B,* respiratory roots (*R*) of *Jussiaea repens.*

their nodes and often from the internodes, by means of which they attach themselves to their support and climb it. To ensure a foothold such roots secrete a kind of sticky juice which quickly dries up in the air, as seen in ivy (see Fig. 1). Often, they form at their apex, a sort of disc or claw for firmer foothold. Such roots are also called **clinging roots**.

(c) *for vital functions*

8. **Sucking Roots** or **Haustoria** (see Fig. 6). Parasites develop certain kinds of roots which penetrate into the living tissue of the host plant and suck it. Such roots are known as sucking roots or haustoria (sing. haustorium). Parasites, particularly non-green ones, have to live by sucking the host plant, i.e. by absorbing food from it with the help of their sucking roots. Common examples are dodder, broomrape, *Balanophora,* mistletoe, etc.

9. **Respiratory Roots** (Fig. 41B). In *Jussiaea repens* (B. & H. KESSRA), an aquatic plant, the floating branches develop a special kind of adventitious roots which are soft, light, spongy and colourless. They usually develop above the surface of the water and serve to store up air. Thus, they facilitate respiration. They are also sometimes used by the plant as buoys or floating organs.

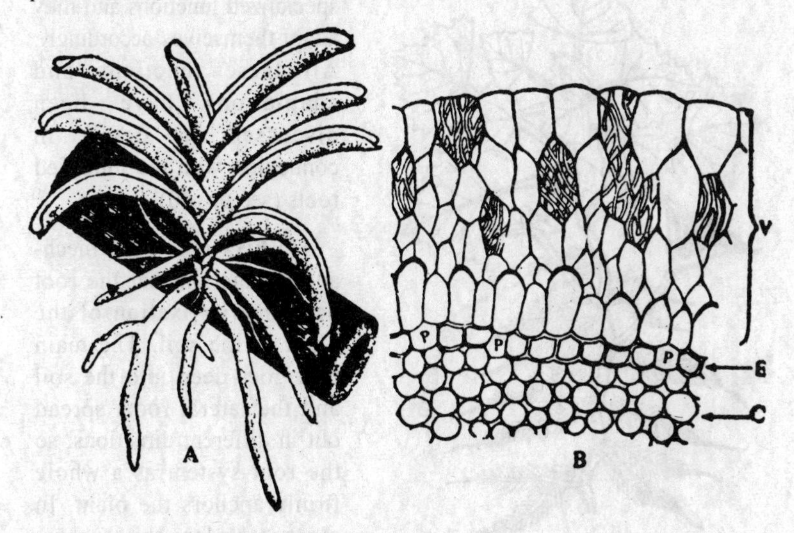

Figure 42. *A,* epiphytic roots of *Vanda* (an orchid); *B,* a root of the same in transection showing *velamen, V,* velamen; *E,* exodermis; *P,* passage cell: and *C,* cortex.

10. **Epiphytic Roots** (Fig. 42). There are certain plants, commonly orchids, which grow on branches of trees. Such plants are known as **epiphytes** (*epi,* upon; *phyta,* plants). They never suck the supporting plant as do parasites. So instead of sucking roots they develop special kinds of aerial roots which hang freely in the air. Each hanging root is surrounded by a spongy tissue, called **velamen**. With the help of this velamen the hanging root absorbs moisture from the surrounding air. *Vanda* (B. & H. RASNA) an epiphytic orchid, is a common example.

11. **Assimilatory Roots.** Branches of *Tinospora* (B. GULANCHA; H. GURCHA) climbing on neighbouring trees produce long, slender, hanging roots which develop chlorophyll and turn green in colour. These green roots are the assimilatory roots. They carry on carbon-assimilation, i.e. they absorb carbon dioxide from the air and manufacture carbohydrate food. The hanging roots of epiphytic orchids (Fig. 42) also often turn green in colour. The submerged roots of water chestnut (*Trapa;* Fig. 43) are green in colour and act as assimilatory roots.

Functions and Adaptations of the Root. The root per-forms manifold functions—*mechanical* such as **fixation**, and *physiological* such as **absorption**, **conduction** and **storage**. These are the normal functions of the roots. Roots also have specialized functions and they adapt themselves accordingly. All these functions and adaptations have been discussed in detail in connexion with the modified roots (see pp. 40).

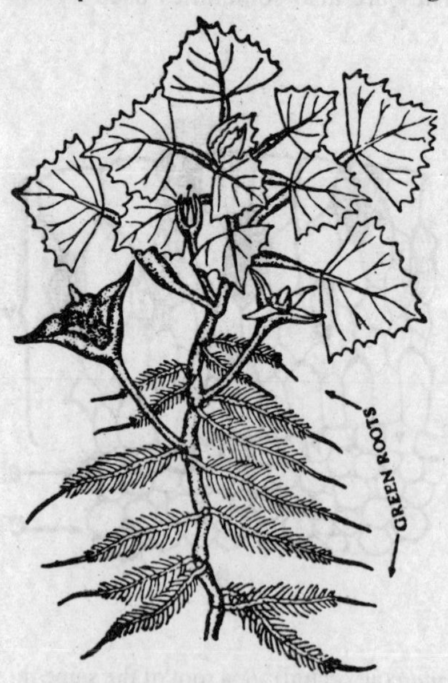

Figure 43. Assimilatory (green) roots of water chestnut *(Trapa natans).*

(1) **Fixation.** The mechanical function that the root performs is fixation of the plant to the soil. The main root goes deep into the soil and the lateral roots spread out in different directions; so the root system as a whole firmly anchors the plant. In monocotyledons this anchorage is afforded by the fibrous roots.

(2) **Absorption.** The most important physiological function is the absorption of water and necessary inorganic salts from the soil. This is done with the help of root-hairs which develop in a cluster at a little distance behind the root-cap. These root-hairs adhere to the soil particles and absorb water and soluble salts from them.

(3) **Conduction.** The root is concerned with the conduction of water and mineral salts, sending them upwards into the stem and ultimately into the leaf.

(4) **Storage.** There is a certain amount of food stored in the root, particularly in its mature region. As the root grows this stored food is utilized.

It may be summarized that anchorage, conduction and storage are carried on normally by the older portions of the root system, and absorption by the root-hairs and the tender portions.

Chapter 5

The Stem

Characteristics of the Stem. The stem is the ascending organ of the plant, and is the direct prolongation of the plumule. It normally bears leaves, branches and flowers, and when young, it is green in colour. The growing apex is covered over and protected by a number of tiny leaves which arch over it (Fig. 44). The stem often bears multicellular hairs of different kinds; it branches exogenously; and it is provided with nodes and internodes which may not be distinct in all cases. Leaves and branches normally develop from the nodes. When the stem or the branch ends in a vegetative bud it continues to grow upwards or sideways. If, however, it ends in a floral bud, the growth ceases.

Nodes and Internodes. The place on the stem or branch where one or more leaves arise is known as the node, and the space between two successive nodes is called the internode. Sometimes nodes and internodes are very conspicuous, as in bamboos and grasses; in others they are not always clearly marked.

THE BUD

A bud (Fig. 44) is a young undeveloped shoot consisting of a short stem and a number of tender leaves arching over the growing apex. In the bud the internodes have not yet developed and the leaves remain closely crowded together forming a compact structure. The lower leaves of the bud are older and larger than those higher up. The bud that grows in the axil of a leaf (**axillary bud**) or at the apex of a stem or branch (**terminal bud)** is regarded as *normal*. Sometimes some extra buds develop by the side of the axillary bud; they are called *accessory buds*. The bud that arises in any other part of the plant body is regarded as adventitious. Adventitious buds may be *radical buds* growing on the roots as in sweet potato (see Fig. 38A) or *foliar buds* growing on the leaf, as in sprout leaf plant (*Bryophyllum;* Fig 45A) and elephant ear plant (*Begonia;* Fig. 45B) or *cauline buds* growing on any part of the stem or branch. When a stem or branch is cut, adventitious buds often appear all round the cut surface. Buds that develop into branches with leaves are called **vegetative buds** and those that develop into flowers are called **floral buds**.

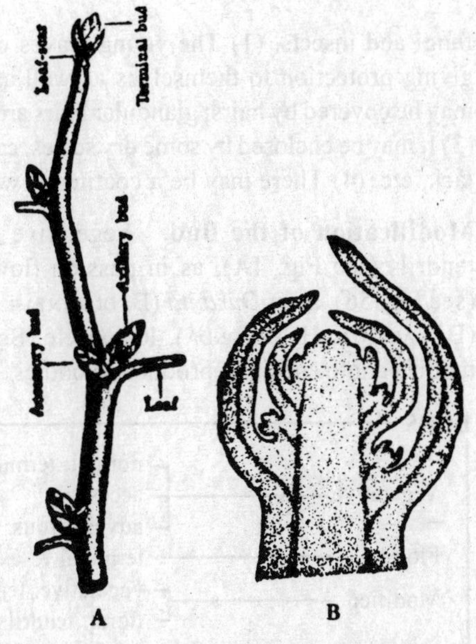

Figure 44. Buds. *A,* a branch showing position of buds; *B,* a bud in longitudinal section.

Protection of the Bud. The bud is protected in various ways against sun, rain,

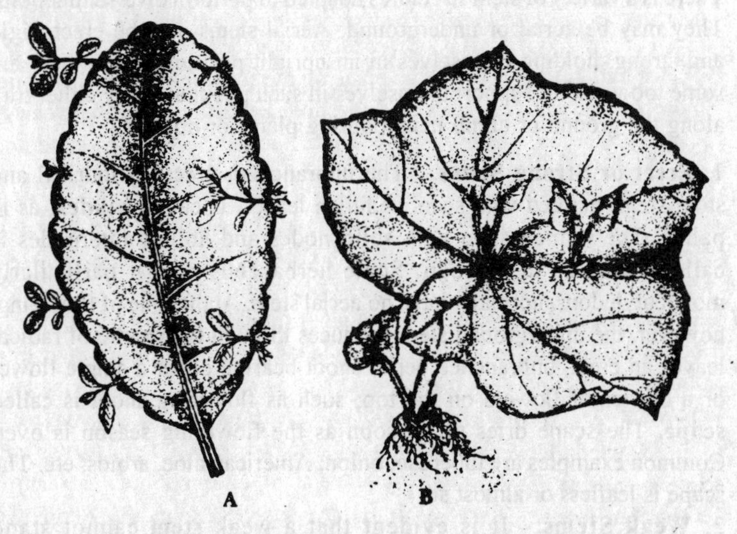

Figure 45. *A,* foliar buds and adventitious roots of sprout leaf plant (*Bryophyllum*); *B,* the same of elephant ear plant (*Begonia*).

fungi and insects. (1) The young leaves of a bud overlap one another giving protection to themselves as well as to the growing apex. (2) It may be covered by hairs; glandular hairs are very effective in this respect. (3) It may be enclosed by some dry scales, called **bud-scales**, as in banyan, jack, etc. (4) There may be a coating of wax or cutin.

Modification of the Bud. Vegetative buds may be modified into tendrils (see Fig. 4A), as in passion-flower and vine, or into thorns (see Fig. 56), as in *Duranta* (B. DURANTA-KANTA; H. NIL-KANTA), *Carissa* (B. KARANJA; H. KARONDA), lemon etc. Sometimes these may become modified into special reproductive bodies, known as **bulbils** (see p. 57).

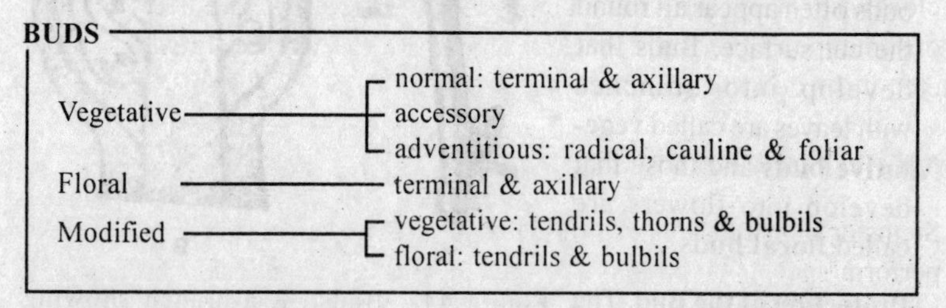

BUDS
Vegetative
normal: terminal & axillary
accessory
adventitious: radical, cauline & foliar
Floral
terminal & axillary
Modified
vegetative: tendrils, thorns & bulbils
floral: tendrils & bulbils

FORMS OF STEMS

There is a variety of stem structures adapted to perform diverse functions. They may be aerial or underground. Aerial stems may be erect, rigid and strong, holding themselves in an upright position; while there are some too weak to support themselves in such position. They either trail along the ground or climb neighbouring plants or objects.

1. **Erect or Strong Stems.** The unbranched, erect, cylindrical and stout stem, marked with scars of fallen leaves, is called **caudex**, as in palms. The jointed stem with solid nodes and hollow internodes is called **culm**, as in bamboo. Some herbaceous plants, particularly monocotyledons, normally have no aerial stem. At the time of flowering, however, the underground stem produces through the rosette of radical leaves an erect, unbranched aerial shoot bearing either a single flower or a cluster of flowers on the top; such as flowering shoot is called **scape**. The scape dries up as soon as the flowering season is over. Common examples are tuberose, onion, American aloe, aroids, etc. The scape is leafless or almost so.

2. **Weak Stems.** It is evident that a weak stem cannot stand upright. When such a stem lies flat on the ground, e.g. wood-sorrel (see Fig. 50), Indian pennywort (see Fig. III/38), dog grass, etc., it is said to

be (1) **prostrate.** When such a stem after trailing for some distance lifts its head, e.g. *Tridax* (see Fig. VII/31), it is said to be (2) **decumbent.** When the stem is much branched and the branches spread out on the ground on all sides, e.g. *Boerhaavia* (B. PUNAR-NAVA; H. SANTH), it is said to be (3) **diffuse.** A weak stem creeping on the ground and rooting at the nodes, e.g. sweet potato (see Fig. 38A), is said to be (4) **creeping.** When the stem bodily twines round a support without any special organ of attachment, e.g. *Clitoria* (B. APARAJITA; H. APARAJIT), *Abrus* (B. KUNCH; H. RATTI), etc., is said to be (5) **twining.** Some twiners by nature move clockwise, while other anti-clockwise. When the stem attaches itself to a nearby support by means of some special device, e.g. betel, cane, rose, pea, passion-flower, gourd, etc., it is said to be (6) **climbing** (see p. 5).

MODIFICATIONS OF STEMS

Stems or branches of certain plants are modified into various shapes to perform special functions. The special functions are: (a) perennation, i.e. surviving from year to year through bad seasons by certain underground stems; (b) vegetative propagation by certain horizontal sub-aerial branches spreading out in different directions; and (c) highly specialized functions of varied nature by certain metamorphosed aerial organs. Thus, in response to the above functions stems undergoes modifications into different and distinct forms, each to meet a special need, as follows.

1. **Underground Modifications of Stems.** For the purpose of perennation stems of certain plants develop underground and lodge there permanently, lying in a dormant, leafless condition for some period and then giving off aerial shoots annually under favourable conditions. They are always thick and fleshy, having a heavy deposit of reserve food material in them. Developing underground they often look like roots but are readily distinguished from them by the presence of (a) nodes and internode, (b) scale-leaves, and (c) buds (axillary and terminal). The main function of this group of modified stems is, as already stated, (a) perennation; but they also (b) store up food material and (c) propagate, i.e. multiply in number vegetatively. The various types met with in this group are as follows:

(1) **Rhizome** (Fig. 46). The rhizome is a thickened, prostrate, underground stem provided with distinct nodes and internodes, scaly

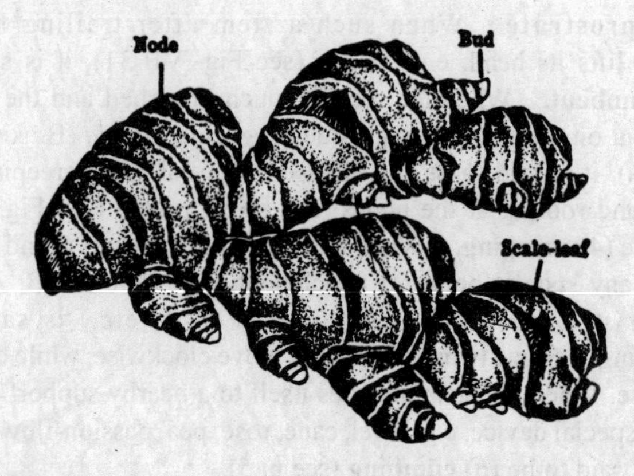

Figure 46. Rhizome of ginger.

leaves at the nodes, a bud in the axil of each such leaf, and a terminal bud. Some slender adventitious roots are given off from its lower side. It may be branched or unbranched. Most of the time it remains underground in a dormant condition, but after a few showers of rain the terminal bud and some of the axillary buds grow up into long or short leafy aerial shoots which again die down after a few months. Common examples are seen in *Canna,* ginger, turmeric, arrowroot, water lily, ferns, etc. Its direction is normally horizontal, but sometimes it grows in a vertical direction (**rootstock**), as in *Alocasia* (B. MANKACHU; H. MANKANDA).

(2) **Tuber** (Fig. 47). This is the swollen end of a special underground branch (tuber means a swelling). The underground branch arises from the axil of a lower leaf, grows horizontally outwards and ultimately swells up at the apex due to accumulation of a large quantity of food there, and becomes almost spherical, e.g. potato (*Solanum tuberosum*). It has on its surface a number of 'eyes' or buds which grow up into new plants. Adventitious roots are usually absent in a tuber. Jerusalem artichoke (*Helianthus tuberosus*) is another example.

Stem-tuber and Root-tuber. Both these structures lie underground and look alike, but (a) a stem-tuber develops exogenously from the node of the stem in the axil of a scale-leaf, has nodes and internodes, and bears buds in the axils of scale-leaves for vegetative propagation; while a root-tuber or tuberous root develops adventitiously from any part of the stem, as in sweet potato. In the sweet-potato, however, a few buds without scale-leaves are seen scattered on the root; (b) a stem-tuber has the internal structure of a stem; while a root-tuber has the internal structure of a root. We conclude, therefore,

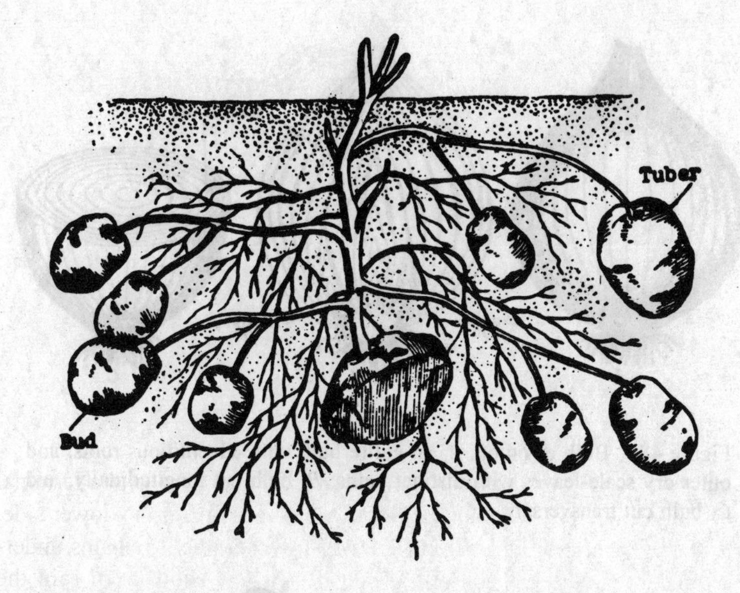

Figure 47. Tubers of potato.

that a stem-tuber is a stem structure, being an underground modification of it, and a root-tuber is a root structure, being a modification of an adventitious root or of a tap root.

(3) **Bulb** (Fig. 48). This is another underground modified shoot (really a single, often large, terminal bud) consisting of a shortened convex or slightly conical stem, a terminal bud and numerous scale-leaves (which are the swollen bases of foliage leaves), with a cluster of fibrous roots at the base. The scale-leaves, often simply called scales, commonly occur surrounding the short stem in a concentric manner (tunicated bulb), rarely they are narrow and just overlap each other (scaly bulb). The inner scales of the bulb are usually fleshy storing water and food (mainly sugar), while the outer ones are dry giving protection. The terminal bud grows into the aerial shoot; some of the axillary buds also do the same and finally form daughter bulbs. Common examples are onion, garlic, tuberose, most lilies, etc.

(4) **Corm** (Fig. 49). This is a condensed form of rhizome and consists of a stout, solid, fleshy, underground stem growing in the vertical direction. It is more or less rounded in shape or often somewhat flattened from top to bottom. It contains a heavy deposit of food material and often grows to a considerable size. It bears one or more buds in the axils

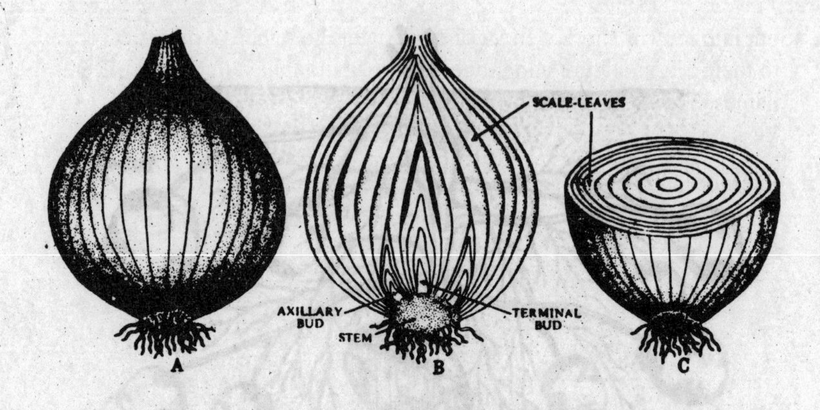

Figure 48. Bulb of onion. *A*, an entire bulb with adventitious roots, and outer dry scale-leaves with distinct veins; *B*, bulb cut longitudinally; and *C*, bulb cut transversely.

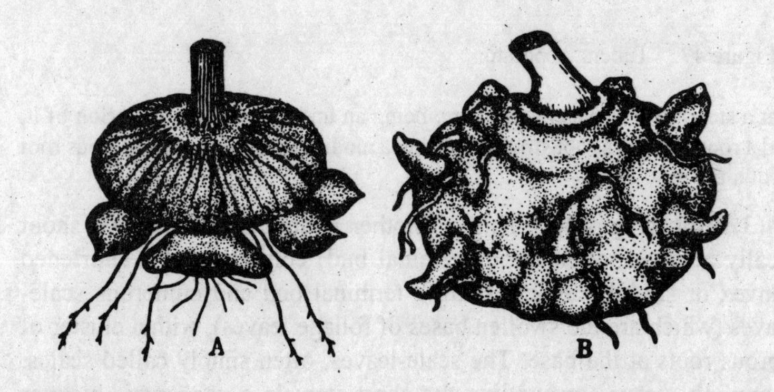

Figure 49. *A*, corm of *Gladiolus; B*, the same of *Amorphophallus* (B. OL; H. KANDA)

of scale-leaves, and some of these buds grow up into daughter corms. Adventitious roots normally develop from the base but sometimes also from the sides. Corm is found in *Amorphophallus* (B. OL; H. KANDA), taro (*Colocasia;* B. KACHU; H. KACHALU), *Gladiolus,* saffron (*Crocus*), etc.

2. **Sub-aerial Modifications of Stems.** For the purpose of vegetative propagation some of the lower buds of the stem in certain plants grow

out into long or short, slender or stout, lateral branches which, according to their origin, nature and mode of propagation, have received different names. These are as follows:

(1) **Runner** (Fig. 50). This is a slender, prostrate branch with long internodes, creeping on the ground and rooting at the nodes. The runner

Figure 50. Runner of wood-sorrel (*Oxalis*).

arises as an axillary bud and creeps some distance away from the mother plant, then strikes roots and grows into a new plant. Many such runners are often produced by the mother plant and they spread out on the ground on all sides. Examples are seen in wood-sorrel (*Oxalis;* Fig. 50); Indian pennywort (*Centella;* see Fig. III/39), *Marsilea* (B. SUSHNISAK), strawberry (*Fragaria*), dog grass (*Cynodon*), etc.

(2) **Stolon** (Fig. 51). This is a slender lateral branch which originates from the base of the stem, bends down on or into the ground. It then strikes roots and develops a bud which soon grows into a new plant. The stolon may continue to grow outwards for a shorter or longer distance, striking roots and producing a bud at each node. It may be provided with long or short internodes. The stolon resembles a runner in many respects, particularly when it straightens out and creeps on the

ground. Wild strawberry (*Fragaria indica*) and *Oenanthe stolonifera* (B. PANTURASI) are good examples.

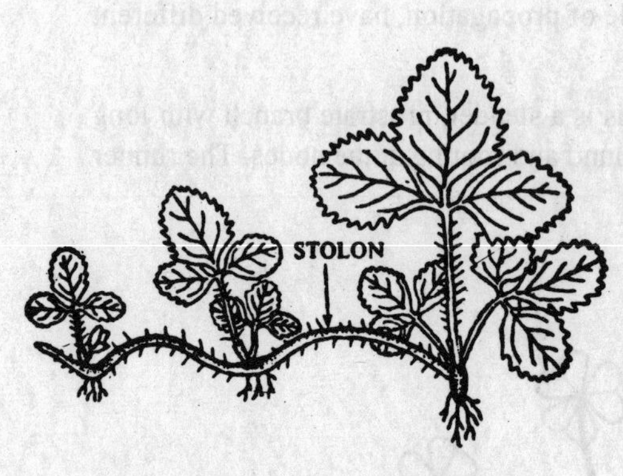

Figure 51. Stolon of wild strawberry (*Fragaria*).

(3) **Offset** (Fig. 52). Like the runner this originates in the axil of a leaf as a short, more or less thickened, horizontal branch. It elongates to some extent only. The apex then turns up and produces a tuft of leaves above and a cluster of roots below. The offset often breaks away from the mother plant into an independent one. Common examples are water lettuce (*Pistia;* Fig. 52) and water hyacinth (*Eichhornia;* see Fig. 62B). An offset is shorter and stouter than a runner, and is found only in the rosette type of plants.

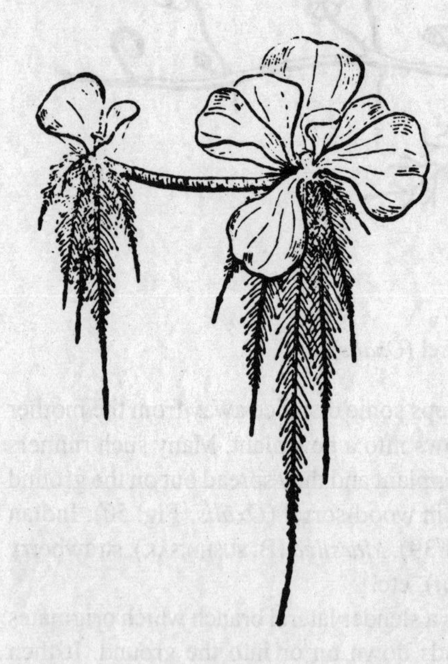

Figure 52. Offset of water lettuce (*Pistia*).

(4) **Sucker** (Fig. 53). Like the stolon the sucker is also a lateral branch develop-ing from the underground part of the stem. But it grows obliquely upwards and gives rise to a leafy shoot or a new plant. It may be a slender branch, or a short stout one, as in banana. A sucker is always much shorter than a stolon. The sucker strikes roots at the base either before it separates from the mother plant or soon after. Examples are seen in *Chrysanthemum,* rose, mint (*Mentha;* B. PUDINA; H. PODINA), pineapple, banana, dagger plant (*Yucca;* see Fig. 92), etc.

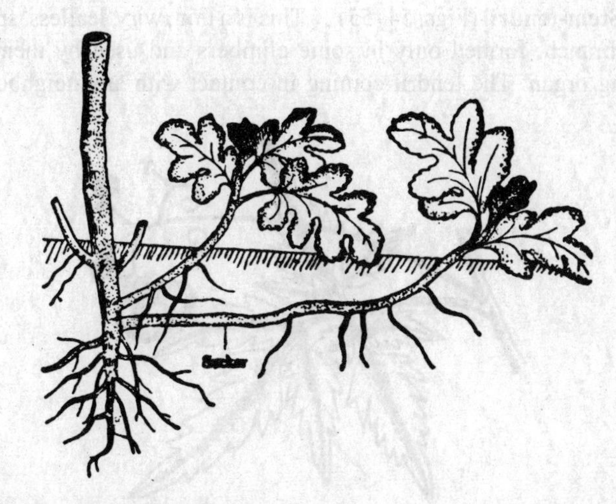

Figure 53. Suckers of *Chrysanthemum.*

3. **Aerial Modifications: Metamorphoses.** Vegetative and floral buds which would normally develop into branches and flowers, often undergo extreme degrees of modification (metamorphosis) in certain plants for definite purposes. Metamorphosed organs are stem-tendril for climbing, thorn for protection, phylloclade for food manufacture, and bulbil for vegetative reproduction.

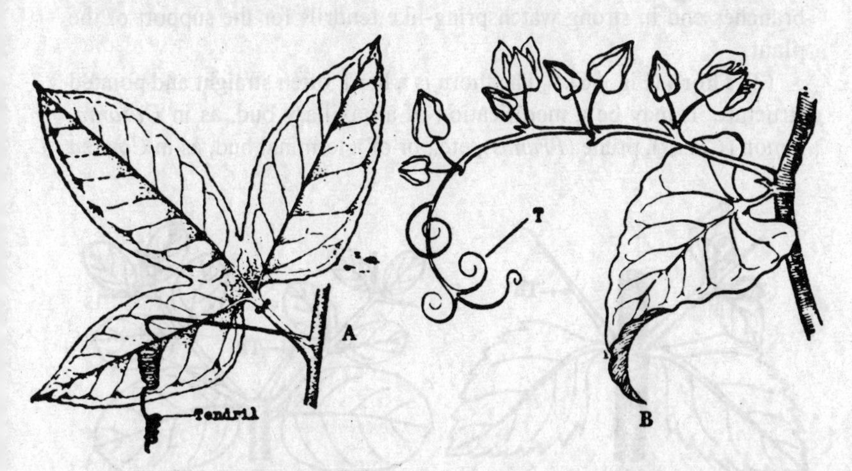

Figure 54. *A,* tendril of passion-flower (*Passiflora*); *B,* tendrils of Sandwich Island climber (*Antigonon = Corculum*); *T,* a tendril

(1) **Stem-tendril** (Figs. 54–55). This is a thin, wiry, leafless, spirally coiled branch, formed only in some climbers and used by them as a climbing organ. The tendril coming in contact with any neighbouring

Figure 55. Tendrils of balloon vine (*Cardiospermum*). *T,* a tendril.

objects, coils round it and helps such plants to climb. The stem-tendril may be a modification of an axillary bud, as in passion-flower (Fig. 54A), or of a terminal bud, as in vine, or even of a flower, as in balloon vine (*Cardiospermum;* Fig. 55) and Sandwich Island climber (*Antigonon* = *Corculum;* Fig. 54B). In *Gouania,* an extensive climber, some of the branches end in strong watchspring-like tendrils for the support of the plant.

(2) **Thorn** (Fig. 56). The thorn is a hard, often straight and pointed structure. It may be a modification of an axillary bud, as in *Duranta,* lemon (*Citrus*), prune (*Prunus*), etc., or of a terminal bud, as in *Carissa*

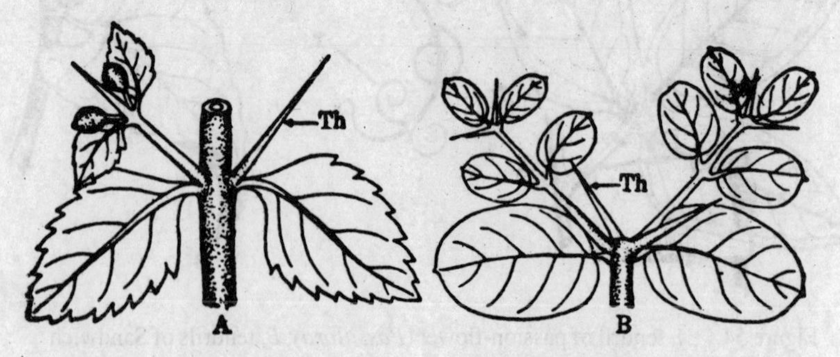

Figure 56. Thorns. *A,* thorns of *Duranta; B,* thorns of *Carissa. Th,* thorn.

(B. KARANJA; H. KARONDA). The thorn may sometimes be branched, and may even bear leaves, flowers and fruits. The thorn is a defensive organ meant to keep off browsing animals. Sometimes, as in glory-of-the-garden (*Bougainvillea;* see Fig. 2A), it is also used as a climbing organ.

Differences between Thorns and Prickles. Both are defensive organs of plants, but prickles being, usually curved, are commonly used for climbing; thorns are seldom used for this purpose. Their differences, however, are as follows:

Thorns	Prickles
• axillary or terminal in position	• irregularly distributed on the plant
• modifications of axillary or terminal buds	• not modifications of any morphological organs
• deep-seated in origin	• mostly superficial in origin
• may bear leaves, flowers and fruits	• never bear them
• usually straight	• usually curved
• may be branched	• often remain unbranched

(3) **Phylloclade** (Fig. 57). This is a *green,* flattened or cylindrical stem or branch *of unlimited growth,* consisting of succession of

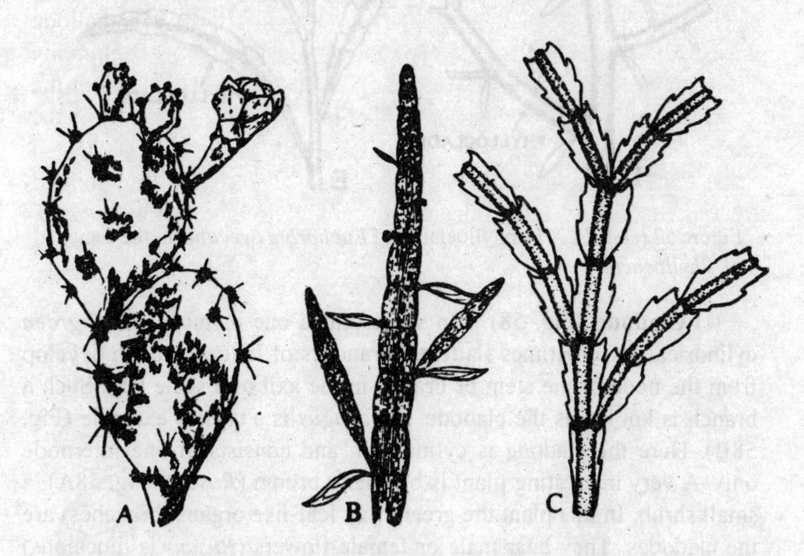

Figure 57. Phylloclades. *A,* prickly pear (*Opuntia*); *B,* cocoloba; *C,* christmas cactus (*Epiphyllum*).

nodes and internodes at long or short intervals. The phylloclade characteristically develops in many xerophytic plants where the leaves often grow out feebly, or fall off early, or are modified into spines, evidently reducing evaporating surfaces. The phylloclade then takes over all the functions of the leaves, particularly photosynthesis. It also often functions as a storage organ, retaining plenty of water and mucilage. Further, because of strong development of cuticle it can reduce transpiration to a considerable extent. Common examples are cacti such as prickly pear (*Opuntia;* Fig. 57A), night-blooming cacti (*Cereus* and *Phyllocactus;* Fig. 57E), Christmas cactus (*Epiphyllum;* Fig. 57C), etc., cocoloba (Fig. 57B), *Casuarina* (B. & H. JHAU), several species of *Euphorbia,* e.g. E. *tirucalli* (phylloclades cylindrical; Fig. 57D), *E. antiquorum* (phylloclades flattened), etc.

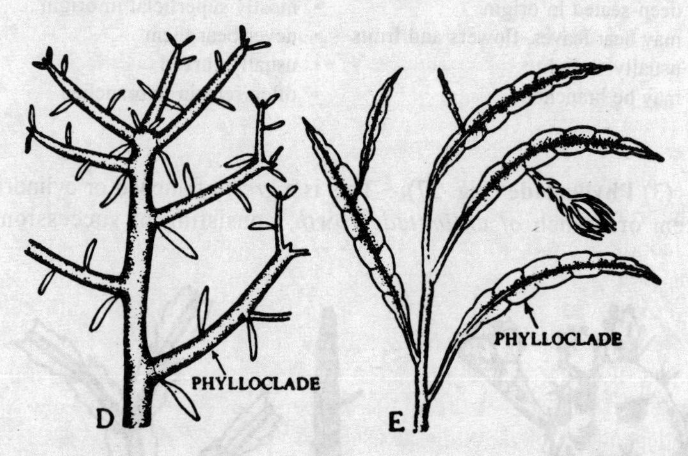

Figure 57 (*contd.*). *D,* phylloclades of *Euphorbia tirucalli; E,* the same of *Phyllocactus*.

(4) **Cladode** (Fig. 58). In some plants one or more short, *green* cylindrical or sometimes flattened branches of *limited growth* develop from the node of the stem or branch in the axil of a scale leaf. Such a branch is known as the cladode. *Asparagus* is a typical example (Fig. 58B). Here the cladode is cylindrical, and consists of one internode only. A very interesting plant is butcher's broom (*Ruscus:* Fig. 58A), a small shrub. In this plant the green, flat, leaf-like organs (branches) are the cladodes. They bear male or female flowers (*Ruscus* is dioccious) from a point (representing a node) half-way up on their surface in the axil of another scale leaf. The female flower subsequently produces a

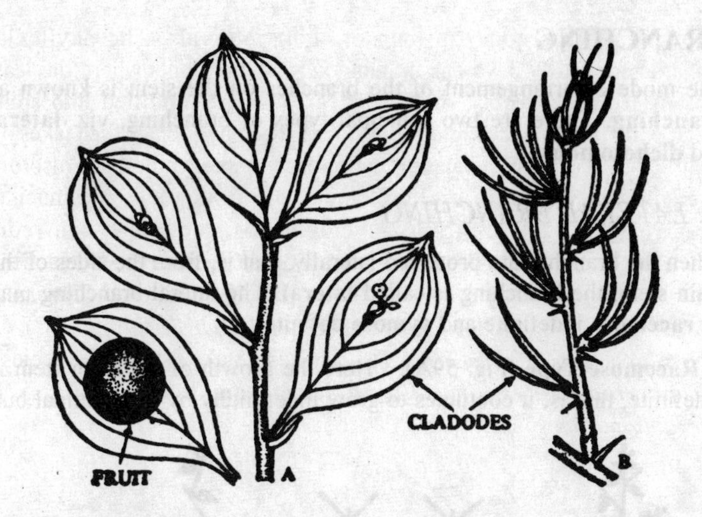

Figure 58. Cladodes. *A,* Ruscus; *B,* Asparagus.

large red berry. The flat, green, floating blade (stem) of duckweed (*Lemna;* see Fig. 31) is also regarded by many as a cladode. Similarly, the frond of *Wolffia,* a minute, rootless, floating plant, is regarded as a cladode. *Wolffia,* it may be noted, is the smallest flowering plant, with its frond as small as a sand grain.

(5) **Bulbil** (see Figs. III/42–4). Bulbil is a special multi cellular reproductive body, i.e. it is essentially meant for the reproduction of the plant. It may be the modification of a vegetative bud or of a floral bud. In either case it sheds from the mother plant and grows up into a new independent one. Bulbils are seen in *Globba,* wild yam (*Dioscorea;* B. GACHH-ALOO; H. ZAMINKHAND), American aloe (*Agave*), wood-sorrel (*Oxalis*), etc.

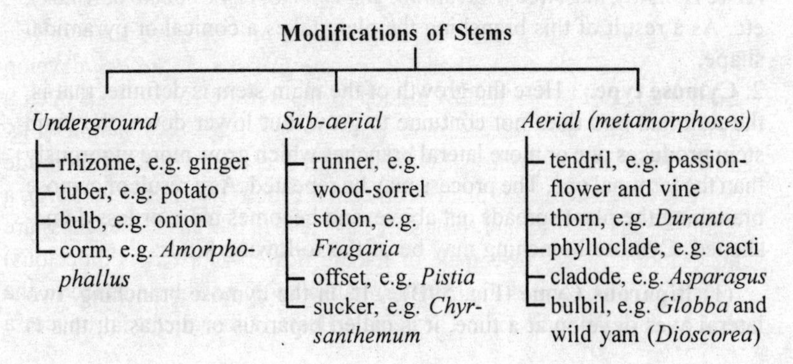

BRANCHING

The mode of arrangement of the branches on the stem is known as **branching**. There are two principal types of branching, viz. **lateral** and **dichotomous**.

A. *LATERAL BRANCHING*

When the branches are produced laterally, that is, from the sides of the main stem, the branching is called **lateral**. The lateral branching may be **racemose** indefinite and **cymose** definite.

1. **Racemose Type** (Fig. 59A). Here the growth of the main stem is indefinite, that is, it continues to grow indefinitely by the terminal bud

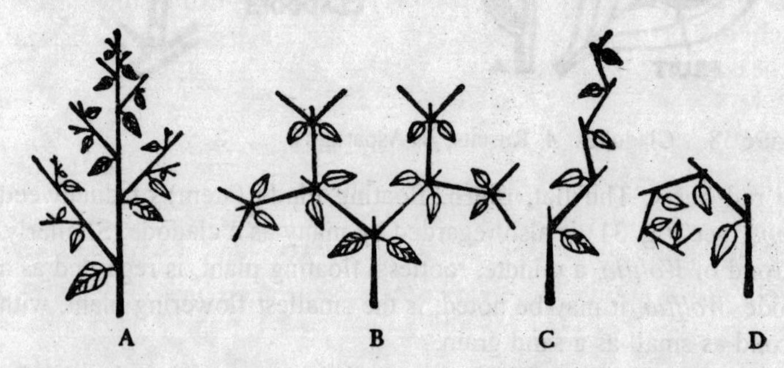

Figure 59. Branching. *A,* racemose type; *B,* true (biparous) cyme; *C,* scorpioid cyme; *D,* helicoid cyme.

and give off branches laterally in *acropetal* succession, i.e. the lower branches are older and longer than the upper ones. The branching of this type is also called **monopodial** (*monos,* single; *podos,* foot or axis) because there is a single continuous axis (**monopodium**), as in *Casuarina* (B. & H. JHAU), mast tree (*Polyalthia;* B. DEBDARU; H. DEVADARU or ASHOK), etc. As a result of this branching the plant takes a conical or pyramidal shape.

2. **Cymose type.** Here the growth of the main stem is definite, that is, the terminal bud does not continue to grow, but lower down, the main stem produces one or more lateral branches which grow more vigorously than the terminal one. The process may be repeated. As a result of cymose branching the plant spreads out above, and becomes more or less dome-shaped. Cymose branching may be of the following kinds:

(1) **Biparous Cyme** (Fig. 59B). If, in the cymose branching, two lateral axes develop at a time, it is called biparous or dichasial; this is

true cyme, and common examples are misteltoe (*Viscum;* see Fig. 8), four o'clock plant (*Mirabilis*), *Carissa* (B. KARANJA; H. KARONDA—see Fig. 56B), temple or pagoda tree (*Plumeria;* B. KATCHAMPA; H. GOLAINCHI), etc. Sometimes it so happens that the terminal bud remains undeveloped or soon falls off, the branching then looks like a dichotomy, often called *false dichotomy.*

(2) **Uniparous Cyme.** If, in the cymose type, only one lateral branch is produced at a time, the branching is said to be uniparous or monochasial (i.e. having one axis at a time). This type of branching is othewise called **sympodial** (*syn,* together or united; *podos,* foot or axis) because there is a succession of daughter axes (false axes) which, as they grow, seemingly become the parent axis. It has two distinct forms: (a) **helicoid** or one-sided cyme (Fig. 59D), when successive lateral branches develop on the same side, forming a sort of helix, as in *Saraca* (B. ASOK; H. SEETA ASHOK), and (b) **scorpioid** or alternate-sided cyme (Fig. 59C), when successive lateral branches develop on alternative sides, forming a zigzag, as in *Vitis* (e.g. vine and wild vine), *Cissus quadragularis* (B & H. HARHJORA), etc. In them, the apparent or false axis (**sympodium**) is a succession of lateral axes, and the lateral axes, and the tendrils are the modified terminal vegetative buds (Fig. 60).

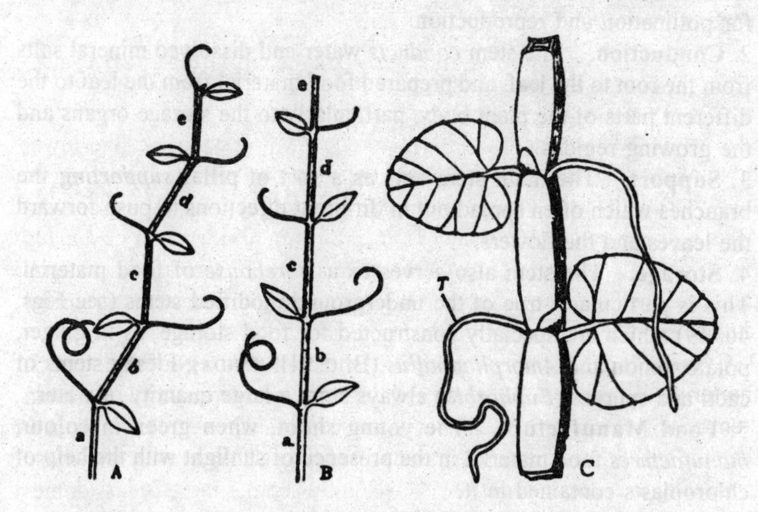

Figure 60. Sympodial branching. *A,* scorpiod type terminal tendrils and lateral axes; *B,* the same straightened out after growth; *a–e* are respective axes of sympodium; *C,* sympodial branching of *Cissus quadrangularis* (B. & H. HARHJORA); *T,* a tendril.

(3) **Multiparous Cyme.** If more than two branches develop at a time, the branching is said to be multiparous or polychasial, as in *Croton sparsiflorus* and some species of *Euphorbia,* e.g. *E. tirucalli* (see Fig. 57D).

B. *DICHOTOMOUS BRANCHING*

When the terminal bud bifurcates, that is, divides into two, producing two branches in a forked manner, the branching is termed dichotomous. Dichotomous branching is common among the 'flowerless' plants, as in *Riccia* (Fig. 60D), *Marchantia, Lycopodium,* etc. Among the 'flowering' plants, doum-palm (*Hyphaene*), screwpine (*Pandanus;* B. KETAKY; H. KEORA), *Clinogyne dichotoma* (B. SITALPATI), etc. show dichotomy.

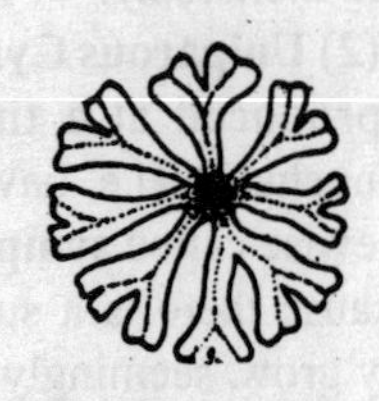

Figure 60D. Dichotomous Branching. *Riccia.*

FUNCTIONS OF THE STEM

1. **Bearing Leaves and Flowers.** The stem and the branches *bear* leaves and flowers, often numerous, and spread them out on all sides for proper functioning—the leaves to get adequate amount of sunlight for manufacture of food, and the flowers to attract insects from a distance for pollination and reproduction.
2. **Conduction.** The stem *conducts* water and dissolved mineral salts from the root to the leaf, and prepared food material from the leaf to the different parts of the plant body, particularly to the storage organs and the growing regions.
3. **Support.** The main stem acts as a sort of pillar *supporting* the branches which often spread out in different directions to push forward the leaves and the flowers.
4. **Storage.** The stem also serves as a *storehouse* of food material. This is particularly true of the underground modified stems (see Figs. 46–49) which are specially constructed for food storage, as in ginger, potato, onion and *Amorphophallus* (B. OL; H. KANDA). Fleshy stems of cacti and spurges (*Euphorbia*) always store a large quantity of water.
5. **Food Manufacture.** The young shoot, when green in colour, *manufactures* food material in the presence of sunlight with the help of chloroplasts contained in it.

In addition to those stated above, metamorphosed stems carry on specialized functions; for example, the tendril helps a plant to climb, and the thorn protects it against grazing animals, and so on.

Chapter 6

The Leaf

The leaf may be regarded as the flattened, lateral outgrowth of the stem or the branch, developing from a node and having a bud in its axil. It is normally green in colour and is regarded as the most important vegetative organ of the plant since food material is prepared in it. Leaves always develop in an *acropetal order* and are *exogenous* in origin.

Parts of a Leaf (Fig. 61). A typical leaf consists of the following parts, each with its own function.

(1) **Leaf-base** is the part attached to the stem. In many plants the leaf-base expands into a **sheath** which partially or wholly clasps the stem. This *sheathing* leaf-base is of frequent occurrence among monocotyledons, and is well developed in grasses; in the banana plant the so-called stem is made up of leaf-sheaths. In dicotyledons, on the other hand, the leaf-base usually bears two lateral outgrowths, known as the **stipules**. In some leaves such as those of gram, pea, tamarind, sensitive plant, rain tree, gold mohur, butterfly pea (*Clitoria;* B. APARAJITA; H. APARAJIT), etc., the leaf-base is swollen, and then it is known as the **pulvinus** (Fig. 62A).

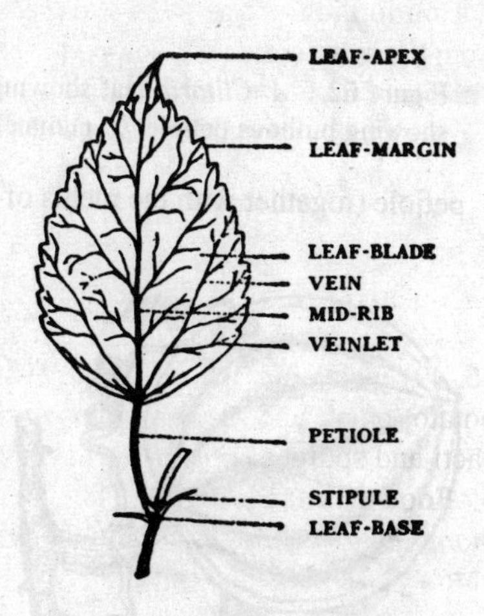

Figure 61. Parts of a leaf.

(2) **Petiole** is the stalk of the leaf. A long petiole pushes out the leaf-blade and thus helps it to secure more sunlight. When the petiole is

absent the leaf is said to be sessile; and when present it is said to be **petiolate** or stalked. In many plants the petiole shows certain peculiarities. Thus in water hyacinth (Fig. 62B) it swells into a spongy bulb, often called pseudobulb, containing innumerable air-chambers for facility of floating; while in orange, pummelo or shaddock, etc., it becomes winged (Fig. 62C). In Australian *Acacia* (see Fig. 80), the

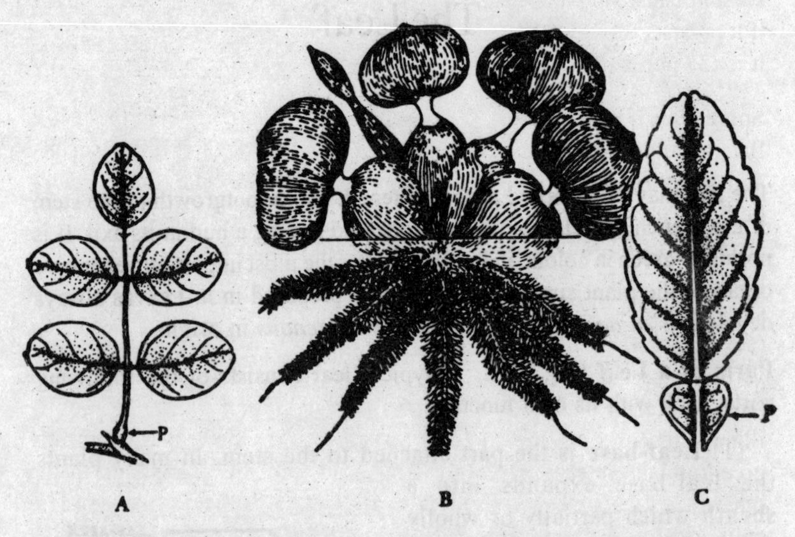

Figure 62. *A, Clitoria* leaf showing pulvinus (P); *B,* water hyacinth leaf showing bulbous petiole; *C,* pummelo leaf showing winged petiole (*P*).

petiole (together with the rachis of the leaf) is modified into a flattened, sickle-shaped, green lamina or blade called the **phyllode.** In *Clematis* (see Fig. 5A) the petiole is tendrillar in nature. In sarsaparilla (*Smilax;* Fig. 63) two strong, closely coiled tendrils, one on each side, develop from the leaf-stalk. They are formed by splitting of the petiole. They help the plant to climb neighbouring shrubs and trees.

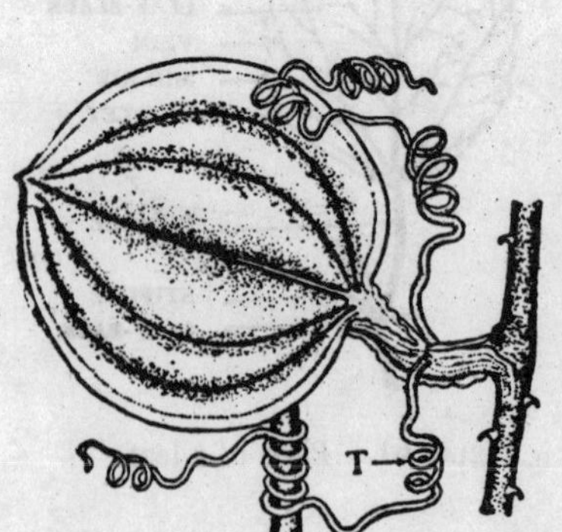

Figure 63. Tendrils (*T*) of *Smilax.*

(3) **Leaf-blade** or **lamina** is the green, expanded portion. A strong vein, known as the mid-rib, runs

centrally through the leaf-blade from its base to the apex; this produces thinner lateral **veins** which in turn give rise to still thinner veins or veinlets. The lamina is the most important part of the leaf since this is the seat of food-manufacture for the entire plant.

Duration of the Leaf. The leaf varies in its duration. It may fall off soon after it appears; then it is said to be (1) **caducous**; if it lasts one season, usually falling off in winter, it is (2) **deciduous** or **annual**; and if it persists for more than one season, usually lasting a number of years, it is (3) **persistent** or **evergreen**.

Some Descriptive Terms. (1) **Dorsiventral Leaf.** When the leaf is flat, with the blade placed horizontally, showing distinct upper surface and lower surface, it is said to be dorsiventral (*dorsum,* back; *venter,* belly or front), as in most dicotyledons. A dorsiventral leaf is more strongly illuminated on the upper surface than on the lower and, therefore, this surface is greener in colour than the lower. In internal structures also there is a good deal of difference between the two sides (see Fig. II/53). (2) **Isobilateral Leaf.** When the leaf is directed vertically upwards, as in most monocotyledons, it is said to be isobilateral (*isos,* equal; *bi,* two; *lateris,* side). An isobilateral leaf is equally illuminated on both the surfaces and, therefore, the leaf is uniformly green and its internal structure is also uniform from one side to the other (see Fig. II/54). (3) **Centric Leaf.** When the leaf is more or less cylindrical and directed upwards or downwards, as in pine, onion, etc., the leaf is said to be centric. A centric leaf is equally illuminated on all sides and, therefore, it is evenly green.

STIPULES

Stipules are the lateral appendage of the leaf borne at its base. They are present in many families of dicotyledons, but they are absent or very rare in monocotyledons. These are often green, but sometimes they have a withered look. They may remain as long as the lamina persists (**persistent**) or may fall off soon after the lamina unfolds (**deciduous**) or sometimes they may shed even before the lamina unfolds (**caducous**). Their function is to protect the young leaves in the bud, and when green they manufacture food material like leaves. When stipules are present the leaf is said to be **stipulate**, and when absent **exstipulate**. Sometimes, as in *Clitoria* (B. APARAJITA; H. APARAJIT), a small stipule may be present at the base of each leaflet. Such a small stipule is otherwise known as a **stipel**.

Kinds of Stipules. According to their shape, position, colour and size, stipules are of the following kinds:

(1) **Free Lateral Stipules** (Fig. 61). These are two free stipules, usually small and green in colour, borne on the two sides of the leaf-base, as in China rose, cotton, etc.

(2) **Scaly Stipules.** These are small dry scales, usually two in number, borne on the two sides of the leaf-base, as in Indian telegraph plant (*Desmodium gyrans*).

(3) **Adnate Stipules** (Fig. 64C). These are the two lateral stipules that grow along the petiole up to a certain height, adhering to it and making it somewhat winged in appearance, as in rose, groundnut, strawberry and lupin.

(4) **Interpetiolar Stipules** (Fig. 64B). These are the two stipules that lie between the petioles of opposite or whorled leaves, thus alternating with the latter. These are seen in *Ixora* (B. RANGAN), *Anthocephalus* (B. & H. KADAM), coffee (*Coffea*), madder (*Rubia;* B. & H. MANJISTHA), *Vangueria* (B. & H. MOYNA), etc.

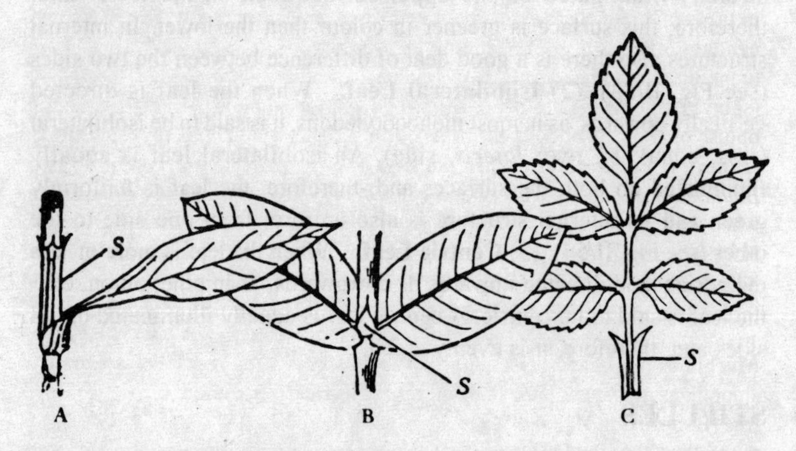

Figure 64. Kinds of Stipules. *A,* ochreate stipule (*S*) of *Polygonum; B,* interpetiolar stipule (*S*) of *Ixora; C,* adnate stipule (*S*) of rose.

(5) **Ochreate Stipules** (Fig. 64A). These form a hollow tube encircling the stem from the node up to a certain height of the internode in front of the petiole, as in *Polygonum*.

(6) **Foliaceous Stipules** (see Fig. 77A–B). These are two large, green, leafy structures, as in pea (*Pisum*), wild pea (*Lathyrus*), and some species of passion-flower (*Passiflora*).

(7) **Bud-scales.** These are scaly stipules which enclose and protect the vegetative buds, and fall off as soon as the leaves unfold. They are seen in banyan, jack, *Magnolia,* etc.

(8) **Spinous Stipules** (Fig. 65). In some plants, as in gum tree (*Acacia*), Indian plum (*Zizyphus*), sensitive plant (*Mimosa*), caper (*Capparis*), etc., the stipules become modified into two sharp, pointed structures known as the spines, one on each side of the leaf-base. Such spinous stipules give protection to the leaf against the attack of herbivorous animals.

Figure 65. Spinous stipules of Indian plum (*Zizyphus*).

LEAF-BLADE

Apex of the leaf (Fig. 66). The apex of the leaf is said to be (A) **obtuse**, when it is rounded, as in banyan; (B) **acute**, when it is pointed in the form of an acute angle, but not stiff, as in China rose; (C) **acuminate** or **candate**, when it is drawn out into a long slender tail, as in peepul and lady's umbrella (*Holmskioldia*); (D) **cuspidate**, when it ends in a long rigid sharp (spiny) point, as in date-palm, screwpine and pineapple; (E) **retuse**, when the obtuse or truncate apex is furnished with a shallow notch, as in water lettuce (*Pistia*); (F) **emarginate**, when the apex is provided with a deep notch, as in *Bauhinia* (B. KANCHAN; H. KACHNAR) and wood-sorrel (*Oxalis*); (G) **mucronate**, when the rounded apex abruptly ends in a short point, as in *Ixora* (B. RANGAN; H. GOTAGANDHAL) and *Ruscus* (see Fig. 58A); and (H) **cirrhose** (*cirrus,* a tendril or a curl), when it ends in a tendril, as in glory lily, or in a slender curled thread-like appendage, as in banana.

Margin of the Leaf. The margin of the leaf may be (1) **entire**, i.e. even and smooth, as in mango, jack, banyan, etc.; (2) **sinuate**, i.e. undulating, as in mast tree (B. DEBDARU; H. ASHOK) and some garden crotons; (3) **serrate**, i.e. cut like the teeth of a saw and the teeth directed upwards, as in China rose, rose, margosa (B. & H. NIM or NIMBA), etc.; (4) **dentate**, i.e. the teeth directed outwards at right angles to the margin of the leaf, as in melon and water lily; (5) **crenate**, i.e. the teeth rounded, as in sprout leaf plant (*Bryophyllum*) and Indian pennywort; and (6) **spinous**, i.e. provided with spines, as in prickly poppy (*Argemone*).

Surface of the Leaf. The leaf is said to be (1) **glabrous**, when the surface of it is smooth and free from hairs or outgrowths of any kind; (2) **rough**, when the surface is somewhat harsh to touch; (3) **glutinous**, when the surface of it is covered with a sticky exudation, as in tobacco; (4) **glaucons**, when the surface is green and shinning; (5) **spiny**, when it is provided with spines; and (6) **hairy**, when it covered, densely or sparsely, with hairs.

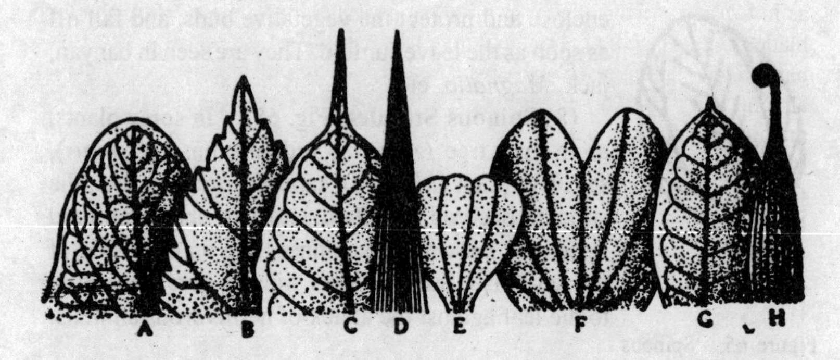

Figure 66. Apex of the leaf. *A*, obtuse; *B*, acute; *C*, acuminate; *D*, cuspidate; *E*, retuse; *F*, emarginate; *G*, mocronate; and *H*, cirrhose.

Shape of the Leaf (Fig. 67). (A) **Acicular**, when the leaf is long, narrow and cylindrical, i.e. needle-shaped, as in pine, onion, etc. (B) **Linear** when the leaf is long, as in many grasses, tuberose, Vallisneria, etc. (C) **Lanceolate**, when the shape is like that of a lance, as in bamboo, olender, mast tree etc.

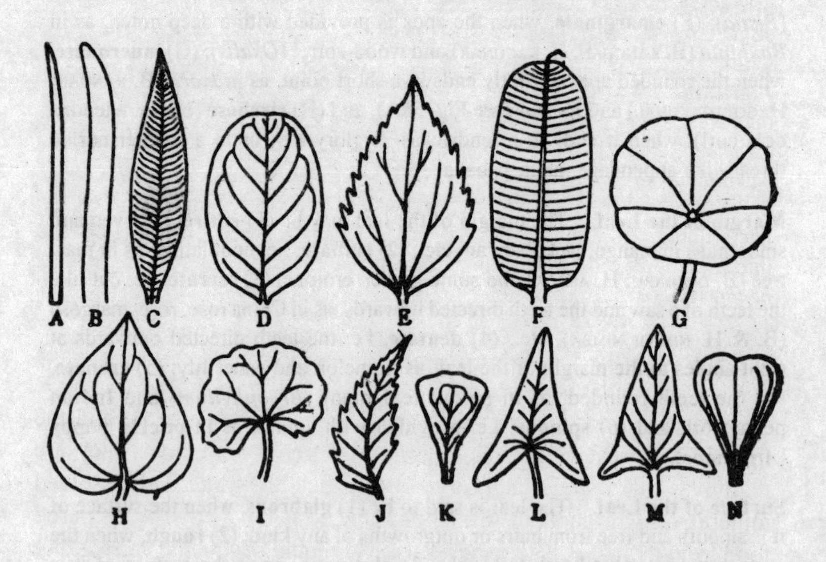

Figure 67. Shape of the Leaf. *A*, acicular; *B*, linear; *C*, lanceolate; *D*, elliptical or oval; *E*, ovate; *F*, oblong; *G*, rotund or orbicular; *H*, cordate; *I*, reniform; *J*, oblique; *K*, spathulate; *L*, sagittate; *M*, hastate; and *N*, cuneate.

(D) **Elliptical** or **oval**, when the leaf has more or less the shape of an ellipse, as in *Carissa,* periwinkle (*Vinca*), guava, roseapple, etc. (E) **Ovate**, when the blade is egg-shaped, i.e. broader at the base than at the apex, as in China rose, banyan, etc.; an inversely egg-shaped leaf is said to be **obovate**, as in country almond and jack. (F) **Oblong**, when the blade is wide and long, with the two margins running straight up, as in banana. (G) **Rotund** or **orbicular**, when the blade is more or less circular in outline, as in lotus, garden nasturtium, etc. (H) **Cordate**, when the blade is heart-shaped, as in betel, *Peperomia,* etc.; (I) **Reniform**, when the leaf is kidney-shaped, as in Indian pennywort. (J) **Oblique**, when the two halves of a leaf are unequal, as in *Begonia;* in margosa (B. & H. NIM) and Indian cork tree (B. & H. AKAS-NIM) and Persian lilac (B. GHORA-NIM) the leaflets are oblique. (K) **Spathulate**, when the shape is like that of a spagula, i.e. broad and somewhat rounded at the top and narrower at the base, as in sun-dew (*Drosera*) and *Calendual* (L) **Sagittate**, when the blade is shaped like an arrow, as in arrowhead and some aroids. (M) **Hastate**, when the two lobes of a sagittate leaf are directed outwards, as in water bindweed (*Ipomoea;* B. & H. KALMI-SAK) and *Typhonium* (B. GHET-KACHU). (N) **Cuneate** when the leaf is wedge-shaped, as in water lettuce (*Pistia*). (O) **Lyrate**, when the shape is like that of a lyre, i.e. with a large terminal lobe and some smaller lateral lobes, as in radish, mustard, etc. (P) **Pedate**, when the leaf is divided into a number of lobes which spread out like the claw of a bird, as in *Vitis pedate* (B. GOALE-LATA).

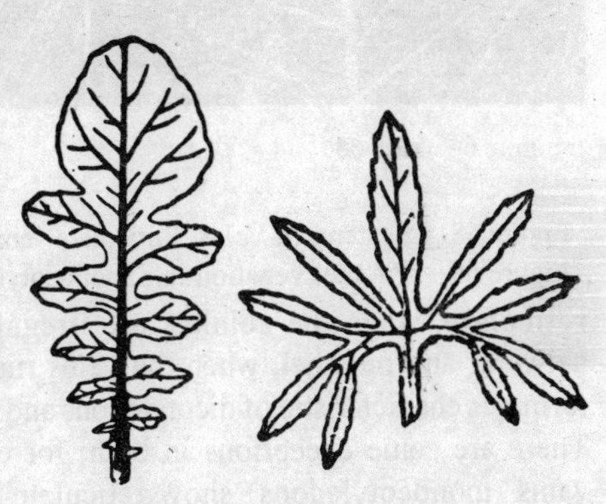

Figure 67 (*contd.*). *O,* lyrate leaf of radish; *P,* pedate leaf of *Vitis pedata.*

VENATION

Veins are rigid linear structures which arise from the petiole and the mid-rib and traverse the leaf-lamina in different directions; they are really vascular bundles and serve to distribute the water and dissolved mineral salts throughout the lamina and to carry away the prepared food from it; they also give the necessary amount of strength and rigidity to the thin, flat leaf-lamina.

The arrangement of the veins and the veinlets in the leaf-blade is known as **venation**. There are two principal types of venation, viz.

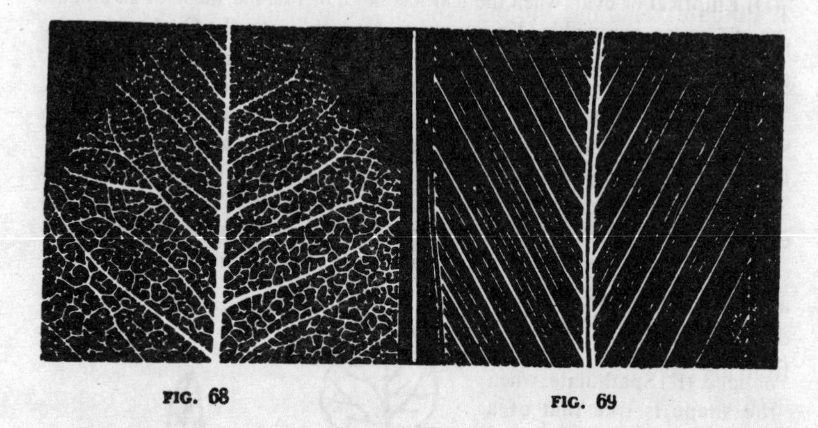

Figure 68. Systems of Veins. Reticulate venation in a dicotyledonous leaf.
Figure 69. Parallel venation in a monocotyledonous leaf.

reticulate, when the veinlets are irregularly distributed, forming a network; and **parallel**, when the veins run parallel to each other. The former is characteristic of dicotyledons and the latter of monocotyledons. There are some exceptions in both; for example, *Smilax,* aroids and yams (monocotyledons) show reticulate venation, and *Calophyllum* (dicotyledon) shows parallel venation.

I. *RETICULATE VENATION*

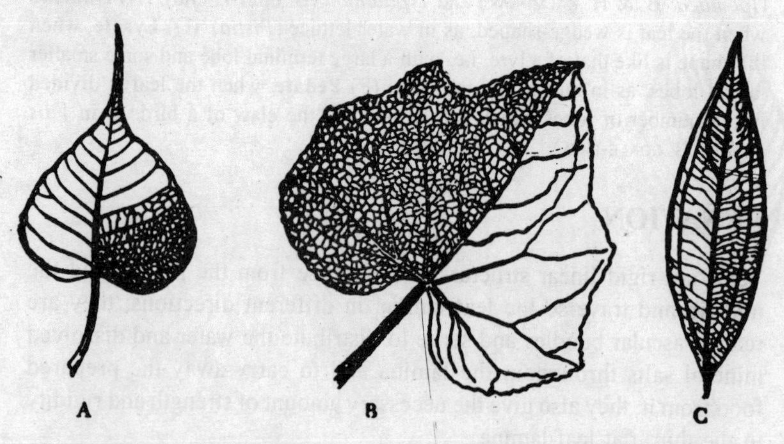

Figure 70. Reticulate Venation. *A,* pinnate type in peepul (*Ficus*) leaf; *B,* palmate (divergent) type in cucumber (*Cucumis*) leaf; *C,* palmate (convergent) type in bay leaf (*Cinnamomum*).

1. **Pinnate Venation.** In this type of venation there is a strong mid-rid; this gives off lateral veins which proceed towards the margin or apex of the leaf, like plumes in a feather (Fig. 70A). These produce still smaller veins and veinlets which pass in all directions and become connected with one another, forming a network as in guava, mango, jack, etc. This is a very common type of venation.

2. **Palmate Venation.** In this type there is a number of more or less equally strong ribs which arise from the tip of the petiole and proceed outwards or upwards. There are two forms: (1) in one the leaf possesses a number of strong veins that arise at the base of the leaf-blade and then diverge from one another towards the margin of the leaf, like the fingers from the palm(**divergent type**; Fig. 70B); these are then connected by a network of smaller veins, as in papaw, gourd, cucumber, castor, China rose, etc.; and (2) in the other, the veins, instead of diverging from one another, run in a curved manner from the base of the blade to its apex (**convergent type**; Fig. 70C), as in *Cinnamomum* (e.g. cinnamon, camphor, bay leaf, etc.), and Indian plum (*Zizyphus;* B. KUL; H. BER).

II. *PARALLEL VENATION*

1. **Pinnate Venation.** In this type of venation the leaf has a prominent mid-rib, and this gives off lateral veins which proceed parallel to each

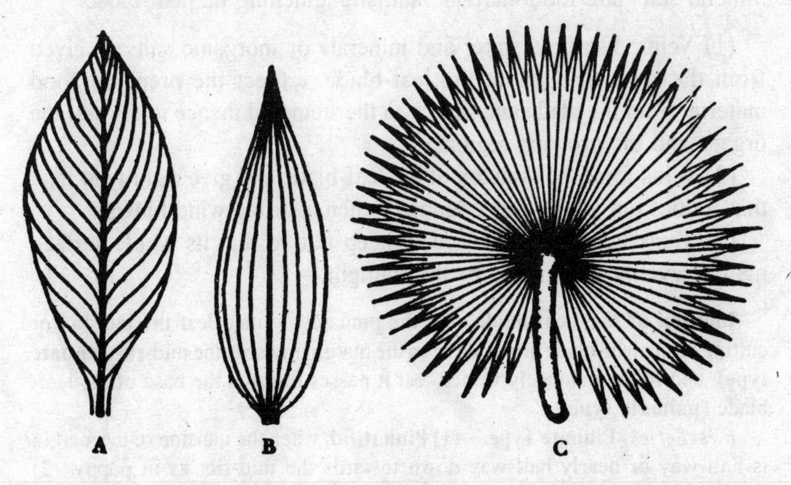

Figure 71. Parallel Venation. *A,* pinnate type in *Canna* leaf, *B,* palmate (convergent) type in bamboo leaf; *C,* palmate (divergent) type in palmyra-pain leaf.

other towards the margin or apex of the leaf-blade (Fig. 71A), as in *Canna,* banana, ginger, turmeric, etc.

2. **Palmate Venation.** Two forms are also met with here: (1) the veins arise from the tip of the petiole and proceed (diverge) towards the margin of the leaf-blade in a more or less parallel manner (**divergent type**; Fig. 71C), as in fan palms such as palmyra-palm; and (2) a number of more or less equally strong veins proceed from the base of the leaf-blade to its apex in a somewhat parallel direction (**convergent type**; Fig. 71B), as in water hyacinth, grasses, rice, bamboo, etc.

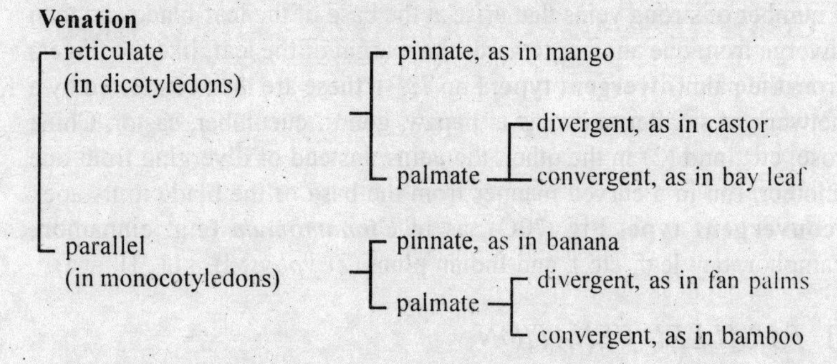

Functions of Veins. Veins are vascular bundles, which ramify through the leaf-blade. Their main functions are conduction of water, mineral salts and food material, and strengthening the leaf-blade.

(1) Veins distribute water and minerals or inorganic salts received from the stem throughout the leaf-blade, collect the prepared food material from the blade and send it to the stem and thence to the storage organs and the growing regions.

(2) Veins form the skeleton of the leaf-blade and give rigidity to it so that it does not get torn or crumpled when a strong wind blows.

(3) Veins help the leaf-blade to keep flat so that its whole surface may be evenly illuminated by the sunlight.

Incision of the Leaf-blade. In the pinnately veined leaf the incision or cutting of the leaf-blade proceeds from the margin towards the mid-rib (**pinnate type**), and in the palmately veined leaf it passes towards the base of the leaf-blade (**palmate type**).

First Series: **Pinnate Type.** (1) **Pinnatifid**, when the incision of the margin is half-way or nearly half-way down towards the mid-rib, as in poppy. (2) **Pinnatipartite**, when the incision is more than half-way down towards the mid-rib, as in radish, mustard, etc. (3) **Pinnatisect**, when the incision is carried down to near the mid-rib, as in some ferns, *Quamoclit* (B. KUNJALATA or TORULATA;

H. KAMALATA), *Cosomos,* etc. (4) **Pinnate compound,** when the incision of the margin reaches the mid-rib, thus dividing the leaf into a number of segments or leaflets, as in pea, gram, etc.

Second Series: **Palmate Type.** (1) **Palmatifid,** as in passion-flower, cotton, etc. (2) **Palmatipartite,** as in castor, papaw, etc. (3) **Palmatisect,** as in tapioca, hemp (*Cannabis;* B. & H. GANJA) and some aroids, e.g., snake plant (see Fig. 94). (4) **Palmate compound,** when the incision is carried down to the base of the leaf-blade, as in silk cotton tree.

COMPOUND LEAVES: PINNATE AND PALMATE

Simple Leaf and Compound Leaf. A leaf is said to be **simple** when it consists of a single blade which may be entire or incised (and, therefore, lobed) to any depth, but not down to the mid-rib or the petiole; and a leaf is said to be **compound** when the incision of the leaf-blade goes down to the mid-rib (rachis) or to the petiole so that the leaf is broken

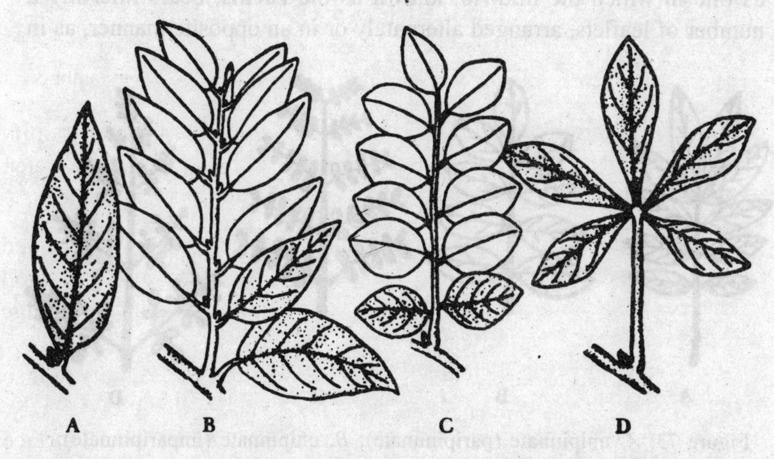

Figure 72. *A,* a simple leaf; *B,* a branch; *C,* a pinnately compound leaf with the leaflets articulated to the mid-rib; *D,* a palmately compound leaf with the leaflets articulated to the petiole. Note the position of the bud in each case.

up into a number of segments, called leaflets, these being free from one another, i.e. not connected by any lamina, and more or less distinctly jointed (articulated) at their base A bud (axillary bud) is present in the axil of a simple or a compound leaf, but it is never present in the axil of the leaflet of a compound leaf. There are two types of compound leaves, viz. **pinnate** and **palmate**.

Compound Leaf and Branch. A compound leaf may sometimes be mistaken for a branch. The following points should be noted by way of distinctions between them (Fig. 72B–C).

Compound leaf	Branch
• This never bears a terminal bud.	• always bears such a bud.
• This bears an axillary bud.	• does not bear such a bud.
• This never arises in the axil of another leaf.	• arises in the axil of a simple or compound leaf.
• Leaflets of a compound leaf have no axillary buds.	• simple leaves of a branch have a bud in their axil.
• Rachis of a compound leaf has no nodes or internodes.	• branch is always provided with nodes or internodes.

1. **Pinnately Compound Leaf.** A pinnately compound leaf is defined as one in which the mid-rib, known as the **rachis**, bears *laterally* a number of leaflets, arranged alternately or in an opposite manner, as in

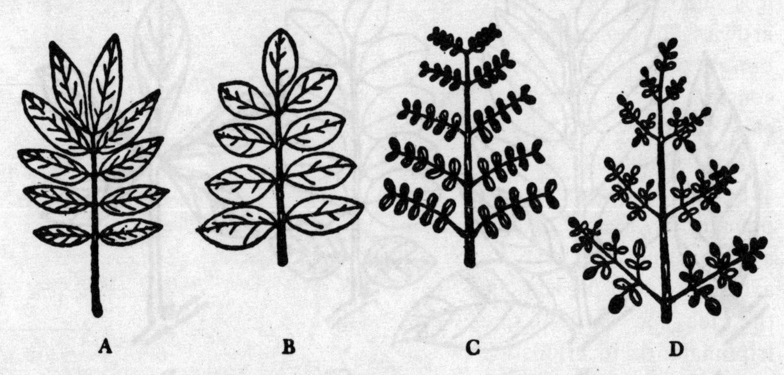

Figure 73. *A*, unipinnate (paripinnnate); *B*, unipinnate (imparipinnate); *C*, bipinnate; *D*, tripinnate.

tamarind, gram, gold mohur, rain tree, sensitive plant, gum tree (*Acacia*), *Cassia* (B. KALKASUNDE; H. KASONDI), etc. It may be of the following types:

(1) **Unipinnate** (Fig. 73A–B). When the mid-rib of the pinnately compound leaf directly bears the leaflets, it is said to be unipinnate. In it the leaflets may be *even* in number (**paripinnate**; Fig. 73A), as in *Cassia, Saraca* (B. ASOK; H. SEETA-ASOK), *Sesbania* (B. BAK-PHUL; H. AGAST), etc. or *odd* in number (**imparipinnate**; Fig. 73B), as in rose, margosa (B. & H. NIM), etc.

The pinnate leaf is said to be unifoliate, when it consists of only one leaflet, as in *Desmodium gangeticum;* bifoliate or unijugate (one pair), when of two leaflets, as in *Balanites* (Fig. 74) and sometimes in rose; trifoliate or ternate, when of three leaflets, as in bean, coral tree (*Erythrina*) and wild vine (*Vitis trifolia;* B. AMAL-LATA; H. AMAL-BEL). It may similarly be quadrifoliate, pentafoliate or multifoliate, when the leaflets are four, five or more in number.

Figure 74. Bifoliat leaf of *Balanites*.

(2) **Bipinnate** (Fig. 73C). When the compound leaf is twice pinnate, i.e. the mid-rib produces secondary axes which bear the leaflets, it is said to be bipinnate, as in dwarf gold mohur (*Caesalpinia*), gum tree (*Acacia*), sensitive plant (*Mimosa*), etc.

(3) **Tripinnate** (Fig. 73D). When the leaf is thrice pinnate, i.e. the secondary axes produce the tertiary axes which bear the leaflets, the leaf is said to be tripinnate, as in drumstick (*Moringa;* B. SAJINA; H. SAINJNA), and *Oroxylum* (B. SONA; H. ARLU).

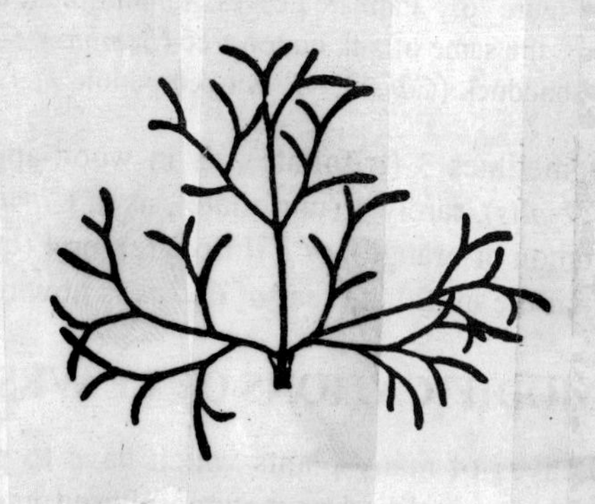

Figure 75. Decompound leaf of coriander (*Coriandrum*).

(4) **Decompound** (Fig. 75). When the leaf is more than thrice pinnate, it is said to be decompound, as in anise, carrot, coriander, *Cosmos,* etc.

2. **Palmately Compound Leaf** (Fig. 76). A palmately compound leaf is defined as the one in which the petiole bears *terminally,* articulated to it, a number of leaflets which seems to be radiating from a common point like fingers from the palm, as in silk cotton tree, lupin, *Gynandropsis* (B. SWET-HURHURE; H. HURHUR), etc. Leaflets are commonly 5 or more (multifoliate or **digitate**), as in silk cotton tree (*Bombax*),

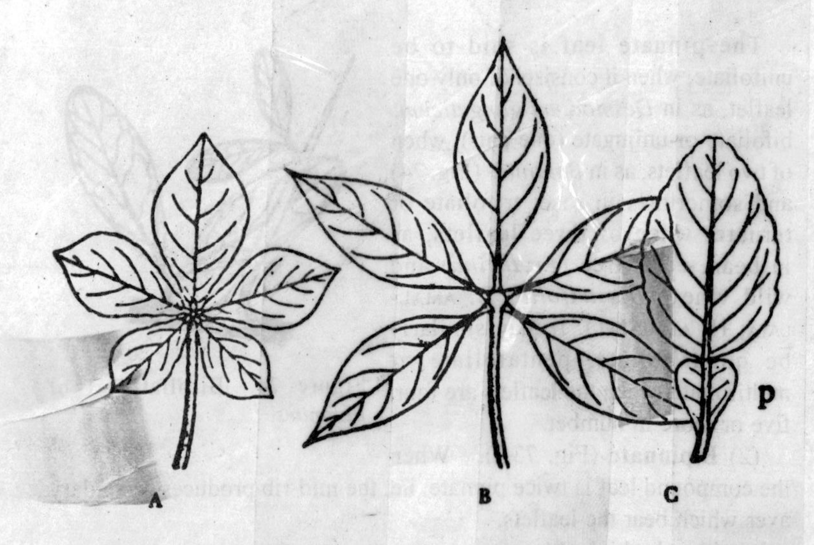

Figure 76. Palmate Leaves. *A*, multifoliate or digitate leaf of *Gynandropsis*; *B*, the same of silk cotton tree (*Bombax*); *C*, unifoliate leaf of pummelo or shaddock (*Citrus*);—*P*, winged petiole.

sometimes 3 (trifoliate), as in wood-apple (*Aegle*), and wood-sorrel (*Oxalis*), rarely 1 (unifoliate), as in *Citrus* (e.g. pummelo or shaddock, lemon or orange), or 2 (bifoliate), or 4 (quadrifoliate). [It may be noted that the unifoliate leaf of *Citrus* is now regarded as a simple leaf.]

MODIFICATIONS OF LEAVES

Leaves of many plants which have to perform specialized functions become modified or metamorphosed into distinct forms. These are as follows:

1. **Leaf-tendrils** (Fig. 77–78). In some plants leaves are modified into slender, wiry, often closely coiled structures known as tendrils. Tendrils are always climbing organs and are sensitive to contact with a foreign body. Therefore, whenever they come in contact with a neighbouring object they coil round it and help the plant to climb. The leaf may be partially or wholly modified. Thus in pea (*Pisum;* Fig. 77A) and *Lathyrus sativus* (B. & H. KHEṢARI) the upper leaflets only are modified into tendrils, while in wild pea (*Lathyrus aphaca;* Fig. 77B) the whole leaf is modified into a tendril. In traveller's joy (*Naravelia;* Fig. 78) the terminal leaflet alone is modified into a tendril, while in

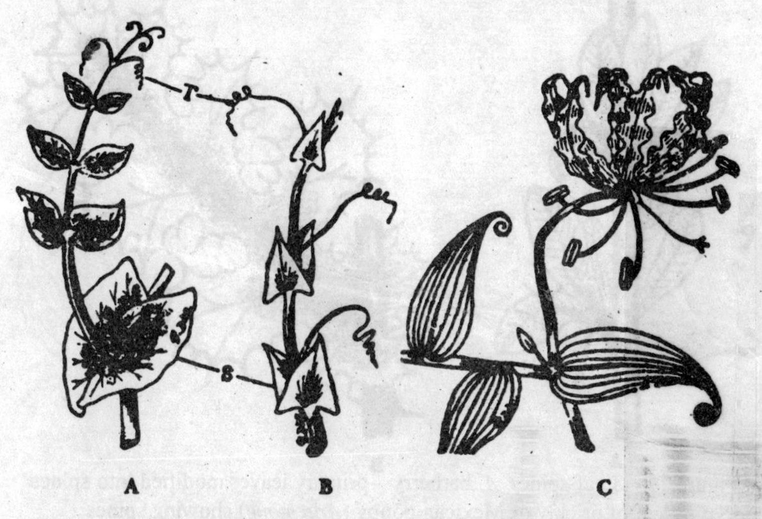

Figure 77. Leaf-tendrils. *A,* pea (*Pisum*) leaf with upper leaflets modified into tendrils; *B,* wild pea (*Lathyrus*) with the entire leaves modified into tendrils; *T,* tendrils; *S,* stipules; *C,* glory lily (*Gloriosa*) with the leaf-apex modified into a tendril.

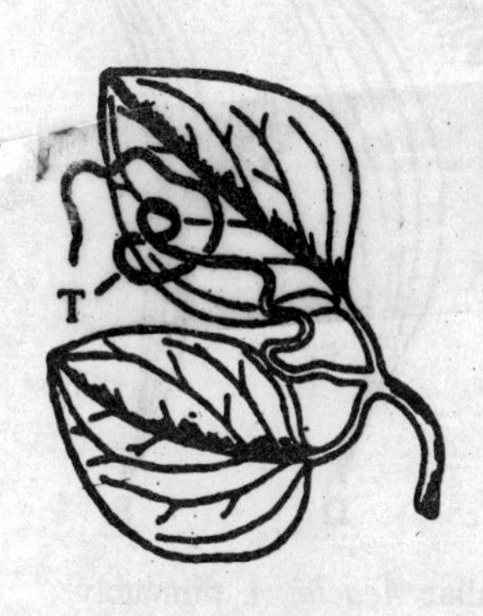

Figure 78. Leaf of Naravelia with the terminal leaflet modified into a tendril (*T*).

glory lily (*Gloriosa;* B. ULAT-CHANDAL; H. KALIARI—Fig. 77C) the leaf-apex only is so modified. In sarsaparilla (*Smilax,* see Fig. 63) the petiole splits into two strongly coiled tendrils one on each side.

2. Leaf-spines (Fig. 79), Leaves of certain plants become wholly or partially modified for defensive purpose into sharp, pointed structures known as **spines**. Thus in prickly pear (*Opuntia;* B. PHANIMANSHA; H. NAGPHANI—see Fig. 57A) the minute leaves of the axillary bud are modified into spines. The leaf-apex in date-palm, dagger plant (*Yucca;* see Fig. 92) etc., is so modified, while in plants like prickly or Mexican poppy (*Argemone;* B. SHEAL-KANTA; H. PILADHUTURA; Fig. 79B), American aloe (*Agave*), Indian aloe (*Aloe*), etc., spines develop on the margin as well as at the apex. In barberry (Fig. 79A) the leaf itself becomes modified into a spine; while the leaves of the axillary bud are normal.

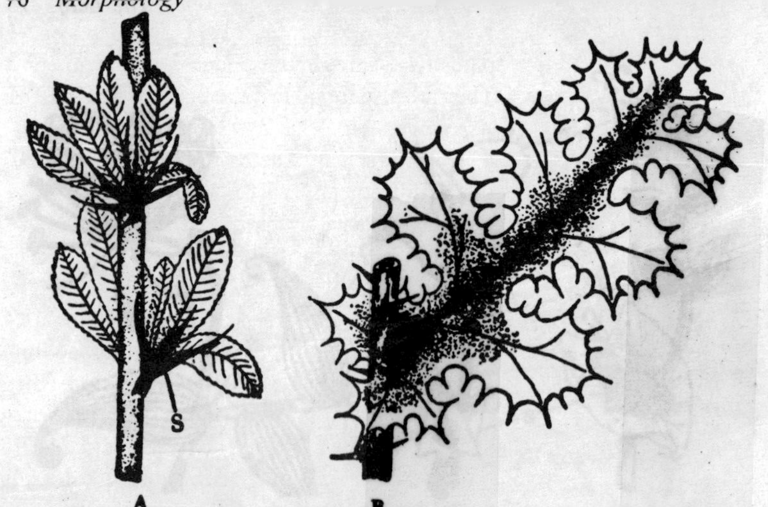

Figure 79. Leaf-spines. *A,* barberry—primary leaves modified into spines (S); *B,* leaf of prickly or Mexican poppy (*Argemone*) showing spines.

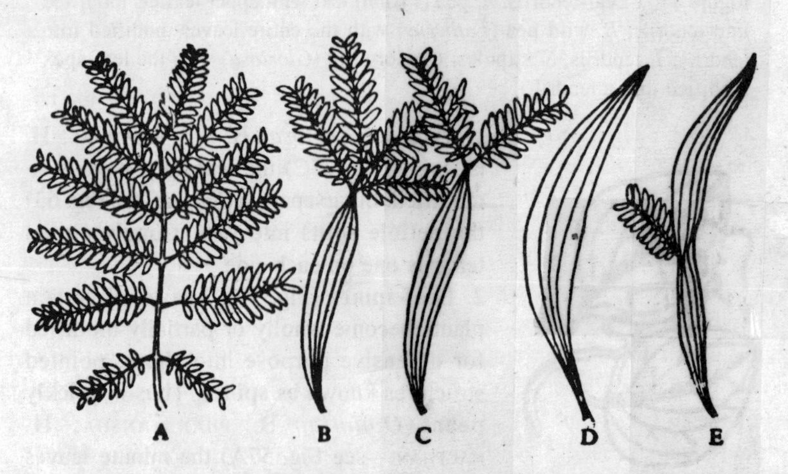

Figure 80. Development of phyllode in Australian *Acacia. A,* pinnately compound leaf; B–C, petiole developing into phyllode; *D,* phyllode; and *E,* petiole and rachis developing into phyllode.

3. **Scale-leaves.** Typically these are thin, dry, stalkless, membranous structures, usually brownish in colour or sometimes colourless. Their function is to protect the axillary bud that they bear in their axil. Sometimes scale-leaves are thick and fleshy, as in onion; then their function is to store up, water and food. Scale-leaves are common in

parasites, saprophytes, underground stems, etc. They are also found in *Casuarina* (B. & H. JHAU), *Asparagus,* etc.

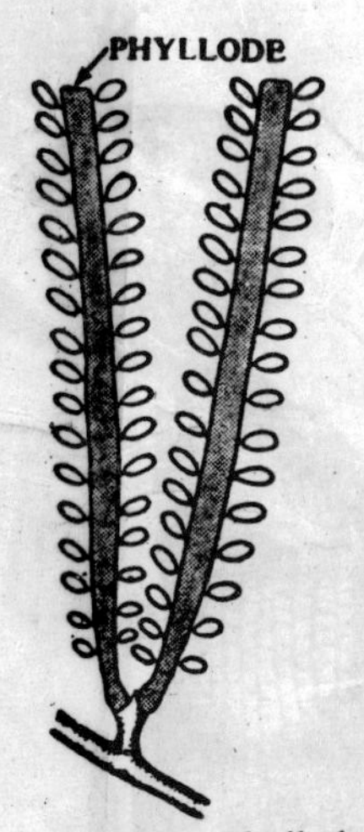

Figure 81. Phyllodes of *Parkinsonia*.

4. **Phyllode.** In Australian *Acacia* (Fig. 80) the petiole or any part of the rachis becomes flattened or winged taking the shape of the leaf and turning green in colour. This flattened or winged petiole or rachis is known as the **phyllode**. The normal leaf which is pinnately compound in nature develops in the seedling stage, but it soon falls off. The phyllode then performs the functions of the leaf. In some species, however, young or even adult plants are seen to bear the normal compound leaves together with the phyllodes. There are about 300 species of Australian *Acacia,* all showing the phyllodes. In Jerusalem thorn (*Parkinsonia;* Fig. 81), a small prickly tree, the primary rachis of the bipinnate leaf ends in a sharp spine, while each secondary rachis is a phyllode being green and flattened. The leaflets are small and fall off soon. The phyllode then performs the functions of the leaflets.

5. **Pitcher** (Fig. 82). In the pitcher plant (*Nepenthes khasiana*) the leaf becomes modified into a **pitcher**. The pitcher may be as big as 20–23 cm in height, sometimes a little more. It has a long, slender but rigid stalk which often coils like a tendril holding the pitcher vertical, and the basal portion is flattened like a leaf. The pitcher is provided with a lid which covers its mouth when the pitcher is young. The function of the pitcher is to capture and digest insects. The morphology of the leaf of the pitcher plant is that the pitcher itself is the modification of the leaf-blade, the inner side of the pitcher corresponding to the upper surface of the leaf; the lid arises as an outgrowth of the leaf-apex. The slender stalk which coils like a tendril is the petiole. The laminated structure which looks like and behaves as the leaf-blade develops from the leaf-base.

6. **Bladder** (Fig. 83). Bladderwort (*Utricularia*) is a rootless, free-floating herb common in many tanks. The leaf of this plant is very much segmented. Some of these segments are modified to form tiny bladders, each with a trap-door entrance which allows aquatic animalcules to pass in, but not to come out. A few species of land

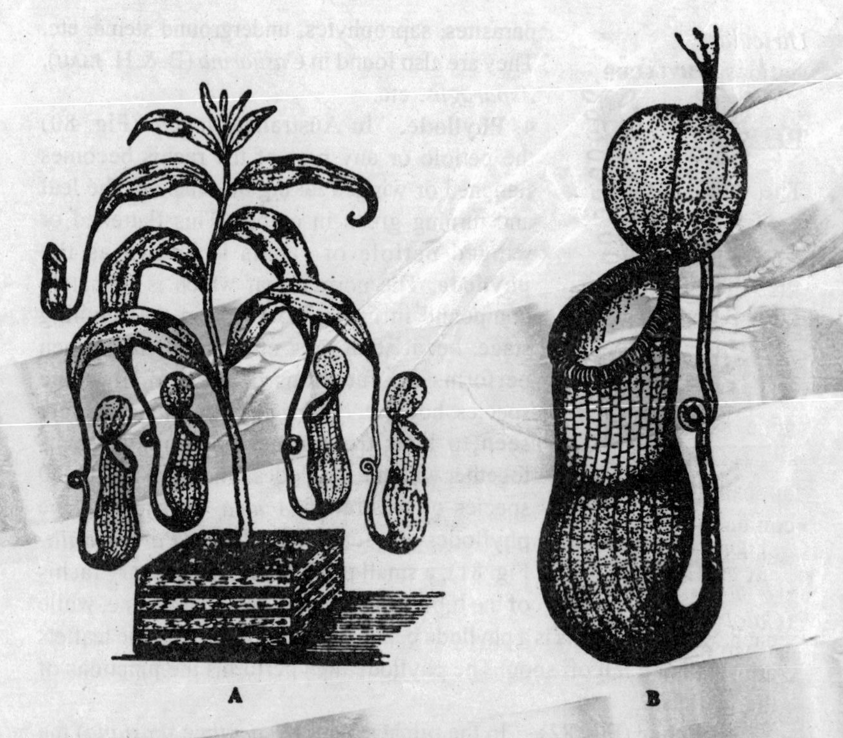

Figure 82. *A*, pitcher plant (*Nepenthes*); *B*, a pitcher.

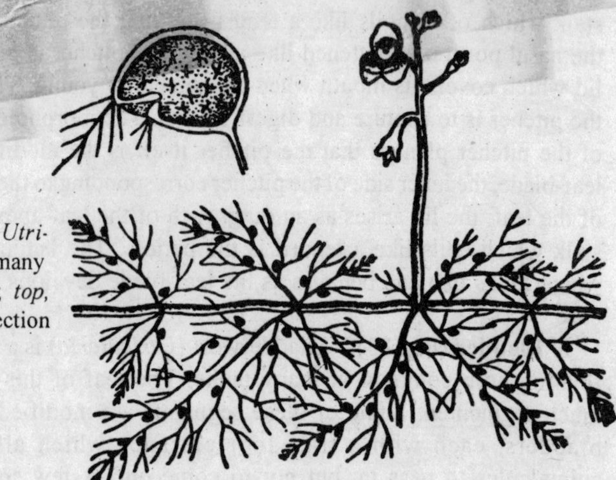

Figure 83. Bladderwort (*Utricularia*) with many small bladders; *top*, a bladder in section (magnified).

Utricularia are seen growing on hill slopes in Shillong and other hill stations.

PHYLLOTAXY

The term **phyllotaxy** (*phyll,* leaves; *taxis,* arrangement) means the various modes in which the leaves are arranged on the stem or the branch. The object of this arrangement is to avoid shading one another so that the leaves may get the maximum amount of sunlight to perform their normal functions, particularly manufacture of food. Three principal types of phyllotaxy are noticed in plants.

(1) **Alternate** or **Spiral** (Fig. 84A), when a single leaf arises at each node, as in tobacco, China rose, mustard, sunflower, garden croton, etc.

(2) **Opposite** (Fig. 84B), when two leaves arise at each node standing opposite each other. In opposite phyllotaxy one pair of leaves is most commonly seen to stand at a right angle to the next upper or lower pair. Such an arrangement of leaves is said to be **decussate**, as in sacred basil (*Ocimum;* B. & H. TULSI), *Ixora* (B. RANGAN; H. GOTAGANDHAL), madar (*Calotropis;* B. AKANDA; H. AK), etc. Sometimes, however, a pair of leaves is seen to stand directly over the lower pair in the same plane. Such an arrangement of leaves is said to be **superposed**, as in guava, Rangoon creeper (*Quisqualis;* B. SAN-DHYAMALATI; H. LALMALATI), etc.

(3) **Whorled** (Fig. 84C–D), when there are more than two leaves at each node and these are arranged in a circle or whorl, as in devil tree

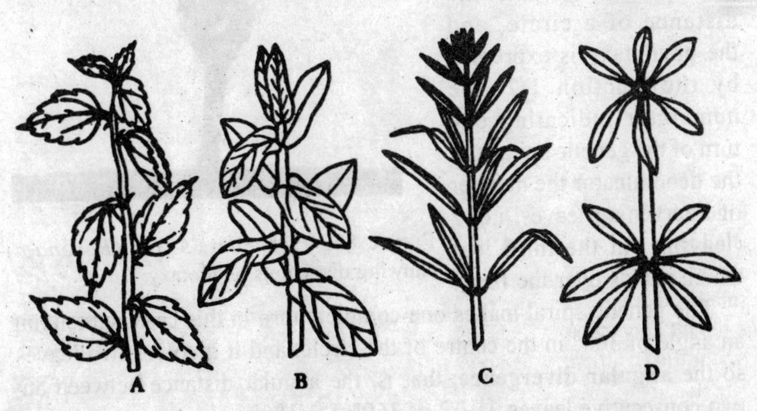

Figure 84. Types of Phyllotaxy. *A,* alternate phyllotaxy of China rose; *B,* opposite phyllotaxy of madar (*Calotropis*); *C,* whorled phyllotaxy of oleander (*Nerium*); *D,* ditto of devil tree (*Alstonia*).

(*Alstonia;* B. CHHATIM; H. CHATIUM), *oleander* (*Nerium;* B. KARAVI; H. KANER), etc.

Alternate Phyllotaxy. The leaves in this case are seen to be spirally arranged round the stem. Now, if an imaginary spiral line starting from any leaf be passed round the stem through the bases of the successive leaves, it is seen that the spiral line finally reaches a leaf which stands vertically over the starting leaf. This imaginary spiral line passing round the stem through the bases of successive leaves is known as the **genetic spiral**, and the vertical line, i.e. the vertical row of leaves is known as the **orthostichy** (*orthos,* straight; *stichos,* line).

(1) **Phyllotaxy 1/2** or **distichous** (Fig. 86). In grasses, traveller's tree (**Ravenala**; Fig. 85), ginger, *Vanda* (see Fig. 42) etc., the *third* leaf always stands over the *first* (starting anywhere). Thus there are only *two* orthostichies, i.e. two rows of leaves, and therefore, the phyllotaxy is distichous. From the starting leaf to the third leaf the genetic spiral makes only *one* turn. Leaves are thus placed at half the distance of a circle, and the phyllotaxy is expressed by the fraction 1/2, the numerator indicating one turn of the genetic spiral and the denominator the number of intervening leaves, i.e. 2 (leaving out the third leaf which stands over the first).

Figure 85. Travellers's tree (*Ravenala*) showing distichous phyllotaxy.

The genetic spiral makes one complete turn in this case, subtending an angle of 360° in the centre of the circle, and it involves two leaves, so the **angular divergence**, that is, the angular distance between any two consecutive leaves, is 1/2 of 360°, i.e. 180°.

(2) **Phyllotaxy** 1/3 or **tristichous** (Fig. 87). In sedges (B. & H. MUTHA) the fourth leaf stands vertically over the first one, and the genetic spiral

makes *one* turn to reach that leaf, and it involves three leaves. Thus there are *three* orthostichies, i.e. three rows of leaves. Leaves are thus placed at one-third the distance of a circle. Phyllotaxy is, therefore, tristichous or 1/3. The angular divergence is 1/3 of 360°, i.e. 120°.

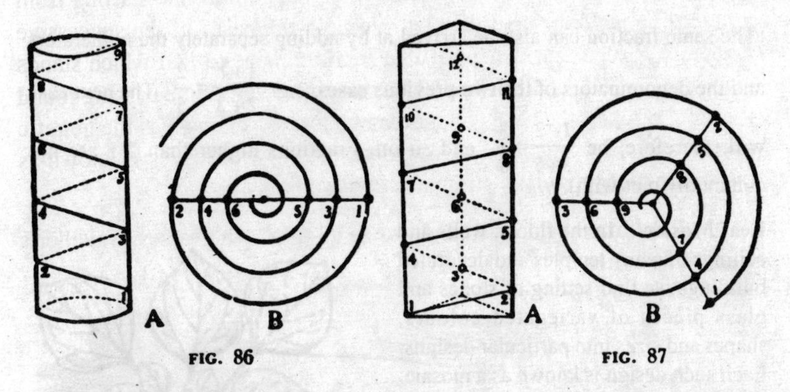

Figure 86. Phyllotaxy and Angular Divergence. *A*, phyllotaxy 1/2; *B*, angular divergence 180°. Figure 87. *A*, phyllotaxy 1/3; *B*, angular divergence 120°.

(3) **Phyllotaxy 2/5** or **pentastichous** (Fig. 88). In China rose the sixth leaf stands over the first, and the genetic spiral completes *two*

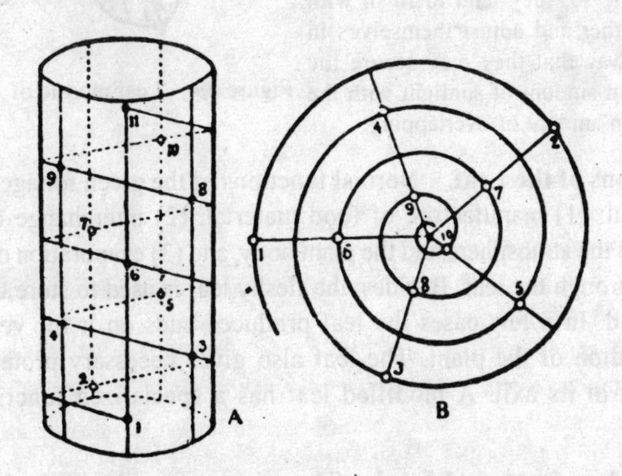

Figure 88. *A*, phyllotaxy 2/5; *B*, angular divergence 144°.

circles to come to that particular leaf. Thus there are five orthostichies, i.e. five rows of leaves, and two turns of the genetic spiral involving five leaves. The latter are thus placed at two-fifths the distance of a circle. Phyllotaxy is, therefore, pentastichous or 2/5. This is the commonest type of alternate phyllotaxy. The angular divergence in this case is 2/5 of 360°, i.e. 144°.

(The same fraction can also be arrived at by adding separately the numerators and the denominators of the two previous cases, e.g. $\frac{1+1}{2+3}=\frac{2}{5}$. The next case will, therefore, be $\frac{1+2}{3+5}=\frac{3}{8}$, and so on, Fractions higher than 2/5 are not commonly met with).

Leaf Mosaic. In the floors, walls and ceilings of many temples and decorated buildings we find setting of stones and glass pieces of variegated colours, shapes and sizes into particular designs. Each such design is known as a mosaic. Similarly in plants we find the setting or distribution of leaves in some definite designs. Each such design of leaf-distribution is known as **leaf mosaic**. Leaves are in special need of sunlight for manufacture of food material, and this being so, they tend to fit in with one another and adjust themselves in such a way that they may secure the maximum amount of sunlight with the minimum amount of overlapping.

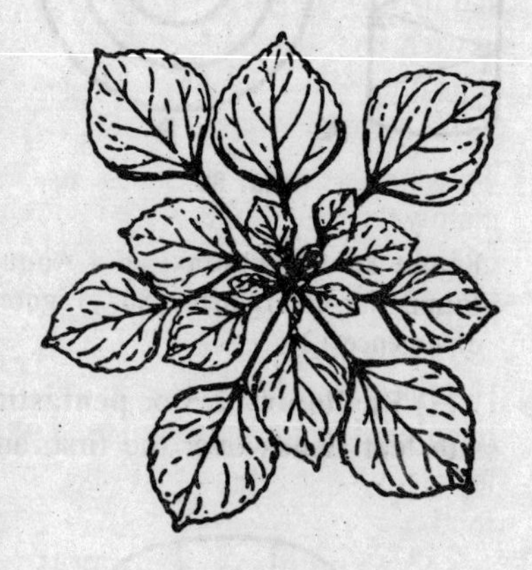

Figure 89. Leaf mosaic of *Acalypha*.

Functions of the Leaf. Normal functions of the green foliage leaf are threefold: (1) manufacture of food material, (2) interchange of gases between the atmosphere and the plant body, and (3) evaporation of excess water through the leaf. Besides, the fleshy leaf is used to store up water and food. In a few cases the leaf produces buds on it for vegetative propagation of the plant. The leaf also gives necessary protection to the bud in its axil. A modified leaf has a specialized function (see pp. 74).

(1) **Manufacture of Food.** The primary function of the leaf is to manufacture food, particularly sugar and starch, during the daytime only, i.e. in the presence of sunlight which is the original source of energy to the plant. The leaf manufactures food with the help of

chloroplasts contained in it out of water and carbon dioxide obtained from the soil and the air respectively. The upper side of the leaf is deeper in colour with more abundant chloroplasts, and the sunlight falls directly on the upper surface, therefore, food manufacture normally takes place in this region.

(2) **Interchange of Gases.** Through the lower surface of the leaf a regular exchange of gases takes place between the atmosphere and the plant body through numerous very minute openings, called **stomata** (see Fig. II/36) which remain open during the daylight only. The gases concerned are oxygen and carbon dioxide. This exchange of gases is mainly for two purposes: *respiration* by all the living cells which absorb oxygen and give out carbon dioxide, and *food manufacture* by green cells only which absorb carbon dioxide and give out oxygen.

(3) **Evaporation of Water.** The excess water absorbed by the root-hairs evaporates during the daytime through the surface of the leaf, mainly through the stomata. At night the excess water sometimes escapes in liquid form through the apices of veins, particularly in herbaceous plants.

(4) **Storage of Food.** Fleshy leaves of Indian aloe (B. GHRITAKUMARI; H. GHIKAVAR), *Portulaca* (B. NUNIA-SAK; H. KULFA-SAG) and fleshy scales of onion store up water and food for their future use. Fleshy and succulent leaves of desert plants always store a quantity of water, mucilage and food.

(5) **Vegetative Propagation.** Leaves of sprout leaf plant (*Bryophyllum;* see Fig. 45A) and elephant-ear plant (*Bergonia;* see Fig. 45B) produce buds on them for vegetative propagation; walking ferns (e.g. *Adiantm caudatum,* etc.; see Fig. III/37) reproduce vegetatively by their leaf-tips. Leaves bow down to the ground, their tips strike roots and form a bud which grows into a new plant.

Heterophylly. Many plants bear different kinds of leaves on the same individual plant. This condition is known as **heterophylly** (*heteros,* different; *phylla,* leaves). Heterophylly is met with in many aquatic plants, particularly in those growing in shallow running water. Here the floating or aerial leaves and the submerged leaves are of different kinds; the former are generally broad, more or less fully expanded, and undivided or merely lobed; while the latter are narrow, ribbon-shaped, linear or much dissected. Heterophylly in water plants is regarded as an adaptation to two different conditions of the environment (aquatic and aerial). Some common examples are *Cardanthera triflora* (Fig. 90A), water crowfoot (*Ranunculus aquatilis*), arrowhead (*Sagittaria;* Fig. 91),

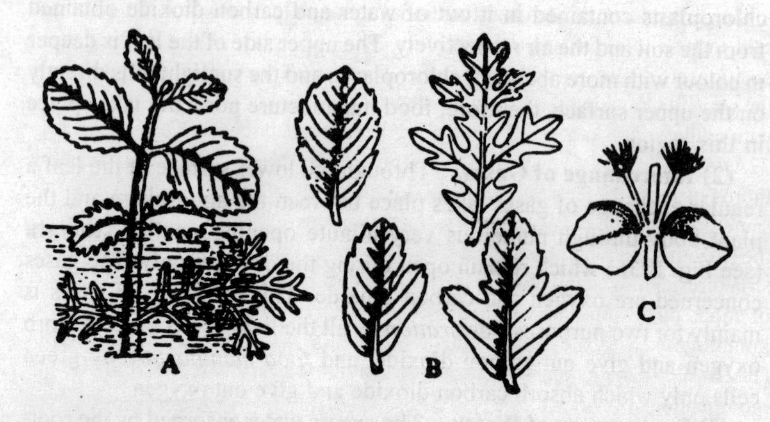

Figure 90. Heterophylly. *A, Cardanthera triflora; B, Artocarpus chaplasha; C, Hemiphragma heterophyllum.*

Limnophila heterophylla, etc. Some land plants also exhibit this phenomenon without any apparent reason. Common examples are *Artocarpus chaplasha* (Fig. 90B), (*A. heterophyllus*) in the young stage only, *Hemiphragma heterophyllum* (Fig. 90C), *Ficus heterophylla,* etc. In *Hemiphragma,* a prostrate herb in Shillong and Darjeeling, the leaves are of two kinds—broad and needle-shaped.

Homology and Analogy. Homology is the morphological study of modified organs from the standpoint of their origin, and analogy is the study of organs from the standpoint of their identical structure and function; or in other words, organs which have the same origin, and are, therefore, morphologically the same, whatever be their structure and function, are said to be *homologous* with one another, and organs which resemble one another in their structure and are adapted

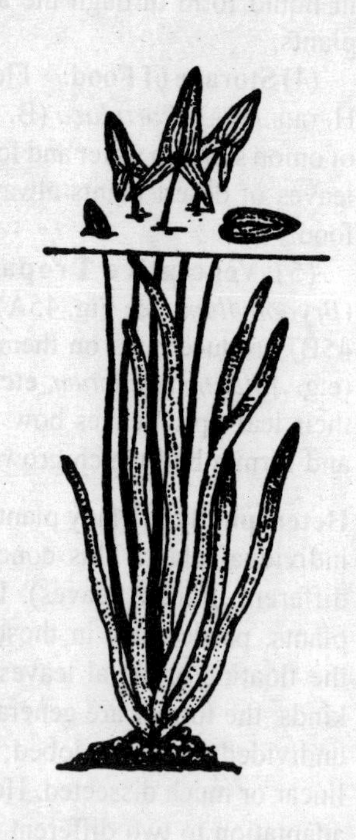

Figure 91. Arrowhead (*Sagittaria*) showing heterohylly.

to the performance of identical functions, although their origin is different, are said to be *analogous* with one another. Thus all tendrils, whatever be their position, are analogous with one another, being structurally the same and having the same function; but tendrils of passion-flower (see Fig. 54A) are homologous with axillary buds, i.e. modifications of the latter, and tendrils of pea (see Fig. 77A) are homologous with leaflets. Similarly tendrils of passion-flower and thorns of *Duranta* (see Fig. 56A) are homologous structures, both having the same origin in the axis of leaves as modifications of axillary buds. Likewise, the rhizome, the tuber, the fusiform root, the napiform root, etc., are analogous structures, being adapted to the performance of identical function, i.e. storage of food; but it must be noted that the former two (rhizome and tuber) are homologous with the stem, being modifications of it, while the latter two (fusiform root and napiform root) are homologous with the root, being modifications of it.

Chapter 7

Defence Mechanisms in Plants

The animal kingdom as a whole is directly or indirectly parasitic upon the plant kingdom, and this being so, plants must either fall a victim to various classes of animals, particularly the herbivorous ones, which live exclusively on a vegetable diet, or they must be provided with special organs or arms of defence, or have other special devices to repulse or avoid the attack of their enemies. Being fixed to the ground they cannot, of course, move when attacked by animals.

1. **Armature.** Various parts of the plant body may take the form of arms or defensive weapons for self-defence against the attack of herbivorous animals. These are as follows:

(1) **Thorns** are modifications of branches, and originate from deeply-seated tissues of the plant body. They are straight, hard and pointed, and can pierce the body of thick-skinned animals. They are usually axillary in position, or sometimes terminal. Plants like *Vangueria* (B. & H. MOYNA), lemon, pomegranate, *Duranta, Carissa* (B. KARANJA; H. KARONDA) and many others are well provided with thorns for self-defence.

(2) **Spines** are modifications of leaves or parts of leaves and serve the purpose of defence. They are seen in pineapple, date-palm, prickly poppy (see Fig. 79B), American aloe, dagger plant, etc. In dagger plant (*Yucca;* Fig. 92) each leaf ends in a very sharp and pointed spine, and is directed obliquely outwards. It acts like a dagger or pointed spike when any grazing animal approaches it.

(3) **Prickles** are also hard and pointed like the thorns, but are usually curved and have a superficial origin, they are further irregularly distributed on the stem, branch or leaf. Prickles are commonly found in cane and rose (see Fig. 3), coral tree (*Erythrina*) silk cotton tree (Bombax), Prosopis (B. & H. SHOMI), etc. Globe thistle (*Echinops*) and

prickly or Mexican poppy (*Argemone;* see Fig. 79B) are armed with both prickles and spines for self-defence.

(4) **Bristles** are short stiff and needle-like hairs, usually growing in clusters, and not infrequently barbed. Their walls are often thickened with a deposit of silica or calcium carbonate, Bristles are commonly met with in prickly pear (*Opuntia;* see Fig. 57A) and in many other cacti.

Figure 92. Dagger plant or Adam's needle (*Yucca*)

(5) **Stinging Hairs.** Nettles (B. BICHUTI; H. BARHANTA) develop stinging hairs on their leaves or fruits or all over their body. Each hair (Fig. 93) has a sharp siliceous apex which readily breaks off even when touched lightly. The sharp point penetrates into the skin, and at once the acid poison of the hair contained in its swollen (bulbous) base is injected into it (the skin) under a sudden pressure. This evidently causes a sharp burning pain, often attended with inflammation. There are various kinds of nettles. e.g. *Laportea* (=*Fleurya* B. LAL-BICHUTI)—an annual weed, *Tragia* (B. BICHUTI; H. BARHANTA)—a twiner, fever or devil nettle (*Laportea*)—a shrub, cowage (*Mucuna;* B. ALKUSHI; H. KAWANCH)—a large twiner, etc. In cowage the stinging hairs develop on the fruits.

(6) **Hairs.** A dense coating of hairs, as in cudweed (*Gnaphalium*) and *Aerua* or presence of stiff hairs, as in some gourd plants (*Cucurbita*), on the body of the plant, is always repulsive to animals as these hairs stick on to their throat and cause an irritation or a choking sensation. Many plants bear **glandular hairs** which secrete a sticky substance. Any animal feeding upon such a plant finds it difficult to brush them off from its mouth. Plants bearing glandular hairs are thus never attacked

by grazing animals, e.g. tobacco (*Nicotiana*), *Boerhaavia* (B. PUNAR-NAVA; H. SANTH; see Fig. 162A), *Jatropha* (B. & H. BHARENDA), *Plumbago* (B. CHITA; H. CHITRAK), etc.

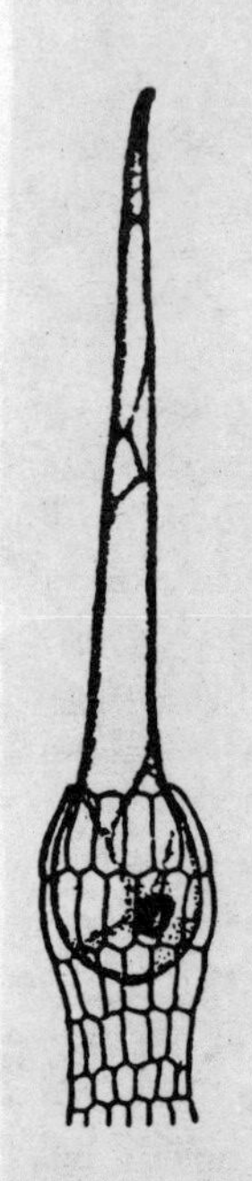

Figure 93.
A stinging hair.

2. **Other Devices of Defence.** Many plants secrete poisonous and irritating substances; such plants are carefully avoided by animals which possess the power of distinguishing between poisonous and non-poisonous ones.

(1) **Latex** is the milky juice secreted by certain plants. It always contains some waste products, and often irritating and poisonous substances so that it causes inflammation and even blisters when it comes in contact with the skin. Plants like madar (*Calotropis*), spurge (*Euphorbia*), oleander (*Nerium*), periwinkle (*Vinca*), papaw (*Carica*), etc., contain latex.

(2) **Alkaloids** are in many cases extremely poisonous, and a very minute quantity is sufficient to kill a strong animal. There are various kinds of them found in plants, e.g. strychnine in nux-vomica, morphine in opium poppy, nicotine in tobacco, daturine in *Datura,* quinine in *Cinchona,* etc.

(3) **Irritating Substance.** Plants like many aroids, e.g. taro (*Colocasia;* B. KACHU; H. KACHALU), *Amorphophallus* (B. OL; H. KANDA), etc., possess needle-like or otherwise sharp and pointed crystals of calcium oxalate, i.e. raphides (see Fig. II/18). These crystals, when such plants are fed upon, prick the tongue and the throat and cause irritation. Therefore, such plants are never attacked by grazing animals.

(4) **Bitter Taste and Repulsive Smell.** These are also effective mechanisms to ward off animals. *Paederia foetida* (B. GANDHAL; H. GANDHALI) emits a bad smell so that no animal likes to go near it. Plants like sacred basil, mint, *Blumea lacera* (B. KUKURSONGA; H. KOKRONDA), *Gynandropsis,* etc., also emit a strong disagreeable odour. The fetid smell of the inflorescence of *Amorphophallus* (see Fig. 36) is very offensive and nauseating. Margosa, bitter gourd, *Andrographis* (B. KALMEGH; H. MAHATITA), etc., have a bitter taste and, therefore, animals avoid them.

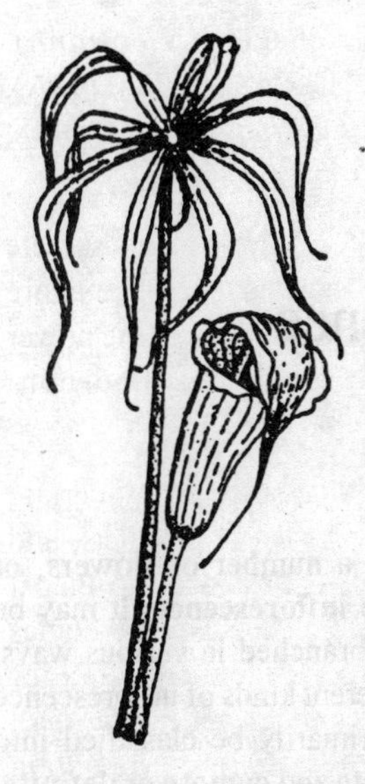

Figure 94. Snake or cobra plant (*Arisaema*).

(5) **Waste Products.** Many plants contain various waste products such as tannin, resin, essential oils, raphides, silica, etc., which keep them free from the attack of animals.

(6) **Mimicry.** Certain plants also protect themselves against grazing animals by imitating the general appearance, colour, shape or any particular feature of another plant or animal; for instance, there are certain aroids (e.g. varieties of *Caladium*) which resemble multi-coloured and variously spotted snakes. Leaves are also variously spotted and striped in many species of bowstring hemp (*Sansevieria;* B. MURGA; H. MARUL) and other allied plants. The inflorescence of devil's spittoon (*Amorphophallus bulbifer*) emerging out of the ground imposes a frightening look. Herbivorous animals, possibly mistaking them for snakes or some other threatening creatures, carefully avoid them. In snake or cobra plant (*Arisaema;* Fig. 94), common in Shillong during the rains, the spathe is greenish-purple in colour and it expands over the spadix like the hood of the cobra. This act of imitating the appearance, colour or any particular feature of another plant or animal is called **mimicry** (*mimikos, imitative*).

Plants have also to protect themselves against the attack of many parasitic fungi and gnawing insects, and also against the scorching rays of the sun; this they do by developing cork and bark.

Chapter 8

The Inflorescence

The reproductive shoot bearing commonly a number of flowers, or sometimes only a single flower, is called the **inflorescence**. It may be terminal or axillary in position, and may be branched in various ways. Thus depending on the mode of branching different kinds of inflorescence have come into existence, and these may primarily be classified into two distinct groups, viz. **racemose** or **indefinite** and **cymose** or **definite**.

1. **Racemose Inflorescence.** Here the main axis of inflorescence does not end in a flower, but it continues to grow and give off flowers laterally. The lower or outer flowers are always older and open earlier than the upper or inner ones, i.e. the order of opening of flowers is *centripetal*. Some of the common types are as follows.

I. *WITH THE MAIN AXIS ELONGATED*

(1) **Raceme** (Fig. 95A). The main axis in this case is elongated and it bears laterally a number of flowers which are all stalked, the lower or older flowers having longer stalks than the upper or younger ones, as in radish, mustard, dwarf gold mohur (*Caesalpinia*), etc. When the main axis of the raceme is branched and the lateral branches bear the flowers, the inflorescence is said to be a compound raceme or **panicle** (Fig. 96), as in gold mohur (*Delonix*).

The main axis of the inflorescence together with the lateral axes, if present, is known as the **peduncle**. The stalk of the individual flower of the inflorescence is called the **pedicel**. In some solitary flowers such as China rose, etc. The peduncle and the pedicel may be clearly marked out due to the presence of an articulation on the floral axis. When the peduncle of an inflorescence is short and dilated forming a sort of convex platform, as in sunflower (see Fig. 99), or becoming hollow and pear-shaped, as in fig. (*Ficus*), it is often called **receptacle** (see Fig. 103).

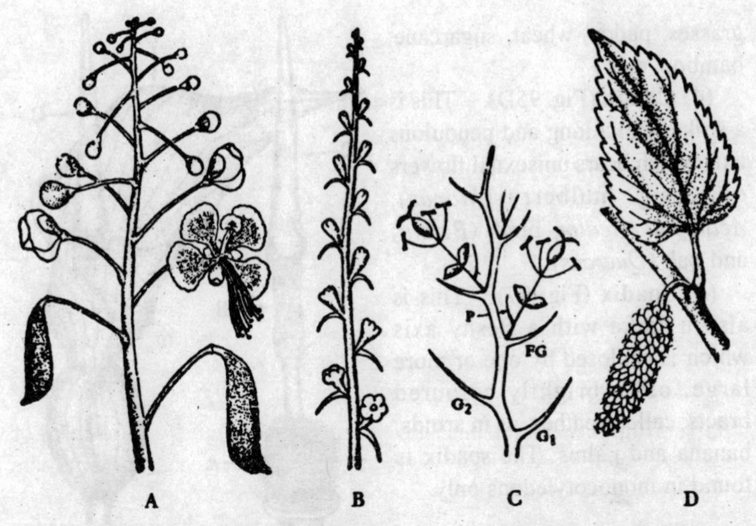

Figure 95. Racemose Inflorescences. *A,* raceme of dwarf gold mohur; *B,* spike (diagrammatic); *C,* spikelet of a grass (diagrammatic); G_1, first empty glume; G_2, second empty glume; *FG,* flowering glume (lemma); and *P,* palea; *D,* female catkin of mulberry.

(2) **Spike** (Fig. 95B). Here also the main axis is elongated and the lower flowers are older, opening earlier than the upper ones, as in raceme, but the flowers are sessile, that is, without any stalk. Examples are seen in tuberose, *Adhatoda* (B. BASAK; H. ADALSA), amaranth (B. NATE-SAK; H. CHULAI), chaff-flower (*Achyranthes;* B. APANG; H. LATJIRA), etc.

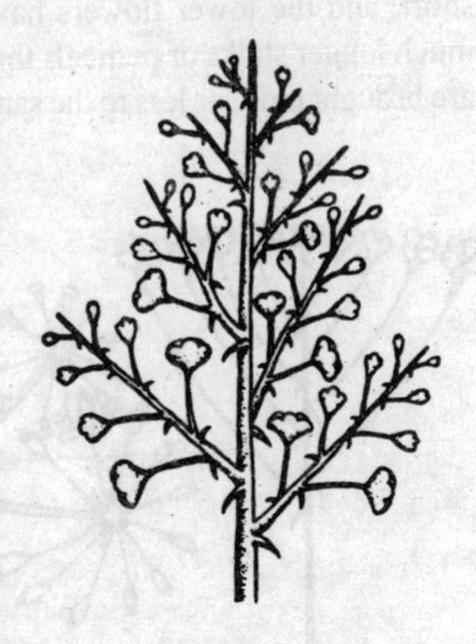

Figure 96. A panicle.

(3) **Spikelet** (Fig. 95C). This is a very small spike with one or a few small flowers (florets). In it there are two small *empty glumes* at the base, and just above them a *flowering glume* called *lemma* with a flower in its axil, and opposite to the lemma there is a small 2-nerved bracteole called *palea.* The flower remains enclosed by the lemma and the palea. Succeeding flowers likewise occur within the lemma and the palea. Spikelet is characteristic of the grass family, e.g.

grasses, paddy, wheat, sugarcane, bamboo, etc.

(4) **Catkin** (Fig. 95D). This is a spike with a long and pendulous axis which bears unisexual flowers only, e.g. mulberry (*Morus*), *Acalypha tricolor,* birch (*Betula*) and oak (*Quercus*).

(5) **Spadix** (Fig. 97). This is also a spike with a fleshy axis which is enclosed by one or more large, often brightly coloured bracts, called spathes, as in aroids, banana and palms. The spadix is found in monocotyledons only.

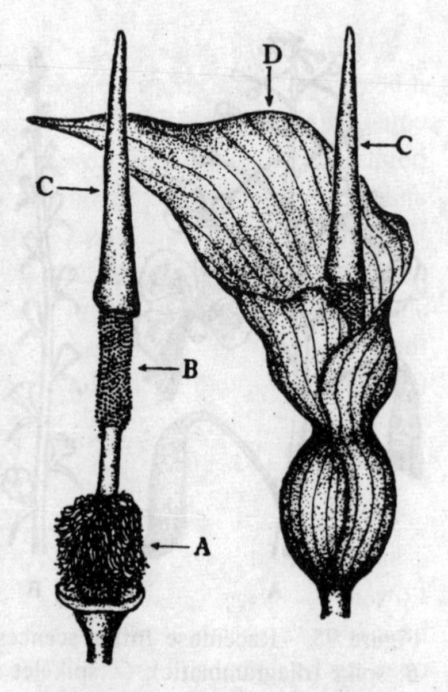

Figure 97. Spadix of an aroid (*Typhonium*); *A,* female flowers; *B,* male flowers; *C,* appendix, *D,* spathe.

II. *WITH THE MAIN AXIS SHORTENED*

(6) **Corymb** (Fig. 98A). Here the main axis is comparatively short, and the lower flowers have much longer stalks or pedicels than the upper ones so that all the flowers are brought more or less to the same level, as in candytuft and wallflower.

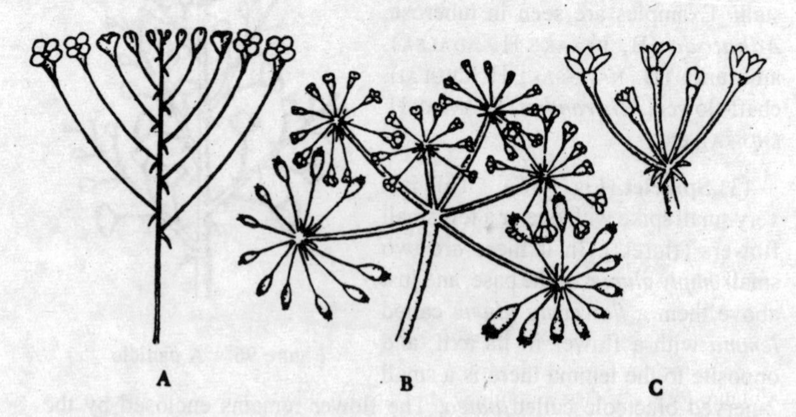

Figure 98. *A,* corymb; *B,* a compound umbel; *C,* a simple umbel.

(7) **Umbel** (Fig. 98B–C). Here the primary axis is shortened, and it bears at its tip a group of flowers which have pedicels of more or less equal lengths so that the flowers are seen to spread out from a common point. In the umbel there is always a whorl of bracts forming an involucre, and each flower develops from the axil of a bract. Commonly the umbel is branched (**compound umbel**) and the branches bear the flowers as in anise or fennel, coriander, cumin, carrot, etc. Sometimes, however, it is simple or unbranched (**simple umbel**), the main axis directly bearing the flowers, as in Indian pennywort (*Centella*) and wild coriander (*Eryngium*). Umbel is characteristic of coriander family.

III. *WITH THE MAIN AXIS FLATTENED*

(8) **Head** or **Capitulum** (Fig. 99). Here the main axis or receptacle is suppressed, becoming almost flat, and it bears masses of small sessile flowers (florets) on its surface, with one or more whorls of bracts at the base forming an *involucre*. In the head the outer flowers are older and open earlier than the inner ones. The florets are commonly of two kinds—**ray florets** (marginal strap-shaped ones) and **disc florets** (central tubular ones). The head may also consist of only one kind of florets. A head or

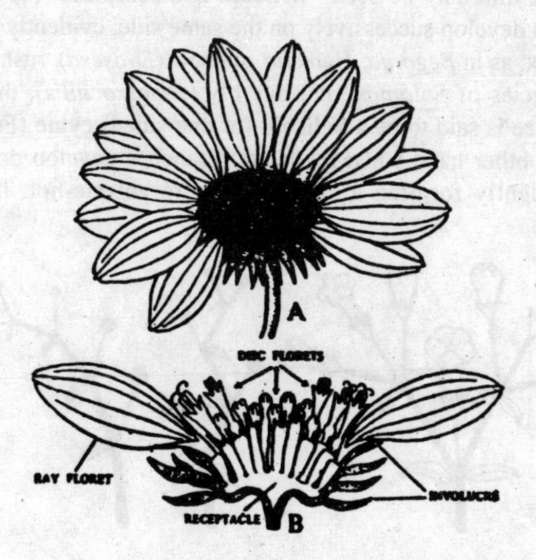

Figure 99. Head or capitulum. *A*, a head (a few ray florets removed to show the involucre); *B*, a head in longitudinal section.

capitulum is characteristic of the sunflower family (e.g. sunflower, marigold, safflower, *Zinnia, Cosmos,* etc.). It is also found in gum tree (*Acacia*), sensitive plant (*Mimosa*), *Anthocephalus* (B. & H. KADAM), *Adina* (B. & H. KELI-KADAM), etc.

The advantages of this kind of inflorescence are that the head as a whole becomes more showy and attractive, and the florets being close together, one or a few insects can pollinate most of them within a short time.

2. **Cymose Inflorescences.** Here the main axis ends in a flower and similarly the lateral axis also ends in a flower. Thus the growth of each axis is checked. In cymose inflorescences the terminal flower is always older and opens earlier than the lateral ones, i.e. the order of opening of flowers is *centrifugal*. Cymose inflorescences may be of the following types.

(1) **Uniparous** or **Monochasial Cyme.** In this type of inflorescence the main axis ends in a flower and it produces only one lateral branch at a time ending in a flower. The lateral and succeeding branches again produce only one branch at a time like the primary one. Two forms of uniparous cyme may be seen—helicoid and scorpioid. (a) When the lateral axes develop successively on the same side, evidently forming a sort of helix, as in *Begonia, Hamelia,* sundew (*Drosera*), rush (*Juncus*), several species of *Solanum,* and day lily (*Hemerocallis*), the cymose inflorescence is said to be a **helicoid** (or one-sided) **cyme** (Fig. 100C). (b) On the other hand when the lateral branches develop on alternate sides, evidently forming a zigzag, as in forget-me-not, heliotrope

Figure 100. Cymose Inflorescences. *A,* biparous cyme; *B,* scorpioid cyme; *C,* helicoid cyme.

(B. HATISUR; H. HATTASURA), *Crassula,* cotton and *Freesia,* the cymose inflorescence is said to be a **scorpioid** (or alternate-sided) **cyme** (Fig. 100B). The scorpioid cyme is otherwise called cincinnus.

In *monochasial* cyme successive axes may be at first zigzag or curved, but subsequently become straightened due to rapid growth, thus forming a central axis or pseudo-axis which is really a succession of branches. This type of inflorescence is otherwise called a **sympodial cyme** (*syn,* together; *podos,* foot). This may be distinguished from the racemose type by examining the position of a bract to a flower; in a sympodial cyme a bract appears opposite to a flower, while in a racemose type a bract appears at the base of a flower.

(2) **Biparous** or **Dichasial Cyme.** In this type of inflorescence the main axis ends in a flower and at the same time it produces two lateral younger flowers, sessile or stalked. The lateral and succeeding flowers develop in the same manner (Fig. 100A). This is **true cyme**. Examples are seen in pink jasmine, teak, night jasmine, *Ixora* (B. RANGAN; H. GOTAGAN-DHAL), etc.

(3) **Multiparous** or **Polychasial Cyme.** In this kind of cymose inflorescence the main axis, as usual, ends in a flower, and at the same time it again produces a number of lateral flowers around. There being a number of lateral flowers developing more or less simultaneously, the whole inflorescence looks like an umbel, but is readily distinguished from the latter by the opening of the middle flower first. This is seen in madar (*Calotropis;* B. AKANDA; H. AK), and blood flower (*Asclepias;* B. KAKTUNDI; H. KAKATUNDI).

Compound and Mixed Forms. In a compound inflorescence the main axis is branched and the branches bear the flowers. Thus raceme may be branched into a compound form, otherwise called **panicle** (see Fig. 96), as in gold mohur, margosa, dagger plant, etc. Similarly other compound forms may be met with; for example, a **compound spike**, as in wheat; **compound spadix**, as in palms; **compound corymb**, as in candytuft; **compound umbel**, as in coriander and anise; and **compound head**, as in globe thistle (*Echinops*). Mixed inflorescences are also not uncommon; thus two racemose types, e.g. raceme and umbel, raceme and spike, may be mixed up; sometimes racemose and cymose types may be combined in the same inflorescence.

3. **Special Types.** The following types may be noted.

(1) **Cyathium** (Fig. 101). This is a special kind of inflorescence found in *Euphorbia,* e.g. poinsettia (B. & H. LALPATA), spurges (B. & H. SIJ), etc., and also in jew's slipper (*Pedilanthus;* B. RANG-CHITA; H. NAGDAMAN). In cyathium there is a cup-shaped involucre, often

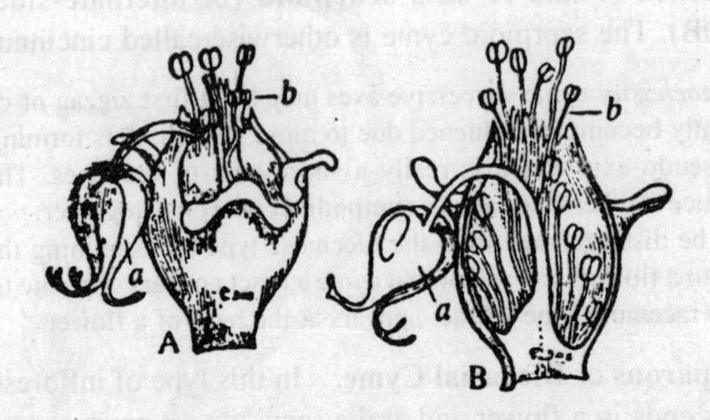

Figure 101. Cyathium of poinsettia. *A,* cyathium; *B,* the same in section *a,* female flower; *b,* male flowers. Note the involucre with nectar glands.

provided with nectar-secreting glands. The involucre encloses a single female flower (reduced to a pistil) in the centre, seated on a comparatively long stalk, and a number of male flowers (each reduced to a solitary stamen) around this, seated on short slender stalks. That each stamen is a single male flower is evident from the fact that it is articulated to a stalk (peduncle) and that it has a scaly bract at the base. The flowers follow centrifugal (cymose) order of development. The female flower in the centre matures first, and then the stamens (male flowers) gradually outwards.

(2) **Verticillaster** (Fig. 102). This is a special form of cymose inflorescence. In it there is a cluster of sessile or almost sessile flowers in the axil of a leaf, forming a false whorl at the node. The first axis ends in a flower; it bears two lateral branches, each ending in a flower; succeeding lateral branches are produced in an alternating manner. This kind of inflorescence is

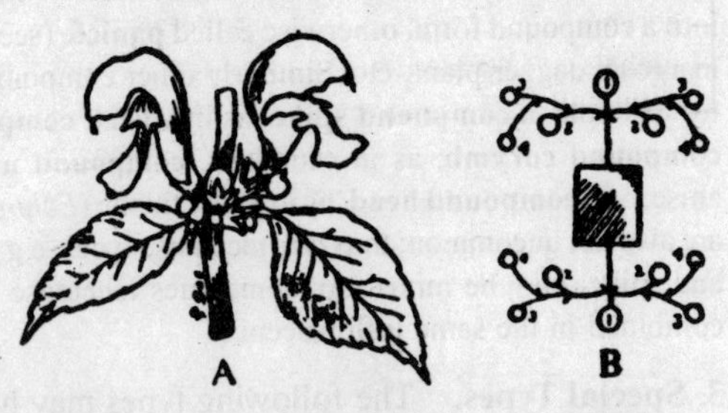

Figure 102. Verticillaster of *Coleus.* *A,* verticillaster; *B,* diagram of verticillaster.

found in several members of basil family, e.g. *Coleus,* mint (*Mentha;* B. PUDINA; H. PODINA), *Leonurus* (B. DRONA; H. HALKUSHA), etc. In sacred basil (*Ocimum sanctum;* B. & H. TULSI) the verticillaster is a condensed cyme, succeeding branches remaining undeveloped.

(3) **Hypanthodium** (Fig. 103). When the fleshy receptacle forms a hollow cavity with an apical opening guarded by scales, and the flowers are borne on the inner wall of the cavity, the inflorescence is a hypanthodium, as in *Ficus* (e.g. banyan, fig, peepul, etc.). Here the female flowers develop at the base of the cavity and the male flowers higher up towards the apical pore.

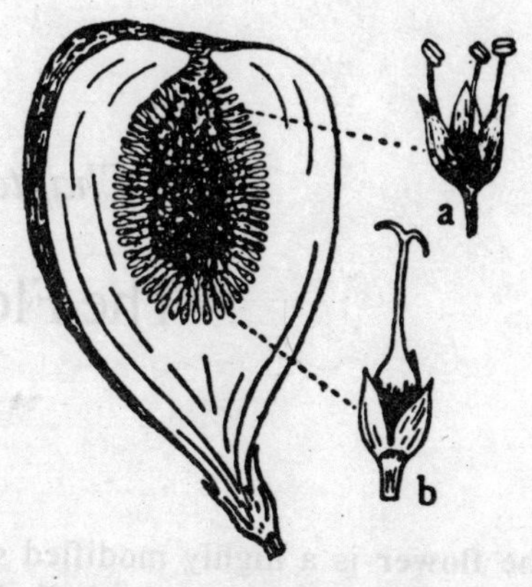

Figure 103. Hypanthodium of fig. (*Ficus*). *a,* male flower, *b,* female flower.

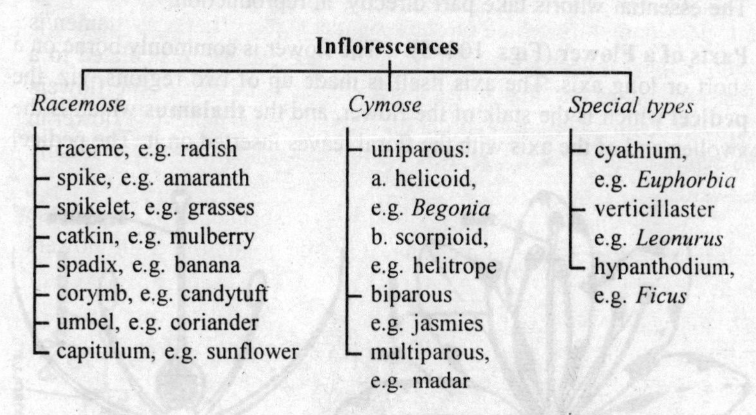

Chapter 9

The Flower

The **flower** is a highly modified shoot meant essentially for the reproduction of the plant. Typically it is a collection of *four* different kinds of floral members arranged in *four* separate whorls or circles in a definite order. Of the four whorls the upper two are called *essential* or *reproductive* whorls, and the lower two *helping* or *accessory* whorls. The essential whorls take part directly, in reproduction.

Parts of a Flower (Figs. 104–5). The flower is commonly borne on a short or long axis. The axis itself is made up of two regions, viz. the **pedicel** which is the stalk of the flower, and the **thalamus** which is the swollen end of the axis with the floral leaves inserted on it. The pedicel

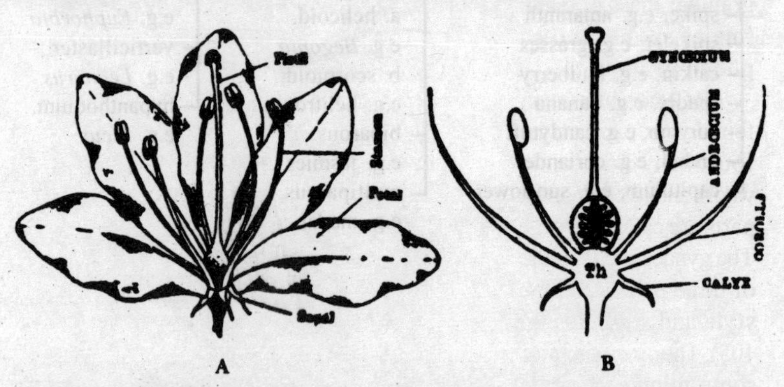

Figure 104. *A*, parts of flower; *B*, a flower in longitudinal section showing the position of the whorls on the thalamus (*Th*).

may be short or long or even absent. A typical flower consists of *four* whorls arranged in a definite order, one just above the other. The whorls and their component parts are as follows.

(1) **Calyx** is the first or the lowermost whorl of the flower, and consists of a number of green leafy **sepals**.

(2) **Corolla** is the second whorl of the flower, and consists of a number of usually brightly coloured **petals**.

(3) **Androecium** (*andros,* male) is the third or the male whorl; its component parts are called **stamens**. Each stamen is made of three parts—**filament**, **anther** and **connective** (Fig. 105). The anther bears four chambers or **pollen-sacs**, each filled with a granular mass of small (male) spores, called **pollen grains**.

(4) **Gynoecium** (*gyne,* female) or pistil is the fourth or the female whorl, and its component parts are called **carpels**. The gynoecium is made of three parts—**ovary**, **style** and **stigma** (Fig. 105). The ovary encloses some minute egg-like bodies called **ovules**. Each ovule encloses a large oval cell called **embryo-sac** (see Fig. 131).

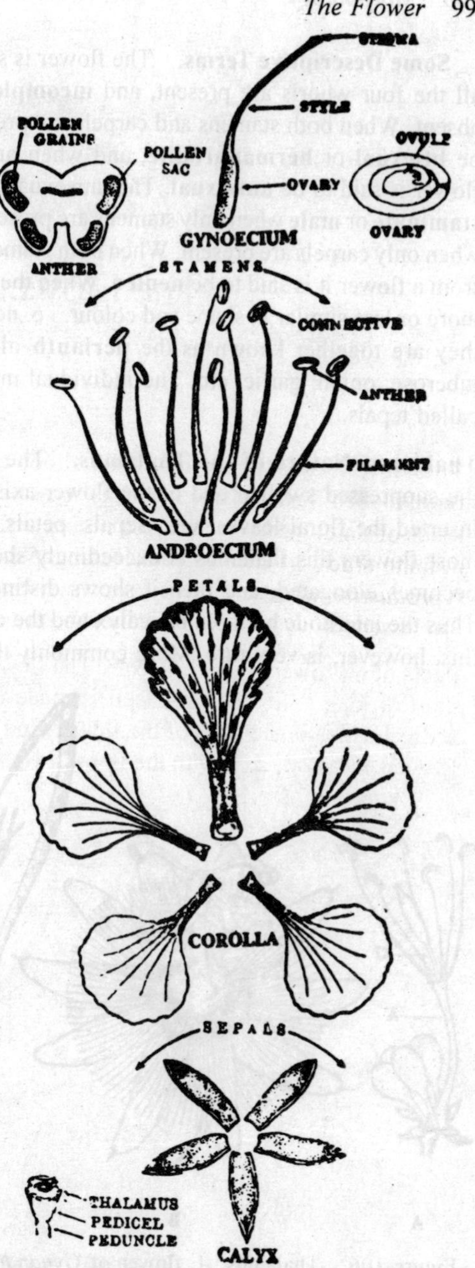

Figure 105. Flower of gold mohur (*Delonix regia*) dissected out.

Some Descriptive Terms. The flower is said to be **complete** when all the four whorls are present, and **incomplete** when any of them is absent. When both stamens and carpels are present the flower is said to be **bisexual** or **hermaphrodite**, and when any of them is absent the flower is said to be **unisexual**. The unisexual flower may, therefore, be **staminate** or **male** when only stamens are present, or **pistillate** or **female** when only carpels are present. When both stamens and carpels are absent from a flower it is said to be **neuter**. When the calyx and the corolla are more or less similar in shape and colour, i.e. not clearly distinguishable, they are together known as the **perianth** of the flower, as in lilies, tuberose, onion, garlic, etc. The individual members of a perianth are called tepals.

Thalamus. Nature of the Thalamus. The thalamus (Fig. 104B) is the suppressed swollen end of the flower-axis (pedicel), on which are inserted the floral leaves, viz., sepals, petals, stamens and carpels. In most flowers this thalamus is exceedingly short; but in a few cases it becomes elongated, and then it shows distinct nodes and internodes. Thus the internode between the calyx and the corolla may be elongated; this, however, is very rare. More commonly the internode between the

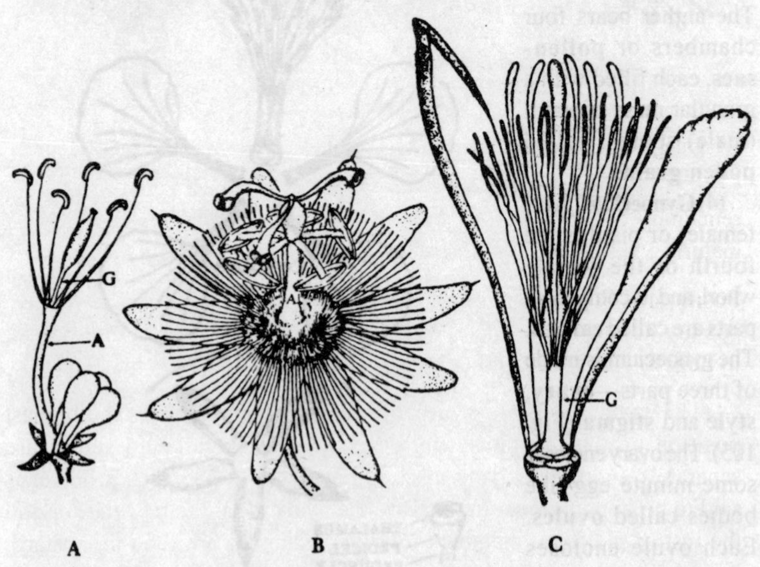

Figure 106. Thalamus. *A*, flower of *Gynandropsis; B*, passion-flower; *C*, flower of *Pterospermum* (with the staminal tube adnate to gynophore). *A*, androphore; *G*, gynophore.

corolla and the androecium is considerably elongated, and is known as the **androphore** (*andros,* male; *phore,* stalk), as in *Gynandropsis* (Fig. 106A) and passion-flower (Fig. 106B). In *Capparis* (Fig. 107A), *Gynandropsis* (Fig. 106A) and *Pterospermum* (B. MOOCHKANDA; H. KANAK-CHAMPA—Fig. 106C) the internode between the androecium and the gynoecium is elongated, and is known as the **gynophore** (*gyne,* female). It may be noted that in *Gynandropsis* both androphore and gynophore develop and they are together known as the **gynandrophore**. In *Magnolia* (B. DULEE-CHAMPA) and *Michelia* (B. CHAMPA; H. CHAMPAK) the thalamus is fleshy and elongated, and bears the floral leaves spirally round it. In rose (Fig. 107B) it is concave and pear-shaped. The thalamus of lotus (Fig. 107C) is spongy and top-shaped. In anise, coriander, balsam, etc., the thalamus elongates upwards into a slender axis with the carpels

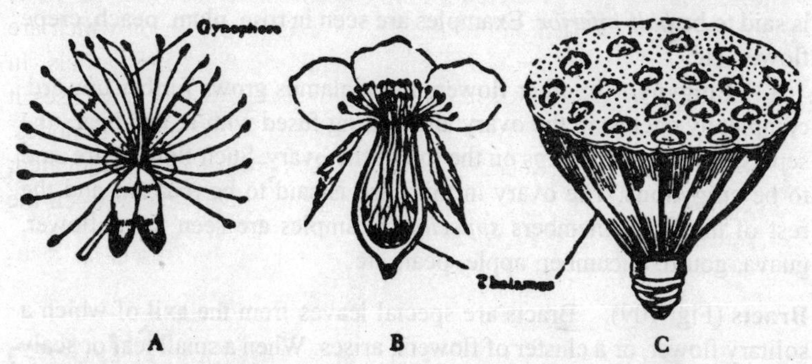

Figure 107. Thalamus (*contd.*). *A,* flower of *Capparis*; *B,* rose (in section); *C,* lotus.

remaining attached to it at first and separating from it afterwards on maturity; such an axis is called **carpophore** (see Fig. 148A).

Position of Floral Leaves on the Thalamus (Fig. 108). The relative positions of the floral whorls with respect to the ovary are of three kinds: **hypogyny**, **perigyny** and **epigyny**.

(1) **Hypogyny.** In a typical flower the ovary occupies the highest position on the thalamus, while the stamens, petals and sepals are separately and successively inserted below the ovary. Such a flower is said to be **hypogynous**. In this case the ovary is said to be *superior* and the rest of the floral members *inferior*. Examples are seen in mustard, brinjal, China rose, *Magnolia,* etc.

(2) **Perigyny.** In some flowers the thalamus grows upward around the ovary in the form of a cup, carrying on its rim the sepals, petals and

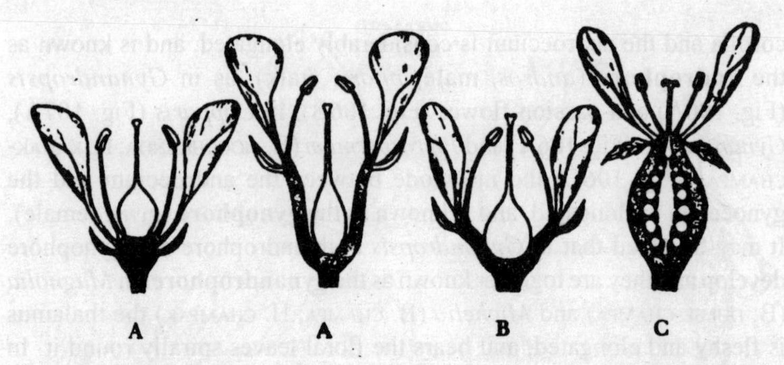

Figure 108. Thalamus (*contd.*). *A,* hypogyny; *B,* perigyny (two types—A & B); *C,* epigyny.

stamens. Such flowers are said to be **perigynous**, and the ovary in them is said to be *half-inferior*. Examples are seen in rose, plum, peach, crepe flower, etc.

(3) **Epigyny,** In other flowers the thalamus grows further upward, completely enclosing the ovary and getting fused with it, and bears the sepals, petals and stamens on the top of the ovary. Such flowers are said to be **epigynous**. The ovary in this case is said to be *inferior,* and the rest of the floral members *superior*. Examples are seen in sunflower, guava, gourd, cucumber, apple, pear, etc.

Bracts (Fig. 109). Bracts are special leaves from the axil of which a solitary flower, or a cluster of flowers, arises. When a small leaf or scaly structure is present on any part of the flower-stalk (pedicel) it goes by the name of **bracteole**. Bracts vary in size, colour and duration, and are commonly of the following kinds.

(1) **Leafy Bracts.** These are green, flat and leaf-like in appearance, as in *Acalypha* (B. MUKTOJHURI; H. KUPPI), *Adhatoda* (B. BASAK; H. ADALSA), *Gynandropsis,* etc.

(2) **Spathe** (*A–B*). This is a large, sometimes very large, commonly boat-shaped and bright coloured bract, enclosing a cluster of flowers or even a whole inflorescence (spadix). It protects the flowers while they are still young, and later attracts insects for pollination by its colour. Examples arc seen in aroids, banana, palms, maize cob, etc.

(3) **Petaloid Bracts** (*C*). These are brightly coloured bracts looking somewhat like petals, as in glory of the garden, (*Bougainvillea;* B. BAGANBILAS) and poinsettia (*Euphorbia;* B. LALPATA).

(4) **Involucre** (*D*). This is a group of bracts occurring in one or more whorls around a cluster of flowers, as in sunflower, marigold.

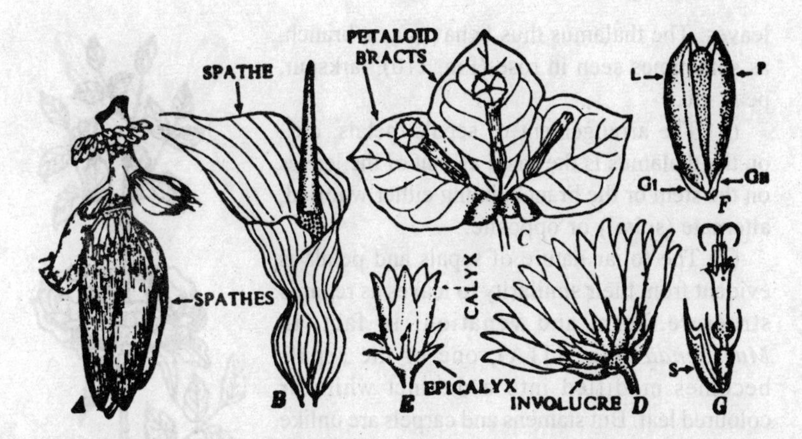

Figure 109. Bracts and Bracteoles. *A,* spathes of banana; *B,* spathe of an aroid (*Typhonium*); *C,* petaloid bracts of *Bougainvillea; D,* involucre of sunflower; *E,* epicalyx (bracteoles) of China rose; *F,* glumes of paddy grain (G_I, G_{II}, empty glumes; *L,* lemma or flowering glume; *P,* palea—a bracteole); *G,* scaly bracteole (S) of a central floret of sunflower.

Cosmos, etc. The involucre often encloses the whole inflorescence and protects the flowers when young. The bracts of the involucre are usually green, and may be free or united.

(5) **Epicalyx** (*E*). This is a whorl of bracteoles developing at the base of the calyx, as in China rose, cotton, lady's finger, etc.

(6) **Glumes** (*F*). These are special bracts, small and dry, found in the spikelets of grass family and sedge family only (see p. 91).

(7) **Scaly Bracteole** (*G*). This is a very small, thin, papery scale occurring at the base of the central floret of sunflower.

FLOWER IS A MODIFIED SHOOT

The following facts may be cited to prove that the *thalamus* is a modified branch; *sepals, petals, stamens* and *carpels* are modified vegetative leaves; and the *flower* as a whole a modified vegetative bud.

(1) In some flowers the thalamus becomes elongated showing distinct nodes, and internodes (see Figs. 106–7), as in *Gynandropsis,* passion-flower, etc. The thalamus may, therefore be regarded as a modified branch.

(2) The thalamus sometimes shows monstrous development, i.e. after bearing the floral members it prolongs upwards and bears ordinary

leaves. The thalamus thus behaves as a branch, as sometimes seen in rose (Fig. 110), larkspur, pear, etc.

(3) The arrangement of sepals, petals, etc., on the thalamus is the same as that of the leaves on the stem or the branch, being either whorled, alternate (spiral) or opposite.

(4) The foliar nature of sepals and petals is evident from their similarity to leaves as regards structure, form and venation: in fact, in *Mussaenda* (Fig. 111A), one of the sepals becomes modified into a distinct white or coloured leaf. But stamens and carpels are unlike leaves in all respects. Their homology with leaves can be made out from certain flowers. Thus water lily flower (Figs. 111B & 112) shows a gradual transition from sepals to petals and from petals to stamens. The cultivated rose shows many petals; while in wild rose (e.g. *Rosa canina*) there are only five. The explanation is that many stamens have gradually become modified into petals. Similarly in *Hibiscus mutabilis* (B. STHAL-PADMA; H. GULA-JAIB) some or many of the stamens have modified into petals. Further the foliar nature of the carpel may be made out from the flowers of pea, bean, etc. in which the single carpel (or pod) may be compared to a leaf which has been folded along its mid rib.

Figure 110. Rose showing monstrous development of the thalamus.

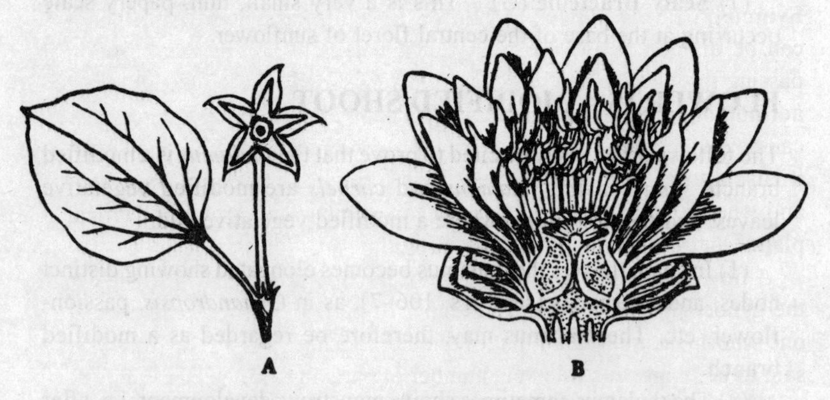

Figure 111. *A, Mussaenda* flower with a sepal modified into a leaf; *B,* water lily flower showing transition of floral parts (see also Fig. 112)

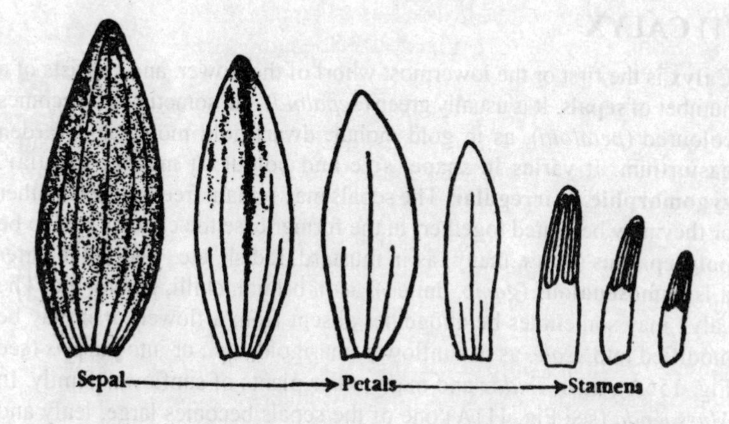

Figure 112. Transition of floral parts in water lily flower.

(5) The inflorescence axis normally bears flowers. Sometimes, as in American aloe (*Agave;* see Fig. III/41B), some of the floral buds become modified into vegetative buds, called bulbils, for vegetative reproduction. In pineapple also, the inflorescence axis bears one or more vegetative buds or bulbils (see Fig. III/42) for vegetative propagation. Such bulbils thus show a reversion to ancestral forms, i.e. the forms from which they have been derived.

(6) A floral bud, like a vagetative bud, is either terminal or axilliary in position. This is so because a floral bud is a modified vegetative bud (shoot).

Symmetry of the Flower. A flower is said to be symmetrical when it can be divided into two exactly equal halves by *any* vertical section passing through the centre. Such a flower is also said to be **regular** or **actinomorphic**, e.g. mustard, potato, brinjal, *Datura,* etc. When a flower can be divided into two similar halves by *one* such vertical section only, it is said to be **zygomorphic**, e.g. pea, bean, gold mohur, *Cassia,* etc., and when it cannot be divided into two similar halves by any vertical plane whatsoever, it is said to be **irregular**.

A flower is also to be symmetrical when its whorls (often leaving out the gynoecium) have an equal number of parts or when the number in one whorl is a multiple of that of another. Such a symmetrical flower is said to be **trimerous** when the number of parts in each whorl is 3 or any multiple of it, as mostly in monocotyledons, and **pentamerous** when the number is 5 or any multiple of it, as mostly in dicotyledons.

(1) CALYX

Calyx is the first or the lowermost whorl of the flower, and consists of a number of **sepals**. It is usually green (*sepaloid*), but sometimes it becomes coloured (*petaloid*), as in gold mohur, dwarf gold mohur and garden nasturtium. It varies in shape, size and colour; it may be **regular**, **zygomorphic**, or **irregular**. The sepals may remain free from each other or they may be united together; in the former case the calyx is said to be **polysepalous** (*polys,* many) as in mustard, radish, etc., and in the latter it is **gamosepalous** (*gamo,* united), as in brinjal, chilli, potato, etc. The calyx may sometimes be altogether absent from a flower, or it may be modified into *scale,* as in sunflower, marigold, etc., or into *pappus* (see Fig. 156A), as in *Tridax* and many other plants of sunflower family. In *Mussaenda* (see Fig. 111A) one of the sepals becomes large, leafy and often yellow or orange, sometimes scarlet or white.

Functions. (1) *Protection,* as in most flowers. (2) *Assimilation,* when green in colour. (3) *Attraction,* when coloured and showy. (4) *Special function,* when modified into pappus (see Fig. 156A); the pappus is persistent in the fruit and helps its distribution by the wind.

Duration. The calyx may fall off as soon as the floral bud opens, as in poppy. More commonly it falls off with the corolla when the flower withers; it is then said to be **deciduous**. Sometimes it persists and adheres to the fruit; then it is said to be **persistent**. A persistent calyx may remain green, as in brinjal, or it may assume a withered appearance, as in cotton, or it may continue to grow and become fleshy, as in *Dillenia* (B. & H. CHALTA).

(2) COROLLA

Corolla is the second whorl of the flower, and consists of a number of **petals**. The petals are often brightly coloured and sometimes scented, and, then their function is to attract insects for *pollination;* they are rarely sepaloid. In the bud stage of the flower the corolla encloses the essential organs, namely, stamens and carpels, and protects them from external heat and rain, and from insect attack.,

Like the calyx, the corolla may also be **regular**, **zygomorphic** or **irregular**. Like the calyx again, the corolla may be **gamopetalous,** i.e. petals united, or **polypetalous**, i.e. petals free. In the former case the petals may be united **partimorphic** or **irregular**. Like the calyx again the corolla may sometimes be narrowed below, forming a sort of stalk, known as the *claw,* and expanded above: this expanded portion is called the *limb,* as in mustard, radish, etc.

Forms of Corollas. The various forms of corollas may be studied under the following four main heads:

I. *REGULAR AND POLYPETALOUS*

(1) **Cruciform** (Fig. 113A). The cruciform corolla consists of four free petals (each differentiated into a claw and a limb) arranged in the form of a cross, as in mustard family. e.g. mustard, radish, cabbage, cauliflower, candytuft, etc.

(2) **Caryophyllaceous** (Fig. 113B). This form of corolla consists of five petals with comparatively long claws, and the limbs of the petals are placed at right angles to the claws, as in pink (*Dianthus*).

(3) **Rosaceous** (Fig. 113C). This form consists of five petals with very short claws or none at all, and the limbs spread regularly outwards, as in rose, tea, prune, etc.

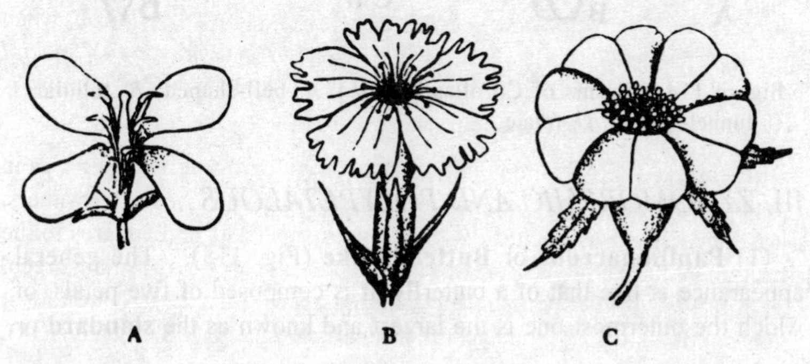

Figure 113. Forms of Corollas. *A*, cruciform; *B*, caryophyllaceous; *C*, rosaceous.

II. *REGULAR AND GAMOPETALOUS*

(1) **Bell-shaped** (Fig. 114A). When the shape of the corolla resembles that of a bell, as in gooseberry, bell flower, wild mangosteen (B. GAB; H. KENDU), etc., it is said to be bell-shaped or campanulate.

(2) **Tubular** (Fig. 114B). When the corolla is cylindrical or tube-like, that is, more or less equally expanded from base to apex, as in the central florets of sunflower, marigold, *Cosmos,* etc., it is said to be tubular.

(3) **Funnel-shaped** (Fig. 114C). When the corolla is shaped like a funnel, that is, gradually spreading outwards from a narrow base, as in *Datura, Ipomoea,* e.g. water bind-weed (B. & H. KALMI-SAK), railway creeper, morning glory, etc., *Petunia,* yellow oleander (*Thevetia*), etc., it is said to be funnel-shaped.

(4) **Rotate** or **Wheel-shaped** (Fig. 114D). When the tube of the corolla is comparatively short and its limb is at a right angle to it, the corolla having more or less the appearance of a wheel, as in night jasmine (*Nyctanthes*), periwinkle (*Vinca*), *Ixora*, etc., it is said to be rotate.

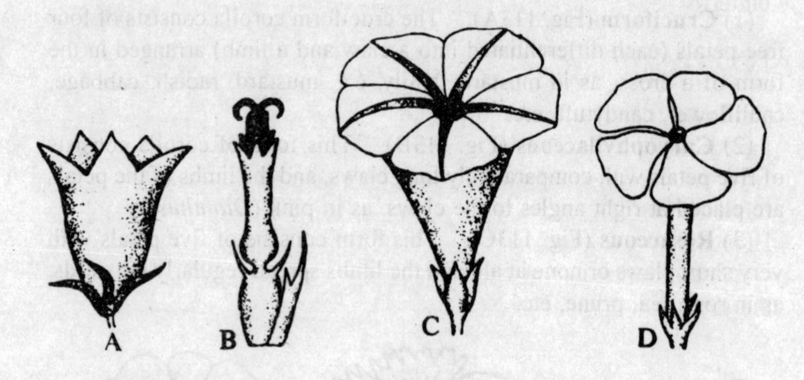

Figure 114. Forms of Corollas (*contd.*). *A*, bell-shaped; *B*, tubular; *C*, funnel-shaped; *D*, rotate.

III. *ZYGOMORPHIC AND POLYPETALOUS*

(1) **Papilionaceous** or **Butterfly-like** (Fig. 115). The general appearance is like that of a butterfly. It is composed of five petals, of which the outermost one is the largest and known as the **standard** or

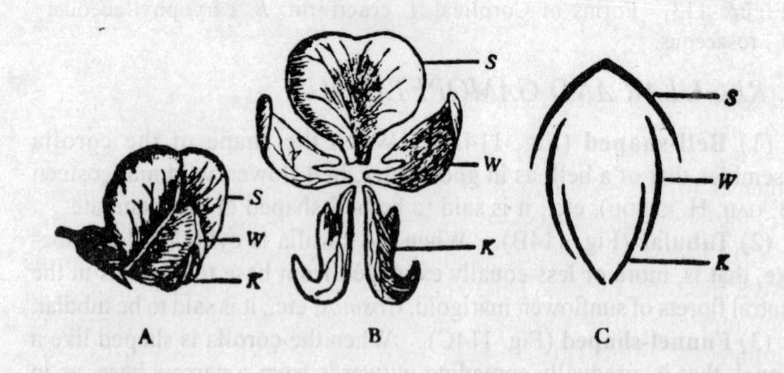

Figure 115. *A*, papilionaceous flower of pea; *B*, petals of the same opened out; *C*, vexillary aestivation of papilionaceous corolla. *S*, standard or vexillum; *W*, wing, *K*, keel.

vexillum, the two lateral ones are known as the **wings** or **alae** and the two innermost ones are the smallest and are together known as the **keel** or **carina**. These two are apparently united to form a boat-shaped cavity. Examples are found in pea family, e.g. pea (Fig. 115), bean, gram, butterfly pea (*Clitoria;* B. APARAJITA; H. APARAJIT or GOKARNA), etc.

IV. *ZYGOMORPHIC AND GAMOPETALOUS*

(1) **Bilabiate** or **Two-lipped** (Fig. 116A). In this form the limb of the corolla is divided into two portions or lips—the upper and the lower, with the mouth gaping wide open. Examples may be seen in sacred basil (*Ocimum sanctum;* B. & H. TULSI), *Leonurus* (B. DRONA; H. HALKUSHA), *Adhatoda* (B. BASAK; H. ADALSA), etc.

(2) **Personate** or **Masked** (Fig. 116B). This is also two-lipped like the previous one, but in this case the lips are placed so near to each

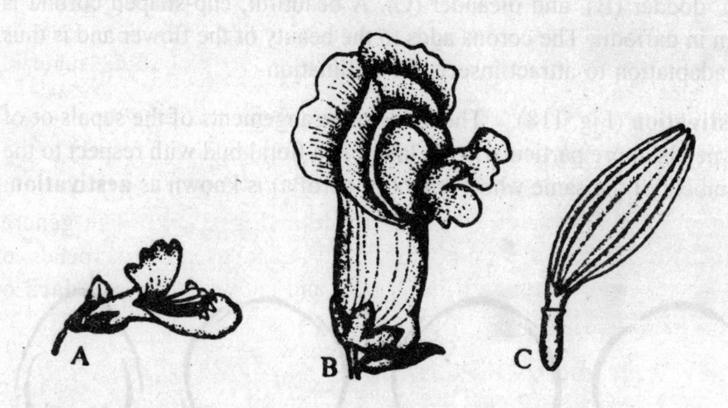

Figure 116. Forms of Corollas (*contd.*). *A,* bilabiate; *B,* personate; *C,* ligulate.

other as to close the mouth of the corolla, as in snapdragon, *Linaria,* etc. Insects sit on the lower lip and push open the mouth of the corolla.

(3) **Ligulate** or **Strap-shaped** (Fig. 116C). When the corolla forms into a short narrow tube below but is flattened above like a strap, as in the outer florets of sunflower, marigold, *Cosmos,* etc., it is said to be ligulate.

Corona (Fig. 117). Sometimes, by a transverse splitting of the corolla, additional whorl may be formed at its throat. This additional whorl may be made up of lobes, scales or hairs, free or united, and is

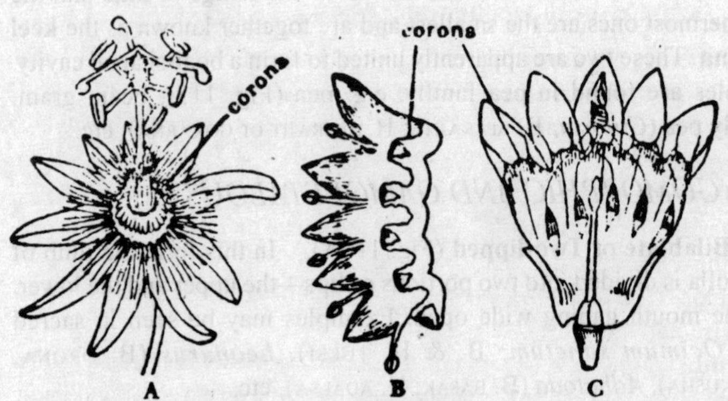

Figure 117. Corona. *A*, passion-flower; *B*, dodder; *C*, oleander.

known as the **corona** (*crown*). The corona may be seen in passion-flower (A), dodder (B), and oleander (C). A beautiful, cup-shaped corona is seen in daffodil. The corona adds to the beauty of the flower and is thus an adaptation to attract insects for pollination.

Aestivation (Fig. 118). The mode of arrangements of the sepals or of the petals, more particularly the latter, in a floral bud with respect to the members of the same whorl (calyx or corolla) is known as **aestivation**.

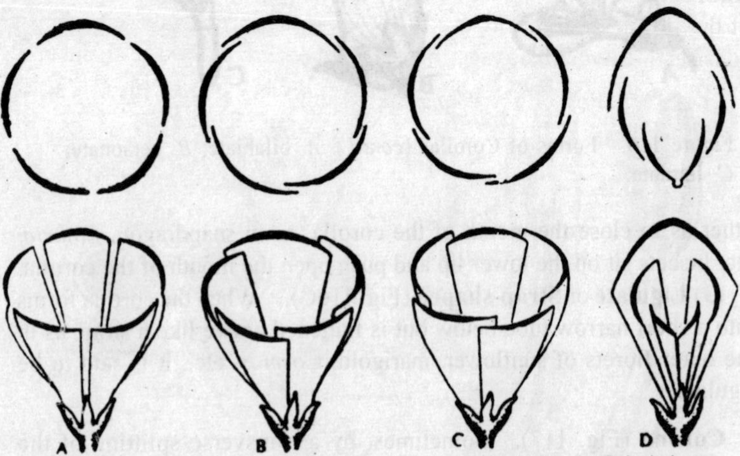

Figure 118. Aestivation of Corolla. *A*, valvate; *B*, twisted; *C*, imbricate; *D*, vexillary. *Top*, corolla in transection; *bottom*, floral bud cut transversely.

Aestivation is an important character from the viewpoint of classification of plants, and may be of the following types:

(1) **Valvate** (A), when the members of a whorl are in contact with each other by their margins, or when they lie very close to each other, but do not overlap, as in custard-apple (*Annona*), madar (*Calotropis*), *Artabortys* (B. & H. KANTALI-CHAMPA), etc.

(2) **Twisted** or **Contorted** (B), when one margin of the sepal or the petal overlaps that of the next one, and the other margin is overlapped, by the third one, as in China rose, cotton, etc. Twisting of the petals may be clockwise or, anti-clockwise. In China rose (*Hibiscus*) both types (clockwise and anti-clockwise) are found.

(3) **Imbricate** (C), when one of the sepals or petals is internal being overlapped on both the margins, and one of them is external and each of the remaining ones is overlapped on one margin and it overlaps the next one on the other margin, e.g. *Cassia,* gold mohur (*Delonix*), dwarf gold mohur (*Caesalpinia*), *Bauhinia* (B. KANCHAN; H. KACHNAR), etc.

(4) **Vexillary** (D), when there are five petals, of which the posterior one is the largest and it almost covers the two lateral petals, and the latter in their turn nearly overlap the two anterior or smallest petals. Vexillary aestivation is universally found in all papilionaceous corollas, as in pea, bean, butterfly pea (*Clitoria*), rattlewort (*Crotalaria*), etc.

(3) ANDROECIUM

Androecium (*andros,* male) is the third or the male reproductive whorl of the flower, and is composed of a number of **stamens**. Each stamen consists of **filament**, **anther** and **connective** (Fig. 119). The filament is

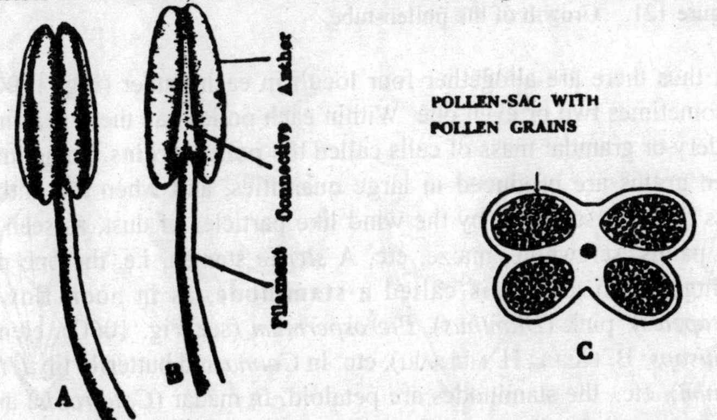

Figure 119. *A,* a stamen—face of the anther showing four pollen-sacs; *B,* the same—back of the anther showing connective; *C,* an anther in section.

the slender stalk of the stamen, and the anther is the expanded head borne by the filament at its tip. Each anther consists usually of two lobes connected together by a sort of midrib known as the **connective**. Each lobe contains within it two chambers or loculi, called the **pollen-**

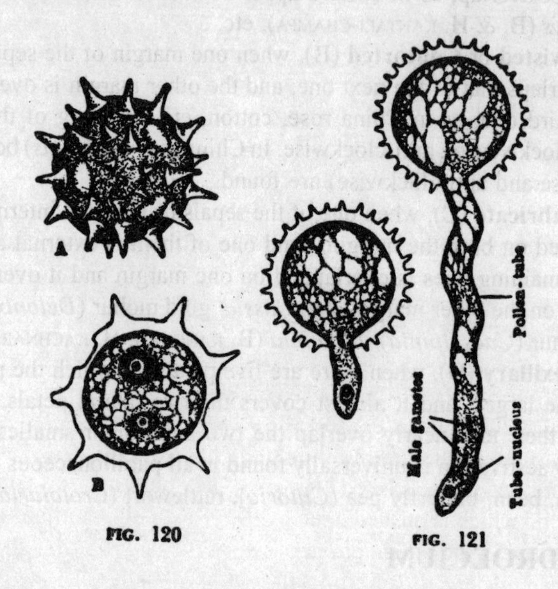

Figure 120. Pollen grains. *A,* an entire grain; *B,* a grain in section showing tube-nucleus (bigger one) and generative nucleus (smaller one).
Figure 121. Growth of the pollen-tube.

sacs; thus there are altogether four loculi in each anther (Fig. 119C), but sometimes two or even one. Within each pollen-sac there is a fine, powdery or granular mass of cells called the **pollen grains**. Sometimes pollen grains are produced in large quantities, and when the anther bursts they are scattered by the wind like particles of dust, as seen in pine, palms, screwpine, maize, etc. A *sterile* stamen, i.e. the one not bearing pollen grains, is called a **staminode**, as in noon flower (*Pentapetes*), pink (*Dianthus*), *Pterospermum* (see Fig. 106C), elengi (*Mimusops;* B. BAKUL; H. MULSARI), etc. In *Canna* and butterfly lily (*Hedychium*), etc., the staminodes are petaloid. In madar (*Calotropis*) and orchids the pollen cells are not free but become united into a mass known as the **pollinium** (Fig. 122). Pollen cells may also be in compound forms,

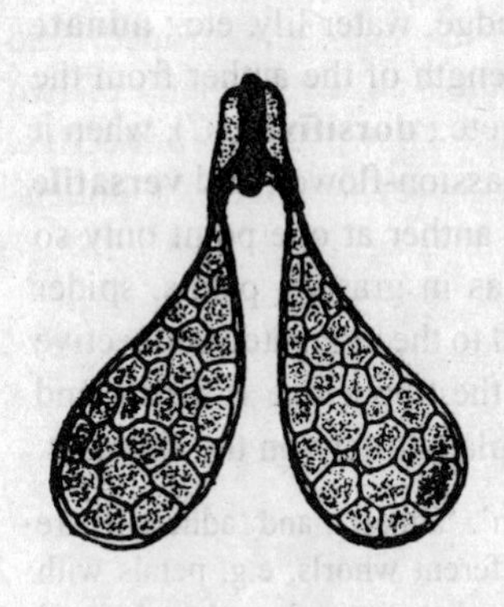

Figure 122. Pollinia of madar (*Calotropis*).

i.e. united in small masses—**pollen masses**—each consisting of 8–32 or more cells, as in sensitive plant (*Mimosa*) and gum tree (*Acacia*).

Pollen Grains. These are the male reproductive bodies of a flower, and are contained in the pollen-sacs. They are very minute in size, usually varying from 10 to 200 microns, and are like particles of dust. Each pollen grain consists of a single microscopic cell, and possesses two coats: the **exine** and the **intine**. The exine is a tough, cutinized layer which is often provided with spinous outgrowths or markings of different patterns, sometimes smooth. The intine, however, is a thin, delicate, cellulose layer lying internal to the exine. In pine, the pollen grain is provided with two distinct wings. When the pollen grain germinates the intine grows out into a tube, called the **pollen-tube** (Fig. 121), through some definite thin and weak slits or pores, called **germ pores** present in the exine (Fig. 120). The pore may be covered by a distinct lid which is pushed open by the growth of the intine. Two nuclei may be seen in the pollen grain, of which the larger one is known as the vegetative nucleus or **tube-nucleus** and the smaller one the **generative nucleus**. As the pollen-tube grows it carries with it, at its apex, both the nuclei. The generative nucleus soon divides into two male reproductive units called **male gametes**, while the tube-nucleus becomes disorganized.

Attachment of the Filament to the Anther (Fig. 123). The anther is said to be **basifixed** or **innate** (*A*), when the filament is attached to the

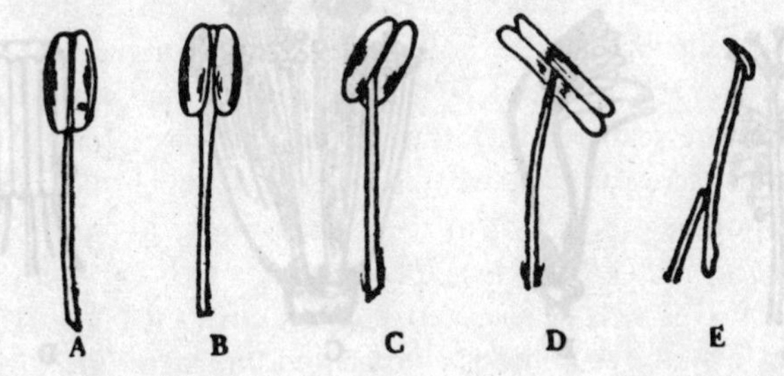

Figure 123. *A*, basifixed; *B*, adnate; *C*, dorsifixed; *D*, versatile *E*, elongated connective of sage (*Salvia*) separating the two anther-lobes

base of the anther, as in mustard, radish, sedge, water lily, etc.; **adnate** (*B*), when the filament runs up the whole length of the anther from the base to the apex, as in *Michelia, Magnolia,* etc.; **dorsifixed** (*C*), when it is attached to the back of the anther, as in passion-flower; and **versatile** (*D*), when it is attached to the back of the anther at one point only so that the latter can swing freely in the air, as in grasses, palms, spider lily, etc. In *Salvia* (*E*) the filament is attached to the elongated connective separating the two anther-lobes, of which the upper one is fertile and the lower one sterile. In it, the connective plays freely on the filament.

Cohesion and Adhesion. The terms 'adhesion', 'adnate', and 'adherent' are used to designate the union of members of different whorls, e.g. petals with stamens, or stamens with carpels; and 'cohesion', 'connate', and 'coherent' are used to designate the union of members of the same whorl, e.g. stamens with each other, and carpels with each other.

Cohesion of Stamens. Stamens may either remain free or they may be united (coherent). There may be different degrees of cohesion of stamens, and these may be referred to as the (a) adelphous condition when the stamens are united by their filaments only, the anthers remaining free; or (b) syngenesious condition when the stamens are united by their anthers only, the filaments remaining free. Accordingly the following types are seen.

(1) **Monadelphous Stamens** (*monos,* single; *adelphos,* brother). When all the filaments are united together into a single bundle but the anthers are free, the stamens are said to be monadelphous (Fig. 124A), as in China rose family, e.g. China rose, lady's finger, cotton, etc. In them, the filaments are united into a tubular structure called *staminal tube,* ending in free anthers.

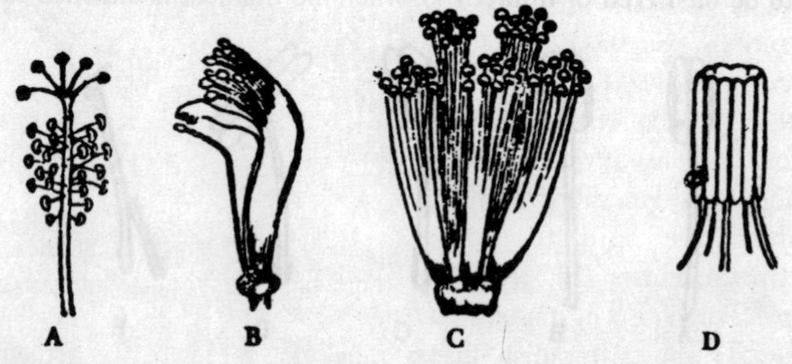

Figure 124. Cohesion of Stamens. *A,* monadelphous; *B,* diadelphous; *C,* polyadelphous; *D,* syngenesious.

(2) **Diadelphous Stgmens** (*di,* two). When the filaments are united into two bundles, the anthers remaining free, the stamens are said to be diadelphous (Fig. 124B), as in pea family, e.g. pea, bean, gram, butterfly pea, coral tree, rattle-wort, etc. In them, there are altogether ten stamens of which nine are united into one bundle and the tenth one is free.

(3) **Polyadelphous Stamens** (*polys,* many). When the filaments are united into a number of bundles—more than two—but the anthers are free, the stamens are said to be polyadelphous (Fig. 124C), as in silk cotton tree, castor, lemon, pummelo or shaddock, etc.

(4) **Syngenesious Stamens** (*syn,* together or united; *genes,* producing). When the anthers are united together into a bundle or tube, but the filaments are free, the stamens are said to be syngenesious (Fig. 124D), as in sunflower family, e.g. sunflower, marigold, safflower, *Tridax,* etc.

Adhesion of Stamens. When the stamens adhere to the corolla wholly or partially by their filaments, anthers remaining free, they are said to be (1) **epipetalous**, as in *Datura,* tobacco, potato, *Ixora,* sunflower, etc. Most of the flowers with a gamopetalous corolla have epipetalous stamens. When the stamens adhere to the carpels, either throughout their whole length or by their anthers only, they are said to be (2) **gynandrous**, as in madar (*Calotropis*), orchids, etc.

Length of Stamens (Fig. 125). The stamens of a flower may be of the same length, or their lengths may vary without any definite relation to each other. But in some cases there is a definite relation between short and long stamens. Thus in *Ocimum* (B. & H. TULSI), *Leonurus* (B. DRONA; H. HALKUSHA), *Leucas* (B. SWET-DRONA; H. CHOTA HALKUSHA), etc., there are four stamens, of which two are long and two short; such stamens are said to be (1) **didynamous** (*di,* two; *dynamis,* strength). In mustard family, e.g. mustard, radish, turnip, rape, etc., there are six stamens, of which four are long and two short; such stamens are said to be (2) **tetradynamous** (*tetra,* four).

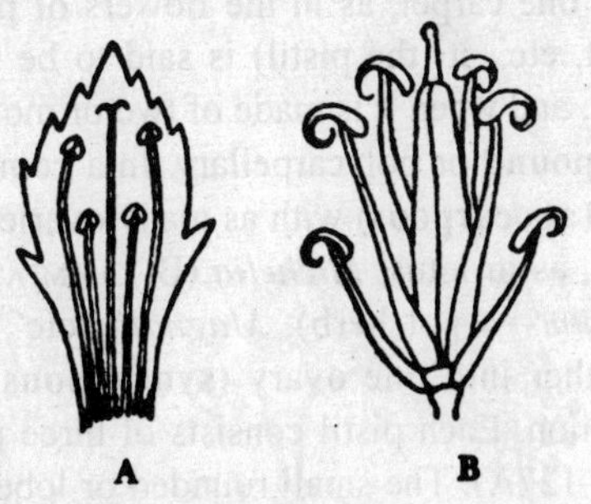

Figure 125. Length of Stamen. *A,* didynamous; *B,* tetradynamous.

(4) GYNOECIUM OR PISTIL

Gynoecium (*gyne,* female) or pistil is the fourth or the female reproductive whorl of the flower, and is composed of one or more **carpels** which are modified leaves meant to bear ovules (Fig. 127B–C) and an embryo-sac within each ovule (see Fig. 131). When the pistil is made of

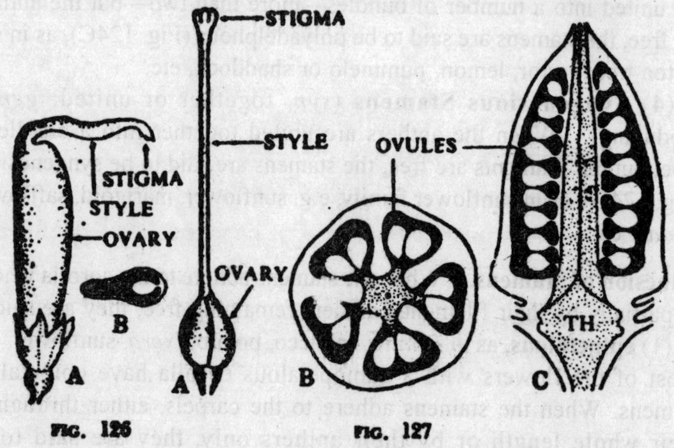

Figure 126. Pistil. *A,* a simple pistil of pea; *B,* one-chambered ovary of the same. Figure 127. *A,* a syncarpous; *B,* ovary of the same in transection; *C,* ovary of the same in longitudinal section. *TH,* thalamus.

only one carpel, as in the flowers of pea, bean, gold mohur, sensitive plant, etc., it (the pistil) is said to be **simple** or monocarpellary (Fig. 126), and when it is made of two or more carpels, the pistil is said to be **compound** or polycarpellary. In a compound pistil the carpels may be free (**apocarpous**) with as many ovaries as the number of carpels (Fig. 128), as in lotus, *Michelia* (B. CHAMPA; H. CHAMPAKA), rose, stonecrop (*Sedum*—a pot herb), *Magnolia* etc., or the carpels may be united together into one ovary (**syncarpous**; Fig. 127); the latter is more common. Each pistil consists of three parts—**stigma**, **style** and **ovary** (Fig. 127A). The small rounded or lobed head of the pistil is known as the **stigma**; the slender stalk supporting the stigma is called the **style**; and the swollen basal part of the pistil which forms one or more chambers is termed the **ovary**. The ovary contains one or more little, roundish or oval, egg-like bodies which are the rudiments of seeds and are known as the **ovules**. Each ovule encloses a large oval cell known as the

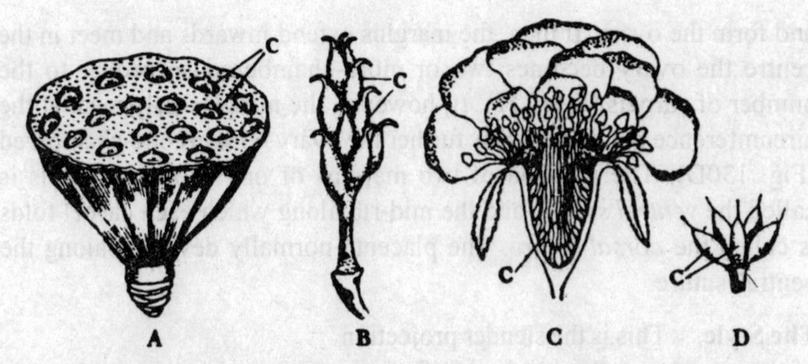

Figure 128. Apocarpous Pistil. *A*, lotus; *B*, *Michelia;* *C*, rose; *D*, stonecrop (*Sedum*). *C*, carpels.

embryo-sac (see Fig. 131). The ovary gives rise to the fruit and the ovules to the seeds. A functionless or sterile pistil is called a **pistillode** as in the ray floret of sunflower.

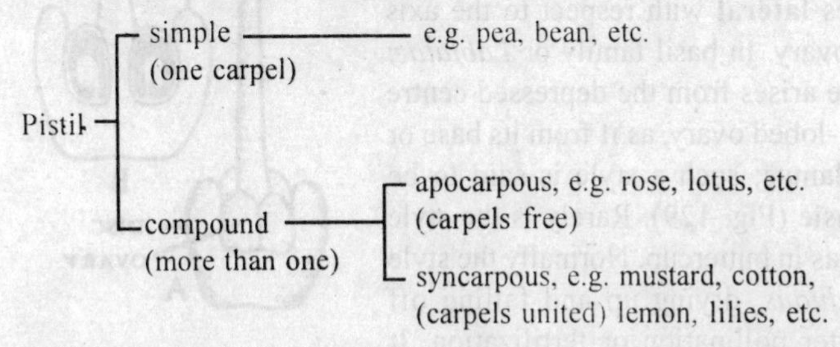

Carpels in Syncarpous Pistil. In a syncarpous pistil it is often difficult to determine the number of carpels. To obviate this difficulty the following points should be noted: 1st, the number of stigmas or of stigmatic lobes; 2nd, the number of styles; 3rd, the number of lobes of the ovary; 4th, the number of chambers (loculi) of the ovary; 5th, the number of placentae in the ovary; and 6th, the number of groups of ovules in the ovary. It is seen that in most cases the number of parts, as mentioned above, corresponds to the number of carpels making up the syncarpous pistil.

The Ovary. The ovary is the closed chamber formed by the union of margins of one or more carpels (which are regarded as metamorphosed leaves). In the simple pistil, as in pea, bean, etc., or in the apocarpous pistil, as in *Ranunculus* the carpel folds along the mid-rib and the two margins meet and fuse together forming the ovary (Fig. 126). In the syncarpous pistil the carpels likewise meet at their respective margins and form the ovary. If then, the margins extend inwards and meet in the centre the ovary becomes two or more-chambered according to the number of carpels (Fig. 127). If, however, the margins only meet at the circumference but do not grow further the ovary remains one-chambered (Fig. 130D). The junction of two margins of one or more carpels is called the *ventral suture,* and the mid-rib along which each carpel folds is called the *dorsal suture*. The placenta normally develops along the ventral suture.

The Style. This is the slender projection of the ovary, usually developing from its top and acting as the stalk of the stigma. When the style grows straight up from the ovary it is said to be **terminal** or **apical**. Sometimes, as in strawberry, the tip of the ovary is bent on one side; the style then becomes **lateral** with respect to the axis of the ovary. In basil family or *Labiatae,* the style arises from the depressed centre of the 4-lobed ovary, as if from its base or the thalamus; such a style is said to be **gynobasic** (Fig. 129). Rarely is the style absent, as in buttercup. Normally the style is *deciduous,* drying up and falling off soon after pollination or fertilization. It is, however, *persistent* in some cases, as in *Clematis* and *Naravelia* (see Fig. 157). A flattened and coloured style (i.e. *petaloid style*) is seen in *Canna* and *Iris*.

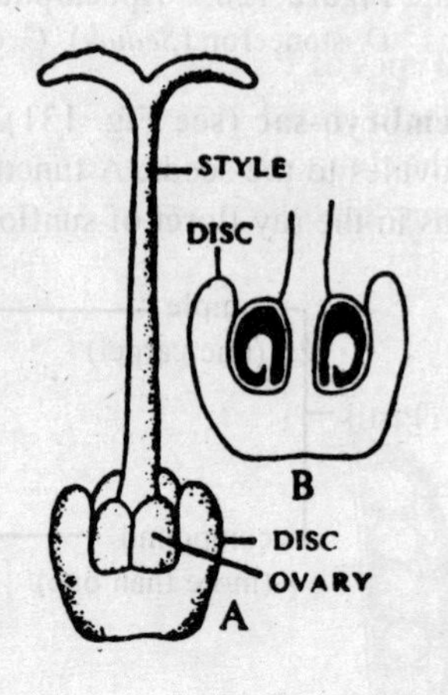

Figure 129. *A,* gynobasic style of basil (*Ocimum*); *B,* the same in longi-section.

The Stigma. This is the terminal end of the style, i.e. the part that receives the pollen grains. Generally it is knob-like, sometimes slightly pointed or even somewhat elongated. In a compound pistil the stigma may be lobed or even radiating. Its surface may be smooth, rough or even hairy. When it matures it becomes sticky to receive the pollen grains.

Cohesion of Carpels. The carpels may be united throughout their whole length, as in most syncarpous pistils; or, they may be united in the region of the ovary alone, styles and stigmas remaining free, as in pink (*Dianthus*), linseed (*Linum*), etc.; or, in the region of the ovary and the style, stigmas remaining free, as in China rose, cotton, etc.; or in the region of the style and the stigma, ovaries remaining free, as in periwinkle (*Vinca*), oleander (*Neritum*), etc.; or in the region of the stigma (with a part of the style), ovaries (and styles partly) remaining free as in madar (*Calotropis*).

PLACENTATION

Placenta is a ridge of tissue in the inner wall of the ovary bearing one or more ovules; and the manner of distribution of the placentae within the ovary is called **placentation**. The placentae most frequently develop on the margins of carpels either along their whole line of union called the **suture**, or at their base or apex.

Types of Placentation (Fig. 130). In the simple ovary (of one carpel) there is one common type of placentation known as **marginal**, and in the compound ovary (of two or more carpels united together) placentation may be **axile**, **central**, **parietal**, **basal** or **superficial**.

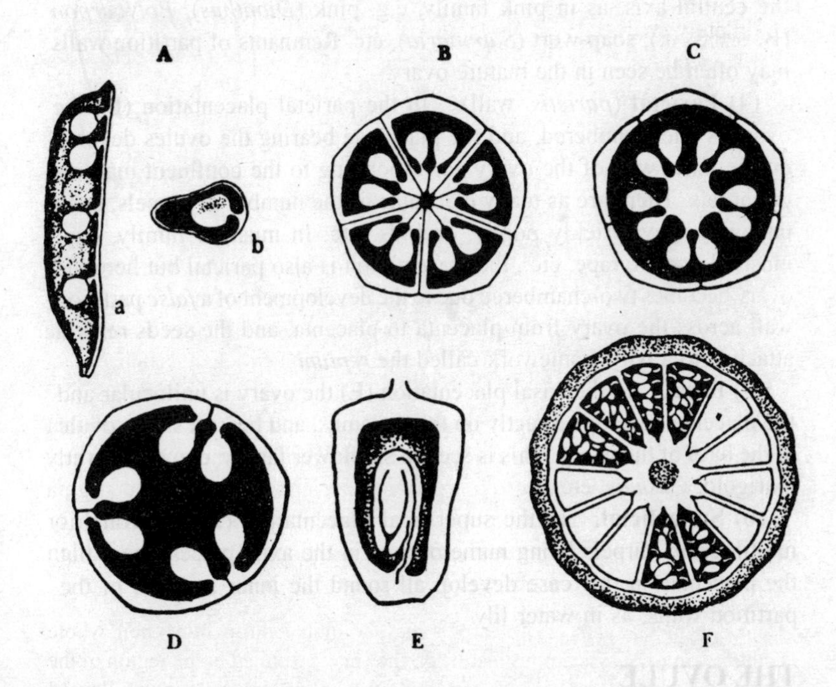

Figure 130. Types of Placentation. *A*, marginal; *a*, longitudinal section; *b*, transverse section; *B*, axile; *C*, central; *D*, parietal; *E*, basal; *F*, superficial.

(1) **Marginal.** In marginal placentation (A) the ovary is one-chambered and the placenta develops along the junction of the two margins of the carpel, called the *ventral suture,* as in pea, wild pea, gram, gold mohur, *Cassia,* sensitive plant, etc. In *Ranunculus* the pistil is apocarpous, and each ovary bears a single pendulous ovule attached to the ventral suture.

(2) **Axile.** In the axile placentation (B) the ovary is two to many-chambered—usually as many as the number of carpels—and the placentae bearing the ovules develop from the central axis corresponding to the confluent margins of carpels, and hence the name axile (lying in the axis), as in potato, tomato, *Petunia, China* rose, lady's finger, hollyhock (*Althaea*), lemon, orange, etc. Three-chambered ovary is common among monocotyledons, as in lily, asphodel, onion, etc.

(3) **Central.** In the central placentation (C) the septa or partition walls in the young ovary soon break down so that the ovary becomes one-chambered and the placentae bearing the ovules develop all round the central axis, as in pink family, e.g. pink (*Dianthus*), *Polycarpon* (B. GIMA-SAK), soap-wort (*Saponaria*), etc. Remnants of partition walls may often be seen in the mature ovary.

(4) **Parietal** (*parietis,* wall). In the parietal placentation (D) the ovary is one-chambered, and the placentae bearing the ovules develop on the inner wall of the ovary corresponding to the confluent margins of carpels. There are as many placentae as the number of carpels, as in papaw, poppy, prickly poppy, orchids, etc. In mustard family, e.g., mustard, radish, rape, etc., the placentation is also parietal but here the ovary becomes two-chambered due to the development of a *false* partition wall across the ovary from placenta to placenta, and the seeds remain attached to a wiry framework called the *replum.*

(3) **Basal.** In the basal placentation (E) the ovary is unilocular and the placenta develops directly on the thalamus, and bears a single ovule at the base of the ovary. This is seen in sunflower family, e.g. sunflower, marigold, *Cosmos,* etc.

(6) **Superficial.** In the superficial placentation (F) the ovary is multilocular, carpels being numerous, as in the axile placentation, but the placentae in this case develop all round the inner surfaces of the partition walls, as in water lily.

THE OVULE

Structure of the Ovule. Each ovule (Fig. 131) is attached to the placenta by a slender stalk known as (1) the **funicle**. The point of

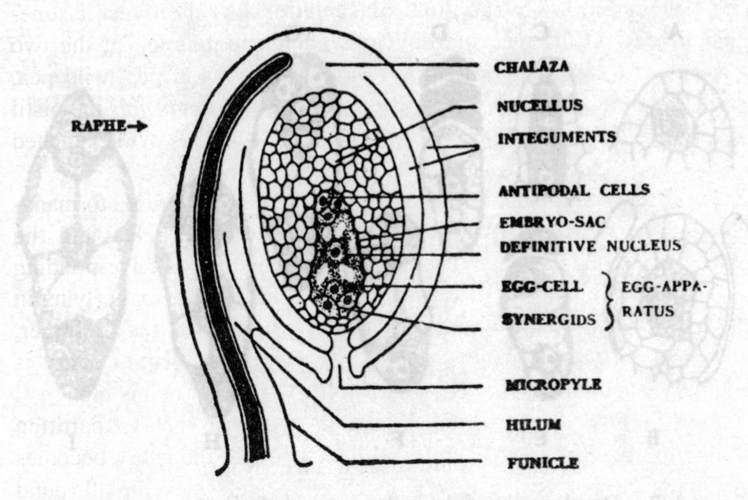

Figure 131. An anatropous or inverted ovule in longitudinal section.

attachment of the body of the ovule to its stalk or funicle is known as (2) the **hilum**. In the inverted ovule, as shown in Fig. 131, the funicle continues beyond the hilum alongside the body of the ovule forming a sort of ridge; this ridge is called (3) the **raphe**. Through the raphe food is carried to the nucellus. The distal end of the raphe or the funicle which is the junction of the integuments and the nucellus is called (4) the **chalaza**. The main body of the ovule is called (5) the **nucellus**, and it is surrounded by *two* coats (or only one in some) termed (6) the **integuments**. A small opening is left at the apex of the integuments; this is called (7) the **micropyle**. Lastly, there is a large, oval cell lying embedded in the nucellus towards the micropylar end; this is (8) the **embryo-sac**, i.e. the sac that bears the embryo, and is the most important part of the ovule.

Parts and Functions of the Embryo-sac (Figs. 131–32). In the mature embryo-sac a group of three cells, each surrounded by a very thin wall, may be seen always lying towards the micropyle; this group is called (1) the **egg-apparatus**. One cell of this group is the female gamete (reproductive unit) known as (a) the **egg-cell** or **ovum**, and the other two known as (b) the **synergids**. The egg-cell on fertilization, i.e. on fusion with a male gamete of the pollen-tube, gives rise to the embryo; this is the most important function of the embryo-sac. The synergids

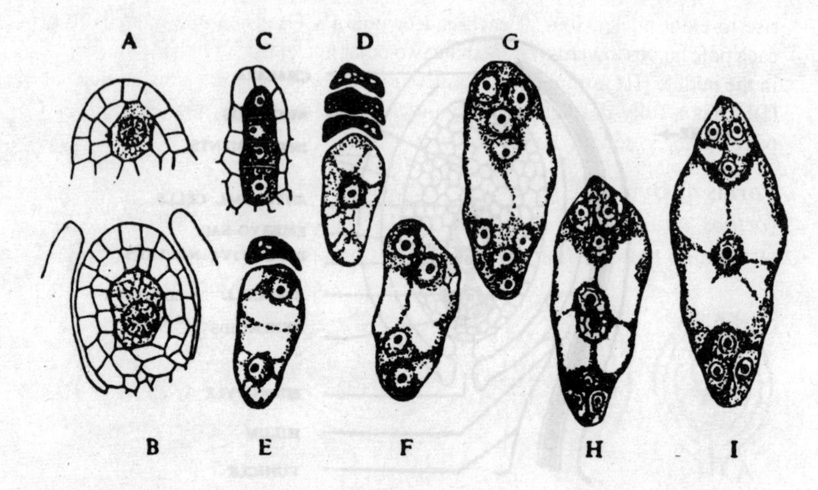

Figure 132. Development of the embryo-sac. *A, B, C,* etc., are stages in its development; *I,* fully developed embryo-sac.

are short-lived structures, and get disorganized soon after fertilization or sometimes even before or during the process. At the opposite end of the embryo-sac there is another group of three cells known as (2) the **anti-podal cells**, each often surrounded by a very thin wall. These have no definite function; so sooner or later they also get disorganized. Somewhere in the middle of the embryo-sac there is a distinct nucleus known as (3) the **definitive nucleus** which is the fused product of the two polar nuclei, i.e. the two nuclei coming from the two poles or ends of the embryo-sac and meeting somewhere in the centre (Fig. 132 G–I). After a second fusion with the remaining male gamete it forms the **endosperm nucleus** (the product of fusion of three nuclei) which may soon grow into the **endosperm** or food storage tissue of the seed; this is the second important function of the embryo-sac.

Ovule—funicle, hilum, raphe, chalaza, integuments, micropyle, nucellus, and embryo sac.
Embryo-sac—egg-apparatus (egg-cell and synergids), definitive nucleus (two polar nuclei fused), and antipodal cells.

Development of Embryo-sac (Fig. 132). At a very early stage in the life of the ovule a particular cell of it—the mother cell of the embryo-sac—enlarges (A–B). It divides twice to produce a row of four megaspores (C). Usually the upper three degenerate and appear as dark caps (D), while the lowest one enlarges and finally forms the embryo-sac; its nucleus divides thrice to give rise to eight nuclei, four at each end or pole (E–G). Then one nucleus from each pole moves inwards (G), and the two polar nuclei fuse together, somewhere in the middle (H) forming the definitive nucleus, also called the fusion nucleus (I). Thus a fully developed embryō-sac consists of the parts as described in page 121.

Forms of Ovules (Fig. 133). The ovule is said to be (A) **orthotropus** (*orthos,* straight; *tropos,* a turn) or **straight** when the ovule is erect or straight so that the funicle, chalaza and micropyle lie on and the same

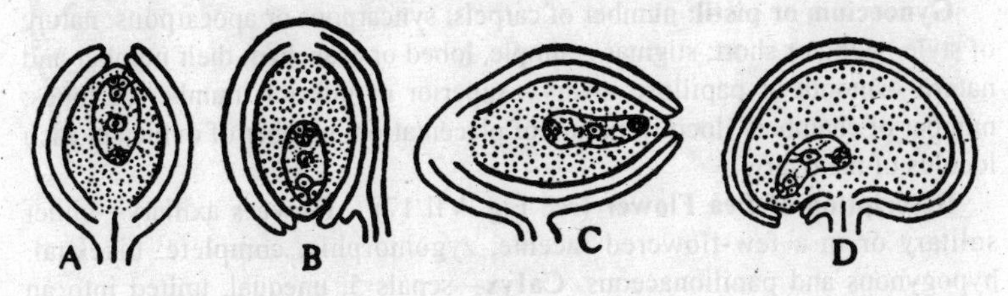

Figure 133. Forms of Ovules. *A,* orthotropous; *B,* anatropous; *C,* amphitropous; *D,* campylotropous.

vertical line, as in *Polygonum,* dock or sorrel (*Rumex*), betel (*Piper*), etc.; (B) **anatropous** (*ana,* backwards or up) or **inverted** when the ovule bends back alongside the funicle so that the micropyle lies close to the hilum; here the micropyle and the chalaza, but not the funicle, lie on the same straight line; this is the commonest form of ovule; (C) **amphitropous** (*amphi,* on both sides) or **transverse** when the ovule is placed transversely at a right angle to its stalk or funicle, as in duckweed; and (D) **campylotropous** (*kampylos,* curved) or **curved** when the transverse ovule is bent round like a horse-shoe so that the micropyle and the chalaza do not lie on the same straight line, as in mustard, radish, caper (*Capparis;* B. KANTA-GURKAMAL; H. KANTHARI), beet (*Beta*), etc.

Features used to describe a Flower

Flower: solitary or in inflorescence (mention the type); sessile or stalked; complete or incomplete; unisexual or bisexual; regular, zygomorphic or irregular; hypogynous, epigynous or perigynous; nature of bracts and bracteoles, if present; shape of the flower, its colour and size.

Calyx: polysepalous or gamosepalous; number of sepals or of lobes; superior or interior; aestivation, shape, size and colour.

Corolla: polypetalous or gamopetalous; number of petals or lobes; superior or inferior; aestivation; shape, size, colour and scent; corona or any special feature. (When there is not much difference between the calyx and the corolla the term **perianth** should be used; it may be sepaloid of petaloid; polyphyllous or gamophyllous).

Androecium: number of stamens—definite (ten or less) or indefinite (more than ten); free or united; nature of cohesion—monadelphous, diadelphous, polyadelphous, syngenesious or synandrous; nature of adhesion—epipetalous or free from the petals; whether alternating with the petals (or corolla-lobes) or opposite them; length of stamens—general length; inserted or exerted; didynamous or tetradynamous; position of stamens—hypogynous, perigynous or epigynous; attachment of the anther.

Gynoecium or pistil: number of carpels; syncarpous or apocarpous; nature of style—long or short; stigmas—simple, lobed or branched; their number and nature—smooth or papillose; ovary—superior or inferior; number of lobes; number of chambers (loculi); nature of placentation; number of ovules in each loculus of the ovary.

Description of Pea Flower (see Fig. VII/17). **Flowers** axillary—either solitary or in a few-flowered raceme, zygomorphic, complete, bisexual, hypogynous and papilionaceous. **Calyx**—sepals 5, unequal, united into an oblique tube, 5-lobed. **Corolla**—petals 5, free, papilionaceous, with vexillary aestivation—the outermost petal known as the *standard* or *vexillum* is broad, the lateral two are the *wings* or *alae* enclosing the two innermost ones—the *keel* or *carina*. **Androecium**—stamens ten, (9) + 1, diadelphous. **Gynoecium**—carpel 1; ovary subsessile, one-chambered and few-ovuled; placentation marginal; style one, inflexed, bearded on the inner side.

Chapter 10

Pollination

Pollination is the transference of pollen grains from the anther of a flower to the stigma of the same flower or of another flower of the same or sometimes allied species. Pollination is of two kinds, viz. (1) **self-pollination** or **autogamy** (*autos,* self; *gamos,* marriage) and (2) **cross-pollination** or **allogamy** (*allos,* different). Pollination taking place within a single flower (evidently bisexual) or between two flowers (bisexual or unisexual) borne by the same parent plant is self-pollination. In this process only one parent plant is concerned to produce the offspring. On the other hand, pollination taking place between two flowers (bisexual or unisexual) borne by two separate parent plants of the same or allied species is cross-pollination. In this process two parent plants are concerned and, therefore, a mingling of two sets of parental characters takes place, and this results in healthier offspring. Both the methods of pollination (self- and cross-) are, however, widespread in nature.

I. *SELF-POLLINATION OR AUTOGAMY*

Self-pollination may, under natural conditions, take place when both the anthers and the stigma of a bisexual flower mature at the same time (**homogamy**). It is likely then that some of the pollen grains are dropped on the stigma of the same flower through insects or wind. There are again many cases where the bisexual flowers never open. The flowers remaining closed, the pollen grains may only pollinate the stigma of the same flower (**cleistogamy**), as in *Commelina bengalensis* (Fig. 134), and also in some species of pansy (*Viola*), balsam (*Impatiens*), wood-sorrel (*Oxalis*), etc.

II. *CROSS-POLLINATION OR ALLOGAMY*

This is brought about by external agents which carry the pollen grains of one flower and deposit them on the stigma of another flower, the two being borne by the same plant or by two separate plants of the same or closely allied species. The agents are insects (bees, flies, moths, etc.), some animals (birds, snails, etc.), wind and water, and to achieve cross-pollination through them the adaptations in flowers are many and varied.

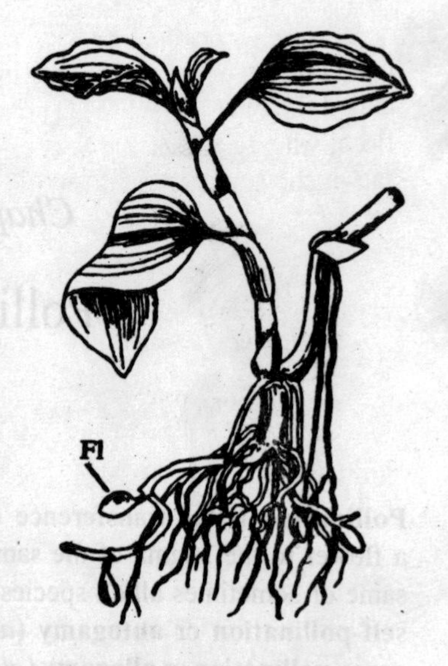

Figure 134. *Commelina bengalensis. Fl,* underground flower (cleistogamous)

1. **Entomophily** (*entomon,* an insect; *philein,* to love). Pollination by insects is of very general occurrence among plants. Entomophilous or insect-loving flowers have various adaptations by which they attract insects and use them as conveyors of pollen grains from one flower to another for the purpose of cross-pollination. Principal adaptations are **colour**, **nectar** and **scent**. There are some special adaptations also in certain flowers.

Colour. One of the most important adaptations is the **colour** of the petals. In this respect, the brighter the colour and the more irregular the shape of the flower the greater is the attraction. Sometimes when the flowers themselves are not conspicuous, other parts may become coloured and showy to attract insects. Thus in *Mussaenda* (see Fig. 111A) one of the sepals is modified into a large white or coloured leafy structure which serves as an 'advertisement' flag to attract insects. In some cases bracts become highly coloured and attractive, as in glory of the garden (*Bougainvillea;* B. BAGANBILAS—Fig. 109C), poinsettia (*Euphorbia;* B. LAL-PATA), etc. The spathes also often become brightly coloured, as in bananas and aroids. In sunflower, marigold, etc., the head or capitulum consisting of a cluster of small florets becomes, as a whole, very attractive.

Nectar. Another important adaptation is the **nectar**. Nearly all flowers with gamopetalous corolla secrete nectar which is a positive attraction to insects like bees. Nectar is contained in a special gland, called *nectary,* and sometimes in a special *sac,* or a tube-like structure, called the *spur* (Fig. 135). The nectary occurs at the base of one of the floral whorls, and as the bees collect the nectar from the nectary or the sac or the spur they incidentally bring about pollination.

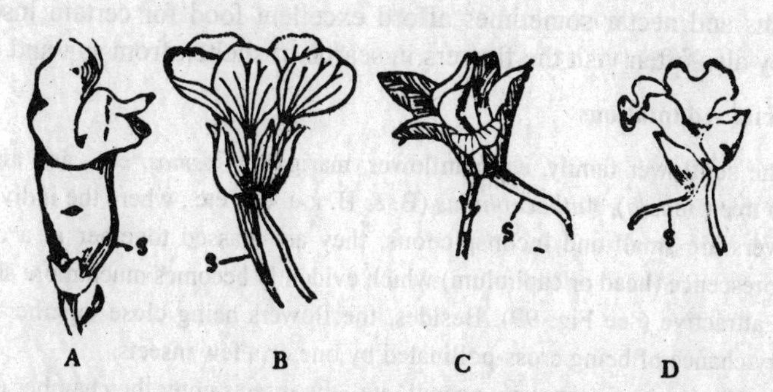

Figure 135. Appendages of Perianth. *A,* saccate corolla (*S'*) of snapdragon; *B,* flower of garden nasturtiums; *C,* flower of larkspur; *D,* flower of balsam; *S,* spur.

Scent. The third adaptation is the **scent**. Most of the nocturnal flowers are insect-loving and they emit at night a sweet scent which attracts insects from a distance. At night, when the colour fails, the scent is particularly useful in directing the insects to the flowers. Thus nocturnal flowers are mostly sweet-smelling. Common examples are night jasmine (*Nyctanthes*), queen of the night (*Cestrum*), jasmines (*Jasminum*), Rangoon creeper (*Quisqualis;* B. SANDHYA-MALATI; H. LALMALTI), etc. On the other hand, the stinking smell that is emitted from the appendix of mature *Amorphophallus* inflorescence (Fig. 136) is immensely liked

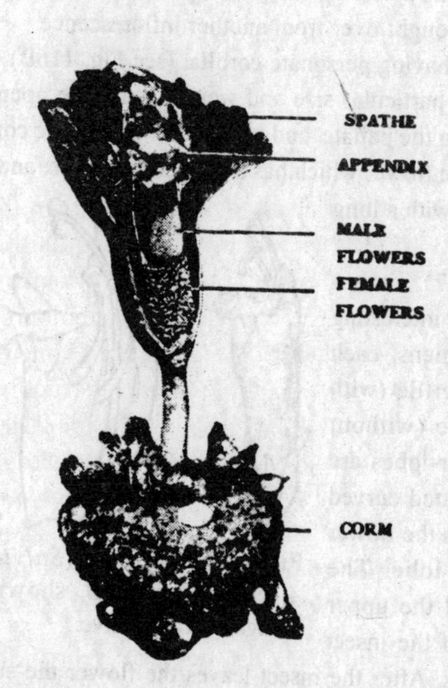

Figure 136. Spadix of *Amorphophallus* (B. OL; H. KANDA).

by certain small flies (carrion-flies), and pollination is achieved through them.

The pollen grains of entomophilous flowers arc either sticky or provided with spinous outgrowths. The stigma is also sticky. Pollen grains and nectar sometimes afford excellent food for certain insects. They also often visit the flowers in search of shelter. from sun and rain.

Special Adaptations

In the sunflower family, e.g. sunflower, marigold, *Cosmos,* etc., and also in gum tree (*Acacia*), *Anthocephalus* (B. & H. KADAM), etc., where the individual flowers are small and inconspicuous, they are massed together in a dense inflorescence (head or capitulum) which evidently becomes much more showy and attractive (see Fig. 99). Besides, the flowers being close together have every chance of being cross-pollinated by one or a few insects.

In *Ficus,* e.g. fig, banyan, peepul, etc., the insects enter the chamber of the inflorescence (hypanthodium) through the apical pore, and as they crawl over the unisexual flowers inside the chamber they bring about cross-pollination (see Fig. 103). The female flowers lie at the base of the cavity and open earlier, while the male flowers lie near the apical opening and open later so that pollen grains have necessarily to be brought over from another inflorescence.

In snapdragon and *Linaria,* having personate corolla, (see Fig. 116B) only certain types of insects having particular size and weight can force open the mouth of the corolla. They sit on the pallate, and under their weight the corolla opens. The insects then enter the flower which has a long corolla-tube, and can only be pollinated by an insect with a long proboscis.

In sage (*Salvia;* Fig. 137), cross-pollination by insects is of a very interesting type. The flower has two stamens, each with two anther-lobes—one fertile (with pollen grains) and one sterile (without pollen grains). The two anther-lobes are widely separated by the elongated curved connective. As the insect enters the flower it pushes the lower sterile lobe. The connective swings round and the upper fertile lobe strikes the back of the insect and dusts it with pollen grains. After the insect leaves the flower the stigma matures and bends down to receive the pollen grains from the back of another insect which has brought them from another flower.

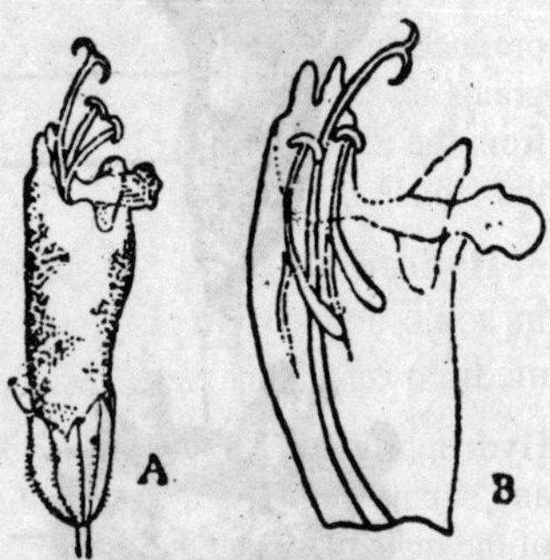

A B

Figure 137. Sage (*Salvia*). *A,* entire flower; *B,* showing elongated connective.

Figure 138. Anemophily in maize plant. Male flowers in a panicle (above) and female flowers in a spadix (below). Note the long hanging styles.

2. **Anemophily** (*anemos,* wind). In some cases pollination is brought about by the wind. Anemophilous or wind-loving flowers are small and inconspicuous. They are never coloured or showy. They do not emit any smell nor do they secrete any nectar. The anthers produce an immense quantity of pollen grains, wastage during transit from one flower to another being considerable. They are also minute, light and dry, and sometimes, as in pine, provided with wings. In this way the pollen grains are easily carried by the wind and distributed over a wide area, evidently helping cross-pollination. Stigmas are comparatively large and protruding, sometimes branched and often feathery. Examples are seen in maize, rice, grasses, sedges, bamboo, sugarcane, pine and several palms. (Wheat, however, is habitually self-pollinated).

Anemophily is well-illustrated by maize or Indian corn plant (Fig. 138). The male flowers (spikelets) of the panicle on the top produce an immense quantity of pollen grains. As the anthers burst, the pollen grains are set adrift by air-currents and many of them, particularly those brought from the neighbouring plants, are caught by the long hanging styles borne by the female flowers (spikelets) of the spadix lower down.

3. **Hydrophily** (*hydro,* water). Pollination may also be brought about in some aquatic plants, particularly the submerged ones, through the medium of water, e.g. *Naias, Vallisneria, Hydrilla,* etc.

Hydrophily may be illustrated by *Vallisneria* (Fig. 139). The plant is dioecious and submerged. The minute male flowers get detached from the small spadix of the male plant and float on water. Each female flower borne on a long stalk by the female plant is brought to the level of water. Then the free-floating male flowers are so adrift by currents, and some of them come in contact with the female flower. The anthers burst and the pollen grains are distributed on the stigma of the female flower. Thus pollination is brought about. The stalk of the female flower then becomes closely coiled and the fruit develops under water.

4. **Zoophily** (*zoon,* animal). Birds, squirrels, bats, snails, etc., also act as useful agents of pollination; for example, birds and sometimes

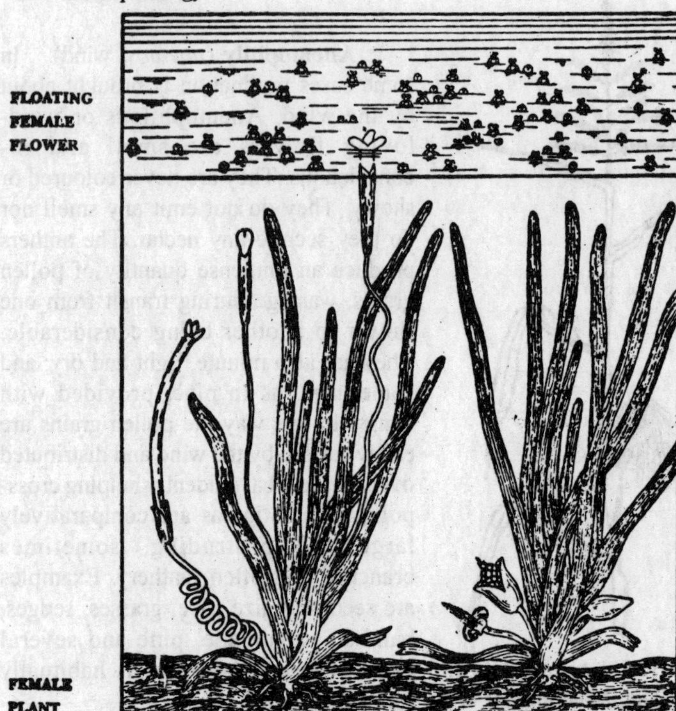

Figure 139. Hydrophily in *Vallisneria*. Female plant with a floating flower, a submerged flower (bud) and a fruit (15 cm long) maturing under water after pollination. Male plant with three spadices—young (covered by spathe, mature (with spathe bursting) and old (after the escape of the male flowers). Male flowers are now seen floating on water.

also squirrels bring about pollination in coral tree (*Erythrina*), silk cotton tree (Bombax), rose-apple (*Syzygium*), etc.; bats in *Anthocephalus* (B. & H. KADAM); and snails in certain large varieties of aroids and in snake or cobra plant (*Arisaema*—see Fig. 94).

Merits and Demerits of Self-pollination and Cross-pollination. Self-pollination has this merit that it is almost certain in a bisexual flower provided that both stamens and carpels of it have matured at the same time. Continued self-pollination, generation after generation, however, results in weaker progeny. The advantages of cross-pollination are many: (a) it always results in much healthier offspring which are better adapted to the struggle for existence; (b) more abundant and viable seeds are produced by this method; (c) germinating capacity is much better;

(d) new varieties may also be produced by the method of cross-pollination; and (e) the adaptability of the plants to their environment is better by this method. The disadvantages of cross-pollination are that the plants have to depend on external agencies for the purpose and, this being so, the process is more or less precarious and also less economical as various devices have to be adopted to attract pollinating agents, and that there is always a considerable waste of material (pollen) when wind is the pollinating agent.

Contrivances for Cross-pollination. By cross-pollination better seeds and healthier offspring are normally produced. Nature, therefore, favours this process and helps it by certain contrivances in flowers, which wholly or sometimes partially prevent self-pollination. It must, however, be noted that in many flowers there is still provision for self-pollination if the other method fails.

(1) **Dicliny** or **Unisexuality.** (a) Unisexual or diclinous flowers, i.e. separate male and female flowers, may be borne by one and the same plant; such a plant is said to be **monoecious** (*monos,* single; *oikos,* house), e.g. gourd, cucumber, castor, maize, etc.; or, (b) these may be borne by two separate plants; such plants are said to be **dioecious** (*di,* two), e.g. palmyra-palm, papaw, mulberry, etc. In monoecious plants the flowers may be self-pollinated or cross-pollinated, while in dioecious plants cross-pollination is indispensable for the production of seeds.

(2) **Self-sterility.** This is the condition in which the pollen of a flower has no fertilizing effect on the stigma of the same flower. Tea flowers, many grasses, some species of passion-flower, some orchids and mallow are self-sterile. Only pollen applied from another plant of the same or allied species is effective in such cases. Cross-pollination is thus the only method in them for the setting of seeds.

(3) **Dichogamy** (*dicha,* in two). In many bisexual flowers the anther and the stigma often mature at different times. This condition is known as **dichogamy**. Dichogamy often stands as a barrier to self-pollination. There are two conditions of dichogamy: (a) **protogyny** (*protos,* first, *gyne,* female) when the gynoecium matures earlier than the anthers of the same flower: here the stigma receives the pollen grains brought from another flower, e.g. *Ficus* (fig, banyan, peepul, etc), four o'clock plant, *Magnolia,* custard-apple, etc.; and (b) **protandry** (*protos,* first; *andros,* male) when the anthers mature (burst and discharge their pollen) earlier than the stigma of the same flower; here the pollen grains are carried over to the stigma of another flower, e.g. cotton, lady's finger,

sunflower, marigold, coriander, rose, etc. Protandry is more common than protogyny.

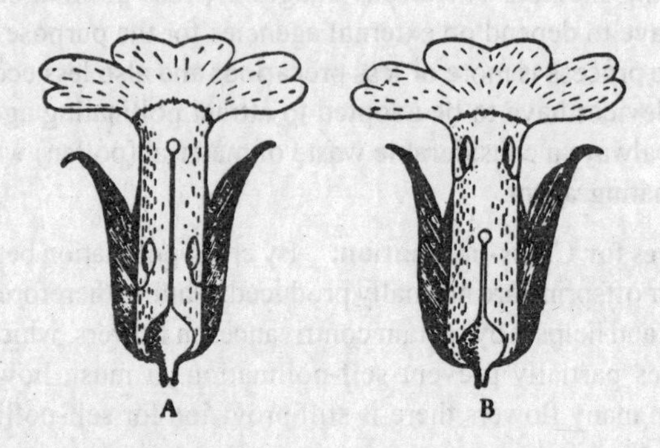

Figure 140. Heterostyly in primrose: dimorphic flowers. *A*, a flower with long style; *B*, a flower with short style.

(4) **Heterostyly** (*heteros*, different). There are some plants which bear flowers of two different forms. One form bears long stamens and a

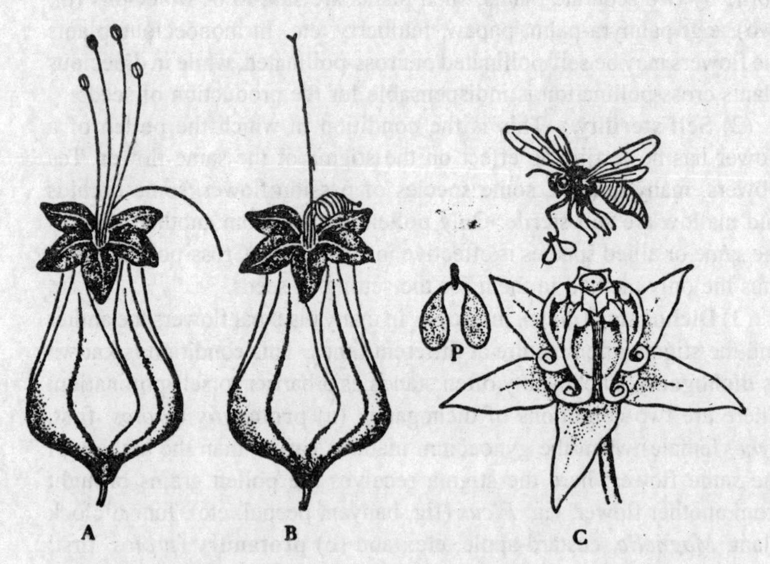

Figure 141. *A–B*, pollination in *Clerodendron; A*, stamens maturing first; *B*, stigma maturing later; *C*, pollination in madar (*Calotropis*); a pair of pollinia being carried away by an insect from a flower; *P*, a pair of pollinia.

short style, and the other form bears short stamens and a long style. This is known as *dimorphic heterostyly*. Similarly there may be cases of *trimorphic* heterostyly, that is, stamens and styles of three different lengths borne by three different forms of flowers. In all such cases cross-pollination is effective only when it takes place between stamens and styles of the same length borne by different flowers (legitimate pollination). Dimorphic heterostyly is seen in primrose (Fig. 140), buckwheat (*Fagopyrum*), wood-sorrel (*Oxalis*), linseed (*Linum*) and *Wood-fordia* (B. DHAINPHUL). Trimorphic heterostyly is found in some species of *Oxalis* and *Linum*.

(5) **Herkogamy** (*herkos,* a fence or barrier). There may be some sort of barrier standing between the stamens and the pistil of the same flower. A hood covering the stigma is a common form of barrier, as in pansy, *Iris,* etc. Anthers and stigmas may lie at some distance from one another—anthers exerted and style inserted or *vice versa.* The stamens and the style may also move away from each other, as in bleeding heart (*Clerodendron thomsonae;* Fig. 141 A–B). The pollinia of madar (*Calotropis*) and orchids, remaining fixed in their position by adhesive discs, can only be carried away by insects, evidently to another flower (Fig. 141C). The mechanism of cross-pollination in sage (*Salvia*) has been already discussed (see p. 128).

Chapter 11

Fertilization

Fertilization is the fusion of two dissimilar sexual reproductive units, called *gametes*. In the 'flowering' plants the process of fertilization, as worked out in detail by Strasburger in 1884, is as follows (Fig. 142). After pollination, that is, after the pollen grains fall on the stigma, the intine of each grows out into a slender tube, called the **pollen-tube** (see Fig. 121), through some thin or weak spot or *germ pore* in the exine

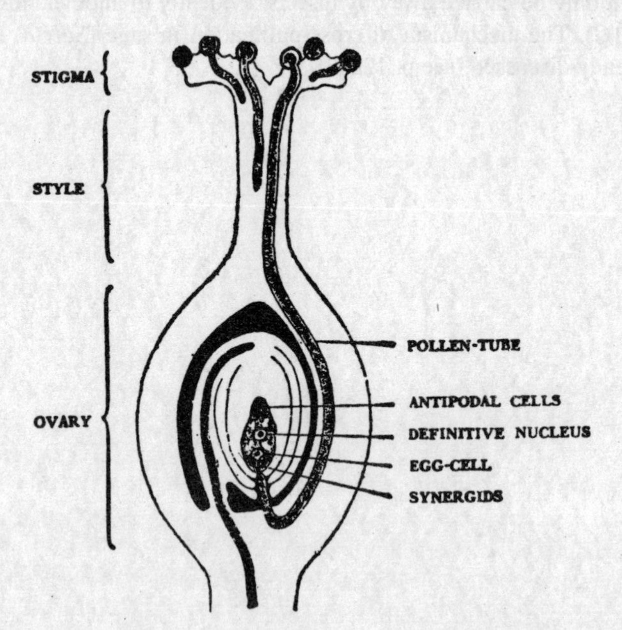

Figure 142. Ovary in longitudinal section showing the process of fertilization. Note the two male gametes at the tip of the pollen-tube.

(see Fig. 120). The tube, as it elongates, penetrates the stigma and pushes its way through the style and the wall of the ovary or alongside it, carrying with it the **tube-nucleus** and the **generative nucleus**. The generative nucleus divides forming two **male gametes**, while the tube-nucleus, gets disorganized sooner or later. The pollen-tube carrying the two male gametes at its tip then turns towards the micropyle and enters into it. As the tube further grows, it penetrates into the nucellus and enters the embryo-sac close to the egg-cell. At this stage the tip of the pollen-tube dissolves and the two male gametes are set free. It may be noted that the growth of the pollen-tube is stimulated by proteins and sugars secreted by the stigma and the style. Now, of the two male gametes already set free, one fuses with the **egg-cell**, while the other pushes further into the embryo-sac and fuses with the **definitive nucleus** (i.e. the fusion product of two polar nuclei). Thus a fusion of three nuclei, called *triple fusion,* takes place, and the product is now called **endosperm nucleus**. Normally the pollen tube enters the ovule through the micropyle when the fertilization process is called *progamic fertilization.* But in some cases, the pollen tube may also enter the ovule penetrating directly either through the chalaza or integuments when the fertilization process is known as *chalazogamic fertilization* and *mesogamic fertilization* respectively. Synergids do not seem to be essential for the process of fertilization. They are, in fact, short-lived bodies, and get disorganized soon after fertilization, or sometimes even before or during the process. Similarly, antipodal cells have no positive function; so they also disappear even before fertilization. After fertilization the egg-cell clothes itself with a cell-wall and becomes known as the **oospore**. The oospore gives rise to the embryo, the ovule to the seed, and the ovary as a whole to the fruit, and the endosperm nucleus to the endosperm. If fertilization fails for some reason or other, the ovary simply withers and falls off. The withering and shedding of the corolla usually indicates that fertilization has been effected. This is readily seen in night jasmine (*Nyctanthes*) where the whole ground is littered with shed flowers (corollas with epipetalous stamens) in the morning, fertilization having taken place the preceding night. Sometimes in banana, papaw, grape, etc. ovary may develop into the fruit (when the fruit is usually seedless) without the act of fertilization. This process is known as parthenocarpy.

Double Fertilization. It must have been noted from the foregoing description that in angiosperms fertilization occurs twice: (a) of the two male gametes of the pollen-tube one fuses with the egg-cell of the embryo-sac, and (b) the other one fuses with the definitive nucleus—the product of fusion of two polar nuclei during the development of the embryo-sac (see p. 122). This process of fusion occurring twice is called double fertilization. It was first discovered in 1898, by Nawaschin in Lilium and Fritillaria and later found to be of universal occurrence among the angiosperms. The significance of double fertilization is not, however, clearly understood.

Chapter 12

The Seed

Development of the Seed. After fertilization a series of changes takes place in the ovule, as a result of which the seed is formed. The fertilized egg-cell or ovum grows and gives rise to the embryo, and the definitive nucleus (endosperm nucleus) to the endosperm; other changes also take place in the ovule.

(1) **Development of the Embryo** (Fig. 143). After fertilization the **egg-cell** or **ovum** secretes a cellulose wall around itself and becomes the **oospore**. The oospore divides into two cells—an *upper* and a *lower*. The lower one lying towards the micropyle further divides in one direction into a row of cell called the **suspensor**. The suspensor absorbs food and feeds the developing embryo. The basal cell of the suspensor often enlarges and acts as an absorbing organ, while its terminal cell, called the **hypophysis cell**, divides and gives rise to the apex of the radicle. The upper cell enlarges, divides repeatedly and gives rise to a mass of cells known as the **embryonal mass**. By further divisions it gives rise to the whole of the embryo with all its differentiated parts, namely, the radicle (except its tip portion), plumule and two cotyledons (in dicotyledonous seeds) or only one cotyledon (in monocotyledonous seeds).

(2) **Development of the Endosperm.** The definitive nucleus (called the endosperm nucleus after fertilization) divides and gives rise to a large number of free nuclei (see Fig. II/22B). Protoplasm collects round each of the nuclei, and finally cell-walls are formed between them. A tissue laden with food material is thus formed by a process of free cell formation, and is known as the **endosperm**. As it grows, it fills up the nucellus. The endosperm is the food storage tissue and when it is present the seed is said to be **endospermic** or **albuminous**. In many seeds,

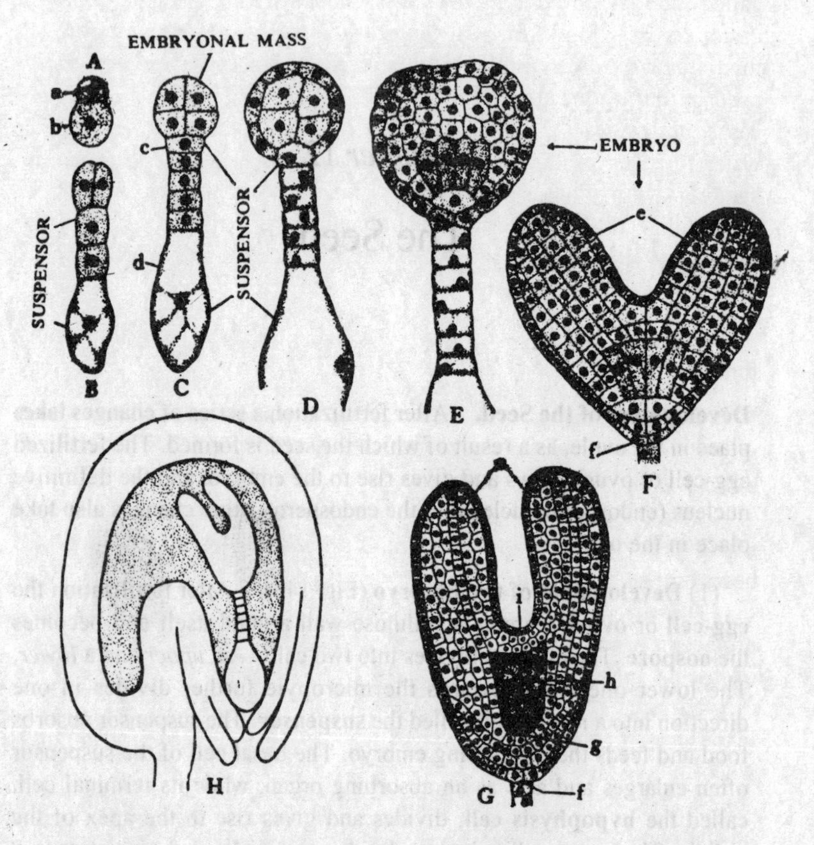

Figure 143. *A–H,* development of dicotyledonous embryo. *a,* embryonal cell; *b,* suspensor cell; *c,* hypophysis cell; *d,* basal cell of the suspensor; *e,* cotyledons; *f,* root-cap; *g,* root-tip; *h,* hypocotyl; and *i,* stem-apex. *H,* embryo within the seed.

however, all the food stored up in the endosperm at its early stage is used up by the developing embryo. The endosperm is thus absent in such seeds, and they are then said to be **non-endospermic** or **ex-albuminous** (see also p. 23).

(3) **Other changes in the Ovule.** A few other changes also take place in the ovule. The two integuments develop into two **seed-coats**, of which the outer one is called the *testa* and the inner one the *tegmen*. In some seeds, as in litchi, nutmeg (B. & H. JAIPHAL), wild mangosteen (B. GAB; H. KENDU), *Baccaurea* (B. LATKAN; H. LUTKO), etc., a fleshy mass

surrounds each seed; this fleshy mass, an outgrowth of the funicle, is called the **aril**. In certain seeds as in balloon vine (see Fig. 55) and castor (see Fig. 15A), a small outgrowth is formed at the micropyle; this is called the **caruncle**. In most seeds the nucellus is completely used up during the development of the ovule into the seed. In some cases, however, as in castor (see Fig. 15B–C), banana, ginger, cubeb, water lily, four o'clock plant, etc., the nucellus persists in the mature seed as a thin compressed tissue lying just within the seed-coat, surrounding the endosperm and the embryo; this tissue is called the **perisperm**. It is, in fact, a remnant of the nucellus and is nutritive in function like the endosperm.

Functions of the Seed. The seed has the following important functions:

(1) *Reproduction.* The 'flowering' plants normally reproduce through the medium of the seeds, and they also multiply in number through them, often profusely.

(2) *Receptacle of Embryo.* The seed is the receptacle or vessel in which the embryo develops and attains full maturity. The seed normally bears only one embryo.

(3) *Protection of the Embryo.* The seed encloses the embryo and protects it from excessive heat, cold and rain, and also from the attack of insects, birds and other animals.

(4) *Storage of Food.* The seed stores up food for the embryo, either, in the endosperm or in the cotyledons. This food is utilized by the embryo when the seed germinates

(5) *Seed Dispersal* (see chapter 14). Many-seeds have special adaptations by which they are easily dispersed by wind, water, and many animals.

Three generations locked up in a seed.

(a) *Past Generation.* Seed coats (testa and tegmen) formed by the two integuments represents the past generation.

(b) *Present Generation.* In endosperms and cotyledons, after the food has been supplied to the growing axis, they gradually dry up and fall off. Therefore, they represent the present generation.

(c) *Future Generation.* Axis which establishes itself into a seedling during germination represents the future generation.

Chapter 13

The Fruit

Development of the Fruit. After fertilization the ovary begins to grow and gradually matures into the fruit. The fruit may, therefore, be regarded as a mature or ripened ovary. If, for some reason or other, fertilization fails the ovary simply withers and falls off. A fruit consists of two portions, viz. the **pericarp** (*peri,* around; *karpos,* fruit) developed from the wall of the ovary, and the **seed** developed from the ovules. In some cultivated varieties of oranges, bananas, grapes, apples, pineapples and some other fruits the ovary may grow into the fruit without fertilization. Such a fruit is seedless or with immature seeds and is known as the **parthenocarpic** fruit. The pericarp may be thick or thin; when thick, it may consist of two or three parts: the outer, called *epicarp,* forms the skin of the fruit; the middle, called *mesocarp,* is pulpy in fruits like mango, peach, plum. etc., and the inner, called *endocarp,* is often very thin and membranous, as in orange, or it may be hard and stony, as in many palms, mango, etc. In many cases, however, the pericarp is not differentiated into these three regions. *Functions of the Fruit.* The fruit gives *protection* to the seed and, therefore, to the embryo. It *stores* food material. It also helps *dispersal* of the seed.

Normally it is only the ovary that grows into the fruit; such a fruit is known as the **true fruit**. Sometimes, however, other floral parts, particularly the thalamus or even the calyx, may grow and form a part of the fruit; such a fruit is known as the **false fruit**. Common examples of false fruits are apple (Fig. 144A), pear, cashew-nut (Fig. 144B), marking nut (Fig. 144C), rose, *Dillenia* (B. & H. CHALTA). etc. In *Dillenia* the calyx becomes thick and fleshy, forming the only edible part of the fruit.

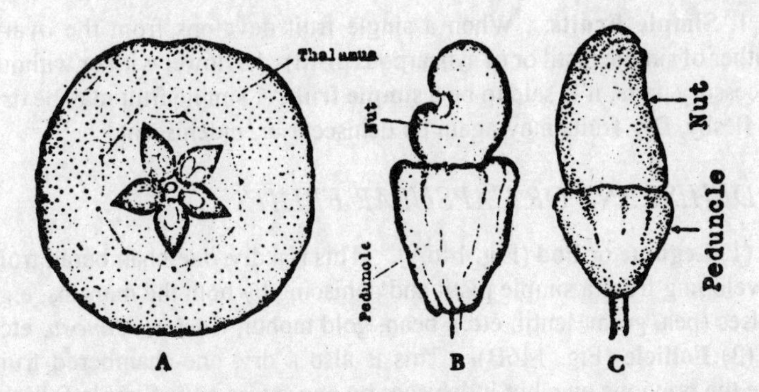

Figure 144. *A*, apple (*Malus*) in transverse section; *B*, cashew-nut (*Anacardium*); *C*, marking nut (*Semecarpus*).

Dehiscence of Fruits (Fig. 145). There are many fruits whose pericarp bursts to liberate the seeds when the former mature; such fruits are said

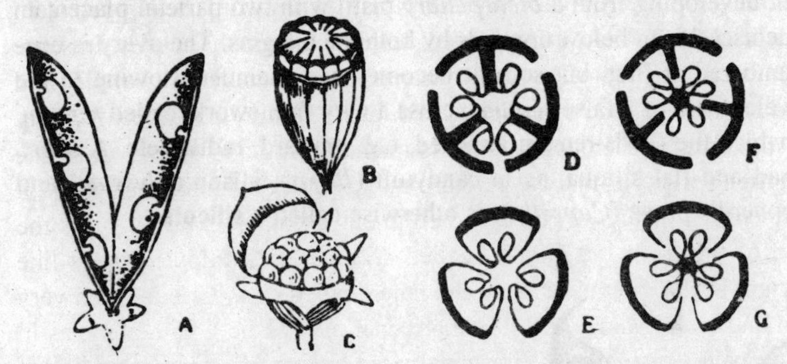

Figure 145. Dehiscence of fruits. *A*, sutural (pea); *B*, porous (poppy); *C*, transverse (cock's comb); *D*, loculicidal; *E*, septicidal; *F–G*, septifragal.

to be **dehiscent**. There are others again whose pericarp does not burst, and consequently the seeds cannot be liberated from the fruits until decay of the latter has set in. Fruits that belong to this category are said to be **indehiscent**. Dehiscent fruits open in various ways, as shown in Fig. 145, and aid in the dispersal of seeds.

CLASSIFICATION OF FRUITS

All the different kinds of fruits may be broadly classified into three groups, viz. **simple**, **aggregate** and **multiple** or **composite**. A few common types are discussed under each group.

1. **Simple Fruits.** When a single fruit develops from the ovary (either of simple pistil or of syncarpous pistil) of a flower with or without accessory parts, it is said to be a **simple fruit**. A simple fruit may be dry or fleshy. Dry fruits may again be dehiscent or, indehiscent.

I. *DEHISCENT OR CAPSULAR FRUITS*

(1) **Legume** or **Pod** (Fig. 146A). This is a dry, one-chambered fruit developing from a simple pistil and dehiscing by both the margins, e.g. pulses (pea, gram, lentil, etc.), bean, gold mohur, *Cassia, Mimosa,* etc.

(2) **Follicle** (Fig. 146B). This is also a dry, one-chambered fruit like the previous one, but it dehisces by one suture only. Simple follicle is rare; it may sometimes be seen in madar (*Calotropis*), blood flower (*Asclepias*), wax plant (*Hoya*), *Rauwolfia,* etc. Usually follicles develop in an aggregate of two, three, or many fruits.

(3) **Siliqua** (Fig. 146C). This is a dry, long, narrow, two-chambered fruit developing from a *bicarpellary* pistil with two parietal placentae. It dehisces from below upwards by both the margins. The ovary is one-chambered at first, but soon it becomes two-chambered owing to the development of a false *septum* across a wiry framework, called *replum,* to which the seeds remain attached, e.g. mustard, radish, etc. A short, broad and flat siliqua, as in candytuft (*Iberis*), alison (*Alyssum*) and shepherd's purse (*Capsella*), is otherwise called a **silicula**.

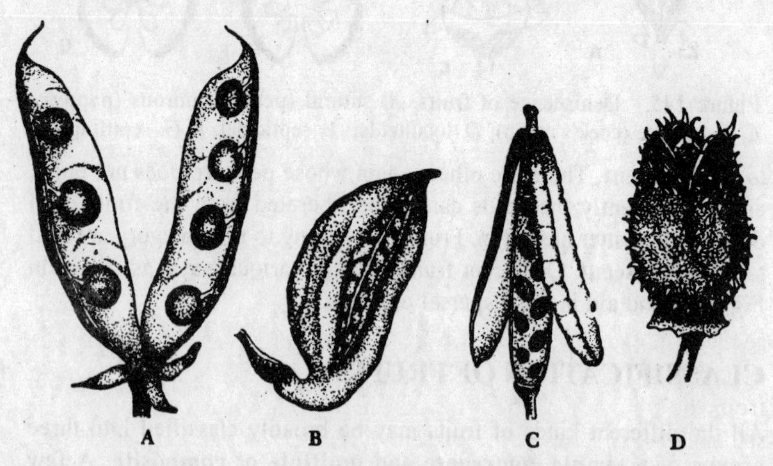

Figure 146. Fruits. *A,* legume or pod of pea; *B,* follicle of madar (*Calotropis*); *C,* siliqua of mustard; *D,* capsule of *Datura.*

(4) **Capsule** (Figs. 146D & 147A). This is a dry, one-to many-chambered fruit developing from a syncarpous pistil, and dehiscing in various ways. All dehiscent fruits developing from a syncarpous pistil are commonly known as capsules, e.g. cotton, lady's finger, *Datura,* cock's comb, poppy, etc.

II. *INDEHISCENT OR ACHENIAL FRUITS*

(1) **Achene** (Fig. 147B). An achene is a small, dry, one-seeded fruit developing from a single carpel; but unlike the next one, the pericarp of this fruit is free from the seed-coat, e.g. four o'clock plant (*Mirabilis*), hogweed (*Boerhaavia*) and buckwheat (*Fagopyrum*). Achenes, however, commonly develop in an aggregate, as in rose, *Clematis, Naravelia* (see Fig. 150B), etc.

(2) **Caryopsis** (147C). This is a very small, dry, one-seeded fruit developing from a simple (or syncarpous) pistil, with the pericarp fused with the seed-coat, e.g. maize, rice, wheat, bamboo, grass, etc.

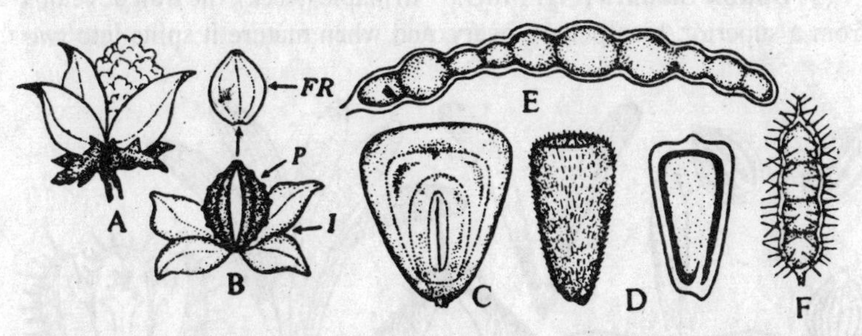

Figure 147. Fruits (*contd.*). *A,* capsule of cotton; *B,* achene of *Mirabilis* (*FR,* fruit; *P,* dry persistenet perianth enclosing the fruit; *I,* involucre); *C,* caryopsis of maize; *D,* cypsela of sunflower (entire and in section); *E,* lomentum of *Acacia; F,* the same of *Mimosa.*

(3) **Cypsela** (147D). This is a dry, one-seeded fruit developing from an inferior *bicarpellary* ovary, e.g. sunflower, marigold, *Cosmos,* etc.

(4) **Samara** (Fig. 148B). This is a dry, indehiscent, one- or two-seeded, winged fruit developing from a superior, bicarpellary or tricarpellary ovary. In samara the wings, one or more, always develop from the pericarp of the fruit, as in *Hiptage* (B. MADHABILATA; H. MADHULATA; Fig. 148B), ash (*Fraxinus*), yam (*Dioscorea;* see Fig. 153B), etc. Fruits of sal tree (*Shorea;* Fig. 148E), wood-oil tree (*Dipterocarpus;* B. & H. GARJAN; see Fig. 154A), *Hopea* (Fig. 148D), etc., are also winged

but in them the wings are the dry, persistent sepals. A winged fruit of this nature is called a **samaroid**.

(5) **Nut** (see Fig. 144B–C). This is a dry, one-seeded fruit developing from a superior *syncarpous pistil, with the pericarp hard and woody,* e.g. cashew-nut, marking nut (see Fig. 144C), chestnut, oak, etc.

III. *SPLITTING OR SCHIZOCARPIC FRUITS*

(1) **Lomentum** (Fig. 147E–F). When the pod is constricted or partitioned between the seeds into a number of one-seeded compartments, it is called a **lomentum**, as in gum tree (*Acacia*), nicker bean (*Entada;* B. & H. GILA), sensitive plant (*Mimosa*), Indian laburnum, (*Cassia fistula*), groundnut (*Arachis*), Indian telegraph plant (*Desmodium*), etc.

(2) **Cremocarp** (Fig. 148A). This is a dry, two-chambered, inferior fruit splitting into two indehiscent, one-seeded pieces, called *mericarps*. Each mericarp remains attached to the forked end of the axis (**carpophore**), as in coriander, anise or fennel, cumin, carrot, etc.

(3) **Double Samara** (Fig. 148C). In maple (*Acer*), the fruit develops from a superior, bicarpellary ovary, and when mature it splits into two

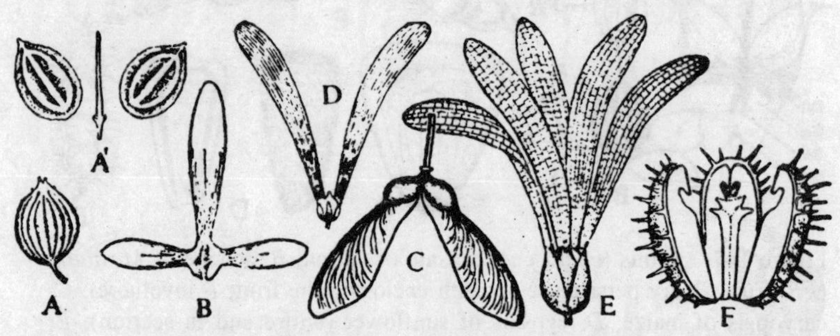

Figure 148. Fruits (*contd.*). *A,* cremocarp of coriander; *A',* fruit split into two mericarps *B,* samara of *Hiptage; C,* double samara of *Acer; D,* samaroid of *Hopea; E,* the same of *Shorea; F,* regma of castor.

samaras, each with a wing and a seed. Such a fruit is called a double samara.

(4) **Regma** (Fig. 148F). This is a dry, 3 to many chambered fruit developing from a syncarpous pistil. It splits away from the central axis into as many parts (cocci) as there are carpels, each, part containing 1 or 2 seeds. Common examples are castor, *Euphorbia* (B. & H. SIJ), *Geranium, Jatropha* (B. & H. BHARENDA), etc.

IV. *FLESHY FRUITS*

(1) **Drupe** (Fig. 149A). This is a fleshy, one or more seeded fruit with the pericarp differentiated into the outer skin or epicarp, often fleshy or sometimes fibrous mesocarp, and *hard and stony* endocarp, and hence this fruit is also known as **stone-fruit**, e.g. mango, plum, coconut-palm, palmyra-palm, country almond, etc.

(2) **Bacca** or **Berry** (Fig. 149B–C). This is a fleshy, superior (sometimes inferior), usually many-seeded fruit, developing commonly from a syncarpous pistil (rarely from a single carpel) with axile or parietal placentation, e.g. tomato, gooseberry, grapes, banana, guava, papaw, etc. With the growth of the fruit the seeds separate from the placentae and lie free in the pulp. It is not infrequent to find a one-seeded berry, e.g. date-palm. In *Artabotrys* (B. & H. KANTALI-CHAMPA) berries develop in an aggregate.

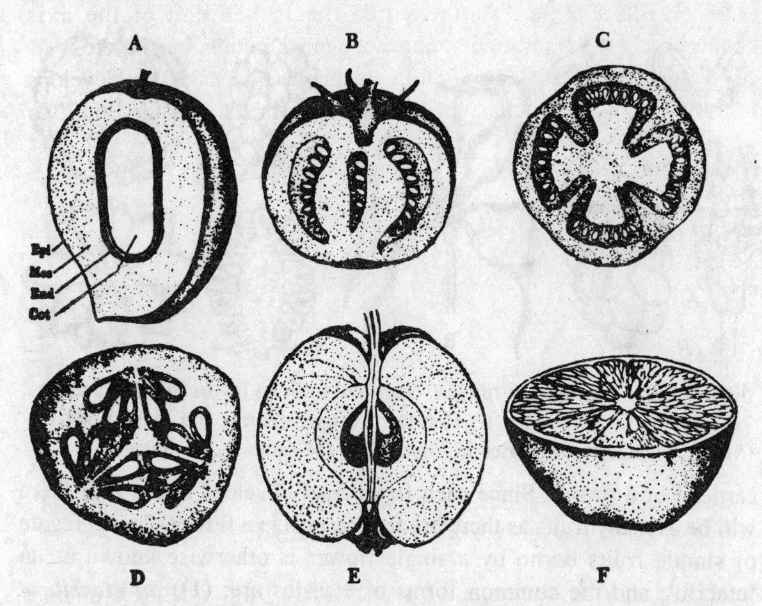

Figure 149. Fruits (shown in sections) *A,* drupe of mango; *Epi,* epicarp; *Mes,* mesocarp; *End,* endocarp; *Cot,* cotyledon; *B–C,* berry of tomato; *D,* pepo of cucumber; *E,* pome of apple (see also Fig. 144A); *F,* hesperidium of orange.

(3) **Pepo** (Fig. 149D). This is also a fleshy, many-seeded fruit like the berry but it develops from all inferior, one-celled or spuriously three-

celled, *syncarpous* pistil with parietal placentation, e.g. gourd, cucumber, melon, water melon, squash, etc. In pepo, the seeds, lying embedded in the pulp, remain attached to the placentae.

(4) **Pome** (Fig. 149E). This is an inferior, two or more celled, fleshy, *syncarpous* fruit surrounded by the thalamus. The fleshy edible part is composed of the thalamus, while the actual fruit lies within, e.g. apple and pear.

(5) **Hesperidium** (Fig. 149F). This is a superior, many celled fleshy fruit with axile placentation. Here the endocarp projects inwards forming distinct chambers, and the epicarp and the mesocarp, fused together, form the separable skin or rind of the fruit, e.g. orange, pummelo or shaddock, lemon etc.

2. **Aggregate Fruits** (Fig. 150). An aggregate fruit is a collection of simple fruits (or fruitlets) developing from an apocarpous pistil (free

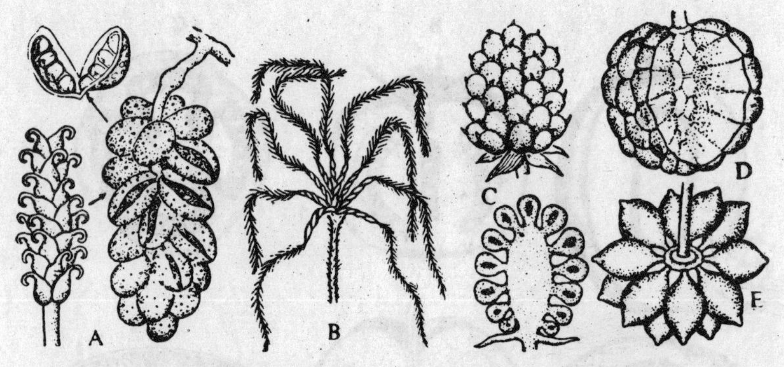

Figure 150. Aggregate Fruits. *A,* etaerio of follicles in *Michelia; B,* etaerio of achenes in *Naravelia; C,* etaerio of drupes (entire and in section) in *Rubus; D,* etaerio of berries in *Annona; E,* the same in *Artabotrys.*

carpels) of a flower. Since each free carpel develops into a fruit there will be as many fruits as there are free carpels in a flower. An aggregate of simple fruits borne by a single flower is otherwise known as an 'etaerio', and the common forms of etaerios are: (1)) *an etaerio of follicles* (A), e.g. Michelia, madar (*Calotropis*), periwinkles (*Vinca*), larkspur (*Delphinium*) etc.; (2) *an etaerio of achenes* (B), e.g. rose, lotus, *Clematis,* strawberry, *Naravelia,* etc.; (3) *an etaerio of drupes* (C), e.g. raspberry (*Rubus*); and (4) *an etaerio of berries* (D–E), e.g. custard-apple (*Annona*), *Artabotrys* (B. & H. KANTALI-CHAMPA), mast tree (*Polyathia*), etc.

3. **Multiple** or **Composite Fruits** (Fig. 151). A multiple or composite fruit is that which develops from an inflorescence where the flowers are crowded together and often fused with one another.

(1) **Sorosis** (Fig. 151 A–B). This is a multiple fruit developing from a spike or spadix. The flowers fuse together by their succulent sepals and at the same time the axis bearing them grows and becomes fleshy or woody, and as a result the whole inflorescence forms a compact mass; e.g. pineapple (A), screwpine, jack-fruit and mulberry (B).

(2) **Syconus** (Fig. 151C). The syconus develops from a hollow pear-shaped, fleshy receptacle which encloses a number of minute, male and female flowers. The receptacle grows, becomes fleshy and forms the so-called fruit. It really encloses a number of true fruits or achenes which

Figure 151. Multiple fruits. *A,* sorosis of pineapple (*Ananas*); *B,* the same of mulberry (*Morus*); *C,* syconus of fig (*Ficus*).

develop from the female flowers lying within the receptacle, as in *Ficus,* (e.g. fig. banyan, peepul, etc.).

Kinds of Fruits

A. Simple Fruits

(*a*) *Capsular Fruits*
1. Legume, e.g. pea
2. Follicle, e.g. madar

(*b*) *Achenial Fruits*
1. Achene, e.g. *Mirabilis*
2. Caryopsis, e.g. maize

3 . Siliqua, e.g. mustard
4. Capsule, e.g. cotton

3. Cypsela, e.g. sunflower
4. Nut, e.g. cashew-nut

(*c*) *Schizocarpic Fruits*
1. Lomentum, e.g. *Acacia*
2. Cremocarp, e.g. coriander
3. Samara, e.g. *Hiptage*
4. Samaroid, e.g. *Shorea*
5. Regma, e.g. castor

(*d*) *Fleshy Fruits*
1. Drupe, e.g. mango
2. Berry, e.g. tomato
3. Pepo, eg. gowd
4. Pome, e.g. apple
5. Hesperidium, e.g. *Citrus*

B. Aggregate Fruits

1. Etaerio of follicles, e.g. *Michelia*
2. Etaerio of achenes, e.g. rose
3. Etacrio of drupes, e.g. raspberry
4. Etaerio of berries, e.g. *Artabotrys*

C. Multiple Fruits

1. Sorosis, e.g. pineapple
2. Syconus, e.g. *Ficus*

Some Common Fruits and their Edible Parts

Apple (pome)—fleshy thalamus. **Banana** (berry)—mesocarp and endocarp. **Cashew-nut** (nut)—peduncle and cotyledons. **Coconut-palm** (fibrous drupe)—endosperm. **Cucumber** (pepo)—mesocarp, endocarp and placentae. **Custard-apple** (etaerio of berries)—fleshy pericarp of individual berries. **Date-palm** (1-seeded berry)—pericarp. *Dillenia* (special)—accrescent calyx. **Fig** (syconus)—fleshy receptacle. **Grape** (berry)—pericarp and placentae. **Guava** (berry)—thalamus and pericarp. **Indian plum** (drupe)—mesocarp including epicarp. **Jack** (sorosis)—bracts, perianth and seeds. **Litchi** (1-seeded nut)—fleshy aril. **Maize**, **oat**, **rice** and **wheat** (caryopsis)—starchy endosperm. **Mango** (drupe)—mesocarp. **Melon** (pepo)—mesocarp. **Orange** (hespiridium)—juicy placental hairs. **Palmyra-palm** (fibrous drupe)—mesocarp. **Papaw** (berry)—mesocarp. **Pea** (legume)—cotyledons. **Pear** (pome)—fleshy thalamus. **Pineapple** (sorosis)—outer portion of receptacle, bracts and perianth. **Pomegranate** (special)—juicy outer coat of the seed. **Pummelo** or **shaddock** (hespiridium)—juicy placental hairs. **Strawberry** (etaerio of achenes)—succulent thalamus. **Tomato** (berry)—pericarp and placentae. **Wood-apple** (special)—mesocarp, endocarp and placentae.

Chapter 14

Dispersal of Seeds and Fruits

If seeds and fruits fall directly underneath the mother plant and the seedlings grow up close together they soon exhaust the soil of its essential

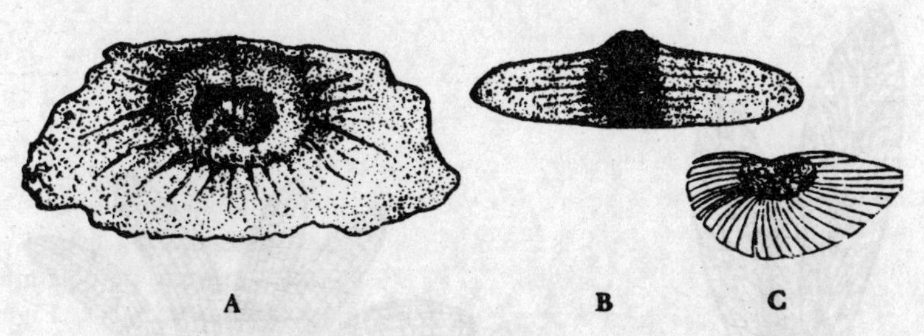

Figure 152. Winged Seeds. *A, Oroxylum; B, Cinchona; C,* crepe tree (*Lagerstroemia*).

food constituents. Besides, the available space, light and air under such a condition fall far short of the demand. A struggle for existence thus ensues, the consequence of which may be fatal to all of them. To guard against this the seeds and fruits have developed various devices for their wide distribution so that some of them at least may meet with favourable condition of germination and normal growth. Thus the risk of a species of plants becoming extinct is practically averted.

1. **Seeds and Fruits dispersed by Wind.** Seeds and fruits have various adaptations like wings, pappus, hairs, etc., which help them to be carried away by the wind to a shorter or longer distance from the parent plant.

(1) **Wings.** Seeds and fruits of many plants develop one or more thin membranous wings for facility of dispersal by the wind. Thus seeds of *Oroxylum* (B. SONA; H. ARLU—Fig. 152A), *Cinchona* (Fig. 152B),

crepe tree (*Lagerstroemia;* B. & H. JARUL—Fig. 152C),drum-stick (*Moringa;* B. SAJINA; H. SAINJNA—Fig. 153A), etc., are provided with wings for this purpose. Similarly many fruits are also provided with one or more wings to achieve the same end, e.g. yam (*Dioscorea;* Fig. 153B), wood-oil tree (*Dipterocarpus;* B. GARJAN—Fig. 154A), *Hiptage* (B. MADHABI-LATA; H. MADHULATA—Fig. 154B), and *Shorea* (B. & H. SAL—Fig. 154C).

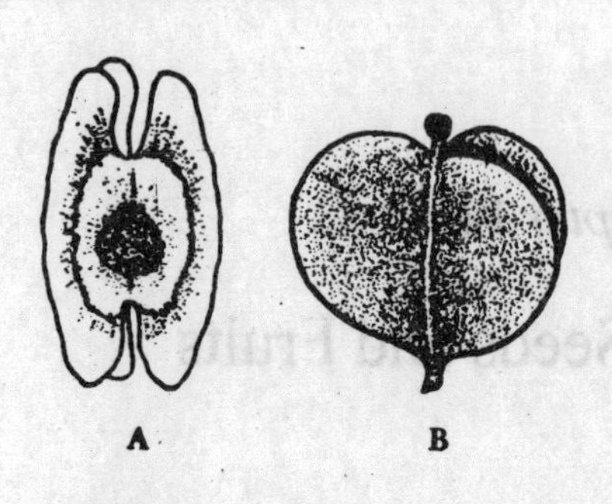

Figure 153. *A,* winged seed of drum-stick (*Moringa*); *B,* winged fruit of yam (*Dioscorea*).

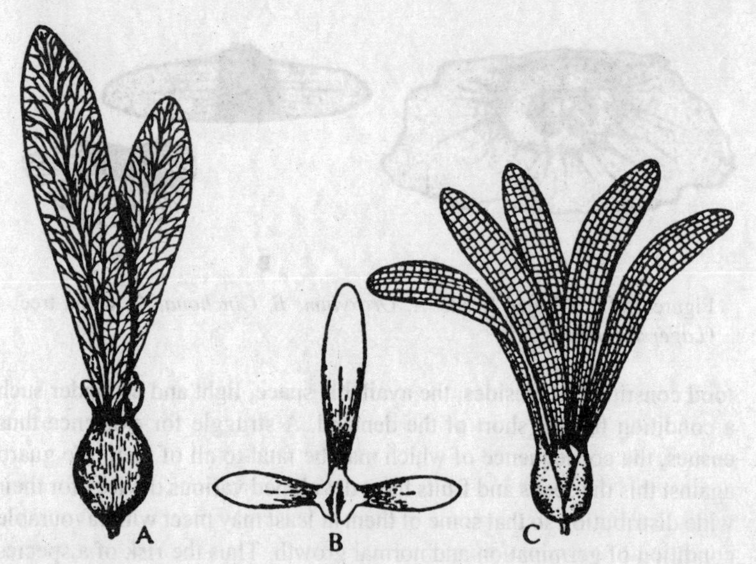

Figure 154. Winged Fruits. *A,* wood-oil tree (*Dipterocarpus*); *B, Hiptage; C,* sal tree (*Shorea*).

(2) **Parachute Mechanism.** In many plants of the sunflower family or *Compositae* the calyx is modified into hair-like structures known as **pappus** (Fig. 156A). This pappus is persistent in the fruit, and opens out in an umbrella-like fashion. Thus, acting like a parachute, it helps the fruit to be carried by air currents to a distance.

(3) **Censer Mechanism.** In some plants, as in poppy, prickly poppy, larkspur, bath sponge or loofah, cock's comb, pelican flower

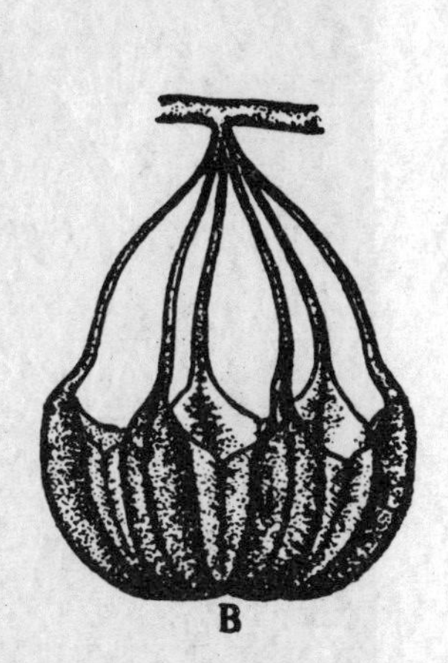

Figure 155. *A*, pelican flower (*Aristolochia*) with duck-shaped flowers; *B*, a fruit of the same like a hanging basket.

(*Aristolochin gigas;* B. HANSA-LATA—Fig. 155), etc., the fruit dehisces, and when it is disturbed by the wind, the seeds are thrown out.

(4) **Hairs.** A tuft of hairs or a dense coating is very useful for distribution of seeds by wind, e.g. madar (*Calotropis;* Fig. 156B), devil tree (*Alstonia;* Fig. 156C), cotton (*Gossypium;* Fig. 156D), etc.

(5) **Persistent Styles.** In *Clematis* (Fig. 157A) and *Naravelia* (Fig. 157B) the styles are persistent and very feathery. The fruits are thus easily carried away by the wind.

(6) **Light Seeds and Fruits.** Some seeds and fruits are so light and minute in size that they may easily be carried away by the gentlest breeze. Thus orchids often bear millions of dust-like seeds (smallest in the vegetable kingdom) in a single fruit (capsule). Seeds of *Cinchona* (the quinine-yielding plant) are also very small, flat, extremely light, and provided with a membranous wing (see Fig. 152B). there are about 2,470 seeds per gramme.

2. **Seeds and Fruits dispersed by Water.** Seeds and fruits to be, dispersed by water usually develop floating devices in the form of spongy

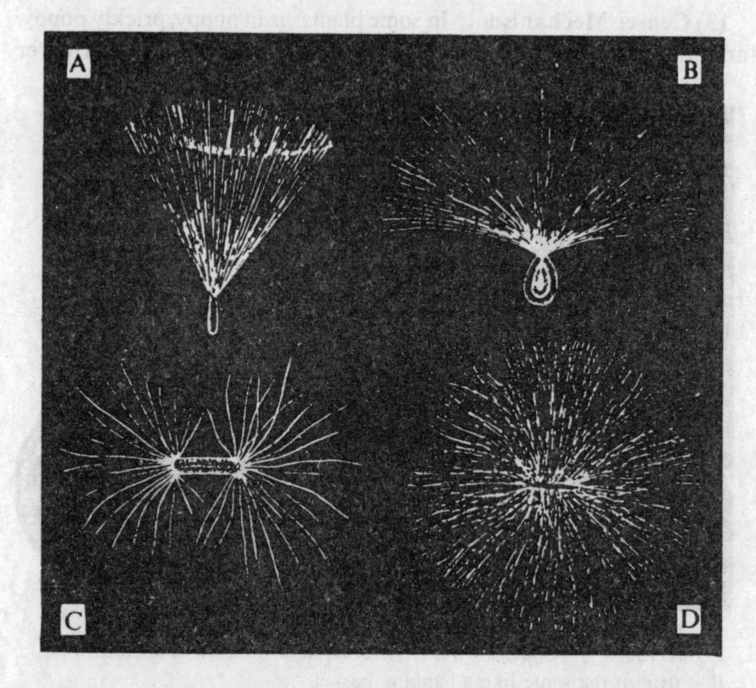

Figure 156. Hairy Fruit and Seeds. *A, pappus of a Compositae fruit; B,* madar (*Calotropis*); *C,* devil tree (*Alstonia*); *D,* cotton (*Gossypium*).

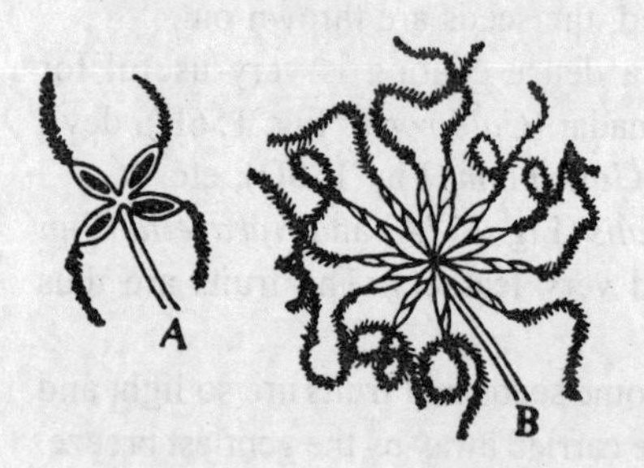

Figure 157. Persistent styles. *A,* fruits of *Clematis, B,* fruits of *Naravelia.*

or fibrous outer coats. The fibrous fruit of coconut is capable of floating long distances in the sea without suffering any injury. Hence coconut forms a characteristic vegetation of sea-coasts and marine islands. The fruit of double coconut (*Lodoicea;* Fig. 158), a native of Seychelles, which bears the largest seed in the world is also distributed likewise by ocean currents (see p. xxxii). The top-shaped spongy thalamus of lotus (see Fig. 107C) bearing the fruits on its surface floats on water and is drifted by water-current or by wind. Seeds of water lily are small and light, and are further provided with an *aril* which encloses air.

3. Seeds dispersed by Explosive Fruits. Many fruits burst with a sudden jerk, with the result that seeds are scattered on all sides. Common examples of explosive fruits are balsam, wood-sorrel, night jasmine, castor, etc. Ripe fruits of balsam burst suddenly. The valves roll up inwards, and the seeds are ejected with great force and scattered in all directions. Dry fruits of *Ruellia* (Fig. 159) coming in contact with water, particularly after a shower of rain, burst suddenly with a noise and scatter the seeds. Further, the seed is provided with a curved hook (jaculator) which straightens out instantly and jerks out the seed. Mature fruits of *Phlox, Andrographis* (B. KALMEGH; H. MAHATITA), *Barleria* (B. JHANTI; H. VAJRADANTI), etc. burst suddenly when the air is dry, particularly at mid-day.

Figure 158. Double coconut seed (Lodoicea Maldivica) .

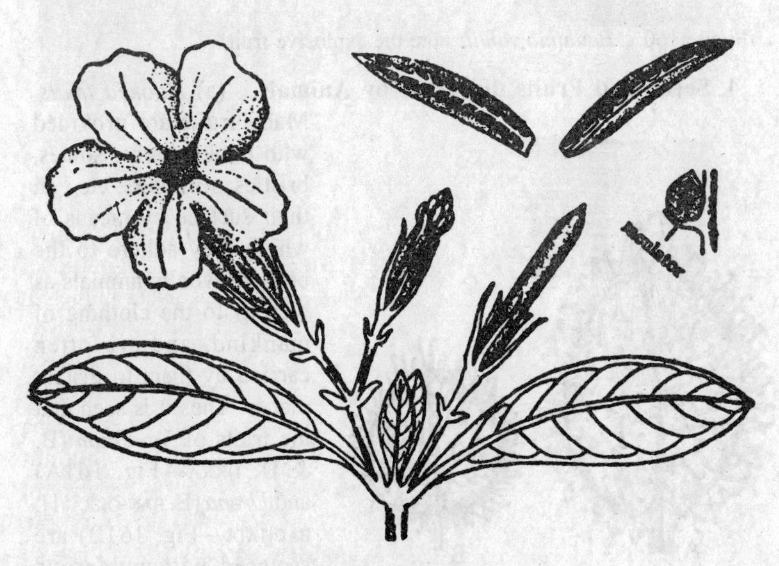

Figure 159. *Ruellia;* note the explosive fruit.

A very interesting example of bursting fruits is found in camel's foot climber (*Bauhinia vahlii,* B. LATA-KANCHAN; H.CHAMBULI). Its long pods, sometimes as long as 30 cm., explode with a loud noise like a cracker, scattering the seeds in all directions (Fig. 160).

Figure 160. *Bauhinia vahlii;* note the explosive fruit.

4. **Seeds and Fruits dispersed by Animals.** (a) *Hooked fruits.* Many fruits are provided with hooks, barbs, spines, bristles, stiff hairs, etc., on their surface by means of which they adhere to the body of woolly animals as well as to the clothing of mankind, and are often carried by them to distant places. Thus it is seen that the fruits of *Xanthium* (B. & H. OKRA—Fig. 161A) and *Urena* (B. BAN-OKRA; H. BACHATA—Fig. 161B) are covered with numerous curved hooks. Seeds (fruits) of spear grass (*Aristida*) and love-thorn (*Chrysopogon;* B. CHORKANTA) have a cluster of stiff hairs pointing upwards. In *Pupalia* (Fig. 162B) the perianth bears clusters of hooked

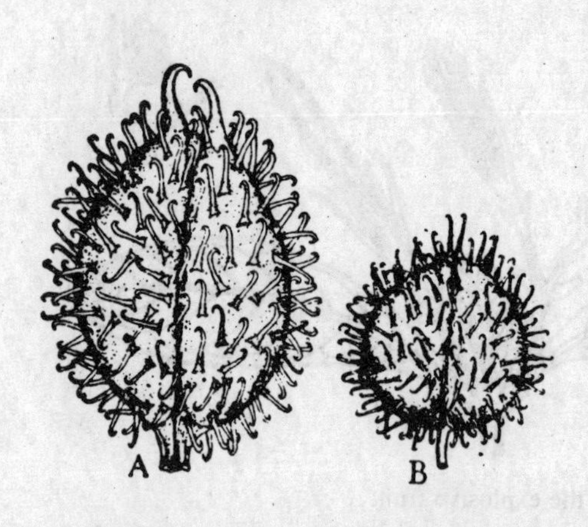

Figure 161. *A,* fruit of *Xanthium* with curved hooks; *B,* fruits of *Urena* with curved hooks.

bristles. Tiger's nail (*Martynia;* B. BAGHNAKHI; H. SHERNUI—Fig. 162C) is a very interesting case. Its seed is provided with two very sharp, pointed, stiff and bent hooks by which it can easily stick to the body of woolly animals. (b) *Sticky fruits.* Fruits of *Boerhaavia* (B. PUNARNAVA; H. SANTH—Fig. 162A) are provided with sticky glands. In mistletoe (*Viscum;* see Fig. 8) the seeds are sticky. (c) *Fleshy fruits.* Many fleshy

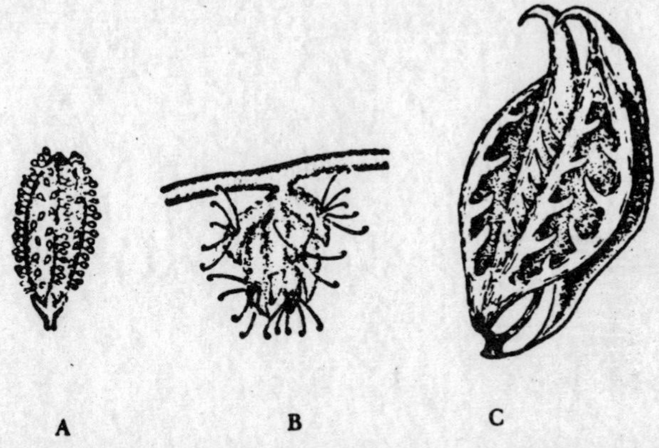

Figure 162. *A,* fruit of *Boerhaavia* with sticky glands (see also Fig. II/33B); *B,* flowers of *Pupalia* with hooked bristles; *C,* seed of tiger's nail (*Martynia*) with a pair of sharp, curved hooks.

fruits, particularly with conspicuous colours, are often carried, by human beings and birds to distant places. (d) *Edible fruits.* Many such fruits are regularly and often widely distributed by animals from one place to another or even from one country to another. Human beings and birds are very useful and active agents in this respect. They feed upon the pulpy or otherwise edible portion of the fruits and reject the seeds which may germinate and grow up into new plants. Common among such fruits are guava, papaw, mango, shaddock, jack, blackberry, Indian plum, custard-apple, rose-apple, etc. Bats, squirrels and jackals are also useful agents in dispersing seeds over wide areas.

Part II

HISTOLOGY

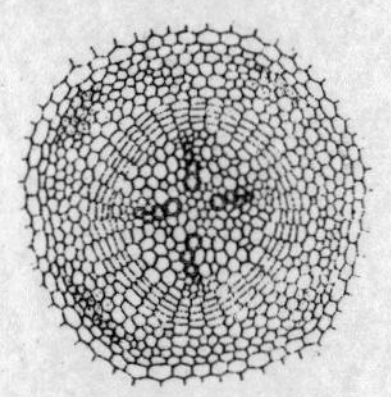

The Cell
The Tissue
The Tissue System
Anatomy of Stems
Anatomy of Roots
Anatomy of Leaves
Secondary Growth in Thickness

Chapter 1

The Cell

An Early History. The study of histology dates from the year 1665, when plant cells were discovered for the first time. It was Robert Hooke, an Englishman, who first studied the internal structure of a thin slice of bottle cork with the help of a microscope improved by himself. He discovered for the first time a honey-comb like structure in it, and to each individual cavity of such a structure he applied the term 'cell'. It was only then that the cell-wall was noticed, this being the prominent part of the cell. Other prominent, workers of that time, who studied plant tissues under the microscope, were Leeuwenhoek, Grew and Malpighi. Jansen, a spectacle-maker of Middleburg in Holland, first invented the compound microscope in 1590. Leeuwenhoek, a cloth merchant of Delft in Holland, at the age of 21 in the year 1653 developed a mania for grinding lenses. He pursued this work with zeal and assiduity, and within 20 years (1653–1673) accomplished marvellous accuracy and perfection in his lenses. He gave a demonstration of his microscope before the Royal Society in 1667. He was first to discover bacteria, protozoa and other forms of life—'the wretched beasties, as he called them—under his own microscope. Grew, an English physician and botanist, published his first paper on plant tissues in 1671. Malpighi, an Italian physician, studied the various tissues of vascular plants, and published his first paper in 1675. In 1838–39 Schleiden, a German botanist, and Schwann, a German zoologist, proved definitely that both plants and animals are cellular in character, and founded the **cell theory**.

Cell-structure. We have already learnt (see Introduction) that the plant body is composed of cells which are its fundamental structural and functional units. A plant cell may be defined as a unit or independent, tiny or microscopic mass of **protoplasm** enclosing in it a denser spherical or oval body, called the **nucleus**, and bounded by a distinct wall, called the **cell-wall**. Protoplasm and nucleus are living, while the cell-wall is non-living; the latter, formed by the protoplasm to maintain its shape and firmness and to afford necessary protection. Cells vary widely in

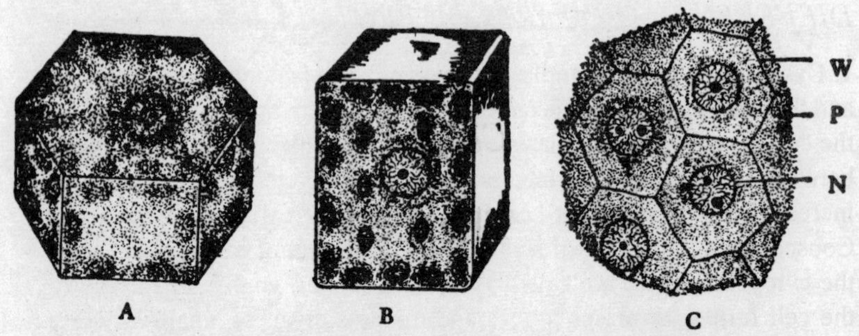

Figure 1. Plant cells. *A*, polygonal cell (three-dimensional diagram); *B*, a cubical cell in section (three-dimensional diagram); *C*, a group of cells in secion. *W*, cell-wall; *P*, protoplasm; *N*, nucleus.

shapes and sizes. In shapes they may commonly be spherical, oval, polygonal, cubical or narrow and elongated. When young, they are often sphericàl or of like-nature. Usually they are very minute in size and invisible to the naked eye. The average size of fully developed rounded or polygonal cells varies between 1/10th and 1/100th of a millimetre. There are, however, many cells far beyond these limits.

LIVING CELL-CONTENTS (THE PROTOPLAST)

Protoplasm (see also pp. xix) is the only substance that is endowed with life; plants and animals containing this substance in their body are, therefore, regarded as living. The living parts of a cell togather constitute what is called protoplast (the term coined by Hanstein in 1880). Protoplasm is the essential living material that comprises the different parts of the protoplast. The protoplasm has to perform manifold vital functions of a cell such as manufacture of food, nutrition, growth, respiration, reproduction, etc., and as such, for the sake of convenience and efficiency of work, it becomes differentiated into distinct living (protoplasmic) bodies, viz. (1) **cytoplasm,** (2) **nucleus** and in special cells (3) **plastids**, of which the first two are constant in all living cells. Such differentiated protoplasmic bodies have certain specialized functions. It must distinctly be noted that these living bodies are never formed afresh in the cells but always develop from pre-existing ones by divisions and that one kind of living body cannot give rise to another kind.

DIFFERENTIATED PARTS OF PROTOPLASM

1. Cytoplasm. The protoplasmic mass of a cell leaving out, the nucleus and the plastids is otherwise called cell-protoplasm or **cytoplasm**. When the cell is young the cytoplasm completely fills its cavity, i.e. the space between the cell-wall and the nucleus (Fig. 2A), but as the cell rapidly increases in size it cannot keep pace with the growth of the cell-wall Consequently, a number of small (non-protoplasmic) cavities appear in the cytoplasm; these are called **vacuoles** (*vacuus,* empty; Fig. 2B), As the cell further increases in size and matures all these small vacuoles fuse together into a large one which then occupies a large part of the

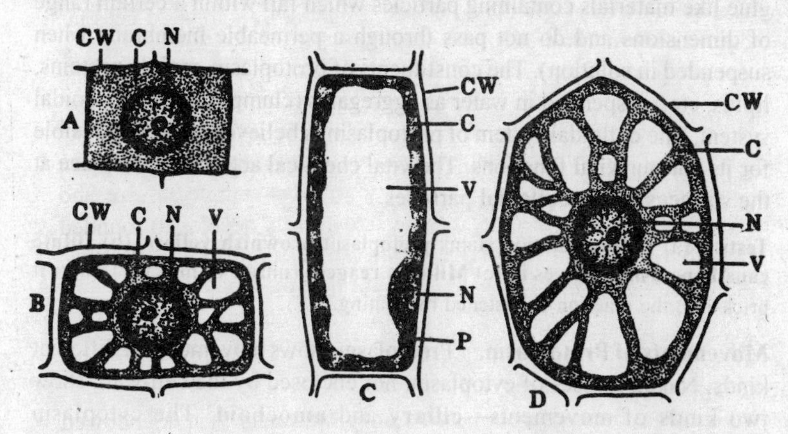

Figure 2. Plant cells. *A,* a very young cell; *B,* a growing cell with many small vacuoles; *C,* a mature cell with a large vacuole; *D,* a mature cell with many vacuoles, *CW,* cell-wall; *C,* cytoplasm; *N,* nucleus; *V,* vacuole; *P,* plastid (chloroplast).

cell, pushing the cytoplasm outwards as a thin lining layer against the cell-wall (Fig. 2C). In some cells comparatively small vacuoles persist and then the cytoplasm forms delicate strands around them (Fig. 2D). The vacuole is filled with a fluid called the **cell-sap** which is water, containing a large number of soluble chemical substances such as inorganic salts, organic acids, soluble carbohydrates, e.g. sugars, soluble proteins, amino-acid, and in certain cells mucilage, antho-cyanins, tannins, latex, alkaloids, etc., in varying proportions. The vacuole is thus a tiny reservoir of the cell from which the cytoplasm draws water and other materials according to its need. Referring to the cytoplasm again we find that it has three distinct parts: (1) its outer surface forms an extremely thin and delicate membrane called the **plasma membrane** or **ectoplasm**; (2) its middle part is granular and is called the **endoplasm**; its fluid portion, however, is called *hyaloplasm;* and (3) its innermost part surrounding the vacuole as a thin membrane is called the **vacuole membrane** or **tonoplasm**. The ectoplasm controls the entrance and exit of water and many chemical substances into and out of the cell, the tonoplasm does the same in respect of the vacuole, while the endoplasm performs the general functions of the cytoplasm.

Physical and Chemical Nature of Protoplasm. (See p. xix) It is also to be noted that protoplasm exists in colloidal condition (colloids are glue like materials containing particles which fall within a certain range of dimensions and do not pass through a permeable membrane when suspended in solution). The constituents of protoplasm, mainly proteins, lipids, etc., suspended in water as aggregates (clumps), form a colloidal system. The colloidal system of protoplasm is believed to be responsible for its various vital functions. The vital chemical activities take place at the surfaces of the colloidal particles.

Tests. (a) **Iodine** solution stains protoplasm **brownish yellow**. (b) **Dilute caustic potash** dissolves it. (c) **Millon's reagent** (nitrate of mercury) stains it brick-red; the reaction is hastened by heating.

Movements of Protoplasm. Protoplasm shows movements of different kinds. Naked masses of cytoplasm, not enclosed by the cell-wall, show two kinds of movements—**ciliary** and **amoeboid**. The cytoplasm enclosed by the cell-wall shows a streaming movement within it, which is spoken of as *cyclosis*. Cyclosis is of two kinds—**rotation** and **circulation**.

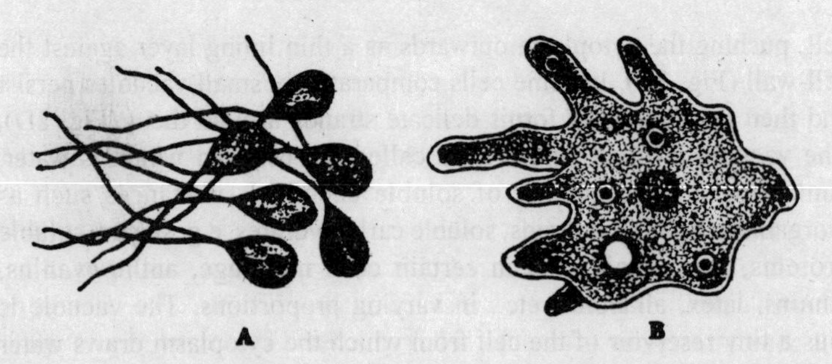

Figure 3. Movements of Protoplasm. *A,* ciliary movement; *B,* amoeboid movement.

(1) **Ciliary Movement** (Fig. 3A) is the swimming movement of free, minute, protoplasmic bodies provided with one or more tail-like structures, called **cilia**. By the vibration of these cilia such ciliary bodies move or swim freely and rapidly in water, e.g. zoospores of many algae and fungi, male gametes of mosses and ferns, etc.

(2) **Amoeboid movement** (Fig. 3B) is the *creeping* movement of naked masses of protoplasm (i.e. not enclosed by cell-wall). They move or creep by the protrusion of one or more parts of their body in the form of false feet or pseudopodia (*pseudos,* false; *podos,* foot), withdrawing the same at the next moment. e.g. some slime fungi, certain zoospores, etc.

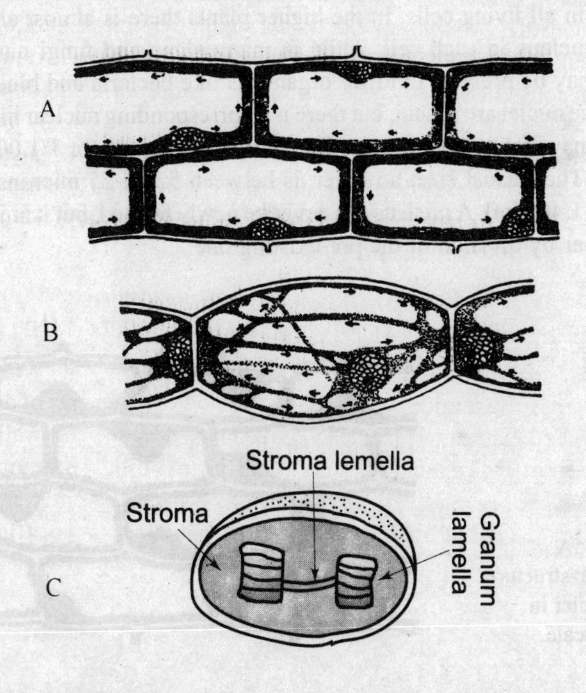

Figure 4. Movements of Protoplasm (*contd.*). *A,* rotation in the leaf of *Vallisneria*; *B,* circulation in the staminal hair of *Commelina obliqua*; *C,* a plastid.

(3) **Rotation** (Fig. 4A). When the protoplasm moves or streams within a cell alongside the cell-wall, clockwise or anti-clockwise, round a large central vacuole, the movement is called rotation. The direction of movement is constant so far as a particular cell is concerned. As the protoplasm rotates, it carries in its current the nucleus and the plastids. Rotation is distinctly seen in *Vallisneria, Hydrilla, Chara* and also in many other aquatic plants.

(4) **Circulation** (Fig. 4B). When the protoplasm moves or streams in different directions within a cell in the form of delicate strands round a number of small vacuoles, the movement is called circulation. Circulation is very distinctly seen in the staminal hairs of *Commelina obliqua,* spiderwort (*Tradescantia*), in the young shoot-hairs of gourd and in many other land plants.

2. Nucleus. The nucleus is a specialized protoplasmic body much denser than the cytoplasm, and is commonly spherical or oval in shape. It always lies embedded in the cytoplasm. The nucleus is universally present in all living cells. In the higher plants there is almost always a single nucleus in each cell, while in many algae and fungi numerous nuclei may be present. In lower organisms like bacteria and blue-green algae true nuclei are absent, but there is a corresponding nuclear material. Nuclei may vary widely in sizes from 1 to 500 microns (or 1/1,000 to 1/2 mm). Their usual size, however, is between 5 and 25 microns (or 1/200 and 1/40 mm). A nucleus can never be newly formed, but it multiplies in number by division of the pre-existing one.

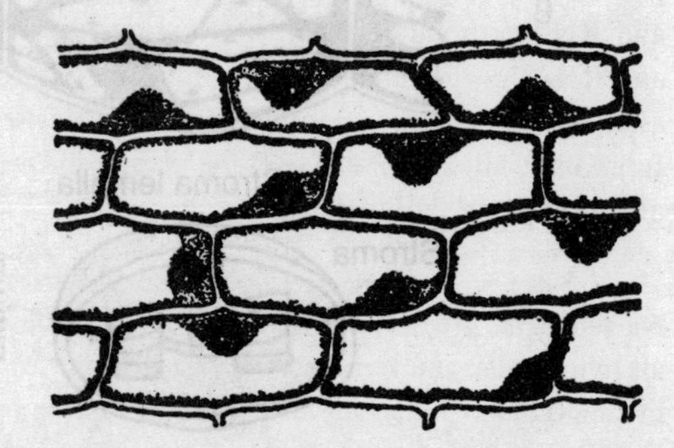

Figure 5A. Cellular structure and nuclei in onion scale.

Structure. Each nucleus (Fig. 5B) is surrounded by a thin, transparent membrane known as (1) **the nuclear membrane** which separates the nucleus from the surrounding cytoplasm. Within the membrane, completely filling up the space, there is a dense but clear mass of protoplasm known as (2) the **nuclear sap** or **nucleoplasm**. Suspended in the nucleoplasm there are numerous fine crooked threads, loosely connected here and there, forming a sort of network, called (3) the **nuclear reticulum** or **chromatin network**. The threads are made of a substance known as chromatin or nuclein which is strongly stainable. The chromatin or nuclein is a nucleoprotein (see below). One or more highly refractive, relatively large and usually spherical bodies may be seen in the nucleoplasm; these are known as (4) the **nucleoli** (sing. nucleolus).

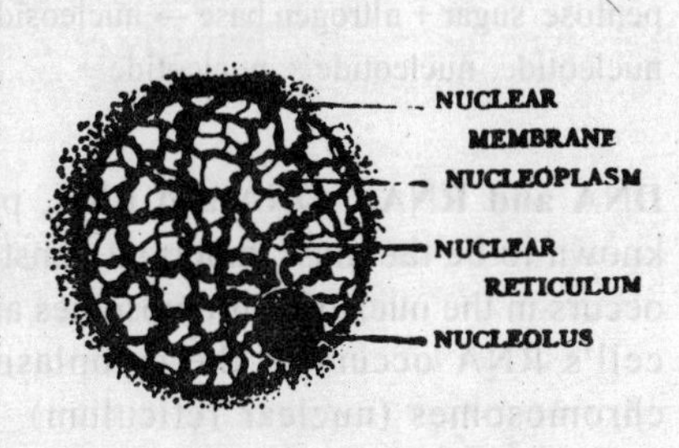

Figure 5B. Nuclear structure.

Chemical Composition. The chemical composition of the nucleus is more or less similar to that of the cytoplasm (see p. xx). The nucleus, however, is predominantly composed of nucleoprotein which is made of phosphorus-containing nucleic acids and certain specific types of proteins. All plant proteins contain carbon (C), hydrogen (H), oxygen (O), nitrogen (N) and sulphur (S), and sometimes phosphorus (P). The two important nucleic acids are DNA (deoxyribonucleic acid) and RNA (ribonucleic acid). DNA occurs in the nuclear reticulum, and RNA, in the nucleolus, chromosome and cytoplasm. The nucleus also contains some amount of lipids, particularly phospholipids. Inorganic salts, such as those of calcium, magnesium, iron and zinc are also present in the nucleus.

Nucleic acids are universally present in the nucleus and in the cytoplasm of all living cells, and form the chemical basis of life. They are very complex organic compounds made of phosphate, 5-carbon (pentose) sugar (ribose in RNA or deoxyribose in DNA) and nitrogen bases (see below). Nucleic acid molecules are very large (macromolecules), often larger than protein molecules, and consist of infinite numbers of *repeating* nucleotide units linked in any sequence into a long chain. They are thus high polymers of nucleotides. A **nucleotide** is a molecular unit (monomer) of a nucleic acid molecule (macro), and consists of three subunits: a phosphate, a pentose sugar and a nitrogen base. Phosphate and sugar alternate as links in the chain, while nitrogen base projects inward from sugar link. A nucleotide is formed when a phosphate group is added to a nucleoside. A **nucleoside** is a compound consisting of two subunits: a pentose sugar and a nitrogen base. There are two kinds of nucleic acids, viz. DNA and RNA. The latter occurs in three forms (see p. 290). An outline of nucleic acid formation may be given thus: pentose sugar + nitrogen base → nucleoside; nucleoside + phosphate group → nucleotide; nucleotide + nucleotide +..................... → nucleic acid.

DNA and RNA. DNA and RNA, particularly the former, are now known to be the most important constituents of the living cells. RNA occurs in the nucleoli, chromosomes and cytoplasm (about 90% of the cell's RNA occurs in the cytoplasm), while DNA occurs in the chromosomes (nuclear reticulum). They are related chemically, consisting of 5-carbon ribose sugar and phosphate but DNA has deoxyribose with one less oxygen atom in its molecule. Both are large molecules or macromolecules but DNA is a double-stranded molecule, while RNA is a single-stranded one. With the advance of knowledge, biologists now hold the view that all secrets of life are embodied in DNA, or in other words it is the chemical basis of life. DNA is the controlling centre of all the vital activities of the cell. DNA is the sole genetic (hereditary) material migrating intact from generation to generation through the reproductive units or gametes, and is responsible for the development of specific characters of plant. It also controls the biosynthetic processes of the cell including protein synthesis. RNA is a chemical messenger and plays a key role in the process of protein synthesis [For details see p. 290].

DNA Molecule (Fig. 6). As already mentioned, DNA molecule is very large and complex (macromolecule), forming the backbone of each chromosome. In 1953 Watson and Crick (co-winners of 1962 Nobel Prize) worked out a model of DNA molecule (Watson-Crick model). According to them DNA occurs as a double-stranded molecule, with the two strands profusely coiled and entwined about each other throughout their whole length. The structure is really like a ladder twisted spirally. Each spiral strand is made of groups of deoxyribose sugar alternating with groups of phosphate. Besides, there are infinite pairs of cross-links (like the rungs of a ladder) connecting the two strands. Each pair is made of two distinct nitrogenous bases—purines and pyrimidines. Altogether there are two purines—adenine and guanine, and two pyrimidines—thymine and cytosine. It is the rule that a specific purine always pairs with a specific pyrimidine, viz. T-A and G.C. In a DNA strand (Fig. 6B) there are altogether four types of nucleotides—PDT, PDA, PDG and PDC, evidently including four types of nucleosides—DT, DA, DG and DC, and four kinds of nitrogen bases—T, A, G and C. Although the bases of the two coiling strands combine in only

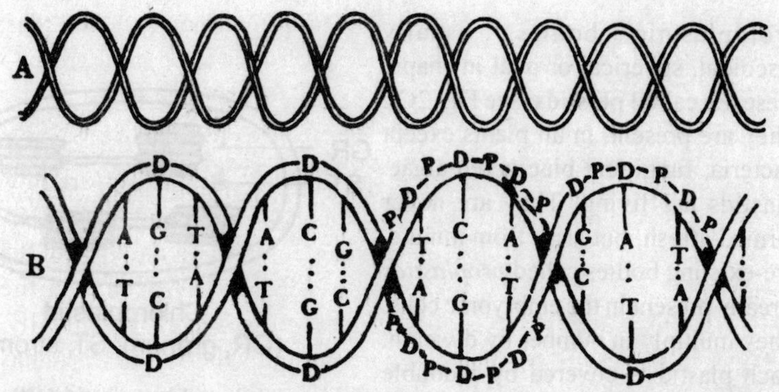

Figure 6 *A,* Watson-Crick model of DNA molecule; the two strands are twisted about each other; *B,* a portion of the same magnified; note the distribution of deoxyribose sugar (*D*), phosphate (*P*) and cross-links—thymine-adenine (*T–A*) and guanine-cytosine (*G–C*).

four specific pairs—T-A, A-T, G-C and C-G they (the pairs) occur in infinite sequences which enable the DNA molecule to coin an infinite number of *chemical codes* (or messages or informations) and transmit appropriate codes to the different parts of the cell for their proper functioning. RNA acts as the chemical messenger or carrier of such codes.

Functions. The nucleus and the protoplasm are together responsible for the life of a cell and the various vital functions performed by it. If they are separated both of them die. The nucleus, however, is regarded as the controlling centre of the vital activities of the cell in many ways. The specific functions performed by the nucleus are as follows:

(1) The nucleus takes a direct part in reproduction. Two reproductive nuclei called gametes (egg-cell and male gamete) fuse together to give rise to an oospore which grows into an embryo. Thus, nuclei are directly concerned in the process of reproduction.

(2) The nucleus takes the initiative in cell division, i.e. it is the nucleus that divides first and this is followed by the division of the cell. This is how the cells multiply in number and the plant body grows.

(3) The nucleus is regarded as the *bearer* of hereditary characters, i.e. it is through the media of two reproductive nuclei that the characteristics of parent plants are transmitted to the offspring. It is to be noted that it is the DNA of the nuclear reticulum that is the sole genetic (hereditary) material of the two reproductive nuclei.

3. Plastids. Besides the nucleus, the cytoplasm of certain cells which have to perform specialized functions, encloses many small specialized protoplasmic bodies, usually discoidal, spherical or oval in shape; these are called **plastids** (see Fig. 2C). They are present, in all plants except bacteria, fungi and blue-green algae. Plastids are living. They are never formed afresh, but arise from minute, pre-existing bodies called *proplastids* already present in the embryonic cells. They multiply in number by division. Each plastid is covered by a double membrane (see Fig. 6C) A proteinaceous material constitutes the colourless ground substance or matrix of the plastid. This ground substance is called stroma.The stroma contains protein grains, starch grains, ribosomes, enzymes, RNA and DNA, etc. Lying in the stroma, there are numerous granules called grana. Each granum consists of flat, disc shaped, small stacks of thick pigmented bodies (like piles of coins) called lamellae (granum lamellae). Lamellae are thin in stroma region (stroma lamellae) and connect the neighbouring grana. According to their colour the plastids are of three types, viz. **leucoplasts**, **chloroplasts** and **chromoplasts**. One form of plastids can change into another; as for example, leucoplasts change into chloroplasts when the former are exposed to light for a prolonged period; similarly, chloroplasts change into leucoplasts in the continued absence of light; similar changes may take place in chromoplasts. In the young tomato fruit the leucoplasts gradually change into chloroplasts which finally turn into chromoplasts as the fruit ripens.

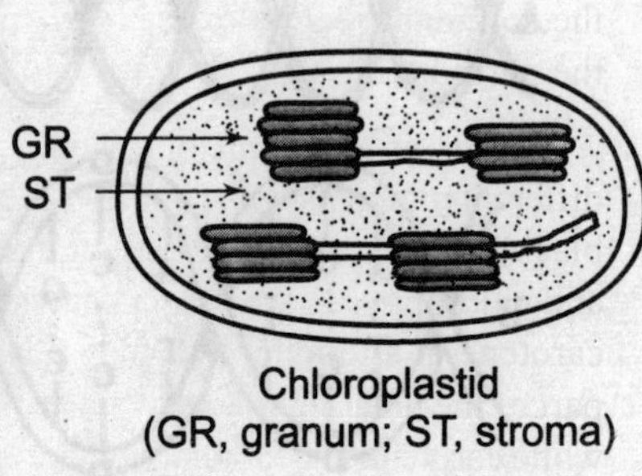

Figure 6C. Chloroplastid (GR, granum; ST, stroma).

(1) **Leucoplasts** (*leucos,* white). These are colourless plastids. Leucoplasts occur most commonly in the storage cells of roots and underground stems; they are also found in other parts not exposed to light. Their function is to convert sugar into starch, an insoluble food substance, for the purpose of storage.

(2) **Chloroplasts** (*chloros,* green). These are green plastids, their colour being due to the presence of a green *pigment* (colouring matter), called **chlorophyll**; sometimes the green colour may be masked by other colours. Lamellae lying in the stroma of a chloroplast contain chlorophyll. Chloroplasts are only found in parts exposed to light and occur abundantly in green leaves. They absorb carbon dioxide from the air, and energy from the sunlight; utilize this energy in manufacturing

sugar and starch from this carbon dioxide, and the water absorbed from the soil; and liberate oxygen (by splitting the water) which escapes to the surrounding air.

Chlorophyll is not one simple substance, but a mixture of four different pigments, viz. chlorophyll *a* (blue-black), chlorophyll *b* (green-black), carotene (orange-red) and xanthophyll (yellow). Chlorophyll *a* and chlorophyll *b* are associated with each other in the chloroplast, but carotene and xanthophyll may also occur without chloroplast in any part of the plant. In old brown leaves chlorophyll becomes decomposed, while carotene and xanthophyll are left intact. Chlorophyll is not soluble in water. It forms about 8% of the dry weight of the chloroplast, while carotene and xanthophyll form about 2%.

Functions. It is definitely known that chlorophyll absorbs energy from the sunlight. It may also help in the chemical process involved in the manufacture of food by the chloroplasts. Chlorophyll, however, does not undergo any chemical change in this process, i.e. it acts as a catalytic agent only.

Extraction of Chlorophyll. Chlorophyll as a whole can be easily extracted from the leaf by boiling it for a minute or so and then dipping it into methylated spirit for some time. When all the chlorophyll is extracted the leaf becomes colourless. The chlorophyll solution examined through transmitted light appears deep green in colour but by reflected light it appears blood-red in colour. This is the physical property of chlorophyll, called *fluorescence*. Then a small quantity of benzene is added to the chlorophyll extract and the whole solution briskly shaken. It is then allowed, to settle for a few minutes. Benzene floats on the top (green solution) carrying chlorophyll, while alcohol settles at the bottom (yellow solution) retaining carotene and xanthophyll.

Chemical composition of chlorophyll

Chlorophyll *a* — $C_{55}H_{72}O_5N_4Mg$ Carotene — $C_{40}H_{56}$
Chlorophyll *b* — $C_{55}H_{70}O_6N_4Mg$ Xanthophyll — $C_{40}H_{56}O_2$

(3) **Chromoplasts** (*chroma,* colour). These are variously coloured plastids—yellow, orange and red. They are mostly, present in the petals of flowers and in fruits, and the colouring matters (pigments) associated with them are **xanthophyll** (yellow) and **carotene** (orange-red) which occur in different proportions. Chromoplasts occurring in the petals of flowers make them showy and attractive to invite insects for the purpose of pollination. (Most other colours of flowers such as violet, purple, blue, brown and often red are due to the presence of a group of colouring matters known as **anthocyanins** which remain dissolved in the cell-sap.)

Anthocyanins. Most colours of flowers such as violet, purple, blue, brown and often red are due to the presence of a group of colouring matters known as anthocyanins which remain *dissolved in the cell-sap*. They are common in the young red leaves of many plants and also in the variegated leaves of garden crotons. They possibly serve as a screen to the chloroplasts. Occurring in flowers, they of course serve to attract insects for pollination. They are also found in coloured roots, as in carrot, and in coloured stems, as in balsam.

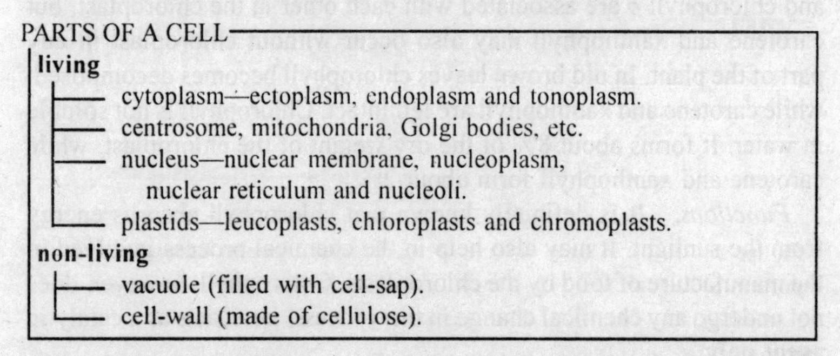

PARTS OF A CELL
living
cytoplasm—ectoplasm, endoplasm and tonoplasm.
centrosome, mitochondria, Golgi bodies, etc.
nucleus—nuclear membrane, nucleoplasm,
nuclear reticulum and nucleoli.
plastids—leucoplasts, chloroplasts and chromoplasts.
non-living
vacuole (filled with cell-sap).
cell-wall (made of cellulose).

Other Cytoplasmic Bodies. Although some of these bodies are seen under a compound microscope, indistinctly though, being very minute in size (Fig. 7A), their true nature (Fig. 7B) has been recently revealed by the electron microscope (see p. 169), and their functions known.

(1) **Centrosome** (Fig. 7A) is a minute body found in animal cells and in those of many lower plants like algae and fungi. It is not found in seed plants. It occurs close to the nucleus and has usually two central bodies called *centrioles*. During nuclear division they pass on to the opposite ends of the cell and organize the nuclear spindle.

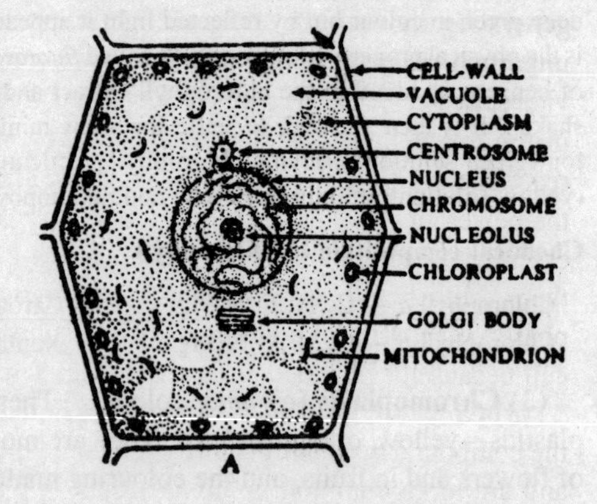

Figure 7. A cell as seen under compound microscope.

(2) **Mitochondria** (Fig. 7A & C) are minute bodies occurring often in very large numbers, in the form of short rods or long filaments or as somewhat spherical or oval bodies in the cytoplasm of all plant and

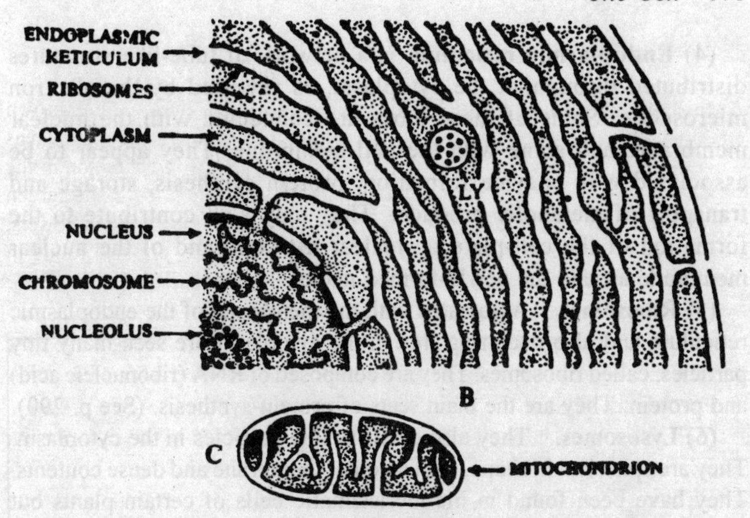

Figure 7. B–C. Parts of a cell as seen under electron microscope; *B*, a portion of the cell (*LY*, lysosome); *C*, a mitochondrion.

animal cells with the exception of bacteria and blue-green algae. They are specially abundant in the young metabolic tissue. Electron microscope has recently revealed their complicated structure: (a) an outer membrane, (b) an inner membrane thrown into folds, and (c) a granular matrix. Mitochondria multiply by division, and are carried down to the next generation through the reproductive units. They are now regarded as important components of a living cell, being centres of *energy* for its vital activities. ATP, an energy rich phosphate compound, is mostly formed in the mitochondria, and is used to activate most of the biochemical reactions in respiration, protein-synthesis, etc. Mitochondria are in fact the powerhouses of the cell, generating *energy*. Besides, all the enzymes essential to respiration, as also several other enzymes, are synthesized in them. Mitochondria are the seat of respiration, particularly its aerobic phase. They are also connected with many other functions of the living cell because of ATP and several enzymes formed in them.

(3) **Golgi bodies** (Fig. 7A) appear as minute net-like structures under the compound microscope. They are found only in certain types of cells, and are more common in animals than in plants. In the gland cells of animals they are associated with secretions of certain enzymes, hormones, etc. In plants their significance is not clear. They may secrete cellulose membrane during nuclear division. Electron microscope has revealed a complicated structure of the Golgi body.

(4) **Endoplasmic reticulum** is a network of tube-like structures distributed throughout the cytoplasm, as revealed by the electron microscope. Some of these tubes are connected with the nuclear membrane, and some with the cell-membrane. They appear to be associated with enzyme formation, protein synthesis, storage and transport of metabolic products. They may also contribute to the formation of the cell-plate in nuclear division, and of the nuclear membrane around the newly-formed nuclei.

(5) **Ribosomes.** Associated with the membrane of the endoplasmic reticulum and also occurring free in the cytoplasm are seen many tiny particles; called ribosomes. They are composed of RNA (ribonucleic acid) and protein. They are the main seats of protein-synthesis. (See p. 290).

(6) **Lysosomes.** They also occur as tiny particles in the cytoplasm. They are spherical in shape, with an outer membrane and dense contents. They have been found in the meristematic cells of certain plants but they are more common in animals. They are rich in several enzymes and are associated with intra-cellular digestion.

Electron microscope was invented by two German scientists, Knoll and Ruska, in 1932. Since 1950 many new interesting aspects of cells began to be revealed by this instrument, magnifying objects as high as 200,000 diameters or even more. Many of the solid grains and rods, as seen under a compound microscope, have now proved to be complex structures under an electron microscope, and also many new discoveries made in the cells. All this has led the biologists to form a new concept of cell-organization.

Formation and Structure of the Cell-wall. All plant cells are bound by a non-living, thick or thin, elastic or semi-rigid wall called the **cell-wall**. It is formed by the protoplast to maintain its form and to protect it from external injury. Besides, the cell-walls form the skeleton of the plant body, and are responsible for, its strength, rigidity and flexibility. The cell-wall is a laminated structure, i.e. it consists of layers laid down by the protoplast one after another. As a whole, it is made of (i) a middle lamella, (ii) a primary wall on each side of the middle lamella, and in many cells, (iii) a secondary wall on each side of the primary wall in their sequence of development. The **middle lamella** is the common middle layer of the cell-wall connecting two adjoining cells, and it is always formed first. It is composed of calcium pectate (calcium salt of pectic acid), and acts as a cementing material firmly holding the adjoining cells together. As the cells enlarge, a thin wall is deposited on each surface of the middle lamella by the protoplast of each cell. This wall is called the **primary wall**, as in most parenchyma, collenchyma,

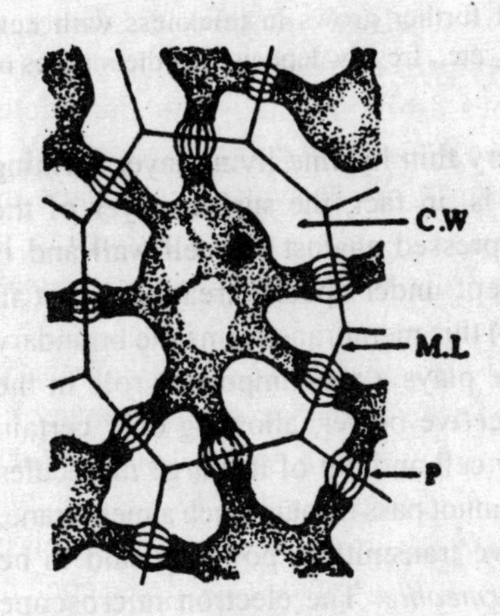

Figure 8. Cells of the endosperm of date seed. *C.W.*, cell-wall (reserve cellulose); *M.L.*, middle lamella; *P*, protoplasmicthreads (plasmodesmata).

cambium, etc., and it consists of cellulose, hemicellulose and pectose in varying proportions. The primary wall is elastic in nature so that it can keep pace with the growth of the cell. In many other cell as they mature, the wall may thicken by the addition of new layers laid down by the protoplast on each surface of the primary wall. This thickened, later-formed wall is called the **secondary wall**, and it is composed of almost pure cellulose. This wall is tough and has a very high tensile strength. The cell-wall as a whole now comprises (i) a middle lamella, (ii) two primary walls, and (iii) two secondary walls. In special cases, as in tracheids, vessels, wood fibres, bast fibres, and stone cells the secondary wall becomes further thickened by the addition of new layers and deposition of new materials, and this thickening may take special patterns (see pp. 172; Fig. 9). It is further seen that the cytoplasm of one cell is connected with that of the adjoining one by fine cytoplasmic strands passing through extremely minute pits in the cell-wall. These cytoplasmic strands are called **plasmodesmata** (sing. plasmodesma; Fig. 8). They are responsible for the transmission of stimuli from cell to cell, and also of nutrient materials, particularly in storage tissues.

The **electron microscope** reveals a complex structure of the cell-wall. The wall develops in successive layers laid down by the protoplast one after another, and each consists of an interwoven network of extremely fine cellulose strands (*microfibrils* visible only under the electron microscope) with pectic compounds deposited in the meshes of the network. The primary wall is made of one layer of microfibrils deposited transversely or somewhat obliquely to the long axis of the cell, forming a loose network. This arrangement helps elongation of the cell-wall. The secondary wall, when formed, commonly consists of three layers, sometimes more, on each side of the primary wall, in which the microfibrils are laid down in different directions in the successive layers, forming a somewhat compact network in each case. This arrangement adds to the greater strength of the cell-wall. At this stage nine layers may be counted in the cell-wall as a whole. In special cases the cell-wall further grows in thickness with new materials such as lignin, cutin, suberin, etc., freshly deposited in the meshes of the network.

Plasma Membrane. This is a very thin hyaline living layer forming the boundary of the cytoplasm. It is, in fact, the surface layer of the cytoplasm. In plant cells it lies adpressed against the cell-wall and is hardly distinguishable from it except under special treatment, but in animal-cells (cell-wall being absent) this membrane forms the boundary of each cell. The plasma membrane plays a very important role in the physiology of the cell. It has a selective power, allowing only certain materials to pass through it into the cell and out of it. Large molecules of proteins, fats and carbohydrates cannot pass through such a membrane. Such a membrane having a selective transmitting power is said to be *semipermeable* or *differentially permeable*. The electron microscope reveals several minute pores in the membrane. These pores may help the diffusion process of materials that have to enter the cell or leave it.

Secondary Thickening of the Cell-wall. The secondary thickening of the cell-wall takes place in vessels (Fig. 9) and tracheids; after they have grown considerably and attained their full dimension their walls begin to thicken. The thickening in them is due to the deposit of a hard and chemically complex substance, called **lignin**, in the interior of the original cell-wall. This deposit of lignin takes the following patterns. All lignified elements are dead.

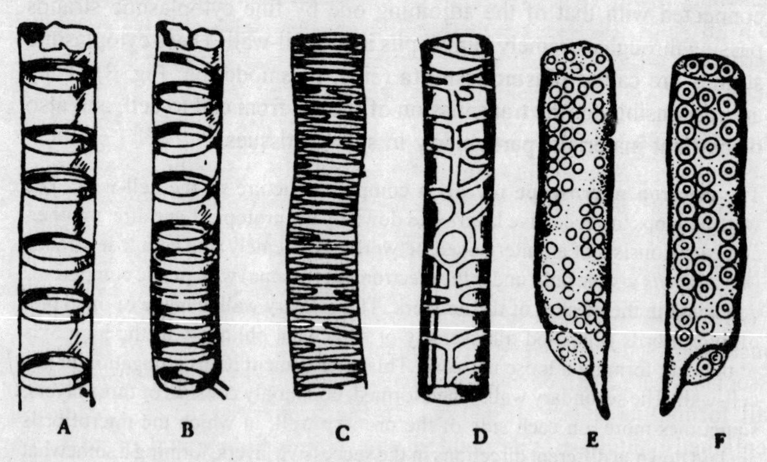

Figure 9. Thickening of the Cell-wall. *A*, annular; *B*, spiral; *C*, scalariform; *D*, reticulate; *E*, pitted (with simple pits); *F*, pitted (with bordered pits).

(A) **Annular or ring-like**, when the deposit of lignin is in the form of rings. (B) **Spiral**, when the thickening takes the form of a spiral band. (C) **Scalariform** or **ladder-like**, when the thickening matter or lignin is deposited transversely in the form of rods or rungs of a ladder, and hence the name scalariform or ladder-like. (D) **Reticulate** or **nettled**, when the thickening takes the form of a network. (E–F) **Pitted**, when the whole inner surface of the cell-wall is more or less uniformly thickened, leaving here and there some small unthickened areas or cavities. These unthickened areas are called **pits**, and are of two kinds, viz. **simple pits** (E) and **bordered pits** (F). Pits are formed in pairs lying against each other on the opposite sides of the wall. When the pit is uniformly wide throughout its whole depth, it is a simple pit (Figs. 10–11); and when the pit is unequally wide, being broader towards the wall and narrower towards the cavity of the cell, more or less like a funnel without the stem, it is a bordered pit (Fig. 12). In the bordered pit the adjoining thickening matter of the wall grows inwards and arches over the pit from all sides forming an overhanging border and hence the name 'bordered' pit. The portion of the middle lamella crossing the pits becomes thickened and is known as the **torus** (Fig. 12B–C).

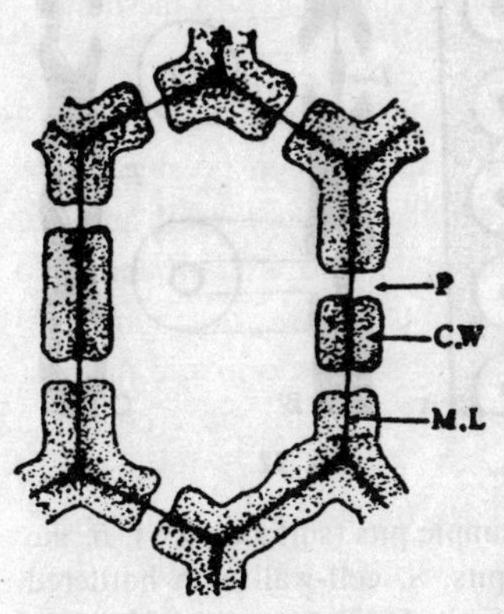

Figure 10. Simple Pits. A cell in section showing simple pits in its walls; *P,* pit: *C.W.,* cell-wall; *M.L.,* middle lamella.

Chemical Nature of the Cell Wall. The cell-wall consists of a variety of chemical substances, of which cellulose is very conspicuous; but as the cell grows older, it undergoes chemical changes and a variety of new substances are formed. Certain mineral matters are also often introduced into the cell-wall.

(1) **Cellulose.** Cellulose, an insoluble carbohydrate, is the chief constituent of the cell-walls of all plants with the exception of fungi. Associated with it are other compounds deposited at different stages of wall formation (see p. 170–171). In the primary wall, as already mentioned, cellulose is associated with hemicellulose and pectose, making the wall soft, elastic and cohesive. In the secondary wall formed later, almost pure cellulose occurs making the wall stiff but flexible.

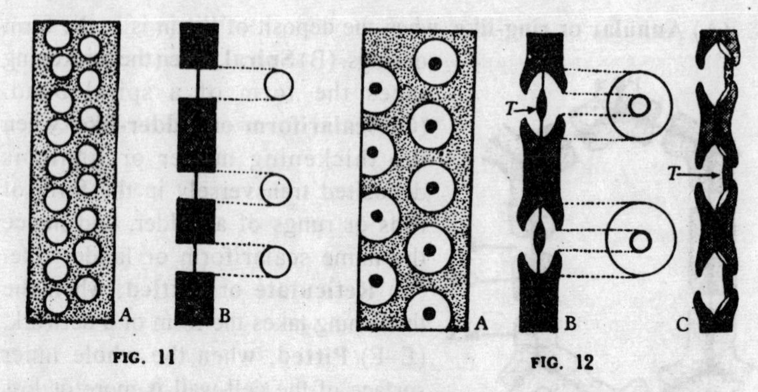

Figure 11. Simple pits. *A*, cell-wall with simple pits (surface view); *B*, the same (sectional view). Fig. 12. Bordered pits, *A*, cell-wall with bordered pits (surface view); *B*, the same (sectional view); *C*, the torus pushed to one side blocking the pit. *T*, torus.

Later still, cellulose is associated with lignin, suberin, cutin, etc. Cellulose is a soft, elastic and transparent substance, and is readily permeable to water and soluble food materials. Seed fibres such as cotton and kapok are made of pure cellulose, while bast fibres and woody tissues are made of a mixture of cellulose and lignin (ligno-cellulose). Cellulose elements are usually thin-walled and living. Cellulose is represented by the formula $(C_6H_{10}O_5)n$. It is directly transformed from glucose in the presence of protoplasm, and its molecules are long chains of glucose units. Many bacteria utilize cellulose as food. They secrete the enzyme *cellulase* to hydrolyse it to glucose for assimilation. It forms a major part of food for the herbivorous animals. Bacteria in the intestines of such animals render necessary help in the digestion of cellulose. Cellulose, however, cannot be digested by human beings. Economically, articles like paper, cellophane, gun-cotton, celluloid, rayon (artificial silk), lacquer, etc., which have worldwide uses, are prepared from cellulose.

(2) **Lignin.** Lignin occurs associated with cellulose in the secondary walls of woody and fibrous tissues, and is responsible for the thickening and strengthening of such tissues. Lignin is deposited in the meshes of the network formed by long chains of cellulose molecules. It is a hard and chemically complex substance; possibly, it is a mixture of several organic compounds. Lignified cells are usually thick-walled and always dead. Although hard and thick, lignin is permeable to water like cellulose. Water-conducting vessels and tracheids, wood fibres and bast

fibres are common lignified structures. Lignified tissues give mechanical strength to the plant body.

(3) **Cutin.** Cutin is a waxy substance. Associated with cellulose and often with pectose it forms a definite layer, sometimes of considerable thickness, called the **cuticle** on the skin (outer surface of the epidermal layer) of the stem, leaf and fruit. Cutin makes the wall impermeable or slightly permeable to water. Its function, therefore, is to prevent or check evaporation of water from the exposed surfaces of the plant body.

(4) **Suberin.** Cell-walls of certain tissues may be charged with another waxy substance, called suberin. It is allied to cutin in many respects but it occurs in the walls of cork cells. Being waxy in nature it makes the cell-wall almost impervious to water, and therefore, like cutin, it also prevents or checks evaporation of water. The bark of cork oak (*Quercus suber*) on the Mediterranean coasts is the source of bottle cork used for this purpose.

(5) **Mucilage.** Mucilage is a slimy substance widely distributed in plants in their different parts. Chemically, it is a complex carbohydrate. Its property is that it absorbs water freely and retains it. When dry, it is very hard and horny, but when wet it forms a viscous mass. Mucilage is abundant in the leaves of Indian aloe (*Aloe*), branches of Indian spinach (*Basella*), flowers of China rose (*Hibiscus*), fruits of lady's finger (*Abelmoschus*), and seeds of linseed (*Linum*), flea seed (*Plantago;* B. ISOBGUL; H. ISOBGUL), etc.

Micro-chemical Tests of the Cell-wall

Reagents	*Cellulose*	*Lignin*	*Cutin & Suberin*	*Mucilage*
1. Iodine solution	pale-yellow	deep yellow	deep yellow	—
2. Chlor-zinc-iodine	blue or violet	yellow	yellowish brown	—
3. Iodine solution + sulphuric acid	blue	brownish	deep brown	violet
4. Aniline sulphate (acid)	—	bright yellow	—	—
5. Phloroglucin (acid)	—	violet red	—	—
6. Caustic potash sol.	—	—	yellow to brown	—
7. Potash + chlor-zinc-iodine	—	—	violet	—
8. Sudan IV	—	—	red	—
9. Methylene blue	—	—		deep blue

Various mineral crystals may also be introduced into the cell-wall; these are usually crystals of silica and calcium oxalate and calcium carbonate (see. Fig. I7–19). In the majority of fungi, and sometimes also in algae, the cell-walls are made of a substance called **chitin**, a complex carbohydrate allied to cellulose to some extent. Chitin, however, is peculiar to animals, particularly insects.

NON-LIVING CELL CONTENTS

A variety of chemical compounds are formed in the plant body and stored up in certain cells. There are three main groups of these, viz. (I) **reserve materials**, (II) **secretory products**, and (III) **waste products**.

I. *RESERVE MATERIALS*

These are substances manufactured by the protoplasm and stored up by it in particular cells, and later utilized by it as *food* for its nutrition and conservation of energy for work. Many of them occur in solution in the cell-sap; others are deposited in solid forms in the cytoplasm. There are three main groups of them. viz. (1) **carbohydrates**, (2) **nitrogenous materials**, and (3) **fats and oils**.

1. **Carbohydrates.** All carbohydrates contain carbon, hydrogen and oxygen. Of these, hydrogen and oxygen occur in the same proportion as they do in water, i.e. H_2O. When these substances are heated the water escapes and the carbon is left behind as a black mass. Some carbohydrates are soluble in water, e.g. sugars and inulin, while others are insoluble, e.g. starch and glycogen.

(a) **Sugars.** There are various kinds of sugars formed in plants. Of these, **grape-sugar** or **glucose** is chiefly found in grapes, and **cane-sugar** or **sucrose** in sugarcanes and beets. Grape-sugar is the simplest of all carbohydrates and is formed in the leaf by choloroplasts *in the presence of sunlight.* Other forms of carbohydrates are derived from it. Commonly, all the glucose formed in the leaf becomes converted into starch, an insoluble carbohydrate. At night this starch is reconverted into sugar which then travels to the storage organs where it is again converted into starch by leucoplasts. The chemical formula of grape-sugar is $C_6H_{12}O_6$ and that of cane-sugar $C_{12}H_{22}O_{11}$. The glucose content of grapes is 12–15% or more; sucrose content of sugarcanes 10–15% and of beet roots 10–20%.

Test for Glucose. Add **Fehling's solution** or an alkaline solution of **copper sulphate** to it and boil, a yellowish red precipitate of cuprous oxide is formed.

Test for Sucrose. Boil sucrose solution with 1 or 2 drops of **sulphuric acid**, and then try the test for glucose.

(b) **Inulin** (Fig. 13). Inulin is a soluble carbohydrate, and occurs in solution in the cell-sap. When required for nutrition it is converted into a form of sugar (fruit-sugar or fructose). Inulin is present in the tuberous roots of *Dahlia* and some other plants.

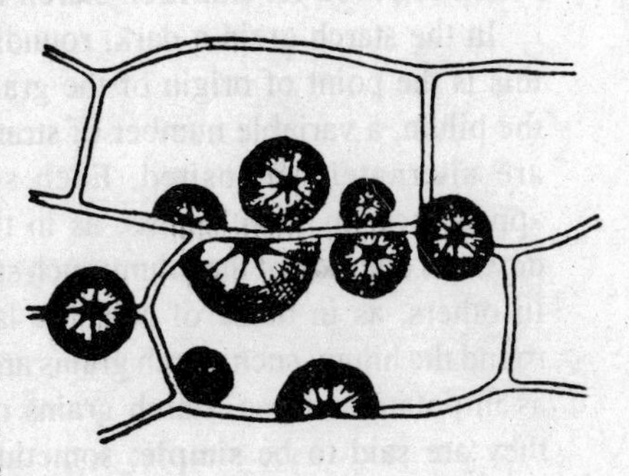

Figure 13. Inulin crystals in the tuberous root of *Dahlia*.

When pieces of *Dahlia* roots are kept in alcohol or glycerine for 6 or 7 days, preferably more, inulin becomes precipitated in the form of spherical crystals (really aggregates of crystals). Under the microscope fully-formed inulin crystals are seen to be star or wheel-shaped, and the half-formed ones more or less fan-shaped. These crystals are deposited mostly across the cell-walls, and occasionally only in the cell-cavity. Sometimes these crystals are so large that they extend through many cells. Inulin has the same chemical composition as many starch, viz. $(C_6H_{10}O_5)n$.

(c) **Starch** (Figs. 14–15). This is an insoluble carbohydrate and occurs in the form of minute grains. Starch grains occur as a reserve food in all green plants in their storage organs. Rice, wheat, maize and

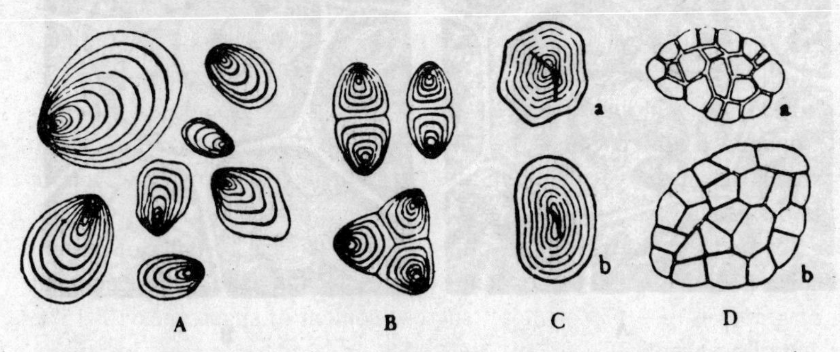

Figure 14. Starch Grains. *A*, simple eccentric grains in potato; *B*, compound grains in the same; *C, a,* simple concentric grain in maize; *b,* ditto in pea; *D, a,* compound grain in rice; *b,* ditto in oat.

millet which constitute the staple food of mankind are specially rich in starch. Starch grains may be *oval, spherical, round* and *flat,* or *polygonal.* They also vary very much in size, the largest known being about 100 microns (or 1/10 mm.) in length, as in the rhizome of *Canna,* and the smallest about 5 microns (or 1/200 mm.) in length, as in rice. In potato they are of varying sizes. Starch is always derived from sugar, either in the leaf by the chloroplasts or in the storage organ by the leucoplasts. When required for nutrition starch is converted into sugar.

In the starch grain a dark, roundish or elongated spot may be seen; this is the point of origin of the grain and is known as **hilum**. Around the hilum, a variable number of strata (i.e. layers) of different densities are alternately deposited. Each starch grain has thus a *stratified* appearance. In some grains, as in those of potato, the layers are laid down on one side of the hilum; such starch grains are said to be **eccentric**. In others, as in those of pea, the layers are deposited concentrically round the hilum; such starch grains are said to be **concentric**. Commonly, as in potato and pea, starch grains occur singly with one hilum, when they are said to be **simple**; sometimes, however, two or more grains occur together in a solid group with as many hila as there are grains in

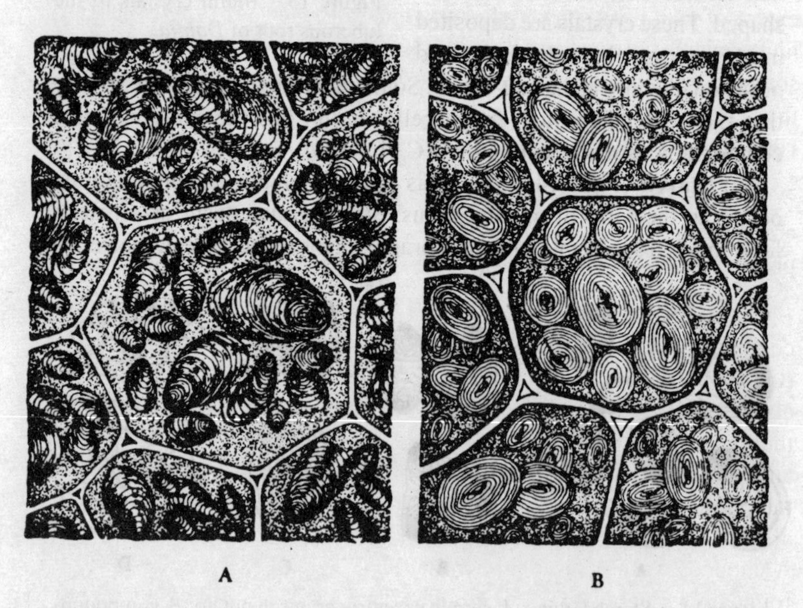

Figure 15 Starch Grains (*contd.*) *A,* eccentric grains in a potato tuber; *B,* concentric grains (and small granules of protein) in pea cotyledon.

it; this group then is said to form a **compound** grain, as in rice and oat (Fig. 14D). A few compound grains are also sometimes formed in potato (Fig. 14B). Starch has the same chemical composition as cellulose and inulin, viz. $(C_6H_{10}O_5)n$. It is insoluble in water and alcohol. Rice contains 70–80%, of starch; wheat about 70%; maize about 68%; barley 60–65%; arrowroot 20–30%; and potato 20%.

Test. It turns **blue** to **black** when treated with **iodine solution**, the density of the colour depending on the strength of the reagent.

Uses of Starch. Apart from its use as food for both plants and animals including human beings starch has a variety of industrial uses. When boiled in water, it forms a thin solution or paste which is extensively used in textile industry, laundry, paper industry, China clay industry, etc., as a sizing and cementing material. Starch is also widely used in the preparation of toilet powders, commercial glucose (by hydrolysis), and industrial alcohol (by fermentation) on a large scale. Sources of commercial starch mainly are potato, maize, tapioca, rice, wheat, sago-palm and arrowroot.

(d) **Glycogen.** This is a very common form of carbohydrate occurring in fungi. In yeast (see Fig. 22C), a unicellular fungus, it occurs to the extent of about 30% of the dry weight of the plant. It is not found in higher plants but is widely distributed among animals and is, therefore, sometimes called 'animal starch'. It occurs in the form of granules in the cytoplasm of the cell. Glycogen dissolves in hot water. It is coloured **reddish brown** with iodine solution. Its chemical formula is $(C_6H_{10}O_5)n$.

2. **Nitrogenous Material.** The nitrogenous reserve materials that are stored up in plants for their use as food are the various kinds of proteins and amino-compounds (amines and amino-acids).

(a) **Proteins.** Proteins are very complex, organic, nitrogenous compounds, essentially containing carbon (C), hydrogen (H), oxygen (O) and nitrogen (N). All plant proteins contain sulphur (S) and some also contain phosphorus (P). Proteins are very important as food, being the source of nitrogen, and they also form an integral part of protoplasm and nucleus, particularly as *nucleoprotein*. There are various kinds of them found in the plant body, particularly in their storage organs. They are mostly insoluble in water but are soluble in strong acids and alkalis. A common form of insoluble or sparingly soluble protein, abundantly found in the endosperm of the castor seed, is the **aleurone grain** (Fig. 16). Each aleurone grain is a solid, ovate or rounded body, and encloses in it a large crystal-like body, known as the **crystalloid**, and a

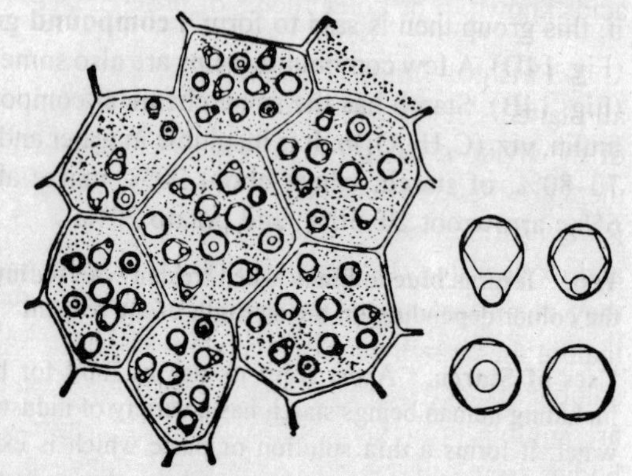

Figure 16. Aleurone grains in the endosperm of castor seed; *right,* a few grains magnified. Note the crystalloid and the globoid in them.

small rounded mineral body, called the **globoid**. The crystalloid is protein in nature, while the globoid is a double phosphate of calcium and magnesium. The occurrence of crystalloid and globoid is not always constant in the aleurone grain. Aleurone grains vary in size. When they occur with starch they are very small, as in pea; but in oily seeds they are very much larger, as in castor.

Fatty seeds usually contain a higher percentage of proteins than starchy seeds, e.g. rice contains only 7% of proteins, wheat 12%, while sunflower seeds contain proteins as high as 30%. Starchy seeds of leguminous plants, however, contain as high a percentage of proteins as fatty seeds, e.g. in the pulses there is an average of 25% of proteins; in soya-bean (*Glycine max*) protein contents are 35% or more.

Average percentage composition of proteins may be given thus: carbon—50–54%: hydrogen—about 7%: oxygen—20–25%; nitrogen—16–18%; sulphur—0.4%; and phosphorus—0.4%.

Test for Proteins. (1) Proteins are coloured yellowish brown with **iodine solution**. (2) **Xanthoproteic reaction**—add some strong nitric acid and a white precipitate is formed; on boiling it turns yellow. After cooling add a little strong ammonia and the yellow colour changes to orange.

(b) **Amino-compounds.** Amino-acids and amines are the simplest forms of all nitrogenous food materials, and occur in solution in the cell-sap. They are found abundantly in the growing regions of plants, less frequently in storage tissues. When translocation is necessary, proteins become converted into amines and amino-acids. They travel to the growing regions where the protoplasm is very active, and are directly assimilated by it. They are also the initial stages in the formation of proteins. They contain carbon, hydrogen, oxygen and nitrogen, and sometimes also sulphur (as in *cystine*). There are 20 different amino-acids known to be constituents of proteins.

3. **Fats and Oils.** Fats and oils occur to a greater or less extent in all plants. They occur in the form of minute globules in the cytoplasm of the living cells. In the 'flowering' plants often special deposits of them are found in seeds and fruits. But in starchy seeds and fruits there is very little of them. Fats and oils are composed of carbon, hydrogen and oxygen, but the latter two do not occur in the same proportion as they do in water—the proportion of oxygen being always much less than in the carbohydrates. They are insoluble in water, but very readily soluble in ether, petroleum and chloroform. Comparatively, few of them are soluble in alcohol, e.g. castor oil. Fats are synthesized in living bodies from fatty acids and glycerine under the action of the enzyme *lipase*. Both these products, viz. fatty acids and glycerine, are derived from carbohydrates (sugar and starch) during respiration. Fats and oils form an important reserve food with a large amount of *energy* stored in them. Their energy value is more than double that of the carbohydrates. When fats are decomposed the energy stored in them is liberated and made use of by the protoplasm for its manifold activities. Digestion of fats into fatty acids and glycerine is also brought about by the enzyme *lipase*. Fats that are liquid at ordinary temperature are known as 'oils'. In plants, fats are usually present in the form of oils.

A large number of them are used for food, for manufacture of soap, and oil-paints, for illumination, lubrication, etc., and are, therefore, of considerable economic importance, e.g. coconut oil, olive oil, sesame or gingerly oil, castor oil; groundnut oil, palm oil, sunflower oil, linseed oil, mustard oil, cotton seed oil, etc.

Tests for Fats and Oils. (1) The dry endosperm of castor or coconut burns when held over a flame. (2) It leaves a permanent greasy (oily) mark on a paper when rubbed on it. (3) Alcoholic solution of Sudan Red stains fats and oils red.

II. *SECRETORY PRODUCTS*

These include various products secreted by the protoplasm within a living cell or outside it, often in minute quantities, to perform some special functions as follows: (1) **Enzymes.** These are certain kinds of proteins acting as organic (biological) catalysts. Commonly they are known as *digestive agents,* and meant to bring about digestion of food, and also thousands of other biochemical reactions, without themselves undergoing any change. (2) **Colouring Matters.** Of the various colouring matters chlorophyll and anthocyanins are specially important. (3) **Nectar.** This is secreted by many flowers in special cells or glands to attract insects for pollination. (4) **Vitamins.** These are certain organic compounds formed in plants as adjuncts to food and used in minute quantities for maintenance of normal health. (5) **Hormones.** These are certain organic products formed in both plants and animals, and have profound influence on the growth and development of the organs of the body.

III. *WASTE PRODUCTS*

These include various substances which are not of any vital use to the protoplasm, nor are they directly secreted by the latter, but are formed as mere *by-products*. There being no excretory system in plants, these waste products are deposited in the bark old leaves, dead wood, and in other special cells away from the sphere of protoplasmic activity.

1. **Tannins.** These are a group of complex compounds widely distributed in plants. They commonly occur in single isolated cells or in small groups of cells in almost all parts of the plant body. They are abundant in the bark, heart-wood, many leaves, young and old, and many unripe fruits. As the fruits ripen, tannins disappear; they are converted to glucose and other substances. They are also abundant in the fruits of myrobalans. Tea leaves contain about 18% of tannin. Catechu, a kind of tannin, is obtained from the heart-wood of *Acacia catechu.* Tannins are bitter substances and that is why 'very strong' tea and fruits of myrobalans taste bitter. They are aseptic, i.e. free from the attack of parasitic fungi and insects. The presence of tannins makes the wood hard and durable. They are extensively used in tanning, i.e. converting hide into leather. They are also used for various medicinal purposes. They turn blue-black with an iron salt such as ferric chloride.

2. **Essential Oils.** These are volatile oils, and occur in *oil-glands* (see Fig. 33A) which appear as transparent spots on the leaves of sacred basil, shaddock, lemon, lemon grass, *Eucalyptus,* etc., in the skin of fruits like orange, lemon, shaddock, etc., and in the petals of flowers of many plants, as in rose, jasmines, etc. In all of them, the essential oils have their own characteristic odours. They differ from fatty oils in their chemical composition as well as in being volatile. They are sufficiently soluble in water to impart to it their taste and odour, but they are readily soluble in alcohol. There are some 200 essential oils of commercial

value. Some of the common ones are lemon-grass oil, eucalyptus oil, clove oil, lavender oil, jasmine oil, sandalwood oil, thyme oil, rose oil (attar), champac oil, etc.

3. **Resins.** These are chemically complex substances mostly found in the stems of conifers (pine, for example) and occur in abundance in special ducts, known as *resin-ducts* (see Fig. 23). They are yellowish solids, insoluble in water but soluble in alcohol, turpentine and methylated spirit. When present in the wood, resins add to its strength and durability. They occur associated with a small quantity of turpentine which is removed by distillation, and the residue is pure resin; they may also occur associated with gums. Resin is the main ingredient of wood varnish.

4. **Gums.** Gums are complex carbohydrates formed in various kinds of plants, being the decomposition products of cell-walls (cellulose). They are insoluble in alcohol but soluble in water, readily swell up in it, and form a viscous mass. They are found in many 'flowering' plants, and are of various kinds. *Acacia senegal* yields the best gum-arabic of commerce. Gums also occur in mixtures with resins.

5. **Mineral Crystals.** The common forms of crystals consist of silica, calcium carbonate and calcium oxalate. They occur either in the cell-cavity or in the cell-wall. Of them, crystals of calcium oxalate are most common, and are very widely distributed among various plants.

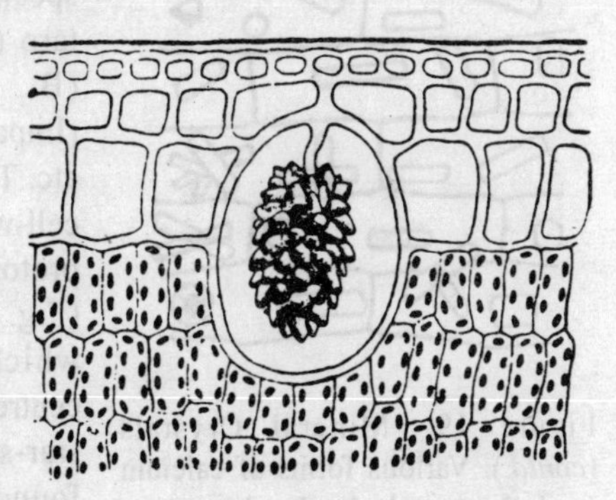

Figure 17. Mineral Crystals. Cystolith in the leaf of India-rubber plant.

(a) **Silica** occurs as an encrustation on the cell-wall or lies embedded in it. It is abundant in horsetail (*Equisetum*) and in the leaves of many grasses. Wheat straw contains about 72%, of silica, rye straw about 50% and Equisetum about 71%.

(b) **Calcium carbonate** occurs as a big mass of small crystals in the leaf of india-rubber plant, banyan, etc. The crystals are deposited on a sort of stalk which is the ingrowth of the inner epidermal wall. Finally, the whole crystalline mass looks like a bunch of grapes suspended from a stalk, and is known as the **cystolith** (Fig. 17).

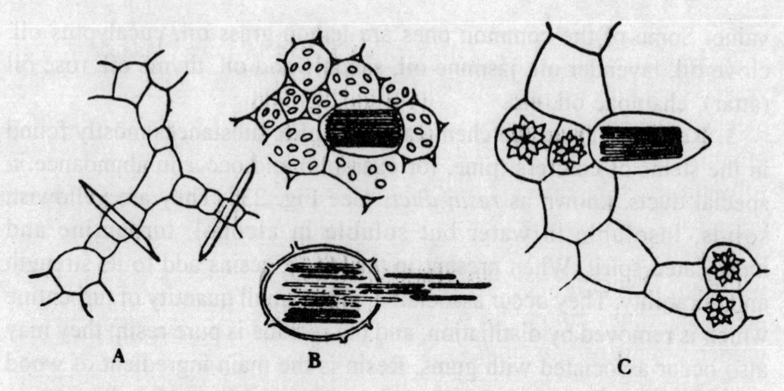

Figure 18. Mineral Crystals (*contd.*). *A,* solitary raphides (two) in the petiole of water hyacinth; *B,* a bundle of raphides in the same; *bottom,* needles (raphides) shooting out; *C,* sphacro-crystals (four) and a bundle of raphides in taro (*Colocasia*).

(c) **Calcium oxalate** occurs as crystals of various forms. (1) **Raphides** (Fig. 18) are needle-like crystals occurring singly or in bundles. They are found in most of the plants in smaller or larger quantities, but are specially common in water hyacinth, taro (*Colocasia*), *Amorphophallus* (B. OL: H. KANDA), balsam (Impatiens; B. DOPATI; H. GULMANDI), etc. They are frequently hidden by a cell-wall to prevent contact with the protoplasm. (2) **Sphaero-crystals** (Fig. 18C) are clusters of crystals which radiate from a common centre, and hence have a more or less star-shaped appearance. They are found in taro (*Colocasia*), water lettuce (*Pistia*), etc. (3) **Octahedral, cubical, prismatic** and **rod-like crystals** (Fig. 19) of calcium oxalate are also common in plants: they can be readily seen in the dry scales of onion.

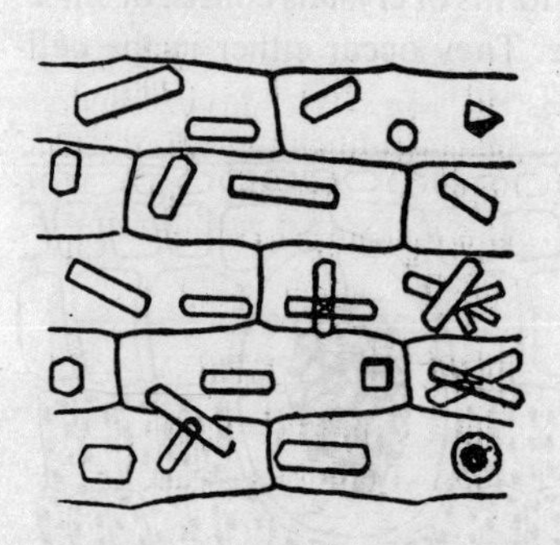

Figure 19. Mineral Crystals (*contd.*). Various forms of calcium oxalate crystals in the dry onion scale.

Tests. (a) 50% nitric acid (or hydrochloric or sulphuric acid) solution dissolves both calcium carbonate and oxalate crystals, but bubbles of carbon dioxide gas are evolved only in the case of carbonate-crystals. (b) 30% acetic acid solution readily dissolves calcium carbonate crystals but not the oxalate crystals.

6. **Latex.** This is the milky juice found in latex cells and latex vessels (see Fig. 32). Latex occurs as an emulsion consisting of a variety of chemical substances. Rod or dumb-bell-shaped starch grains may often be found in the latex. It also sometimes contains some poisonous substances, as in yellow oleander (*Thevetia*). The function of latex is not clear; perhaps in some way it is associated with nutrition, healing of wounds and protection against parasites and animals. Latex is often white and milky, as in banyan, peepul, jack, madar, oleander, *Euphorbia*, etc., sometimes coloured (yellow, orange or red) as in opium poppy, garden poppy, prickly poppy, etc.

7. **Alkaloids.** These are complex nitrogenous substances, and occur combined with some organic acids, mostly in seeds and roots of some plants. They have an intensely bitter taste and many of them are extremely poisonous. There are over 200 known alkaloids found in plants, e.g. quinine in *Cinchona*, nicotine in tobacco, morphine in opium poppy, caffeine in coffee and tea, strychnine in nux-vomica, etc. The role played by the alkaloids in the physiology of plants is not known.

FORMATION OF NEW CELLS

Plants begin their existence as a single cell. This divides and forms two cells; these again divide, and the process continues, resulting in the development of the body of the plant. There are different methods by which new cells are formed in plants by division of the pre-existing cell. In all such cases it is the nucleus that divides first, and this is followed by the division of the cell.

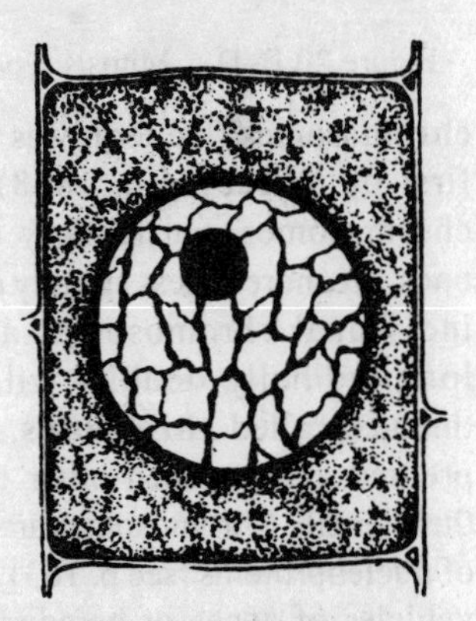

Figure 20 A. Mitosis. Metabolic nucleus.

1. **Somatic Cell Division.**[1] Cell division leading to the development of the vegetative body (soma) of the plant is known as somatic cell division. It includes the division of the nucleus, called **mitosis** (*mitos*, thread) or **karyokinesis** (*karyon*, nut or nucleus; *kinesis*, movement) or **indirect nuclear division,** and the division of the cytoplasm, called

[1]Fig. 20A–J redrawn after Fig. 40 in Fundamentals of Cytology by L.W. Sharp by permission of McGraw-Hill Book Company. Copyright 1943.

cytokinesis; (*kytos,* cell). It occurs in the growing regions, as in the root-tip and the stem-tip. The time taken for complete division is usually between 1/2 and 3 hours.

Mitosis (Fig. 20). In this process the nucleus (A) passes through a complicated system of changes which may conveniently be divided into four phases. The process was first worked out by Strasburger, a German botanist, in 1875. Mitosis can be studied in the root tip or the stem tip.

First Phase or Prophase. The first sign of the prophase is the appearance of a number of separate, slender, crooked threads, called

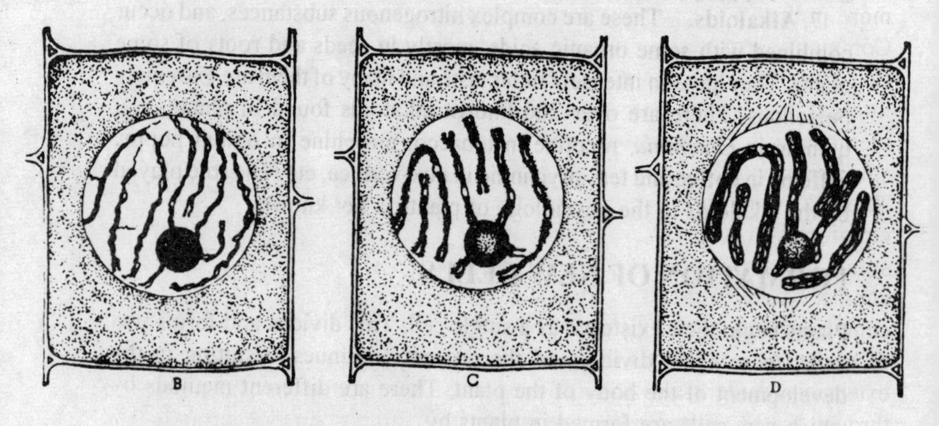

Figure 20 B–D. Mitosis (*contd.*). Prophase

chromosomes (the term was introduced first by Waldeyer in 1888) (B). The chromosomes, particularly the longer ones, are more or less spirally coiled. The individual chromosomes are always longitudinally double, with the two threads, called **chromatids**, remaining pressed against each other throughout their length. Chromosomes are composed of nucleoproteins (see p. 163) and are the vehicles of genes or hereditary factors. As prophase proceeds the chromosomes relax their coils and thicken somewhat (C). Their double nature becomes more apparent. As prophase advances a

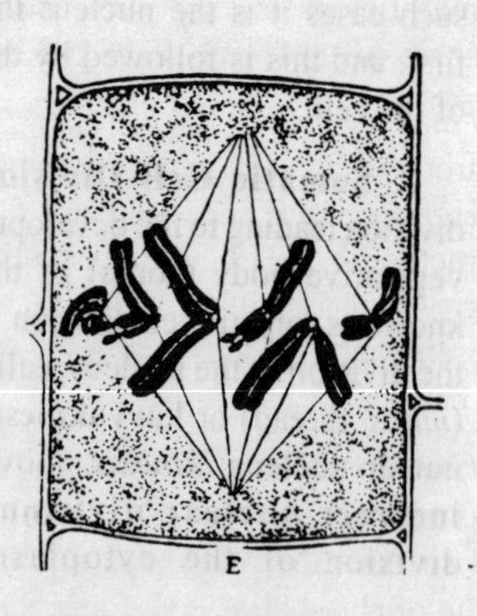

Figure 20 E. Mitosis (*contd.*). Metaphase.

chromosomal substance accumulates in a sheath or matrix round each chromosome and the chromatids become closely coiled in it (D). In well-fixed chromosomes some unstained gaps or constrictions are seen; these are the attachment regions, called **centromeres**. The nucleoli lose their staining power and disappear completely. The nucleus then rapidly passes into the next stage, the metaphase.

Second Phase or Metaphase. The nuclear membrane disappears and a spindle-like body known as the **nuclear spindle** is formed (E). The spindle may be of nuclear origin, or more probably of cytoplasmic origin. Commonly, in root-tips it appears as two opposite polar caps outside the nuclear membrane (as in D). The membrane then disappears and the spindle extends into the nuclear area. The chromosomes move to the equatorial plane of the spindle and stand there clearly apart from one another. At this stage the chromatids come even closer. From the centromeres of each pair of chromatids fibre-like extensions, called **tractile fibres**, are formed towards the opposite poles through the nuclear spindle. The number of chromosomes is normally constant for a particular species of plants and this number which is normally even, is expressed as $2n$ (or 2x) or **diploid**. Chromosome numbers cover a wide range but 24 seem to be a common figure.

Figure 20E. Mitosis (*contd.*). Anaphase.

Third Phase or Anaphase. At the end of the metaphase the centromeres of each pair of chromatids appear to repel each other. They diverge and move ahead towards the two opposite poles along the course of tractile fibres (F). The chromatids soon become separated from each other. Anaphase covers the shortest period in mitosis and includes only the separation of chromatids of each chromosome. The movement of the chromatids is autonomous.

Fourth Phase or Telophase. At each pole the chromatids (daughter chromosomes) form a close group (G). The nuclear spindle disappears and so does the matrix. A nuclear membrane is formed round each group of chromatids (H). Each chromatid splits longitudinally and then behave as a full chromosome which re-organize themselves within the nuclear membrane. Nucleoli re-appear at definite points on certain

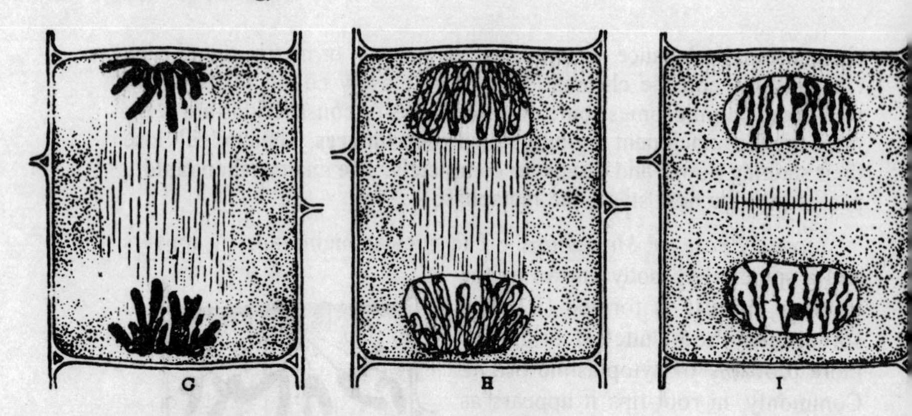

Figure 20*G–1*. Mitosis (*contd.*). Telophase.

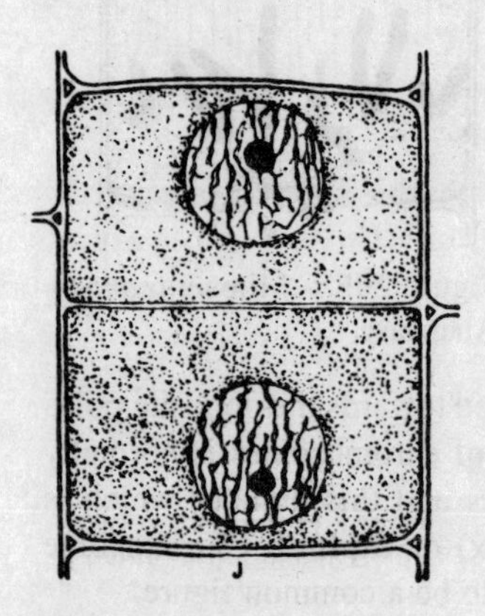

Figure 20. *J.* Mitosis (*contd.*). Cytokinesis.

chromosomes. The nuclear sap reappears and each nucleus increases in size (I). It passes into the metabolic stage or prepares for the next division.

Cytokinesis. This is the division of the cytoplasm by the formation of a new cell-wall in the equatorial region. This process, known as the cell-plate method, is the usual one in the vegetative cell. It usually begins in the telophase when new cellulose particles are gradually deposited in the equatorial zone, and soon these particles fuse together to form a delicate membrane, dividing the cytoplasm into two new cells (J).

Importance. The importance of karyokinesis lies in the fact that by this complicated process of nuclear division, the constituents of the chromosomes are equally distributed to the two daughter nuclei and thus they become qualitatively and quantitatively similar to the mother nucleus. Chromosomes are the bearers of hereditary characters and because of even distribution of chromosomal substance, particularly *DNA* (see p. 163–165), the two daughter nuclei possess all the characteristics and qualities of the mother nucleus.

Chemistry of Chromosomes. Chromosomes are made of nucleoprotein. The two components of the nucleoprotein are: nucleic acids and certain special types of proteins. Nucleic acids occur in the chromosomes mostly in the form of DNA and to some extent only as RNA. The nature and quantity of DNA present in the diploid cells are constant for a particular species of plants or animals, and in the haploid cells just half this quantity occurs.

Structure of the Chromosome. Most of the chromosomes lie within a range of 1–20μ in lengths. Each chromosome consists of two parts (see Fig. 20 E–F): (a) two spiral threads, called **chromatids**, twisted about each other, sometimes very closely, and (b) a chromosomal **matrix**. There is a series of granules arranged in a linear order along the whole length of the chromatid; they look like beads in a chain and are called **chromomeres**. The attachment regions of the chromatids, which appear as unstained gaps or constrictions, are called **centromeres**. Their position is constant in the chromosome, as seen in successive divisions. The portions of the chromosome, lying on the sides of the centromere are called arms. The arms may be equal or unequal, depending on the position of the centromere. Further, one **arm** has commonly a small segment at its distal end; this is called the **satellite**.

2. **Meiosis or Reduction Division** (Fig. 21). Meiosis (*meiosis,* diminition) is a complicated process of nuclear division by which the chromosome number is reduced to half in the four daughter nuclei so formed by this process. The reduced chromosome number is expressed as **haploid** or *n* (or x). For example, if there are 12 chromosomes in the mother nucleus there will be only half this number, i.e. 6 chromosomes in the daughter nuclei. Meiosis is completed in two stages. In this process the mother nucleus with 2*n* chromosomes divides twice to give rise to four nuclei in a group, each nucleus with *n* chromosomes. Of the two successive divisions it is only the first one that is a reduction division, while the second one is mitotic.

Meiosis takes place in all plants reproducing sexually at a certain time in their life-cycle, most often in the formation of spores as in all

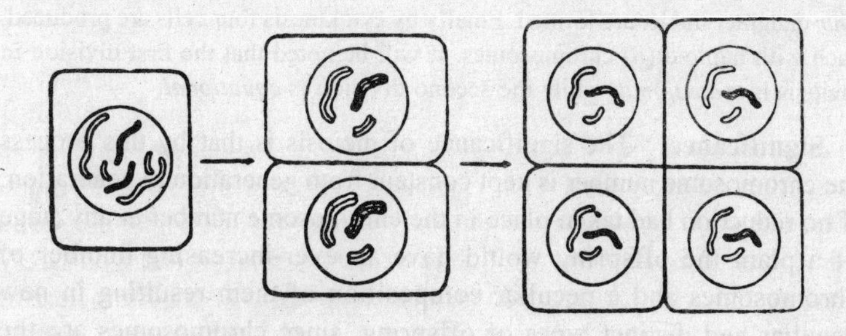

Figure 21. Meiosis (diagramatic). Note that the first division is reductional while the second one is mitotic (equational).

higher cryptogams and 'flowering' plants. In green algae and many fungi meiosis occurs immediately after fertilization or on the germination of the zygote. In some algae meiosis occurs during the formation of the gametes. Wherever meiosis may occur in the life-cycle of a plant, it is universally true that all gametes have half the usual number of chromosomes. Then, when sexual reproduction takes place, i.e. when two gametes (each with n chromosomes) fuse together the chromosome number becomes doubled ($n + n = 2n$) in the zygote.

Process (Fig. 21). **Meiosis** consists of two successive divisions of the mother nucleus ($2n$), of which *division I* is reduction division whereby the chromosome number ($2n$) is reduced to half (n), and, *division II* is mitotic in nature. This being so, the four nuclei (cells) so formed by this process have the same reduced number (n) of chromosomes.

Division I. **Prophase.** Identical or homologous chromosomes (one paternal and one maternal) come together in pairs throughout their whole lengths. This pairing, called **synapsis** (a feature of meiosis) is in the nature of close association (but not actual fusion). Each chromosome pair splits longitudinally and thus four threads or **chromatids** are formed in each chromosome pair. The paired chromosomes, each with two chromatids, now begin to separate from each other except in one or few points, called **chiasmata**. At each chiasma an exchange of genes, called **crossing-over**, takes place; this is a special feature of meiosis. **Metaphase.** The paired chromosomes move to the equator of the spindle and get attached to spindle fibres at a certain point called **centromere**. **Anaphase.** Homologous chromosomes (paternal and maternal) of each pair now begin to move towards the opposite poles. The reduction of chromosome number from diploid ($2n$) to haploid (n) is evident; this is a special feature of meiosis. **Telophase.** The chromosomes form a compact group at each pole. The two daughter nuclei, thus formed, evidently have haploid (n) chromosomes, each with two chromatids.

Division II. This is mitotic in nature. With the second division altogether *four* daughter nuclei are formed. Finally by cytokinesis four cells are produced, each with haploid (n) chromosomes. It will be noted that the first division in meiosis is *reductional,* while the second division is *equational.*

Significance. The significance of meiosis is that by this process the chromosome number is kept constant from generation to generation. If no reduction had taken place in the chromosome number at any stage of a plant the offspring would have an ever-increasing number of chromosomes and a peculiar composition of them resulting in new peculiar and distinct types of offspring, since chromosomes are the bearers of hereditary characteristics, and meiosis, is the mechanism for their transmission to the offspring. By meiosis the DNA which is the sole genetic material is carried down to the offspring from generation to generation through the gametes in sexual reproduction.

Differences between Mitosis and Meiosis.

1. Mitosis is the somatic cell division, in which the chromosome number remains constant, expressed as $2n$ or *diploid;* while meiosis is the reduction division, in which the chromosome number is reduced to half the somatic number, expressed as n or *haploid*.

2. Mitosis occurs in vegetative cells and continues almost indefinitely in them, retaining the same number of chromosomes, i.e. $2n$, in all successive divisions; while meiosis occurs in the formation of reproductive units (spores or gametes) and ends with two divisions (the first being reduction division), resulting in a group of four cells or nuclei (spores or gametes), each with n chromosomes. Mitosis is connected with vegetative growth of the plant body; while meiosis is concerned with reproduction.

3. In both the processes the chromosomes appear in specific numbers. In mitosis, however, they appear in double threads (chromatids); while in meiosis they appear in single threads but in identical pairs (one paternal and one maternal).

4. Pairing of identical (homologous) chromosomes soon occurs in the prophase of meiosis but no such pairing takes place in mitosis. Further, the prophase of mitosis is short, while it is a prolonged one in meiosis.

5. In mitosis each chromosome splits into two sister chromatids which move to the two opposite poles, while in meiosis the homologous chromosomes separate from each other and then move to the opposite poles.

6. Chiasma (pl. chiasmata, i.e. one or more points of contact of the two homologous chromosomes) and crossing-over (exchange of genes) are special and exclusive features of meiosis.

7. Haploid gametes formed by meiosis normally fuse in pairs in sexual reproduction, and their product, i.e. the zygote, becomes diploid ($n + n = 2n$). The diploid zygote divides by mitosis and gives rise to the vegetative body with diploid cells, as in all higher plants. Normally no such fusion of diploid vegetative cells (nuclei) takes place. Meiosis is really a mechanism to keep the chromosome number constant from generation to generation.

3. **Amitosis or Direct Nuclear Division** (Fig. 22A). In this case the nucleus elongates to some extent and then it undergoes constriction, i.e. it becomes narrower and narrower in the middle or it one end, and finally it splits into two. The nuclei so formed may be of equal or unequal sizes. The direct nuclear division may or may not be followed by the division of the cell. Amitosis commonly occurs in certain lower algae and fungi. In the higher plants it is seen to occur in certain old cells here and there.

4. **Free Cell Formation** (Fig. 22B). This is a modification of indirect nuclear division. It differs from the latter in that the cell-wall is not formed immediately after the division of the nucleus. In this process by repeated mitotic divisions a large number of nuclei are formed. When the divisions of the nuclei cease, cytoplasm aggregates round them, and a cell-wall is formed round each nucleus. The formation of the cell-wall gradually proceeds from one side to the other, resulting in a regular tissue (combination of cells). The endosperm, i.e. the food storage tissue of the seed is formed by this method.

5. **Budding** (Fig. 22C). This is seen in yeast—a unicellular fungus. In this plant the cell forms one or more tiny outgrowths on its body. The nucleus undergoes direct division (amitosis) and splits up into two. One of them passes on to one outgrowth. The outgrowth increases in size and is ultimately cut off from the mother yeast as a new independent cell (a new yeast plant). This process of cell formation is known as **budding**. Often budding continues one after the other so that chains and even sub-chains of cells are formed. Ultimately, all the cells separate from one another.

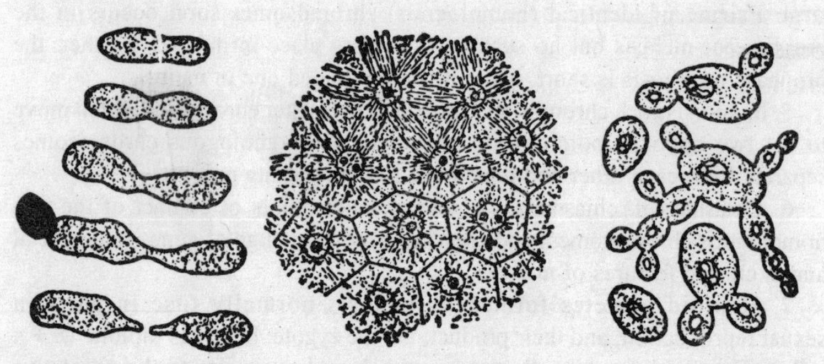

Figure 22. *A,* amitosis or direct nuclear division; *B,* free cell formation in the development of endosperm; *C,* budding in yeast.

INTERCELLULAR SPACES AND CAVITIES

Intercellular Spaces. When the cells are young they remain closely packed without any empty space or cavity between them; but as they grow, their walls split at certain points, giving rise to small cavities or empty spaces; called intercellular spaces. They remain filled with air or water.

Schizogenous Cavities. Bigger cavities are also often formed by the splitting up of common walls and the separation of masses of cells from one another; these are schizogenous (*schizein,* to split) cavities. Intercellular spaces and these cavities form an intercommunicating system so that gases and liquids can easily diffuse from one part of the plant body to the other. Most resin-ducts in plants are schizogenous cavities (Fig. 23).

Figure 23. Schizogenous Cavity. *A*, resin-duct of pine stem with resin.

Lysigenous Cavities. Sometimes, during the development of a mass of cells, their walls break down and dissolve, and as a consequence large irregular cavities appear; these are known as lysigenous (*lysis,* loosening) cavities. These cavities are meant for storing up water, gases, essential oils, etc., and thus act as glands (see Fig. 33A).

Chapter 2

The Tissue

Cells grow and assume distinct shapes to perform definite functions. Cells of the same shape grow together and combine into a group for the discharge of a common function. Each group of mature cells gives rise to a tissue. *A* **tissue** *is thus a group of cells of the same type or of the mixed type, having a common origin and performing an identical function.* Tissues may primarily be classified into two groups: **meristematic** and **permanent**.

Meristematic Tissues (*meristos,* divided). These are composed of cells that are in a state of division or retain the power of dividing. These cells are essentially alike, being either spherical, oval or polygonal in shape without any intercellular spaces; their walls thin and homogeneous; the protoplasm abundant and active with large nuclei; and the vacuoles small or absent. Meristematic tissues may be apical and lateral: (a) the apical meristem lies at the apex of the stem and the root (see Figs. 34–35) and gives rise to primary permanent tissues, while (b) the lateral meristem, e.g. cambium (see Fig. 40), lies among masses of permanent tissues and gives rise to secondary permanent tissues. Besides, in pine, many grasses, Equisetum, etc. some short lived meristematic tissues called intercalary meristem is seen to be present which soon disappear or becomes transformed into permanent tissues.

Permanent Tissues. These are composed of cells that have lost the power of dividing, having attained their definite form and size. They may be living or dead and thin-walled or thick-walled. Permanent tissues are formed by differentiation of the cells of the meristems (apical and lateral), and may be **primary** and **secondary**. The primary permanent tissues are derived from the apical meristems of growing regions and the secondary permanent tissues from the lateral meristems.

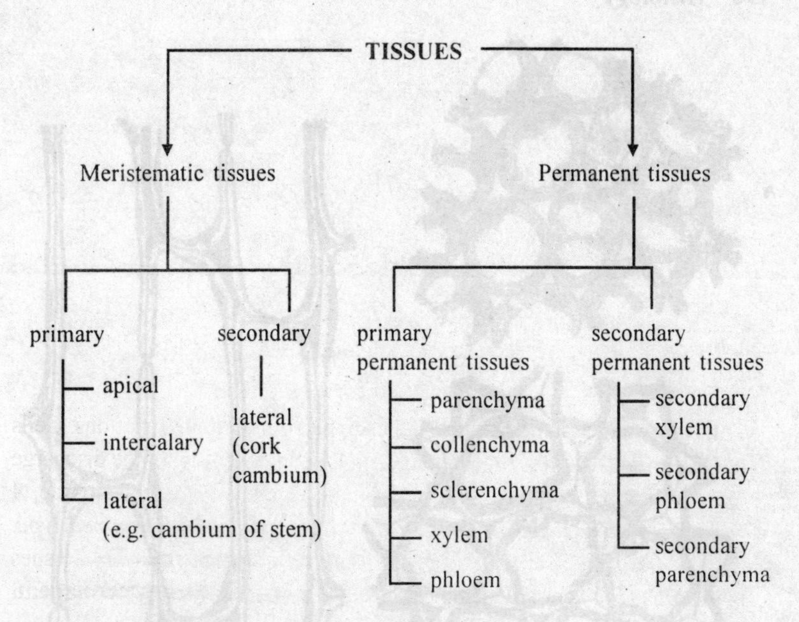

PRIMARY PERMANENT TISSUES

Classification. Primary permanent tissues may, be classified as *simple* and *complex*. A simple tissue is made up of one type of cells forming a homogeneous or uniform mass, and a complex tissue is made up of more than one type of cells working together as a unit. To these may be added another kind of tissue—the secretary tissue.

I. *SIMPLE TISSUES*

1. **Parenchyma** (Fig. 24A). Parenchyma consists of a collection of cells which are more or less isodiametric, that is, equally expanded on all sides. Typical parenchymatous cells are oval, spherical or polygonal in shape. Their walls are thin and made of cellulose; they are usually living. Parenchyma is of universal occurrence in all the soft parts of plants. Its function is mainly storage of food material. Parenchyma containing chloroplasts, often called *chlorenchyma,* manufactures sugar and starch. Star-like parenchyma with radiating arms, leaving a lot of air-cavities, is called *aerenchyma,* as in the petiole of banana and *Canna* (Fig. 25A–B) and also in many aquatic plants.

2. **Collenchyma** (Fig. 24B–C). This tissue consists of somewhat elongated cells with the corners or intercellular spaces much thickened

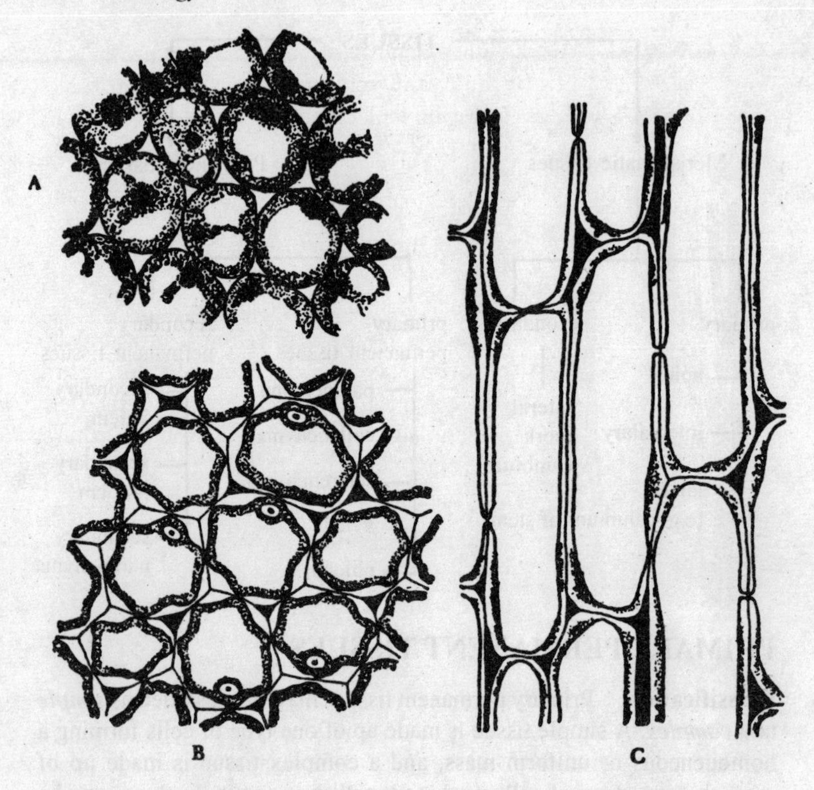

Figure 24. *A,* parenchyma; *B,* collenchyma in transverse section; *C,* collenchyma in longitudinal section.

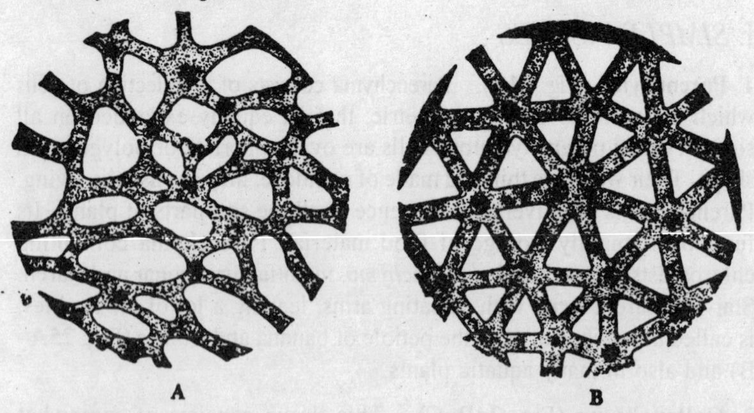

Figure 25. *A,* aerenchyma in the petiole of banana; *B,* the same in the petiole of *Canna.*

with a deposit of cellulose and pectin. In a transverse section of the stem the cells, however, look circular or oval. Their walls are provided with simple pits here and there. Collenchyma occurs in a few layers under the skin (epidermis) of herbaceous dicotyledons, e.g. sunflower, gourd, etc. (see Figs. 42–45). It is absent from the root and the monocotyledon except in special cases. The cells are living and often contain some chloroplasts. Being flexible in nature collenchyma gives tensile strength to the stem. Containing chloroplasts it also manufactures sugar and starch. Its functions are, therefore, both mechanical and vital.

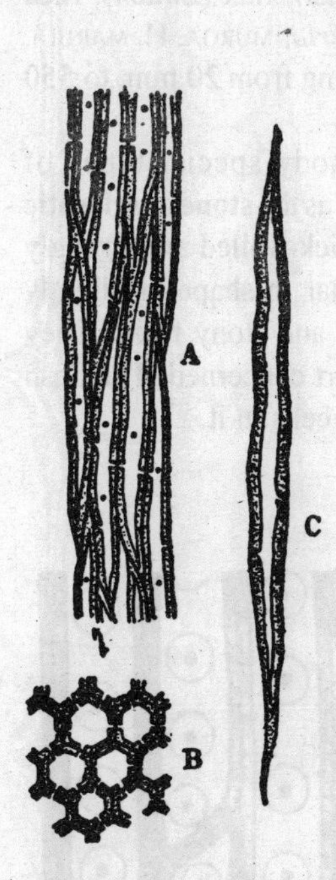

Figure 26. Sclerenchyma. *A,* fibres as seen in longisection; *B,* the same as seen in transection; and *C,* a single fibre.

3. **Sclerenchyma** (Fig. 26). Sclerenchyma (*scleros,* hard) consists of very long, narrow, thick-walled and lignified cells, usually pointed at both ends. They are fibre-like in appearance, and hence they are also called sclerenchymatous fibres, or simply **fibres**. They have simple, often oblique, pits in their walls. The middle lamella is conspicuous in sclerenchyma. Sclerenchymatous cells are found abundantly in plants, and occur in patches or definite layers. They are dead cells and serve a purely mechanical function, that is, they give strength and rigidity to the plant body and thus enable it to withstand various strains. Their average length is 1 to 3 mm.

Figure 27. Stone cells—two types.

but in the fibre-yielding plants such as hemp (*Cannabis;* B. and H. GANJA), Indian hemp (*Crotalaria;* B. SHONE; H. SAN), Deccan hemp (*Hibiscus;* B. NALITA; H. AMBARI), jute (*Corchorus*), flax (*Linum*), rhea or ramie (*Boehmeria*), bowstring hemp (*Sansevieria;* MURGA; H. MARUL), etc. these cells may be of excessive lengths ranging from 20 mm. to 550 mm. Such fibres are of commercial importance.

Sometimes, here and there in the plant body special types of sclerenchyma may be developed. These are known as the **stone** or **sclerotic cells** or **sclereids** (Fig. 27). The cells are very thick-walled and strongly lignified, and are mostly isodiametric or irregular in shape or slightly elongated. Stone cells occur in hard seeds, nuts and stony fruits. They contribute to the firmness and hardness of the part concerned. The flesh of pear is gritty because of the presence of stone cells in it.

II. *COMPLEX TISSUES*

1. **Xylem.** Xylem or wood is a conduct-ing tissue and is composed of elements of different kinds, viz. (a) **tracheids**, (b) **vessels** or **tracheae** (sing. trachea), (c) **wood fibres** and (d) **wood parenchyma**. Xylem as a whole is meant to conduct water and mineral salts upward from the root to the leaf, and to give mechanical strength to the plant body. Except wood parenchyma all other xylem elements are lignified, thick-walled and dead.

Figure 28. Tracheids with bordered pits. *A,* stem in radial section; *B,* the same in tangential section.

(a) **Tracheids** (Fig. 28). These are elongated, tube-like dead cells with hard, thick and lignified walls and a large cell-cavity. Their ends

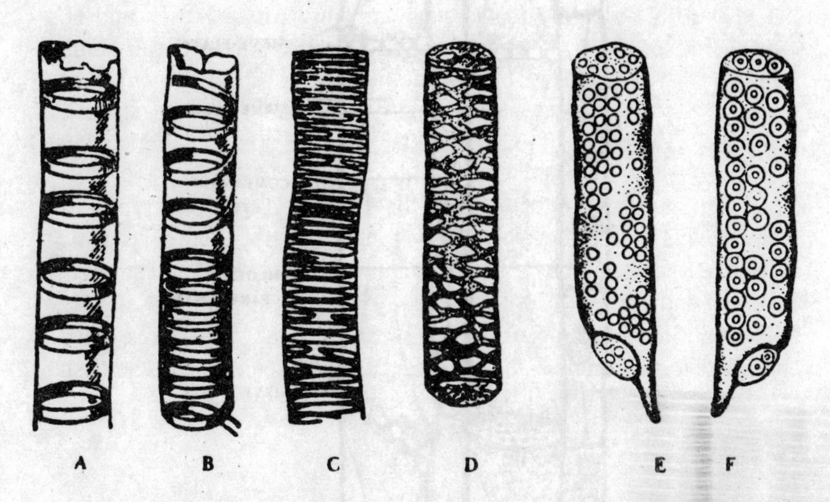

Figure 29. Kinds of Vessels. *A,* annular; *B,* spiral; *C,* scalariforin; *D,* reticulate; *E,* a vessel with simple pits; *F,* a vessel with bordered pits.

are commonly tapering or oblique. Their walls are usually provided with one or more rows of bordered pits. Tracheids may also be annular, spiral, scalariform or pitted (with simple pits). In transverse section they are mostly angular, either poly-gonal or rectangular. Tracheids (and not vessels) occur alone in the wood of ferns and gymnosperms, whereas in the wood of angiosperms they occur associated with the vessels. Being lignified and hard, tracheids give strength to the plant body, but their main function is conduction of water and mineral salts from the root to the leaf.

(b) **Vessels** or **Tracheae** (Fig. 29). Vessels are rows of elongated tube-like dead cells, placed end to end, with their transverse or end-walls dissolved. A vessel or trachea is thus very much like a series of water-pipes forming a pipeline. Their walls are thickened in various ways, and according to the mode of thickening, vessels have received their names such as **annular, spiral, scalariform, reticulate** and **pitted**. Associated with the vessels are often found some tracheids. Vessels and tracheids form the main elements of the wood or xylem of the vascular bundle (see Fig. 40). They have large cell-cavities which serve for conduction of water and mineral salts from the roots to the leaves. They are dead, thick-walled and lignified, and as such they also serve the mechanical function of strengthening the plant body.

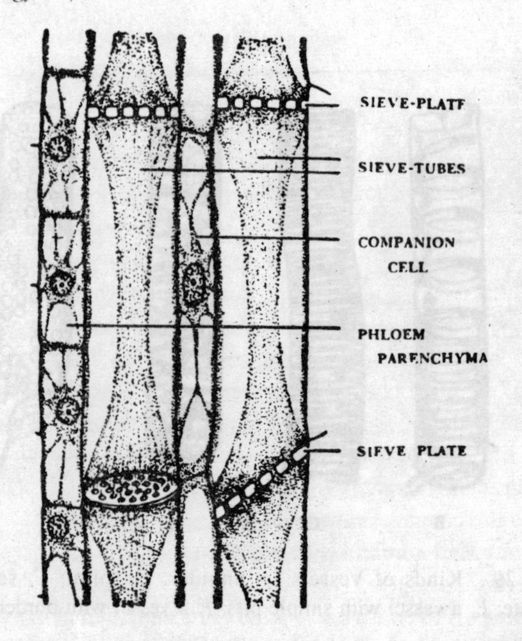

Figure 30. Sieve tissue in longitudinal section.

(c) **Wood Fibres.** Sclerenchymatous cells associated with wood or xylem are known as wood fibres. They occur abundantly in woody dicotyledons and add to the mechanical strength of the xylem and the plant body as a whole.

(d) **Wood Parenchyma.** Parenchymatous cells associated with xylem together form the wood parenchyma. The cells are alive, thin-walled and generally abundant. The wood parenchyma assists, directly or indirectly, in the conduction of water upwards through the vessels and the tracheids; it also serves for food storage.

2. **Phloem.** Phloem or bast is another conducting tissue, and is composed of the following elements: (a) **sieve-tubes**, (b) **companion cells**, (c) **phloem parenchyma**, and (d) **bast fibres** (rarely). Phloem as a whole is meant to conduct prepared food materials from the leaf to the storage organs and the growing regions.

(a) **Sieve-tubes** (Figs. 30–31). Sieve-tubes are slender, tube-like structures composed of elongated cells, placed end on end. Their walls are thin and made of cellulose; each transverse wall is, however, perforated by a number of pores. It then looks very much like a sieve, and is called the **sieve-plate**. In winter, the sieve-plate is covered by a thin pad, called

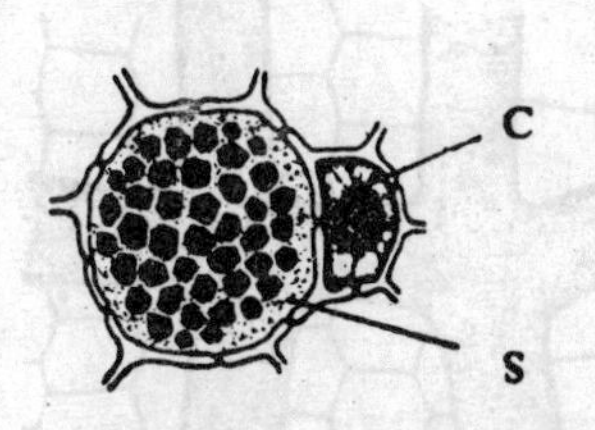

Figure 31. Sieve-tube in transection. *C*, companion cell; *S*, sieve-tube.

callus or **callus pad**. In spring, when the active season begins, the callus gets dissolved. In old sieve-tubes the callus forms a permanent deposit. The sieve-tube contains no nucleus, but has a lining layer of cytoplasm which is continuous through the pores. Sieve-tubes carry prepared food materials—soluble proteins and carbohydrates—from the leaves to the storage organs and later from the storage organs to the growing regions of the plant body. A heavy deposit of food material is found on either side of the sieve-plate with a narrow median portion.

(b) **Companion Cells.** Associated with each sieve-tube and connected with it by simple pits, there is a thin-walled, elongated cell, known as the companion cell. It is living, containing protoplasm and a large elongated nucleus. The companion cell is present only in angiosperms.

(c) **Phloem Parenchyma.** There are some parenchymatous cells in the phloem. These are living, and in shape often cylindrical. Phloem parenchyma, however, is mostly absent in monocotyledons.

(d) **Bast Fibres.** Sclerenchymatous cells occurring in the phloem or bast are known as bast fibres. These are generally absent in the primary phloem but are of frequent occurrence in the secondary phloem.

III. *SECRETORY TISSUES*

1. **Laticiferous Tissue.** This consists of thin-walled, greatly elongated and much-branched ducts (Fig. 32) containing a milky juice known as latex (see p. 184). Laticiferous ducts are of two kinds: latex vessels and latex cells. They contain numerous nuclei which lie embedded in the thin lining layer of protoplasm. They occur irregularly distributed in the mass of parenchymatous cells. The function of laticiferous ducts is lot clearly understood. They may act as food-storage organs or as reservoirs of waste products. They may also act as translocatory tissues.

Latex vessels (Fig. 32B) are rows of more or less parallel ducts, connected with one another *by the fusion of their branches, forming a network*. Latex vessels are found in the poppy family, e.g. opium poppy, garden poppy and prickly poppy and also in some species of the sunflower family, e.g. *Sonchus*.

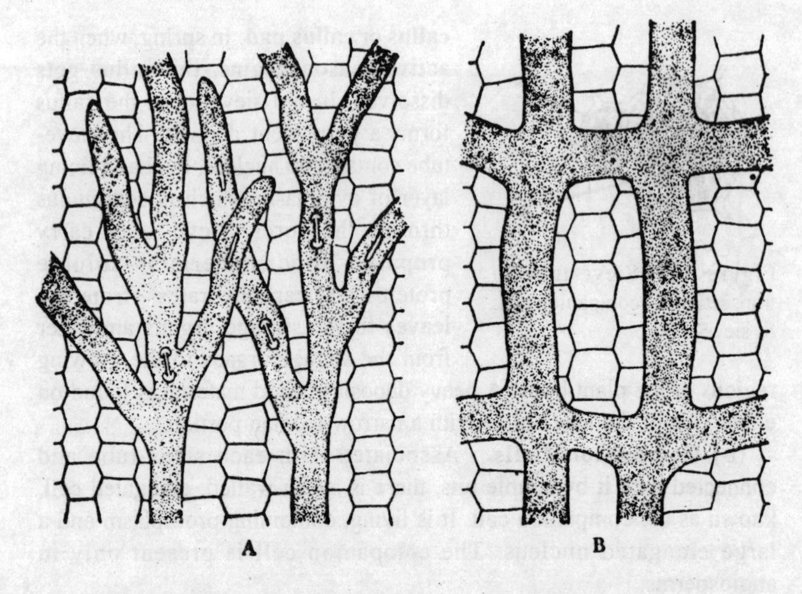

Figure 32. Laticiferous Tissue. *A,* latex cells; *B,* latex vessels.

Latex cells (Fig. 32A), on the other hand, although much-branched like the latex vessels, are really single or independent units. They branch profusely through the parenchymatous tissue of the plant, but *without fusing together to form a network.* Latex cells are found in madar, *Euphorbia* (B. & H. SIJ), oleander (*Nerium*), periwinkle (*Vinca*), *Ficus* (e.g. banyan, Fig. peepul), etc.

2. **Glandular Tissue.** This tissue is made of glands which are special structures containing some sccretory or excretory products. Glands may consist of single isolated cells or small groups of cells with or without a central cavity. They are of various kinds and may be internal or external. *Internal glands* are (i) oil-glands (Fig. 33A) secreting essential oils, as in the fruits and leaves of orange, lemon, pummelo, etc.; (ii) mucilage-secreting glands, as in the betel leaf; (iii) glands secreting gum, resin, tannin, etc.; (iv) digestive glands secreting enzymes or digestive agents; and (v) special water-secreting glands at the tips of veins.

External glands are commonly short hairs tipped by glands. They are: (i) water-secreting hairs or glands; (ii) glandular hairs (Fig. 33B) secreting gummy substances, as in tobacco, *Plumbago* (B. CHITA; H. CHITRAK), and *Boerhavvin* (B. PUNARNAVA; H. THIKRI); (iii) glandular hairs secreting irritating, poisonous substances as in nettles (see Fig. I/93);

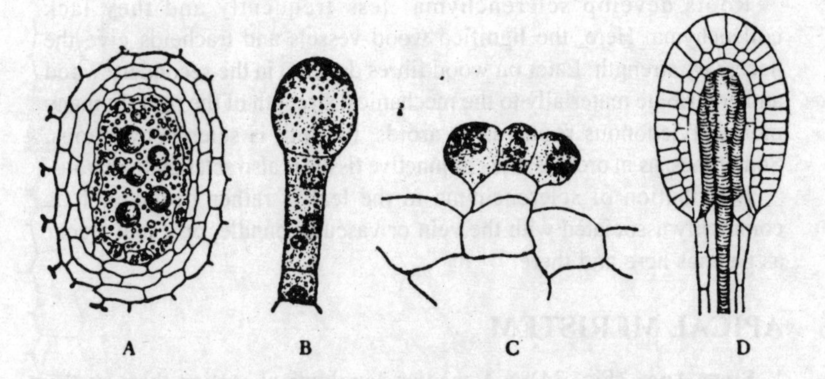

Figure 33. Glands. *A*, an oil-gland of orange skin; *B*, a glandular hair of *Boerhaavia* fruit; *C*, a digestive gland of butterwort (insectivorous); *D*, a digestive gland of sundew (insectivorous).

(iv) honey glands or nectaries, as in many flowers; and (v) enzyme-secreting glands (Figs. 33C–D), as in carnivorous plants.

Permanent tissues
- simple, e.g. parenchyma, collenchyma and sclerenchyma
- complex, e.g. xylem and phloem
- secretory, e.g. laticiferous, and glandular

Distribution of Strengthening or **Mechanical Tissues.** The distribution of mechanical tissues in the plant body is determined by several factors. From a purely mechanical stand-point the principle of distribution is as follows: Stems have to bear the weight of the upper parts, and are swayed back and forth by the wind. They are, therefore, subjected to alternate stretching and compressing. The best position for strengthening tissues in stems, therefore, is close to the periphery, either in the form of a cylinder or in patches. Roots, on the other hand, are subjected to the pulling force exerted by the swaying stem and also to the compressing force exerted by the surrounding soil. These forces are met by roots, by the development of a solid wood cylinder in or around the centre.

Collenchyma and sclerenchyma, including wood fibres and bast fibres, are the two most important tissues concerned with the strengthening of the plant body. Their distribution may be studied with reference to the sunflower stem (see Fig. 43) and the maize stem (see Fig. 46).

Roots develop sclerenchyma less frequently and they lack collenchyma. Here, the lignified wood vessels and tracheids give the necessary strength. Later on wood fibres develop in the secondary wood and contribute materially to the mechanical strength of the root. In many monocotyledonous roots, as in aroids, the pith is sclerenchymatous. Sometimes, as in orchids, the conjunctive tissue is also sclerenchymatous.

Distribution of sclerenchyma in the leaf is rather irregular. It is commonly associated with the vein or vascular bundle, or it may occur as patches here and there.

APICAL MERISTEM

1. **Stem Apex** (Fig. 34). A median longitudinal section through the apex of a stem, when examined under the microscope, shows that the

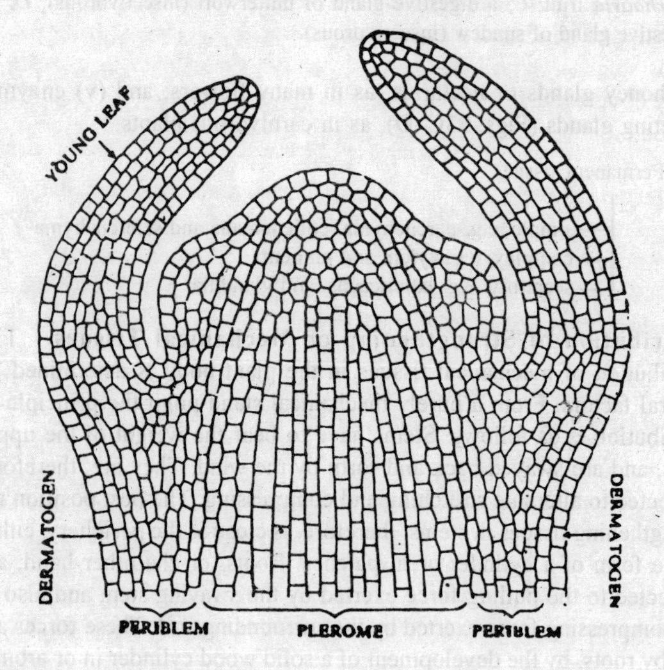

Figure 34. Stem apex in longitudinal section.

apical meristem or growing region is composed of a mass of small, usually rounded or polygonal cells which are all essentially alike and are in a state of division; these meristematic cells constitute the **promeristem**. The cells of the promeristem soon differentiate into three regions, viz., dermatogen, periblem and plerome. The cells of these three regions grow and give rise to primary permanent tissues in the mature portion of the stem. The section further shows on either side a number of outgrowths which arch over the growing apex; these are the young leaves of the bud, which cover and protect the tender growing apex of the stem.

(1) **Dermatogen** (*derma,* skin; *gen,* producing). This is the single outermost layer of cells. It passes right over the apex and continues downwards as a single layer. The cells divide by *radial* walls only, i.e. at right angles to the surface of the stem, and increase in circumference, thus keeping pace with the increasing growth in volume of the underlying tissues. The dermatogen gives rise to the skin layer or epidermis of the stem.

(2) **Periblem** (*peri,* around; *blema,* covering). This lies internal to the dermatogen, and is the middle region of the apical meristem. At the apex it is single-layered but lower down it becomes multi-layered. It forms the cortex of the stem, which is often, particularly in dicotyledons, differentiated into hypodermis, general cortex and endodermis (see p. 209).

(3) **Plerome** (*pleres,* full). This lies internal to the periblem, and is the central region of the stem apex. At a little distance behind the apex certain groups or strands of cells show a tendency to elongate. These groups or strands of elongated cells are said to form the **procambium**. In a transverse section of the stem each procambium appears as a small group of cells which soon become differentiated into the elements of xylem on the inner side and of phloem on the outer, together forming into a vascular bundle. A portion, however, may remain un-differentiated, and it forms the **cambium** of the vascular bundle, lying in between xylem and phloem. Plerome as a whole gives rise to the central cylinder or **stele** (see p. 217), as it is called, which in a dicotyledonous stem is differentiated into the pericycle, medullary rays, pith and the vascular bundles (see Fig. 43).

2. **Root Apex** (Fig. 35). A median longitudinal section through the apex of the root shows that it is covered over and protected by a many-layered tissue which constitutes the **root-cap**. The apical meristem or growing region lies within and behind the root-cap (see Fig. I/33). The promeristem, as in the stem, differentiates into three regions, viz. (1) dermatogen (2) periblem and (3) plerome. In many roots, however, these three regions are not clearly marked.

(1) **Dermatogen.** As in the stem, this also is single-layered but at the apex it merges into the periblem; outside this the dermatogen cuts off many new cells, forming a small-celled tissue known as the **calyptrogen** (*calyptra,* cap; *gen,* producing). The calyptrogen is also meristematic, and by repeated divisions of its cells gives rise to the **root-cap**. As the root passes through the hard soil, the root-cap oftens wears away but then it is renewed by the underlying calyptrogen. The walls of outer cells of the root-cap may be modified into mucilage which helps push the root forward in the soil more easily. At a little distance from the root-tip the outermost layer bears a large number of *unicellular root-hairs*. The dermatogen continues upwards as a single outermost layer called **epiblema**.

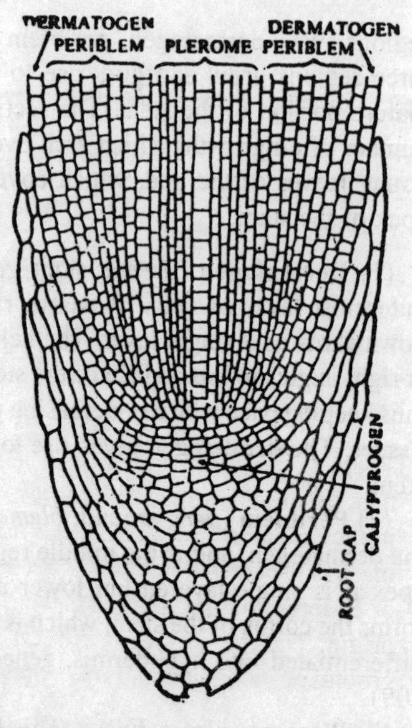

Figure 35. Root apex in longitudinal section.

(2) **Periblem.** As in the stem, this is also is single-layered at the apex and many-layered higher up. Periblem forms the middle region or cortex of the root (see Fig. 49).

(3) **Plerome.** Its structure and function are practically the same as those of the stem. But here some procambial strands give rise to bundles of vessels (xylem) and others to bundles of sieve-tubes (phloem) in an alternating manner (see Fig. 41A).

In many roots and stems, however, the apical meristem is not sharply separable into the three regions mentioned above, particularly into the periblem and the plerome. In such cases the division is as follows: (1) **Protoderm** which corresponds to the dermatogen, (2) **procambium** which forms isolated groups or strands of cells, as in the previous classification, and (3) **ground** or **fundamental meristem**, which fills up the remaining spaces. It is a combination of the periblem and the plerome (excluding the procambium).

Chapter 3

The Tissue System

There exists in the higher plant a division of labour, and in response to this, tissues are arranged into three systems, each taking a definite share in the common life-work of the plant. Each system may consist of only one tissue or a combination of tissues which may structurally be of similar or different nature, but perform a common function and have the same origin. The three systems are: (I) The epidermal tissue system, (II) the ground or fundamental tissue system, and (III) the vascular tissue system.

1. **The Epidermal Tissue System.** The epidermal tissue system consists mainly of a single outermost layer called the **epidermis** (*epi,* upon or outer; *derma,* skin) which extends over the entire surface of the plant body. At surface view the cells of the epidermis are somewhat irregular in outline (Fig. 36), but closely fitted together without intercellular spaces. They, however, appear more or less rectangular in transection. Epidermal cells are parenchymatous in nature with colourless cell-sap. In the leaves and young green shoots the epidermis possesses numerous minute openings called **stomata** (Fig. 36). The outer walls of the epidermis are often thickened and cutinized. The cutinized layer or the **cuticle** checks evaporation of water. In many plants the epidermis bears hairs of different kinds—soft, stiff, sharp, stinging, glandular, etc. *Functions.* The epidermis functions as a protective tissue. It protects the plant body against excessive evaporation of water, attacks of herbivorous animals, parasitic fungi and bacteria, and excessive heat or cold.

The outermost layer of the root is called the **epiblema** or **piliferous layer**. It is mainly concerned with the absorption of water and mineral salts from the soil. Thus, to increase the absorbing surface which may be 5 to 20 times greater, the outer walls of most of its cells a little behind the apex (see Fig. I/33) extend outwards and form tubular unicellular root-hairs. The epiblema is neither cutinized nor is it provided with stomata.

Stomata. *Structure and Behaviour*. **Stomata** (*stoma,* a mouth) are very minute openings (Fig. 36) formed in the epidermal layer in green

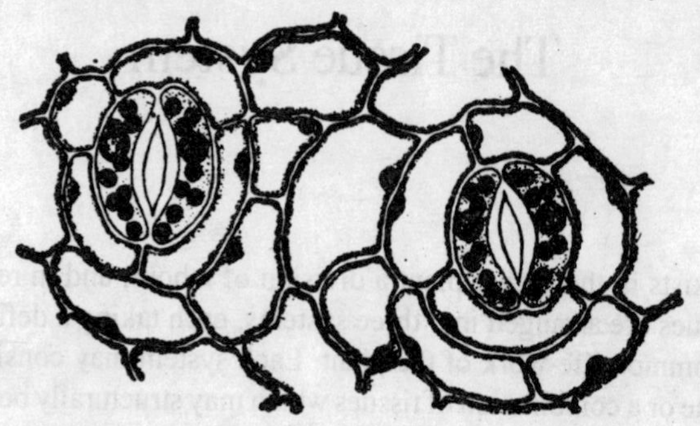

Figure 36 Stomata in epidermal layer (surface view).

aerial parts of the plant, particularly the leaves. Roots and non-green parts of the stem are free from them. Each stoma is surrounded by two semi-lunar cells known as the *guard cells*. The term 'stoma' is often applied to the stomatal opening plus the guard cells. The guard cells are living and always contain chloroplasts, and their inner walls are thicker, and outer walls thinner. They guard the stoma or the passage, i.e. they regulate the opening and closing of it like lips. Under normal conditions the stomata remain closed at night, i.e. in the absence of light, and they remain open during the daytime, i.e. in the presence of light. They may close up at daytime when very active transpiration (evaporation of water) takes place from the surface of the leaf under certain conditions such as high temperature, dryness of the air, blowing of dry wind and deficient supply of water in the soil. The opening and closing of the stomata are due to the movement of the guard cells, and the movement is mainly connected with two factors—light and water. In the presence of light the guard cells absorb water from the neighbouring cells, expand and bulge in an outward direction and the stoma opens. In the absence of light the guard cells lose water and become flaccid and the stoma closes. The intensity of light also directly affects the degree of stomatal opening.

The expansion or contraction of the guard cells is due to the presence of sugar or starch in them. In the presence of light the sugar manufactured by the chloroplasts of the guard cells accumulates in them and, being soluble, increases the concentration of the cell-sap. Under this condition the guard cells absorb water from the neighbouring cells and become turgid, and the stoma opens. In darkness, on the other hand, the sugar present in the guard cells becomes converted into starch—an insoluble compound. The concentration of the cell-sap is, therefore, lower than that of the neighbouring cells. Under this condition the guard cells lose water and shrink, and the stoma closes.

Functions and Distribution. **Stomata** are used for interchange of gases between the plant and the atmosphere—oxygen for respiration

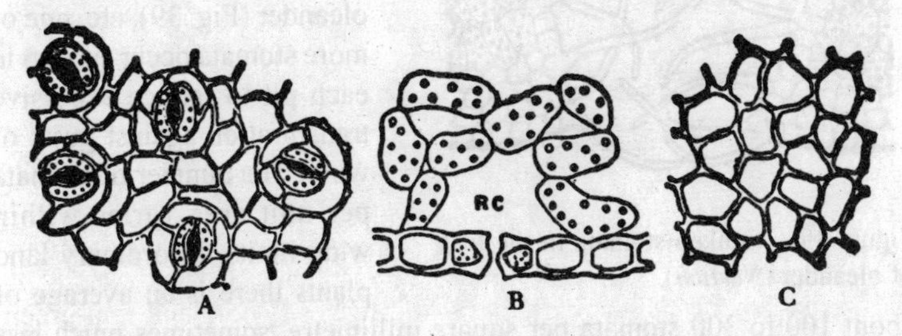

Figure 37. Stomata in betel leaf. *A,* lower epidermis with numerous stomata; *B,* section of leaf (a portion of the lower side); *RC,* respiratory cavity internal to a stoma; *C,* upper epidermis with no stoma.

and carbon dioxide for manufacture of carbohydrates. For the facility of diffusion of these gases each stoma opens internally into a small cavity, known as the **respiratory cavity** (Figs. 37B and 38) which in its turn communicates with the system of intercellular spaces and air-cavities. Stomata are also the organs through which evaporation of water takes place; in this way the plant gets rid of the surplus water. Stomata are most abundant on the lower epidermis (Fig. 37A) of the dorsiventral leaf; none (or sometimes comparatively few) are present in the upper (Fig. 37C), In the isobilateral leaf, stomata are more or less evenly distributed on

Figure 38. Sunken stoma in the leaf of American aloe (*Agave*).

all sides (see Fig. 54), In the floating leaves, as in those of the water lily, stomata remain confined to the upper epidermis alone; in the submerged leaves no stoma is present. In plants growing in deserts or dry regions, e.g. American aloe (Fig. 38), oleander (Fig. 39), etc. one or more stomata occur sunken in each pit to reduce excessive transpiration against gusts of wind. The number of stomata per unit area varies within wide limits. In ordinary land plants there is an average of about 100 to 300 stomata per square millimetre, sometimes much less or many more. In desert plants they may be only 10 to 15 in the same area.

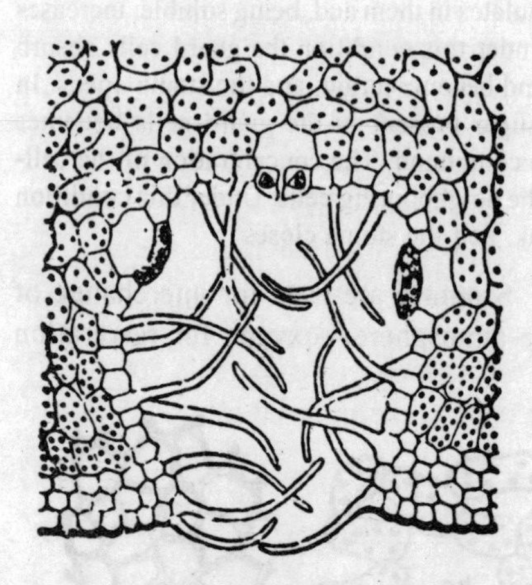

Figure 39. Sunken stomata in the leaf of oleander (*Nerium*).

2. **The Ground** or **Fundamental Tissue System.** This system forms the main bulk of the body of the plant, and extends from below the epidermis to the centre (excluding the vascular bundles), This system consists of various kinds of tissues, of which parenchyma is the most abundant. It is differentiated into the following zones and sub-zones.

(a) **Cortex.** This is the zone that lies between the epidermis and the pericycle, varying in thickness from a few to many layers. In dicotyledonous stems (see Figs. 42–45) it is usually differentiated into the following sub-zones: (a) **hypodermis**—a few external layers of collenchyma or sometimes sclerenchyma; (b) **general cortex** or cortical parenchyma—a few middle layers of thin-walled cells with or without chloroplasts and (c) **endodermis**—a single internal layer, often wavy; it is also called *starch sheath* as it often contains numerous starch grains. In monocotyledonous stems (see Fig. 46), owing to the scattered arrangement of the vascular bundles, there is no such differentiation into sub-zones. In roots (see Fig. 49) the cortex consists of (a) many layers of thin-walled parenchyma and (b) a distinct circular layer of endodermis.

Functions. In stems the cortex primarily functions as a protective tissue; its secondary functions are manufacture and storage of food. In roots the cortex is essentially a storage tissue. It is also the pumping station of the root where the individual cells, by their alternate expansion and contraction, act as pumps forcing water absorbed by the root-hairs into the xylem vessels.

(b) **Pericycle.** This forms a multi-layered zone between the endodermis and the vascular bundles and occurs as a cylinder encircling the vascular bundles and the pith, as in dicotyledonous stems. It may consist wholly of sclerenchyma forming a continuous zone, as in the gourd stem (see Fig. 45), but more commonly it is made of both parenchyma and sclerenchyma, the latter forming isolated strands in it. Each such strand associated with the phloem or bast of the vascular bundle in the form of a cap is known as the **hard bast,** as in the sunflower stem (see Fig. 43), In roots, the pericycle consists of a single layer of small, very thin-walled, more or less barrel-shaped cells.

Functions. In all roots the pericycle is the seat of origin of lateral roots (see Fig. 52), In dicotyledonous roots it further gives rise to lateral meristems—a portion of the cambium (see Fig. 59) and later the whole of the cork cambium (see Fig. 61), In all stems the pericycle is the seat of origin of adventitious roots. Otherwise its function is mechanical or storage.

(c) **Pith and Pith Rays.** The **pith** or **medulla** forms the central core of the stem and the root and is usually made of large-celled parenchyma with abundant intercellular spaces. In the dicotyledonous stem the pith is often large and well-developed. In the dicotyledonous root the pith is either small or absent, bigger vessels having met in the centre; while in the monocotyledonous root a distinct large pith is present. It is often parenchymatous, but sometimes sclerenchymatous. In the dicotyledonous stem the pith extends outwards to the pericycle between the vascular bundles. Each such extension, which is a strip of parenchyma, is called the **pith ray** or **medullary ray**. It is not present as such in the root.

Functions. They serve to store food material. The function of the sclerenchymatous pith is, of course, mechanical. The medullary ray further transmits water and food material outwards to the peripheral tissues, and is the seat of origin of a strip of cambium (see Fig. 55) prior to secondary growth.

3. **The Vascular Tissue System.** This system consists of a number of vascular bundles which are distributed in the **stele**. The stele is the

central column of the dicotyledonous stems and all roots surrounded by the endodermis, and consists of pericycle, vascular bundles, medullary rays and pith. Each bundle may be made up of both **xylem tissue** and **phloem tissue** with a cambium, as in dicotyledonous stems, or without a cambium, as in monocotyledonous stems, or of only one kind of tissue—xylem or phloem, as in roots. The function of this system is to conduct water and raw food materials from the roots to the leaves, and prepare food materials from the leaves to the storage organs and the growing regions. The vascular bundles may be regularly arranged in a ring, as in the stems of most dicotyledons and in all roots, or they may be scattered in the ground tissue, as in the stems of monocotyledons.

Elements of a Vascular Bundle (Fig. 40). A vascular bundle of a dicotyledonous stem, when fully formed, consists of three well-defined

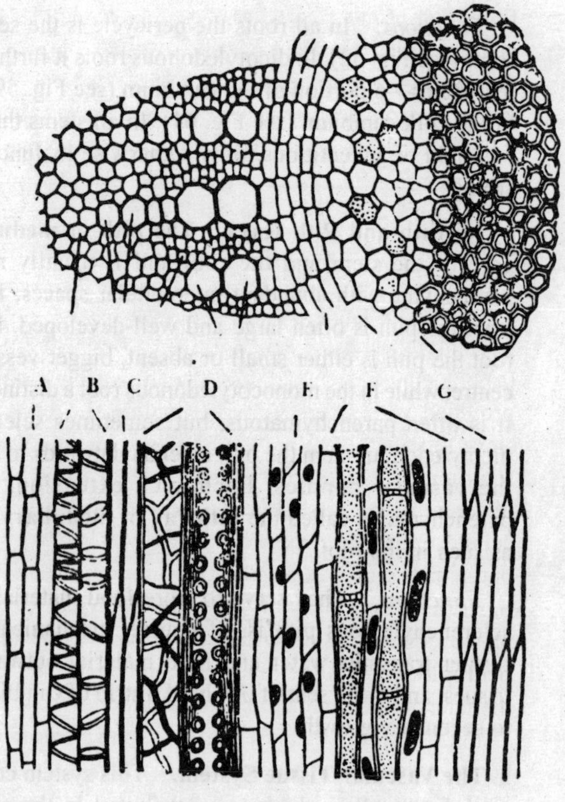

Figure 40. Vascular bundles of sunflower stem in transverse and longitudinal sections. *A,* wood parenchyma; *B,* protoxylem (annular and spiral vessels); *C,* tracheids and wood fibres; *D,* metaxylem (reticulate and pitted vessels); *E,* cambium; *F,* phloem (sieve-tubes companion cells and phloem parenchyma); *G,* sclerenchyma (hard bast).

tissues: (1) xylem or wood, (2) phloem or bast, and (3) cambium. They have different kinds of tissue elements.

(1) **Xylem** or **Wood** (see pp. 197). This lies towards the centre, and is composed of the following elements: (a) tracheae or vessels, (b) some tracheids, (c) a number of wood fibres, and (d) a small patch of wood parenchyma. Vessels are of various kinds (see Fig. 29) such as *spiral, annular, scalariform, reticulate* and *pitted* (with simple or bordered pits). Some tracheids also lie associated with the vessels. Wood fibres and wood parenchyma are ordinary sclerenchymatous and parenchymatous cells lying associated with the wood or xylem, and provided with simple pits in their walls. Xylem vessels and tracheids are used for the *conduction* of water and mineral salts from the roots to the leaves and other parts of the plant; xylem parenchyma assists them in their task and also serves the purpose of food storage, and wood fibres give proper rigidity to the xylem. Except for the wood parenchyma all other xylem elements are thick-walled, lignified and dead, and hence they also give *mechanical strength* to the plant body. The first-formed xylem or **protoxylem** consists of *annular, spiral* and *scalariform* vessels; it lies towards the centre of the stem and its vessels have smaller cavities. The later-formed xylem or **metaxylem** consists of *reticulate* and *pitted* vessels and some *tracheids;* it lies away from the centre and its vessels have much bigger cavities. Xylem is *endarch* in stems, and its development is *centrifugal*.

(2) **Phloem** or **Bast** (see pp. 198). This lies towards the circumference, and consists of (a) sieve-tubes, (b) companion cells, and (c) phloem parenchyma. Companion cells and phloem parenchyma are provided with simple pits, particularly in the walls lying against the sieve-tubes. Phloem as a whole is used for translocation of prepared food materials from the leaves to the storage organs and also to the different growing regions. All the elements of phloem are made of cellulose, and are living. Primary phloem hardly ever contains bast fibres, but it may be capped by a patch of pericyclic sclerenchyma, called **hard bast,** as seen in the sunflower stem (see Fig. 43). The outer portion of phloem consisting of narrow sieve-tubes is the first-formed phloem or **protophloem** and the inner portion consisting of bigger sieve-tubes is the later-formed phloem or **metaphloem**.

(3) **Cambium.** This is a thin strip of primary meristem lying in between xylem and phloem. It usually consists of a few layers of thin-walled and roughly rectangular cells. Although cambial cells look

rectangular in transverse section, they are much elongated, often with oblique ends. They become flattened tangentially, i.e. at right angles to the radius of the stem. Cambium is responsible for secondary growth in thickness of the plant body.

Types of Vascular Bundles (Fig. 41). According to the arrangement of xylem and phloem, the vascular bundles are of the following types:

(1) **Radial** (A), when xylem and phloem form separate bundles and these lie on different radii alternating with each other, as in roots.

(2) **Conjoint,** when xylem and phloem combine into one bundle. There are different types of conjoint bundles.

(a) **Collateral** (B–C), when xylem and phloem lie together on the same radius, xylem being internal and phloem external.

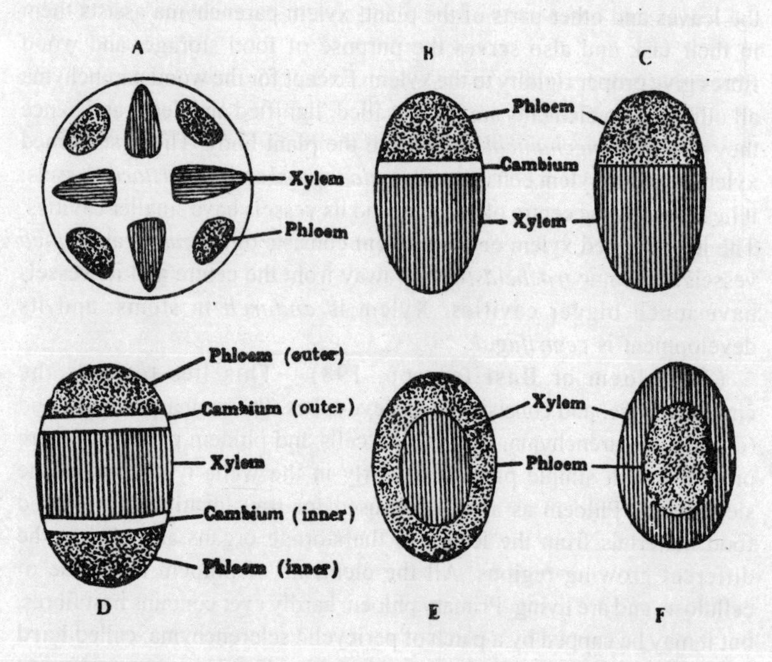

Figure 41. Types of Vascular Bundles. *A,* radial; *B,* collateral (open); *C,* collateral (closed); *D,* bicollateral; *E,* concentric (xylem central); *F,* concentric (phloem central).

When, in a collateral bundle, the cambium is present, as in dicotyledonous stems, the bundle is said to be *open* (B), and when the cambium is absent it is said to be *closed* (C), as in monocotyledonous stems.

(b) **Bicollateral** (D), when in a collateral bundle both phloem and cambium occur twice—once on the outer side of xylem and then again on its inner side. The sequence is: outer phloem, outer cambium, xylem, inner cambium and inner phloem. Bicollateral bundle is characteristic of the gourd family. It is always open.

(c) **Concentric** (E–F), when xylem lies in the centre and is surrounded by phloem (E), as in ferns, or phloem lies in the centre and is surrounded by xylem (F); the latter type is found only in some monocotyledons, e.g. sweet flag (*Acorus;* B. & H. BOCH), dragon plant (*Dracaena*) and dagger plant (*Yucca*). A concentric bundle is always closed.

Apical Meristems and Tissue systems

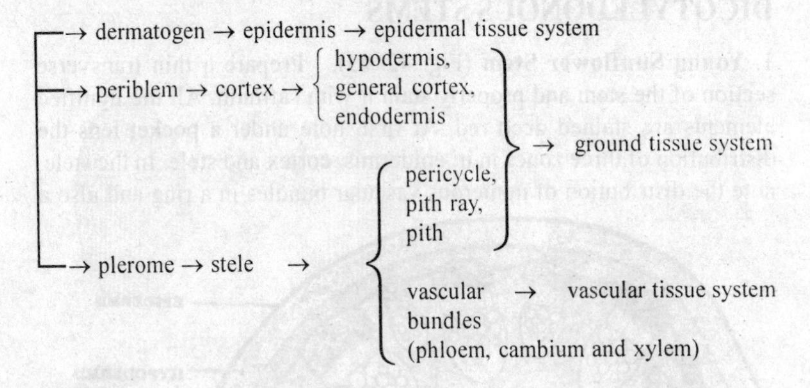

Chapter 4

Anatomy of Stems

DICOTYLEDONOUS STEMS

1. **Young Sunflower Stem** (Fig. 42–43). Prepare a thin transverse section of the stem and properly stain it with safranin. All the lignified elements are stained deep red. At first, note under a pocket lens the distribution of three zones in it: epidermis, cortex and stele. In the stele, note the distribution of numerous vascular bundles in a ring and also a

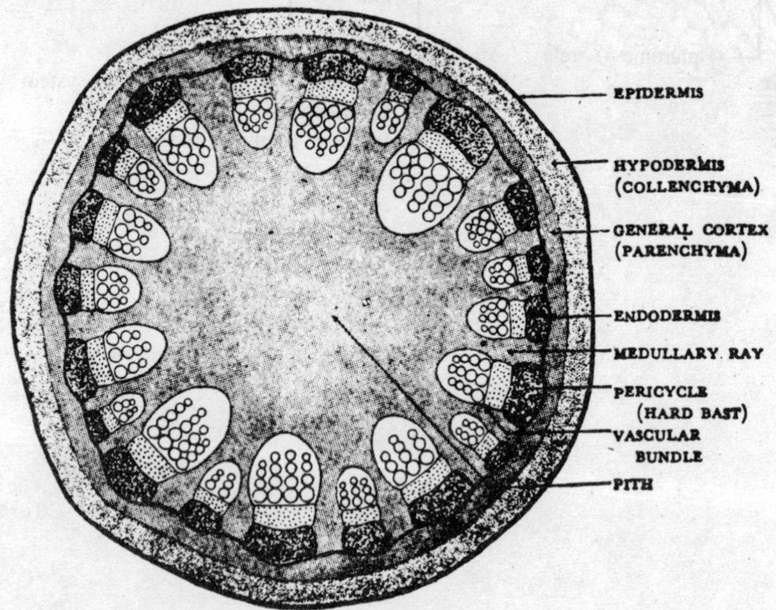

Figure 42. Young sunflower stem in transverse section, as seen under a pocket lens.

large pith. Then, under a microscope, study the internal structure of a sector in detail.

(1) **Epidermis.** This forms the outermost layer, and consists of a single row of cells, flattened tangentially and fitting closely along their radial walls, with a well-defined cuticle extending over it. Here and there it bears some multicellular hairs and a few stomata but no chloroplasts, except in the guard cells.

(2) **Cortex.** This is the zone that lies between the epidermis and the pericycle, and consists of hypodermis externally, general cortex centrally, and endodermis internally.

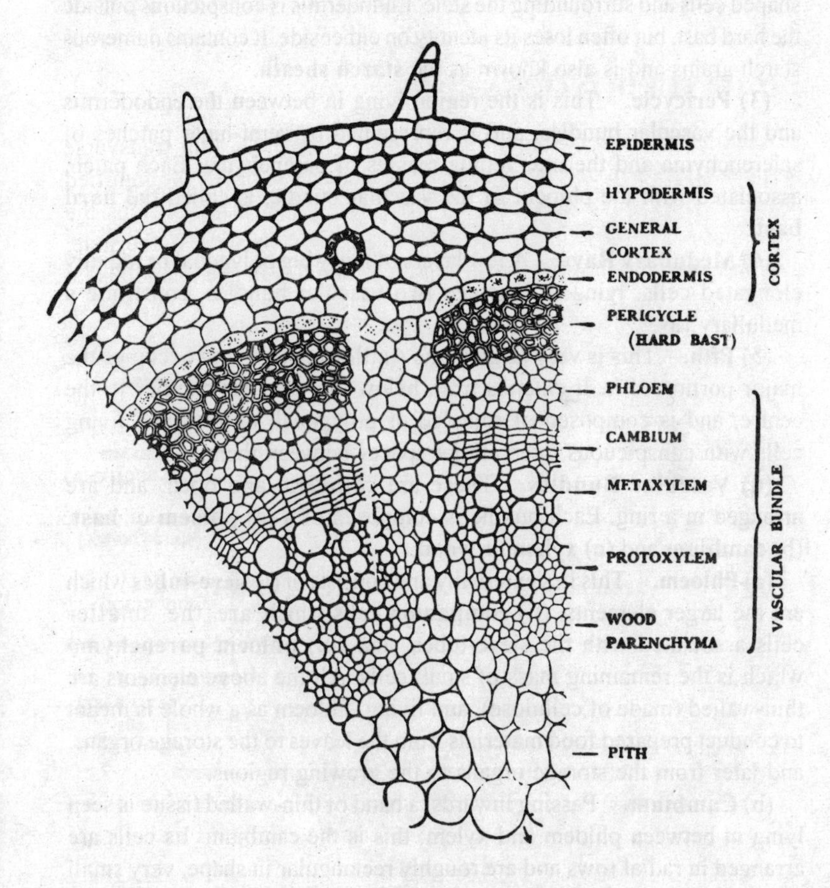

Figure 43. Young sunflower stem (a sector) in transverse section.

(a) **Hypodermis** (collenchyma)—this lies immediately below the epidermis, and consists of some 4 or 5 layers of collenchymatous cells. These cells are specially thickened at the corners against the intercellular spaces owing to a deposit of cellulose and pectin. The cells are living and contain a number of chloroplasts. (b) **General cortex**—this lies internal to the hypodermis and consists of a few layers of thin-walled; large, rounded or oval parenchymatous cells. It may be reduced to 1 or 2 layers outside the vascular bundle. There are conspicuous intercellular spaces in it. Some isolated resin-ducts are also seen here and there in it. (c) **Endodermis**—this is the innermost layer of the cortex consisting of more or less barrel-shaped cells and surrounding the stele. Endodermis is conspicuous outside the hard bast. but often loses its identity on either side. It contains numerous starch grains and is also known as the **starch sheath**.

(3) **Pericycle.** This is the region lying in between the endodermis and the vascular bundles, and is represented by semi-lunar patches of sclerenchyma and the intervening masses of parenchyma. Each patch, associated with the phloem of the vascular bundle, is called the **hard bast**.

(4) **Medullary Rays.** A few layers of fairly big polygonal or radially elongated cells, lying in between two vascular bundles, constitute a medullary ray.

(5) **Pith.** This is very large in the sunflower stem, and occupies the major portion of it. It extends from below the vascular bundles to the centre, and is composed of rounded or polygonal, thin-walled, living cells with conspicuous intercellular spaces between them.

(6) **Vascular Bundles.** These are collateral and open, and are arranged in a ring. Each bundle is composed of a (a) **phloem** or **bast**, (b) **cambium** and (c) **xylem** or **wood**.

(a) **Phloem.** This lies externally and consists of (i) **sieve-tubes** which are the larger elements; (ii) **companion cells** which are the smaller cells associated with the sieve-tubes; and (iii) **phloem parenchyma** which is the remaining mass of small cells. All the above elements are thin-walled (made of cellulose), and living. Phloem as a whole is meant to conduct prepared food materials from the leaves to the storage organs, and later from the storage organs to the growing regions.

(b) **Cambium.** Passing inwards, a band of thin-walled tissue is seen lying in between phloem and xylem; this is the cambium. Its cells are arranged in radial rows and are roughly rectangular in shape, very small in size and very thin-walled. Cambium is responsible for secondary growth in thickness of the plant body.

(c) **Xylem** or **Wood.** This lies internally and consists of the following elements. (i) **Wood vessels** are the large, lignified, thick-walled elements distributed in a few radial rows. The smaller vessels lying towards the centre constitute the *protoxylem,* and the bigger ones lying away from the centre constitute the *metaxylem*. Protoxylem consists of annular, spiral and scalariform vessels, and metaxylem of reticulate and pitted vessels. (ii) **Tracheids** and (iii) **wood fibres** are the smaller thick-walled and lignifted cells lying around the metaxylem vessels and in between them. In a transverse section of the stem these two can hardly be distinguished from each other. (iv) **Wood Parenchyma** is the patch of thin-walled parenchyma lying on the inner side of the bundle surrounding the protoxylem. Its cells are living.

2. **Young Gourd** (***Cucurbita***) **Stem** (Figs. 44–45). Prepare a thin transverse section of the stem and stain it properly with safranin. Note under a pocket lens the three zones in it—epidermis, cortex and stele. Further note the five ridges and five furrows, ten vascular bundles in

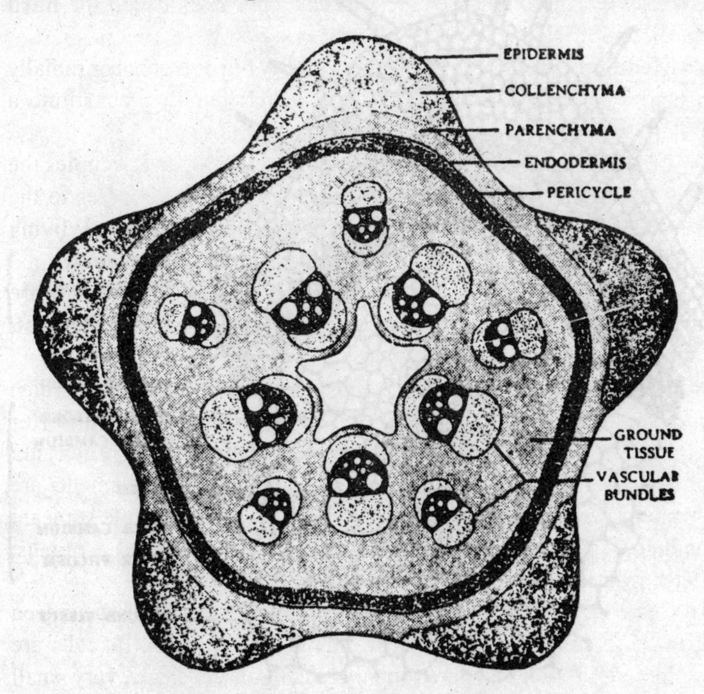

Figure 44. Young gourd (*Cucurbita*) stem in transverse section, as seen under a pocket lens.

two rows, the outer row corresponding to the ridges and the inner row to the furrows, and the central cavity (the stem being hollow), Then under a microscope study the internal structure of a sector in detail.

(1) **Epidermis.** This is the single outermost layer passing over the ridges and furrows; it often bears many long and narrow multicellular hairs.

(2) **Cortex.** This consists of hypodermis externally, general cortex in the middle, and endodermis internally. (a) **Hypodermis** (collenchyma) lies immediately below the epidermis, and consists of six or seven (sometimes more) layers of collenchymatous cells in the ridge. In the furrows the number of layers is reduced to two or three, sometimes none; in the furrows the underlying parenchyma may be seen to pass right up to the epidermis. Collenchyma contains some chloroplasts. (b) **General cortex** forms a narrow zone of parenchyma, 2 or 3 layers thick. In the furrows it often passes outwards right up to the epidermis.

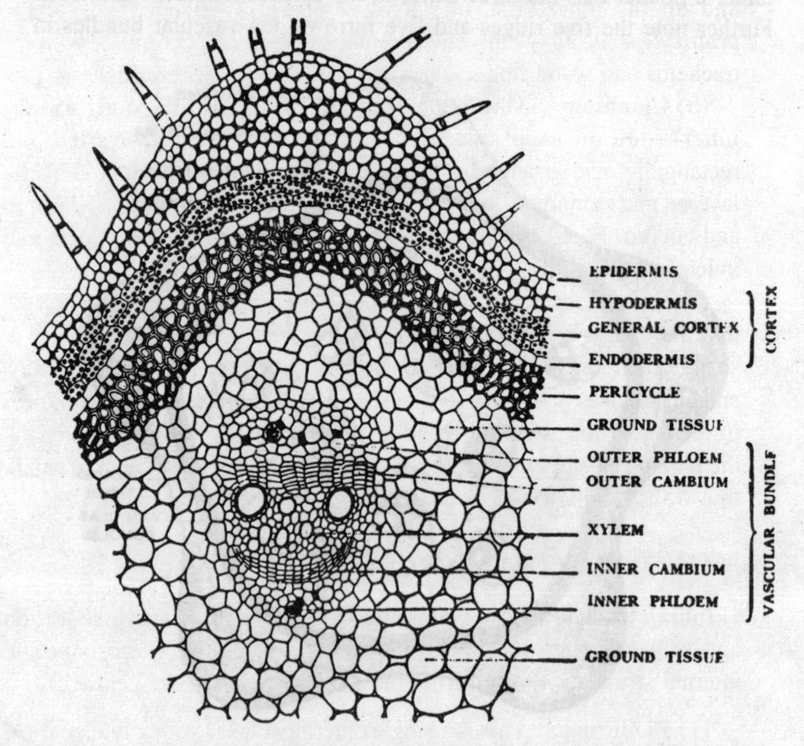

Figure 45. Young gourd (*Cucurbita*) stem (a sector) in transverse section.

Chloroplasts are abundant in the cortex. (c) **Endodermis** is the innermost layer of the cortex, lying immediately outside the pericycle. This layer is wavy in outline and contains starch grains.

(3) **Pericycle.** Below the endodermis there is a zone of sclerenchyma which represents the pericycle. This zone consists of 4 or 5 layers of thick-walled, lignified cells which are polygonal in shape.

(4) **Ground Tissue.** This is the continuous mass of thin-walled parenchyma extending from below the sclerenchyma to the pith cavity; in this tissue lie embedded the vascular bundles.

(5) **Vascular Bundles.** These are *bi-collateral,* usually, ten in number, and are arranged in two rows. Each bundle consists of (a) **xylem**, (b) **two strips of cambium**, and (c) **two patches of phloem**.

(a) **Xylem** occupies the centre of the bundle and consists, on the outer side, of very wide vessels (pitted) which constitute the *metaxylem,* and on the inner side, of narrower vessels which constitute the *protoxylem*. Protoxylem vessels remain scattered. There may be some tracheids and wood fibres, but wood parenchyma is abundant.

(b) **Cambium.** This tissue occurs in two strips—the outer and the inner—one on each side of xylem. Its cell are thin-walled and rectangular, and arranged in radial rows. The outer cambium is many-layered and is more or less flat, while the inner cambium is few-layered and curved. Each strip of cambium gradually merges into phloem and xylem.

(c) **Phloem** occurs in two patches—the outer and the inner. Note that the outer phloem is plano-convex and the inner one semi-lunar in shape. Each patch of phloem consists of sieve-tubes, companion cells, and phloem parenchyma. Sieve-tubes are very conspicuous in the phloem of the *Cucurbita* stem. Here and there sieve-plates with perforations in them may be distinctly seen. The rest of the phloem is made up of small, thin-walled cells which constitute the phloem parenchyma.

MONOCOTYLEDONOUS STEMS

1. **Indian Corn** or **Maize Stem** (Fig. 46). Cut a thin transverse section and properly stain it with safranin. Note under the microscope the internal structure in detail from the circumference to the centre.

(1) **Epidermis.** This is a single outermost layer with a thick cuticle on the outer surface. Here and there in the epidermis a few stomata may be seen.

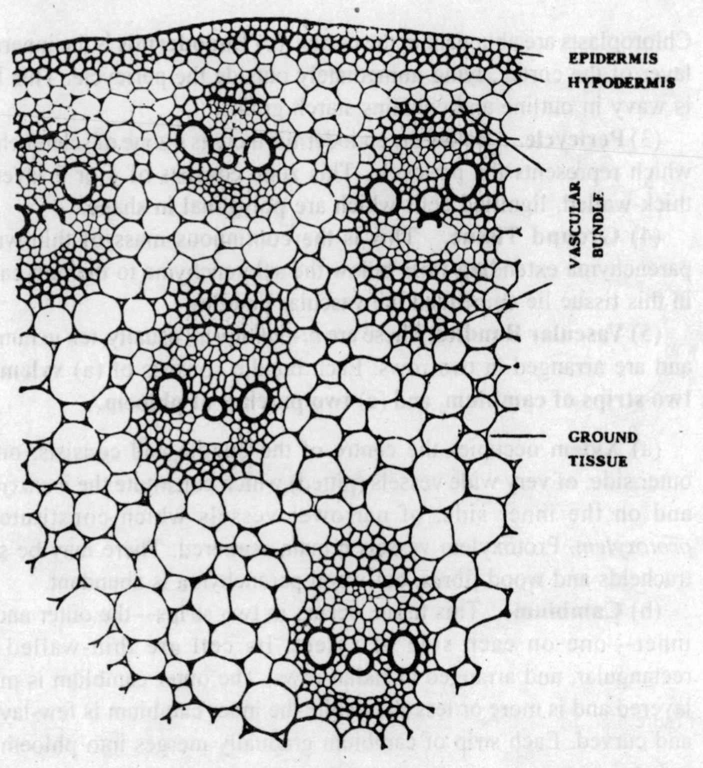

Figure 46. Maize or Indian corn stem (a sector) in transverse section.

(2) **Hypodermis** (sclerenchyma). This forms a narrow zone of sclerenchyma, usually two or three layers thick lying below epidermis.

(3) **Ground Tissue.** This is the continuous mass of thin-walled parenchyma, extending from below the sclerenchyma to the centre. It is not differentiated into cortex, endodermis, pericycle, etc., as in a dicotyledonous stem. The cells of the ground tissue enclose numerous intercellular spaces.

(4) **Vascular Bundles** (Fig. 47). These are collateral and closed, and lie scattered in the ground tissue; they are more numerous, and lie closer together nearer the periphery, than the centre. The peripheral ones are also seen to be smaller in size than the central ones. Each vascular bundle is somewhat oval in general outline and is more or less completely surrounded by a **sheath** of sclerenchyma which is specially developed on the two sides—upper and lower. The bundle consists of (a) xylem and (b) phloem only; cambium is altogether absent.

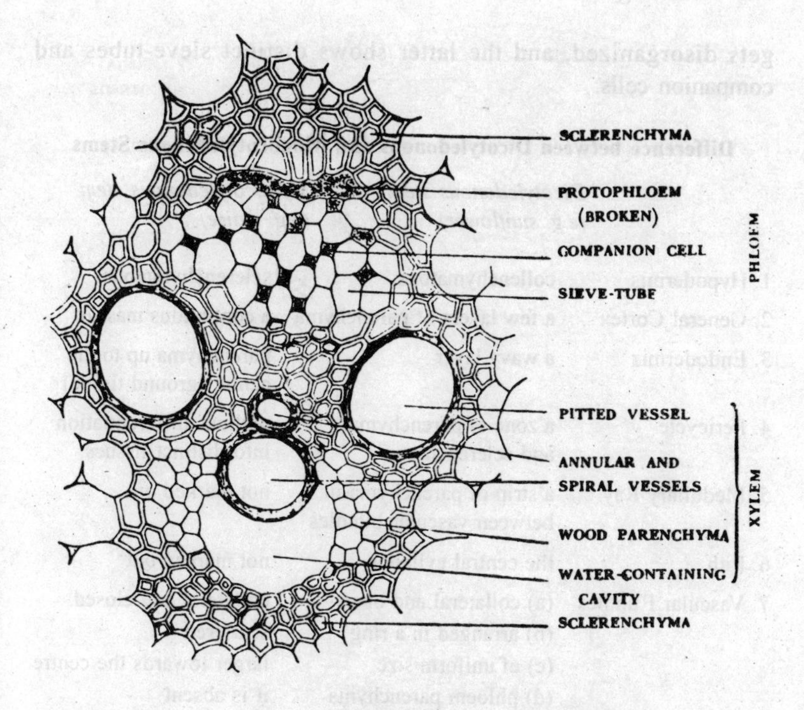

Figure 47. A vascular bundle of maize stem (magnified).

(a) **Xylem** mainly consists of usually four distinct vessels arranged in the form of a Y, and a small number of tracheids arranged irregularly. The two smaller vessels (annular and spiral) lying radially towards the centre constitute the *protoxylem,* and the two bigger vessels (pitted) lying laterally together with the small pitted tracheids lying in between them constitute the *metaxylem*. Besides, thin-walled wood (or xylem) parenchyma almost surrounding a conspicuous water-containing cavity is present in the protoxylem, and a few wood fibres occur associated with the tracheids in between the two big pitted vessels. The water-containing cavity forms lysigenously, i.e. by the breaking down of the inner protoxylem vessel and the contiguous parenchyma during the rapid growth of the stem.

(b) **Phloem** consists exclusively of sieve-tubes and companion cells; no phloem parenchyma is present in the monocotyledonous stem. The outermost portion of the phloem, which is a broken mass, is the *protophloem,* and the inner portion is the *metaphloem*. The former soon gets disorganized, and the latter shows distinct sieve-tubes and companion cells.

Difference between Dicotyledonous and Monocotyledonous Stems

	Dicotyledonous stem (e.g. sunflower)	*Monocotyledonous stem (e.g. maize)*
1. Hypodermis	collenchymatous	sclerenchymatous
2. General Cortex	a few layers of parenchyma	a continuous mass of
3. Endodermis	a wavy layer	parenchyma up to the centre (ground tissue)
4. Pericycle	a zone of parenchyma and sclerenchyma	without differentiation into distinct tissues
5. Medullary Ray	a strip of parenchyma in between vascular bundles	not marked out
6. Pith	the central cylinder	not marked out
7. Vascular Fundles	(a) collateral and open	collateral and closed.
	(b) arranged in a ring	scattered
	(c) of uniform size	larger towards the centre
	(d) phloem parenchyma present	it is absent
	(e) usually wedge-shaped	usually oval
	(f) bundle sheath absent	strongly developed

2. **Flowering Stem (Scape) of *Canna*** (Fig. 48). A thin transverse section stained with safranin shows the following internal structure under a microscope.

(1) **Epidermis.** This is the single outermost layer of very small, polygonal cells flattened tangentially. Its outer walls are cutinized.

(2) **Ground Tissue System.** From below the epidermis to the centre the whole mass of tissues, leaving out the vascular bundles, constitutes the ground tissue system. It is differentiated into (a) **cortex** consisting of two layers of fairly large polygonal cells, (b) **chlorophyllous tissue** consisting of one or two layers of chloroplast-bearing cells, intruding inwards here and there, (c) several patches of **sclerenchyma** of different sizes, lying against the chlorophyllous tissue, and (d) **ground tissue** consisting of a continuous mass of large, thin-walled, parenchymatous cells, containing starch grains and enclosing numerous intercellular spaces between them.

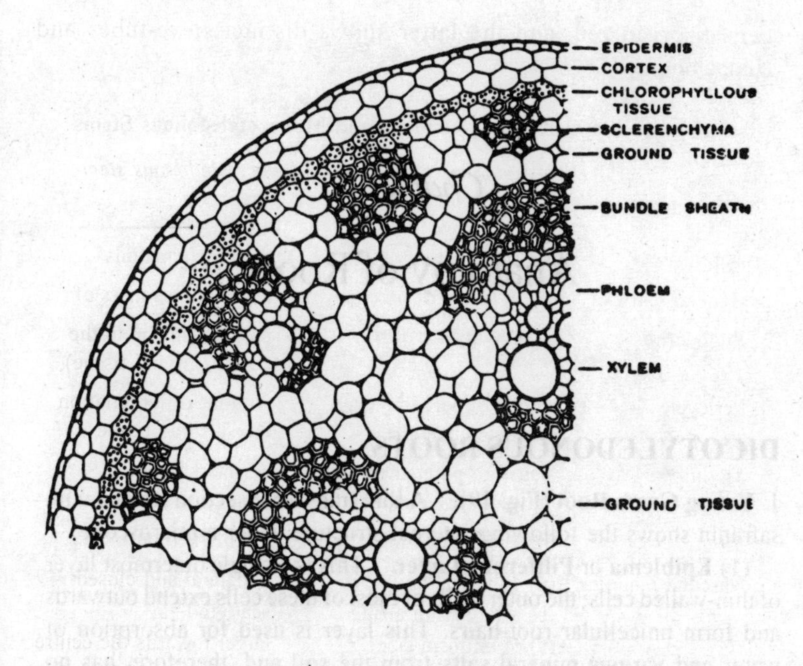

Figure 48. Flowering stem (scape) of *Canna* (a sector) in transverse section.

(3) **Vascular Bundles.** These are numerous and of different sizes, lying scattered in the ground tissue. Each bundle is collateral and closed. It is incompletely surrounded by a sheath of sclerenchyma (**bundle sheath**), with a distinct patch of it on the outer side in the form of a cap, and a thin strip on the inner side; seldom is a regular and complete sheath formed encircling the vascular bundle. Each bundle consists of (a) xylem on the inner side, and (b) phloem on the outer. **Xylem** consists of a large prominent spiral vessel, often with one or two smaller ones, also spiral in nature, lying usually on its outer side, and some parenchyma. **Phloem** consists of sieve-tubes and companion cells.

Chapter 5

Anatomy of Roots

DICOTYLEDONOUS ROOTS

1. **Young Gram Root** (Fig. 49). A thin transverse section stained with safranin shows the following internal structure under a microscope.

(1) **Epiblema** or **Piliferous Layer.** This is a single outermost layer of thin-walled cells; the outer walls of most of these cells extend outwards and form unicellular root-hairs. This layer is used for absorption of water and various mineral salts from the soil and, therefore, has no cuticle. Root-hairs increase the absorbing surface of the root.

(2) **Cortex.** This consists of many layers of thin-walled rounded cells, with numerous intercellular spaces between them. The cells of the cortex contain leucoplasts and store starch grains.

(3) **Endodermis.** This is a single ring-like layer of barrel-shaped cells which are closely packed without intercellular spaces. The radial walls of this layer are often thickened, and sometimes this thickening extends to the inner walls also. The endodermis is the innermost layer of the cortex and surrounds the stele as a cylinder.

(4) **Pericycle.** This lies internal to the endodermis and, like it, is a single ring-like layer; its cells, however, are much smaller and thinner-walled but with abundant protoplasm.

(5) **Conjunctive Tissue.** The parenchyma lying in between xylem and phloem bundles constitutes the **conjunctive tissue**.

(6) **Pith.** This occupies only a small area in the centre of the root. Sometimes the pith is nearly obliterated owing to the wood vessels meeting in the centre.

(7) **Vascular Bundles.** These are arranged in a ring, as in the dicotyledonous stem but here xylem and phloem form an equal number of separate bundles, and their arrangement is *radial* (see p. 213). The

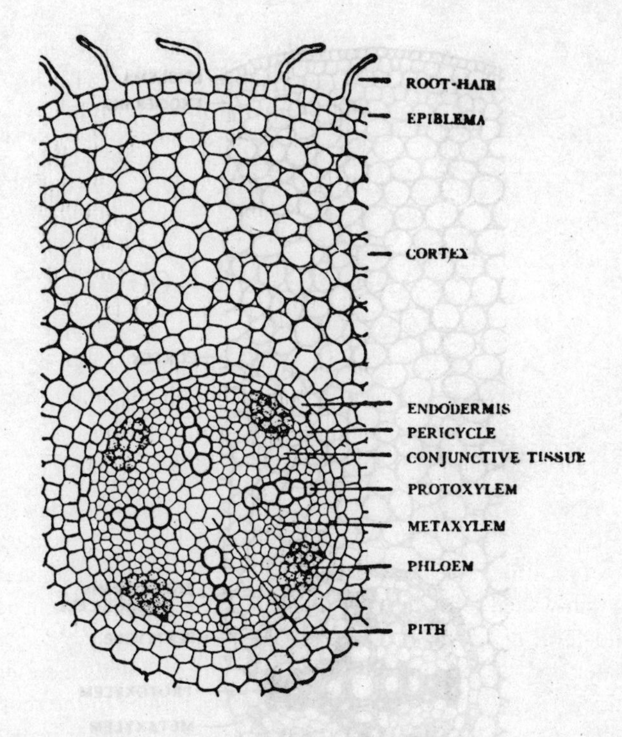

Figure 49. Young gram root in transverse section.

number of xylem or phloem bundles varies from two to six, very seldom more. The cambium is absent in the young root but soon makes its appearance. **Phloem bundle** consists of sieve-tubes, companion cells and phloem parenchyma. **Xylem bundle** consists of protoxylem which lies towards the circumference abutting on the pericycle, and metaxylem towards the centre. Xylem is *exarch* in roots, and its development is *centripetal*. Protoxylem is composed of small vessels (annular and spiral) and metaxylem of bigger vessels (reticulate and pitted). The metaxylem groups often meet in the centre, and then the pith gets broken.

2. **Young Buttercup (*Ranunculus*) Root** (Fig. 50). (1) **Epiblema**—the single outermost layer. (2) **Exodermis**—a few layers internal to the epiblema, representing the outer zone of the cortex. (3) **Cortex**—several layers of rounded or oval cells, leaving a lot of intercellular space beween them. (4) **Endodermis**—a distinct layer of thick-walled cells, representing the inner zone of the cortex; passage cells, however, are thin-walled and mostly lie against the protoxylem. (5) **Pericycle**—a

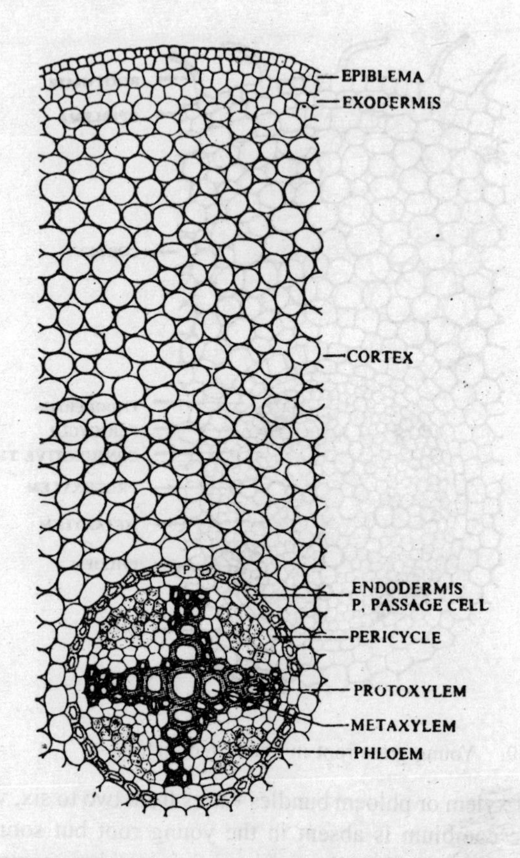

Figure 50. Young *Ranunculus* root in transverse section.

single layer of small and thin-walled cells, lying internal to the endodermis. (6) **Conjunctive tissue**—the parenchyma in between xylem and phloem. (7) **Vascular bundles**—radial, with 4 or 5 xylem bundles and as many phloem bundles; xylem is exarch and the metaxylem vessels meet in the centre. The pith is absent.

MONOCOTYLEDONOUS ROOT

Amaryllis Root (Fig. 51). A thin transverse section stained with safranin reveals the following internal structure under a microscope.

(1) **Epiblima** or **Piliferous Layer.** This is the single outermost layer with a number of unicellular root-hairs.

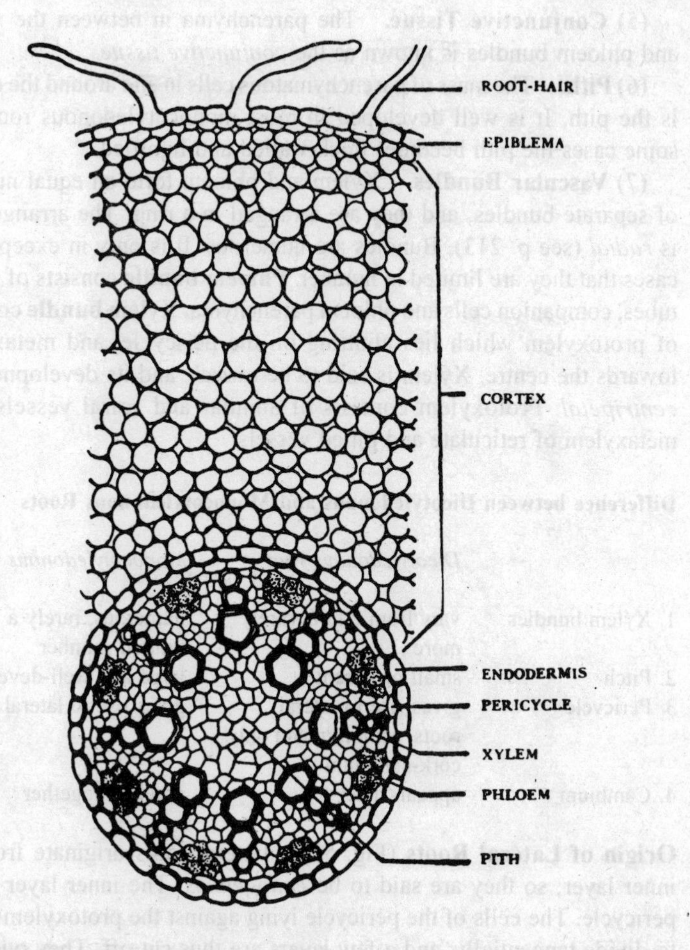

Figure 51. *Amaryllis* root in transverse section.

(2) **Cortex.** This is a many-layered zone of rounded or oval cells with intercellular spaces between them.

(3) **Endodermis**. This is the innermost layer of the cortex and forms a definite ring around the stele. Radial walls and often the inner walls of the endodermis are considerably thickened. Cells of the endodermis are barrel-shaped.

(4) **Pericycle.** This is the ring-like layer lying internal to the endodermis. Its cells are very small and thin-walled, but contain abundant protoplasm.

(5) **Conjunctive Tissue.** The parenchyma in between the xylem and phloem bundles is known as the *conjunctive tissue*.

(6) **Pith.** The mass of parenchymatous cells in and around the centre is the pith. It is well developed in most monocotyledonous roots. In some cases the pith becomes thick-walled and lignified.

(7) **Vascular Bundles.** Xylem and phloem form an equal number of separate bundles, and they are arranged in a ring. The arrangement is *radial* (see p. 213), Bundles are numerous. It is only in exceptional cases that they are limited in number. **Phloem bundle** consists of sieve-tubes, companion cells and phloem parenchyma. **Xylem bundle** consists of protoxylem which lies abutting on the pericycle, and metaxylem towards the centre. Xylem is said to be *exarch,* and its development is *centripetal*. Protoxylem consists of annular and spiral vessels, and metaxylem of reticulate and pitted vessels.

Difference between Dicotyledonous and Monocotyledonous Roots

	Dicotyledonous root	*Monocotyledonous root*
1. Xylem bundles	vary from 2 to 6, rarely more	numerous, rarely a limited number
2. Pitch	small or absent	large and well-developed
3. Pericycle	gives rise to lateral roots, cambium and only cork-cambium	gives rise to lateral roots
4. Cambium	appears later	absent altogether

Origin of Lateral Roots (Fig. 52). Lateral roots originate from an inner layer; so they are said to be *endogenous*. The inner layer is the pericycle. The cells of the pericycle lying against the protoxylem begin to divide tangentially, and a few layers are thus cut off. They push the endodermis outwards and tend to grow through the cortex. At this stage the three regions of the root-apex, namely, dermatogen (or calyptrogen), periblem and plerome, become marked out. The endodermis and some of the cells of the cortex form a part of the root-cap, but as the root passes through the soil this portion soon wears off, and the root-cap is renewed by the calyptrogen.

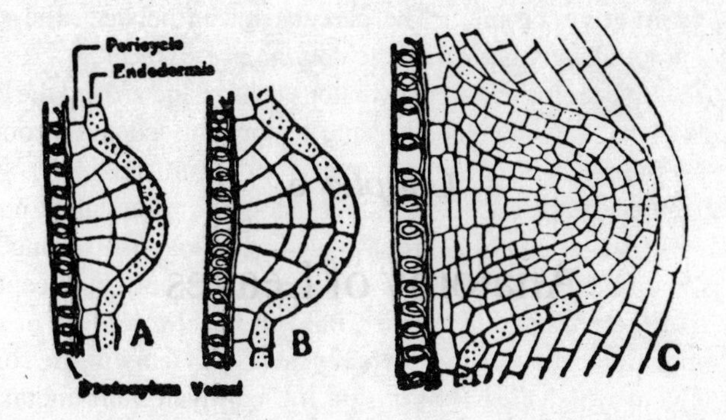

Figure 52. Origin of a lateral root. *A, B* and *C* are stages in its formation from the pericycle.

Chapter 6

Anatomy of Leaves

1. **Dorsiventral Leaf** (Fig. 53). A dorsiventral leaf (see p. 58) is more strongly illuminated on the upper surface than on the lower. This unequal illumination induces a difference in the internal structure between the

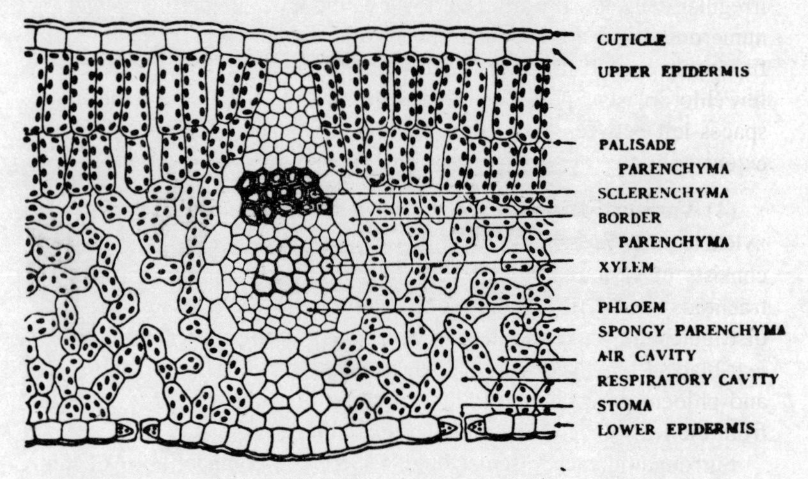

Figure 53. A dorsiventral leaf in section.

upper and the lower sides. A section made at a right angle to one of the bigger veins reveals the following internal structure.

(1) **Upper Epidermis.** This is a single layer of cells with a thick cuticle which checks excessive evaporation of water from the surface. It does not contain chloroplasts. Stomata are also usually absent.

(2) **Lower Epidermis.** This is also a single layer but with a thin cuticle. It is, however, interspersed with numerous stomata, the two

guard cells of which contain some chloroplasts; none are present in the epidermal cells. Internal to each stoma a large cavity, known as the *respiratory cavity,* may be seen. The lower epidermis of the leaf is meant for the exchange of gases (oxygen and carbon dioxide) between the atmosphere and the plant body. Excess water also evaporates from the plant body mainly through the lower epidermis.

(3) **Mesophyll.** The ground tissue lying between the upper epidermis and the lower one is known as the mesophyll. It is differentiated into (a) palisade parenchyma and (b) spongy parenchyma.

(a) **Palisade parenchyma** consists of usually one to two or three layers of elongated, more or less cylindrical cells, closely packed with their long axes at right angles to the epidermis. The cells contain numerous chloroplasts and manufacture sugar and starch in the presence of sunlight.

(b) **Spongy parenchyma** consists of oval, rounded, or more commonly irregular cells, loosely arranged towards the lower epidermis, enclosing numerous, large, intercellular spaces and air-cavities. They, however, fit closely around the vein or the vascular bundle. The cells contain a few chloroplasts. Spongy cells help diffusion of gases through the empty spaces left between them; they manufacture sugar and starch to some extent only.

(4) **Vascular Bundles.** Each vascular bundle (vein) consists of xylem towards the upper epidermis and phloem towards the lower. **Xylem** consists of various kinds of vessels (particularly annular and spiral), tracheids, wood fibres and wood parenchyma. Xylem conducts and distributes the water and the raw food material to different parts of the leaf-blade. **Phloem** consists of some narrow sieve-tubes, companion cells and phloem parenchyma. Phloern carries the prepared food material from the leaf-blade to the growing and storage regions.

Surrounding each vascular bundle there is a compact layer of thin-walled cells, containing a few chloroplasts or none at all; this layer is known as the **border parenchyma** or **bundle sheath**. It may extend radially towards the upper and the lower sides.

Frequently **sclerenchyma** occurs as a *sheath,* complete or incomplete, surrounding a bigger bundle, or as patches associated with xylem and phloem. Otherwise its distribution in the leaf is somewhat irregular.

2. **Isobilateral Leaf** (Fig. 54). An isobilateral leaf (see p. 58) is more or less equally illuminated on both sides. A section at a right angle to one or more veins reveals the following internal structure.

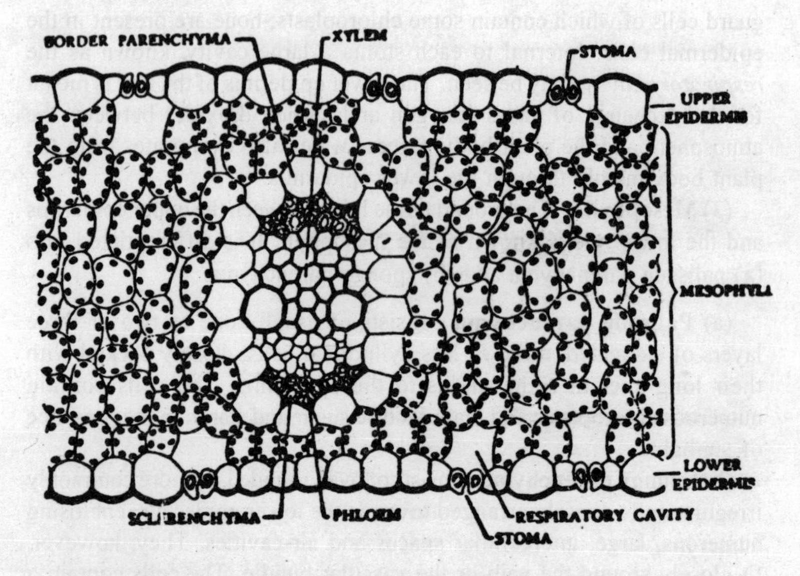

Figure 5·. An isobilateral leaf (a lily leaf) in section.

The structure is more or less uniform from one surface to the other. The epidermis on either side bears more or less an equal number of stomata, and is also somewhat uniformly thickened and cutinized. The mesophyll is often not differentiated into palisade and spongy parenchyma, but consists of spongy cells only, in which chloroplasts are evenly distributed.

Chapter 7

Secondary Growth in Thickness

1. **Dicotyledonous Stem.** In sturdy herbs and in all shrubs and trees secondary growth takes place as a result of the formation of new (secondary) tissues in them. Secondary tissues are formed by two meristems—**cambium** in the stelar region and **cork-cambium** formed later in the extra-stelar or cortical region. *The increase in thickness due to the addition of secondary tissues cut off by the cambium and the cork-cambium in the stelar and extra-stelar regions respectively is spoken of as* **secondary growth**.

A. *ACTIVITY OF THE CAMBIUM*

Cambium Ring. At first a portion of each medullary ray in a line with the cambium becomes meristematic and forms a strip of cambium called the **interfascicular cambium**. This joins on to the cambium proper on either side and forms a complete ring known as the **cambium ring** (Fig. 55). Secondary growth begins with the activity of this cambium ring. **Secondary Tissues.** The cambium ring as a whole begins to cut

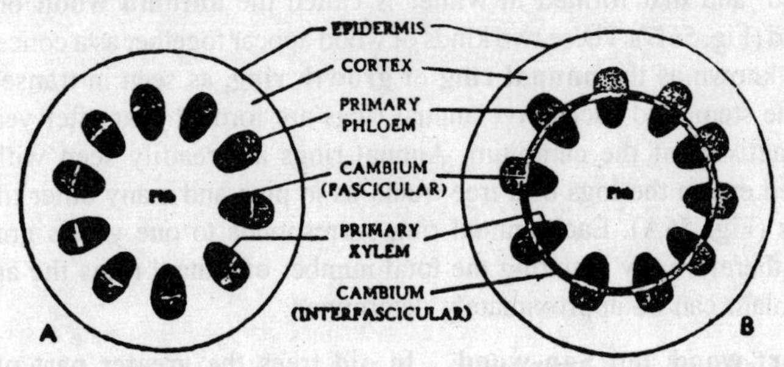

Figure 55. Formation of cambium ring.

off new cells both externally and internally. Those cut off on the outer side are gradually modified into the elements of phloem; these constitute the **secondary phloem**. The secondary phloem consists of sieve-tubes, companion cells and phloem parenchyma and often also some bands or patches of bast fibres. Many of the textile fibres of commerce such as jute, hemp, flax, rhea (or ramie), etc., are the bast fibres of secondary phloem.

The new cells cut off by the cambium on its inner side are gradually modified into the various elements of xylem; these constitute the **secondary xylem**. The secondary xylem consists of scalariform and pitted vessels, tracheids, numerous wood fibres arranged mostly in radial rows, and some wood parenchyma. The cambium is always more active on the inner side than on the outer. Consequently xylem increases more rapidly in bulk than phloem, and soon forms a hard compact mass, occupying the major portion of the stem. As xylem increases in bulk the peripheral tissues become stretched and some of them even get crushed. Primary xylem, however, remains intact.

Here and there the cambium forms some narrow bands of parenchyma, running across the stem in the radial direction through the secondary xylem and the secondary phloem; these are the **secondary medullary rays**. They are one, two or a few layers in thickness, and one to many layers in height.

Annual Rings (Fig. 56). The activity of the cambium increases or decreases according to favourable or unfavourable climatic conditions. Thus it is seen that in spring the cambium becomes more active and forms a greater number of vessels with wider cavities (large pitted vessels); while in winter it becomes less active and forms elements of narrower dimensions (narrow pitted vessels, tracheids and wood fibres) The wood thus formed in the spring is called the **spring wood** or early wood, and that formed in winter is called the **autumn wood** or late wood (Fig. 56B), These two kinds of wood appear together as a concentric ring known as the **annual ring** or **growth ring,** as seen in transection of the stem, and successive annual rings are formed year after year by the activity of the cambium. Annual rings are readily seen with the naked eye in the logs of a tree trunk, as in pine and many other timber trees (Fig. 56A). Each annual ring corresponds to one year's growth, and therefore, by counting the total number of annual rings the age of the plant can be approximately determined.

Heart-wood and **Sap-wood.** In old trees the greater part of the secondary wood is filled up with tannins, resins, gums, essential oils,

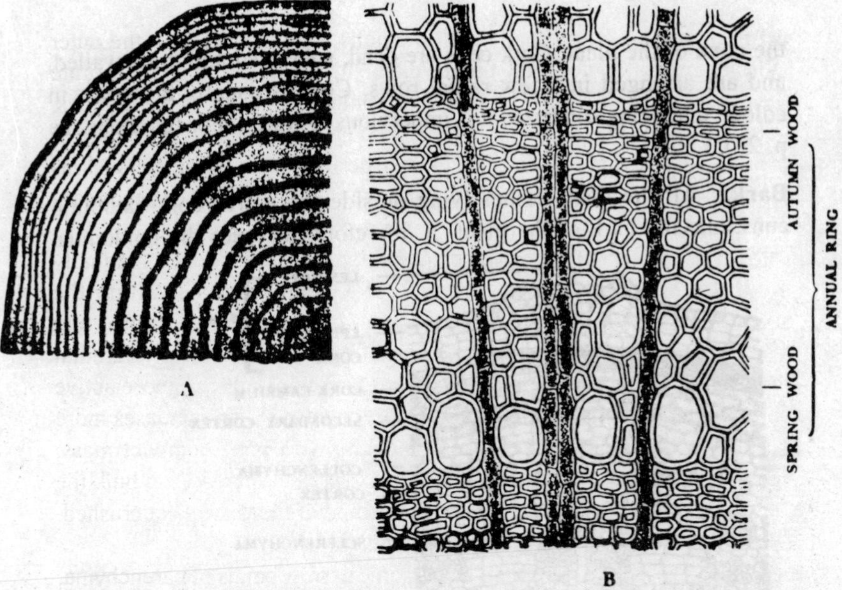

Figure 56 Secondary growth in stem. *A,* cut surface of a stem showing annual rings; *B,* transverse section of a stem showing an annual ring (magnified).

etc., which make it hard and durable. This region is known as the **heartwood** or **duramen**. It looks dark or brown. The heart-wood does no longer conduct water, but it simply gives mechanical support to the stem. The outer region of the secondary wood which is of lighter colour is known as the **sap-wood** or **alburnum**, and this alone is used for conduction of water and salt solutions from the root to the leaf.

B. *ORIGIN AND ACTIVITY OF THE CORK-CAMBIUM*

Sooner or later another meristematic tissue, i.e. the **cork-cambium** (or phellogen), makes its appearance in the cortical region. Commonly it originates in the outer layer of collenchyma. It may also arise in the epidermis itself, or in the deeper layers of the cortex. It is a few layers in thickness and consists of narrow, thin-walled and roughly rectangular cells. It begins to divide and give off new cells on both sides—**cork** on the outer side and **secondary cortex** on the inner. The cells of the secondary cortex are parenchymatous in nature and often contain chloroplasts.

Cork. The new cells cut off by the cork-cambium on its outer side are roughly rectangular in shape and soon become suberized. They form the **cork** of the plant. Cork cells are dead, suberized and thick-walled, and are arranged in a few radial rows. Cork is usually brownish in colour, and being suberized it is impervious to water. (For functions see p. 241.)

Bark. All the dead tissues lying outside the active cork-cambium constitute the bark of the plant. It, therefore, includes the epidermis,

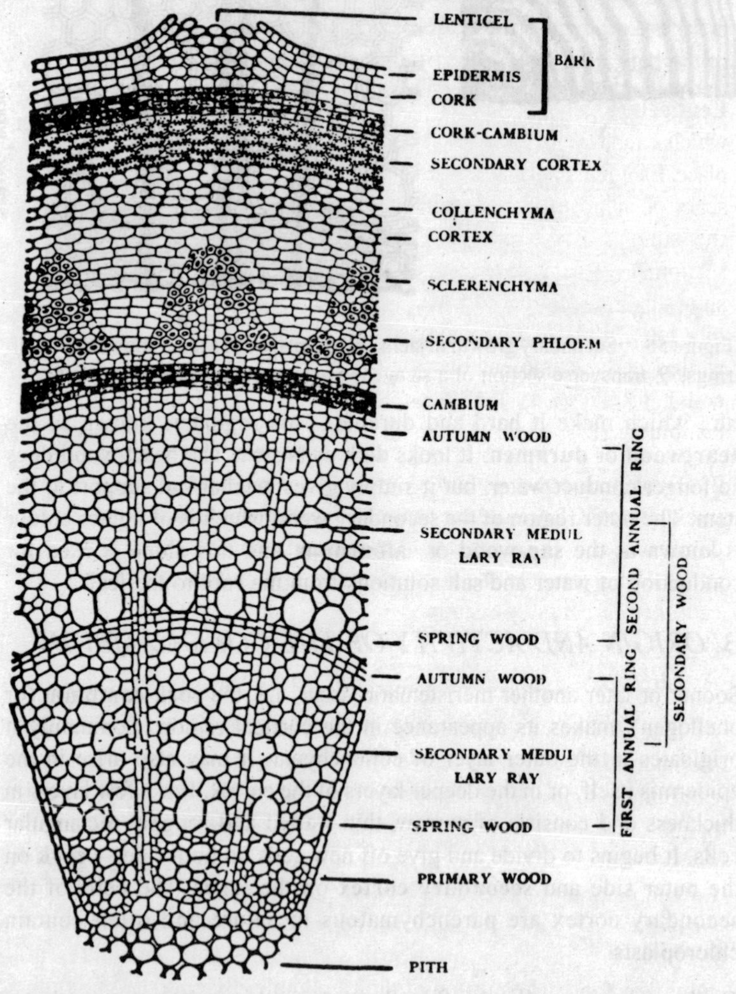

Figure 57. A two-year old dicotyledonous stem (a sector) in transverse section showing secondary growth in thickness.

lenticels and cork, and sometimes also hypodermis and a portion of the cortex, depending on the position of the cork-cambium, that is, the deeper the origin of the cork-cambium, the thicker the bark.

When the cork-cambium appears in the form of a complete ring the bark that is formed comes away in a sheet; such a bark is known as the **ring-bark**, as in *Betula* (B. BHURJJA-PATRA); and when it appears in strips the resulting bark comes away in the form of scales; such a bark is, therefore, known as the **scale-bark**, as in guava and pine. The function of the bark is to give protection (see p. 241).

Lenticels (Fig. 58). These are aerating pores formed in the bark, through which exchange of gases takes place. Externally they appear as scars or small protrusions on the surface of the stem. A section through one of the scars shows that the lenticel consists of a loose mass of small thin-walled cells (complementary cells), produced by the cork cambium.

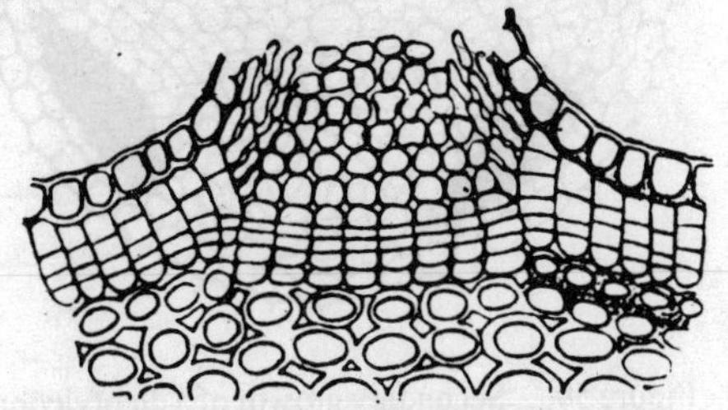

Figure 58. A lenticel, as seen in transverse section.

2. **Dicotyledonous Root.** As in the stem the secondary growth in thickness of the root is due to the addition of new tissues cut off by the cambium and the cork-cambium in the interior as well as in the peripheral region, In the root the secondary growth commences a few centimetres behind the apex.

A. *ORIGIN AND ACTIVITY OF THE CAMBIUM*

At first the conjunctive tissue on the inner side of phloem becomes meristematic and gives origin to a strip of cambium (Fig. 59). The cambium extends outwards between phloem and xylem. Then the portion of the pericycle just outside the protoxylem becomes meristematic; it divides and forms a strip of cambium there, joining with the earlier-formed cambium strips on either side of the xylem. Thus a continuous wavy band of **cambium** is formed, extending over the xylem and down the phloem. The secondary growth commences with the activity of this cambium band. The portion of the cambium adjoining the inner phloem becomes active first. It begins to cut off new cells more profusely on the inside. As a result the cambium and the phloem are gradually pushed

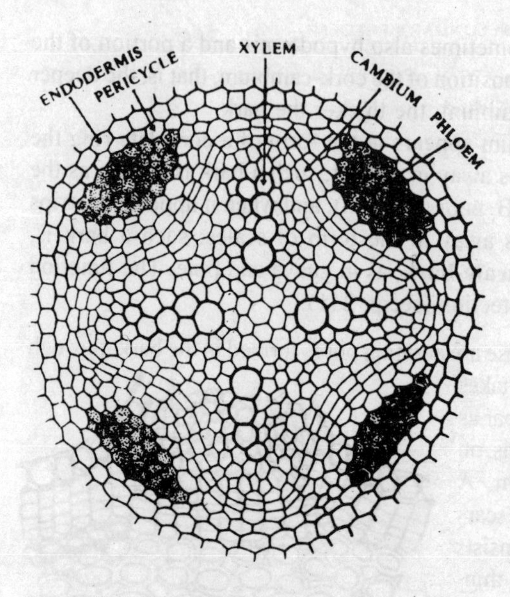

Figure 59. Secondary growth of a dicotyledonous root (early stage) showing the origin of the cambium.

outwards. The wavy band of cambium soon becomes circular or ring-like (Fig. 60). The whole of the cambium ring then becomes active, more so on the inner side than on the outer.

Secondary Xylem. The new cells cut off by the cambium ring on the inner side gradually become differentiated into the elements of xylem and all these new elements together constitute the secondary xylem. The secondary wood increases rapidly and soon forms the main bulk of the root. It is made of numerous large vessels with comparatively thin walls, abundance of wood parenchyma, but few wood fibres. As more wood is added, the cambium and phloem are gradually pushed further out. As the root lies underground it is not subjected to variations of aerial conditions; consequently annual rings, which are so characteristic of woody stems, are rarely formed in the root. Even when the root has increased considerably in thickness the primary xylem bundles still remain intact and can be recognized under the microscope in several cases. Against the protoxylem the cambium forms distinct and widening radial bands of parenchyma, which constitute the **medullary rays**. These extend up to the secondary phloem. Other smaller and thinner medullary rays are also formed later by the cambium. Medullary rays are larger and more prominent in the root than in the stem.

Secondary Phloem. The new elements cut off by the cambium on the outer side become gradually modified into the elements of phloem, and all these together constitute the secondary phloem. It consists of sieve-tubes with companion cells and abundant parenchyma, but less bast

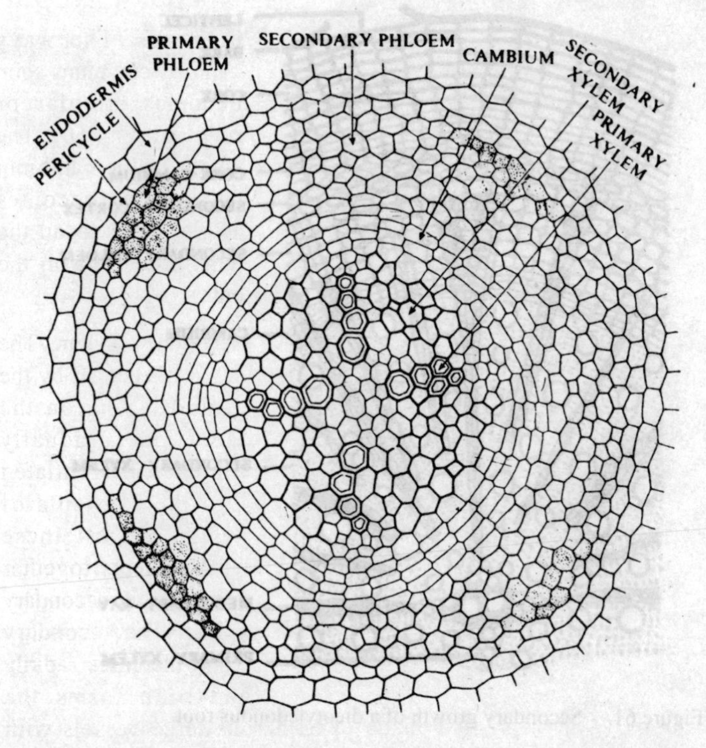

Figure 60. Secondary growth of a dicotyledonous root (later stage) showing the activity of the cambium ring with the formation of secondary xylem and secondary phloem.

fibres (except in special cases). The secondary phloem is much thinner than the secondary xylem. The primary phloem soon gets crushed.

B. *ORIGIN AND ACTIVITY OF THE CORK-CAMBIUM*

When the secondary growth has advanced to some extent, the single-layered pericycle as a whole becomes meristematic and divides into a few rows of thin-walled, roughly rectangular cells; these constitute the **cork-cambium** or **phellogen**. As in the stem, it produces a few browinish layers of **cork** on the outside, and **secondary cortex** on the inside. The secondary cortex of the root does not contain chloroplasts. The **bark** of the root is not extensive; it forms only a thin covering. The cortex, being thin-walled, gets disorganized. Such is also the fate of the endodermis. Epiblema dies out earlier. Here and there **lenticels** may be developed, as in the stem.

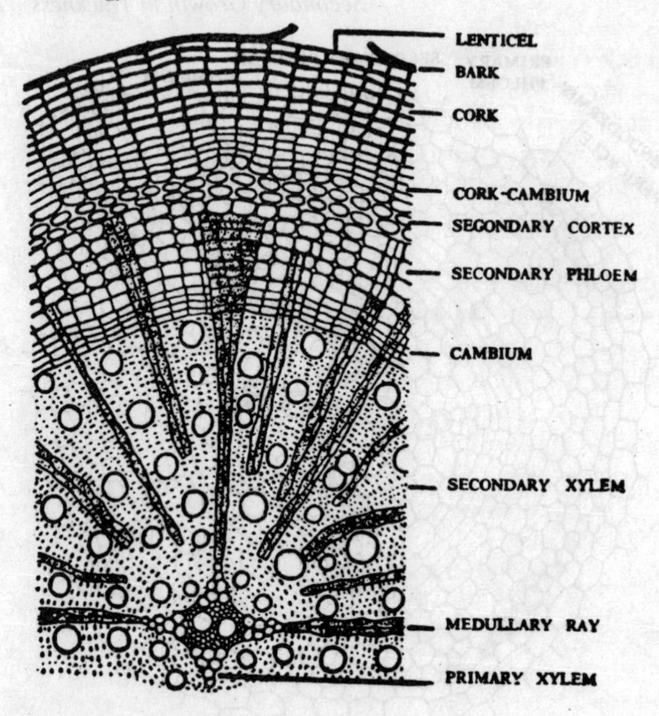

Figure 61. Secondary growth of a dicotyledonous root.

Functions of Cork and Bark. Cork and bark are protective tissues formed in shrubs and trees. Both are much thicker than the epidermis, but the cork has an additional advantage: it can be renewed by the underlying cork-cambium.

(1) **Cork.** (a) All the cork cells are suberized, and thus the cork acts as a waterproof covering to the stem. Loss of water by evaporation is, therefore, prevented or greatly minimized. (b) The cork tissue also protects the plant against the attacks of parasitic fungi and insects. (c) Cork cells, being dead and empty, containing air only, are bad conductors of heat. This being so, a sudden variation in outside temperature does not affect the internal tissues of the plant. (d) Cork is also made use of by the plant for the healing of wounds.

(2) **Bark.** Since bark is a mass of dead tissues lying externally as a hard dry covering, its function is protection. It protects the inner tissues against the attack of fungi and insects, against loss of water by evaporation, and against variation of external temperature.

Part III

PHYSIOLOGY

General Considerations
Soil

Chapter 1

General Considerations

Physiology deals with the various functions of life, such as the manufacturing of food, nutrition of the protoplasm, building up of the body, respiration, metabolism, reproduction, growth, movements, and so on. All these vital functions are performed by the protoplasm. To maintain the life and activity of the protoplasm the primary requirements are **water**, **air**, **food**, **heat** and **light**.

Water. Water is indispensable to the protoplasm for its manifold activities. There is always a high percentage of water—75–95%—associated with the protoplasm in its active state. Besides, inorganic materials are absorbed from the soil in a state of dilute solution; soluble food travels in the plant through the medium of water; similarly gases reach the protoplasm in solution, and many chemical changes are also carried out in solution in the plant body.

Air. Air is another necessity of the plant. Of the gases present in the air the plant normally utilizes only oxygen and carbondioxide. The plant requires oxygen for respiration and carbon dioxide for manufacture of food.

Food. Protoplasm also requires food for its nutrition. This is of primary importance to all living organisms; but unlike animals, plants manufacture their own food from raw food materials—water and inorganic salts absorbed from the soil, and carbon dioxide absorbed from the air. Food furnishes the necessary materials for body-building and is the source of energy.

Heat. A certain amount of heat is necessary to maintain the activity of the protoplasm and for all the vital processes carried on by it. Within certain limits, the higher the temperature, the greater the activity of the protoplasm. Generally speaking, the maximum temperature may be stated to be 45–50°C, with the optimum lying at about 30°C.

Light. Sunlight is the original source of energy and has a stimulating effect on growth; it makes the plant sturdy. It is not, however, essential in the early stages of growth. Light is an important factor responsible for green colouration of the plant, utilization of carbon dioxide of the air, and manufacture of sugar and starch. It is also responsible for some kinds of movement in plant organs.

Chapter 2

Soil

Since water and mineral salts are almost exclusively obtained from the soil for their utilization later in the plant body, a knowledge of soil science in different aspects is an essential prerequisite to the study of plant physiology.

Soil Formation. Soils are formed by the disintegration and decomposition of rocks due to weathering (action of rain-water, running streams, glaciers, wind, alternate high and low temperatures, etc.) and the action of soil organisms such as many bacteria, fungi, protozoa, earthworms, etc., and also interactions of various chemical substances present in the soil. Although soils are normally formed from underlying rocks in a particular region, they may be transported long distances by agencies such as rivers, glaciers, strong winds, etc.

Physical Nature. Physically the soil is a mixture of mineral particles of varying sizes—coarse and fine—of different degrees, some angular and others rounded, with a certain amount of decaying organic matter in it. The soil has been graded into the following types according to the size of the particles:

Coarse particles	...	2–.2 mm form coarse sand
Smaller particles	...	.2–.02 mm form sand
Finer particles	...	.02–.002 mm form silt
Very fine particles	...	less than .002 mm form clay

Types of Soil and their Properties. (1) **Sandy soil** contains more or less 60% of sand particles with a small proportion of clay and silt, usually not exceeding 10% of each. It is well aerated, being very porous; but as it allows easy percolation of water through the large pore spaces, it quickly dries up and often remains dry. This soil is loose and light and has no cohesive power. Capillarity decreases in this soil and it can hold only 25% of water of its own weight, when saturated. It contains very little plant food. It can, however, be improved by the addition of clay, lime or humus. Being loose and porous, it helps seed germination and root growth but is not suitable for subsequent growth. (2) **Clay soil** contains over 50% of clay particles. It is compact and heavy. It easily becomes water-logged, and is badly aerated. Drainage is difficult, and its workability equally so. It is hard and often cracks when dry, but becomes soft and sticky when wet. Capillarity increases considerably in this soil, and it has a great capacity for holding water (40% or more of its own weight), particles being very fine. This soil always contains a considerable amount of plant food, but the root cannot easily penetrate it. The addition of lime or sand improves it and makes it suitable for normal plant growth. (3) **Loam** contains 30–50% of silt and a small amount of clay (5–25%), the rest being sand. It is the best soil for vigorous plant growth and is most suitable for agricultural crops because all the important physical conditions are satisfied—porosity for proper aeration and for percolation (downward movement) of excess water, and capillarity for upward movement of sub-soil water. It can hold 50% or a little more of water of its own weight. At the same time it is rich in plant food.

The proportion of the above constituents of the soil can be approximately determined by stirring a small lump of soil in a beaker to which an excess of water has been added, and then pouring the contents into a measuring cylinder. When the mixture is allowed to settle, it is seen that sand particles collect at the bottom, silt higher up, and clay on the top, in three distinct layers. A fine portion of the clay, however, remains suspended in water. Their proportions are then determined and percentages calculated. Humus mostly floats on water.

There are other kinds of soils also. (1) **Calcareous soil** contains over 20% of calcium carbonate which is useful in neutralizing organic acids formed from humus. It is commonly whitish in colour. (2) **Laterite soil** contains a high percentage of iron and aluminium oxides. It is reddish, brownish or yellowish in colour. (3) **Peat soil** contains a high percentage (even up to 80% or 90%) of humus. It is dark in colour, porous and light. The floating garden of Kashmir is made of peat soil.

Soil Water and Soil Air. Ordinarily two-thirds of the pore space occupied by water and one-third occupied by air are found to be suitable for normal growth of most crop plants. An excess of water in the soil chokes its pore space and is, therefore, harmful to plants. Conversely a very low percentage of water in the soil results in the wilting of plants. The water loosely held by the small soil particles by *capillary force,* with mineral salts dissolved in it, is the water absorbed by the root-hairs.

Water Content of the Soil. To find out the water content of the soil the following procedure may be adopted. Collect from a depth of 0.3–1 m. a small sample of soil by digging the earth, and keep it in a stoppered jar. Take out a small lump from it and weigh it. Heat it at 100°C for a while, stirring the mass occasionally till all the water evaporates. After cooling take the weight of the soil again. To make sure that all the water has been driven out, heat the same soil over again. A constant weight of the soil will indicate the loss of all the water from it. The difference in weight is the quantity of water originally present in the soil. Then calculate the water content on a percentage basis.

Chemical Nature. Chemically the soil contains a variety of *inorganic salts* such as nitrates, sulphates, phosphates, chlorides, carbonates, etc., of potassium (K), calcium (Ca), magnesium (Mg), sodium (Na) and iron (Fe), and of the 'trace' elements like boron (B), manganese (Mn), copper (Cu), zinc (Zn), aluminium (Al), molybdenum (Mo), etc. A certain amount of *organic compounds,* chiefly proteins and their decomposition products, are also present in the soil. *Humus* (see below) is also present in many soils as a source of *organic food. Acidity* or *alkalinity* of the soil is no less important for plant growth than the availability of plant food in the soil. Soils containing a high amount of lime (calcium carbonate) are alkaline, and soils containing a high quantity of humus are acid. These conditions can, however be altered by the addition of one or the other, as the case may be. Most of the field crops prefer a slightly acid soil. The acid or alkaline nature of the soil may be tested by special chemical indicators. The soil containing a certain amount of lime (calcium carbonate) is said to be calcareous soil. The presence of calcium carbonate in such a soil can be detected by adding strong hydrochloric acid to a small sample of it when effervescence is noticed in it either with the naked eye or under pocket lens.

Humus. Humus is the decomposed vegetable matter formed in the soil from dead roots, trunks, branches and leaves under the action of soil bacteria and fungi. It forms a dark-coloured surface layer on the ground, often occurring to some depth in forests and swamps. It is very useful both physically and chemically. It is rich in plant food, particularly nitrogen. Humus absorbs and retains water to a considerable extent (about 190 parts of its own weight). Added to sandy soil it increases its water-holding capacity, and added to clay soil it loosens its compactness and makes it porous for better aeration. Soil containing 5–15% of humus is suitable for agricultural crops. It is mostly the seat of bacterial activities in the soil.

Humus Content of the Soil (ignition method). After heating a lump of soil at 100°C, to drive out the water, cool it in a desiccator and take its weight. Then in a platinum crucible burn the dehydrated soil at a high temperature for about an hour, occasionally stirring the mass. During ignition fumes are seen to escape. (Organic matter becomes converted into ammonia, oxides of nitrogen or free nitrogen, sulphur dioxide and carbon dioxide, and escapes as such). After complete combustion cool it in a desiccator and then weight it again. The loss in weight approximately represents the quantity of humus originally present in the soil sample. Then calculate the humus content of the soil on a percentage basis. The residue left after combustion is the incombustible or inorganic matter present in the soil.

Soil Organisms. Various kinds of bacteria are present in the soil, sometimes to the extent of a few million individuals per gram of soil, particularly in the region of organic matter, and many of them are useful agents of soil fertility. Thus nitrifying bacteria convert proteins of dead plants and animals into nitrates, and it is a fact that but for the activity of such bacteria the proteins would ever remain locked up in the soil as such without any further use. Then there are nitrogen-fixing bacteria, ammonifying bacteria, sulphur bacteria and a host of other types in the soil. Fungi are also abundant in the soil, particularly in acidic soil, often replacing bacteria. Like the bacteria they are also useful agents in decomposing proteins. Many algae are also present in the soil. It is now definitely known that many of the blue-green algae fix atmospheric nitrogen in the soil. Among animals the soil-dwellers such as protozoa, earthworms, rats, etc., are useful agents in altering the soil. The burrowing animals make the soil loose for better aeration and percolation of water.

Fertilizers. Ordinarily the soil contains the necessary salts required by plants. In the cultivated soil, however, deficiency occurs in one or more of them, particularly in nitrogen, phosphorus, potassium and calcium, and to make good this deficiency the use of chemical fertilizers and manures becomes a necessity. Fertilizers are certain chemical substances which when properly added to the soil make it fertile, i.e. enable it to produce more abundantly. Manuring the field for better crop production may be done by any of the following three methods.

(1) Artificial manuring is done by introducing into the soil certain chemical fertilizers such as ammonium sulphate, ammonium nitrate, super-phosphate, urea, leaf compost, bonemeal, oil-cakes, etc., in suitable proportions. It may be mentioned in this connection that the Fertilizer Corporation of India has set up a number of 'fertilizer units' in the country so that her required need of fertilizers may be adequately met internally, i.e., without foreign import. (2) Farmyard manuring is done by adding cowdung and organic refuse to the soil. (3) Green (natural) manuring is done by rotation of crops (see p. 262) and by growing leafy herbs, including certain nodule-bearing leguminous plants (see Fig. 2) and finally ploughing the whole lot into the field.

Biofertilizer. A good soil, rich in inorganic and organic nutrients, is better suited for cultivation. These nutrients are cycled back in nature in the soil and thus fertility of the soil is maintained. Sometimes when the soil becomes deficient in these nutrients due to over cultivation, heavy drainage, etc., the general practice is to add fertilizers or manures to the soil to meet such deficiencies. In recent years, with the increasing demand for food supply, the use of the inorganic fertilizers has also been increased to a greater extent in the field of agriculture. The addition of inorganic salts in excess and frequently to the soil results in an adverse effect on the soil. Dung and agricultural wastes are also used to increase soil fertility in different ways but these are not available in sufficient quantities. Therefore, as an alternative, biofertilizers have been brought into use which cost very little and they have no adverse effect on the soil and are self generating and self sustaining. Biofertilizers include Legume fertilizers like various kinds of nitrogen fixing soil legume bacteria, (*Rhizobium* bacteria) which directly supply amino acids to the Leguminous plants and the most nodules of the latter then directly increase soil fertility. Algal fertilizers such as the blue green algae (*Nostoc, Anabaena*), and Bacterial fertilizers like *Clostrodium, Azotobactor, Bacillus* which help in fixing free nitrogen of the air into the soil and thereby increase the fertility of the soil. These biofertilizers can be grown in culture in the laboratory. Besides, there are also some other indirect sources of biofertilizers. For example, rapidly and easily grown, unwanted aquatic weed (water hyacinth) is rich in potassium and phosphorus and this weed can decompose organic matters present in the refuse, drained in from various sources. Similarly, mycorrhizal fungus, which establishes a symbiotic relationships with the roots of higher plants, can be cultured and introduced into the soil. Likewise aquatic Fern, *Azolla,* which establishes symbiotic relationship with *Anabaena,* can be used as biofertilizer.

Physiology may be divided into:

A. **Physiology of nutrition (or chemical physiology)**
B. **Physiology of growth and movements**
C. **Physiology of reproduction**

A. Physiology of Nutrition

aeration of the roots. It is desirable that the culture solution should be renewed fortnightly. Later the following *observations* are made.

In bottle *A* (with normal culture solution) the growth of the seedling is normal. In bottle *B* (the same minus potassium salts) the growth becomes checked and leaves lose their colour. In bottle *C* (the same minus calcium salts) roots do not develop properly and leaves become yellowish, spotted and deformed. In bottle *D* (the same minus magnesium salts) chlorophyll does not develop and the seedlings is stunted in growth. In bottle *E* (the same minus iron salts) the seedling becomes chlorotic. In bottle *F* (the same minus nitrogen compounds) the seedling is weak and straggling, and leaves yellowish.

Inference. The inference that may be finally drawn from the water culture experiments is that the following elements in suitable soluble compounds are essential for normal growth of a plant: K, Ca, Mg, Fe among metals, H, O, N, S, P among non-metals, and certain 'trace' elements, e.g. Mn, Zn, Cu, Mo and B, making a total of 15 elements including C; that free oxygen and carbon dioxide are obtained from the air (and not from the soil); that free nitrogen of the air is of no use to the plant.

Sand Culture Experiments. To obviate many difficulties in water culture experiments it has become the growing practice with scientists to take to sand or charcoal culture. Charcoal is thoroughly washed and powdered. In the case of sand, it is washed, dried and then ignited to remove organic impurities. Normal culture solution is added to any of the two media and growth of the seedling studied. The effect produced on the seedling under the exclusion of a particular elements is studied in the same way as in water culture experiments.

Classification of Elements

Essential:	metals—K, Ca, Mg and Fe
	non-metal—C, H, O, N, S and P
Non-essential:	metal—Na
	non-metals—Cl and Si
Trace (essential):	metals—Mn, Zn, Cu and Mo
	non-metal—B

Although the other elements occur in easily detectable quantities, the trace elements, as is evident, occur in merc traces. Even then those listed above have been found to be essential for certain physiological processes. They are, however, poisonous in higher concentrations. They are obtained from the soil in the form of suitable salts.

Hydroponics. Hydroponics or soilless cultivation is the technique of growing plants directly in *normal culture solution* including the essential 'trace' elements without the use of soil, or in pure sand irrigated with this solution. Waterproof earthen vessels, troughs, semi-pucca beds, etc., filled with the solution are commonly used for the purpose, and these are laid out in verandahs, backyards, house roofs, etc. Rocky beds, barren areas, etc., where cultivation is not possible, are also profitably utilized for hydroponic culture. Hydroponics was established by Geriche of California University in the year 1929. By this method he was able to grow tomato plants to a height of 8 m. Hydroponics is now regarded as an established science. At present there are about twenty hydroponic research centres in the world

ROLE PLAYED BY THE ELEMENTS IN THE PLANT BODY

(1) **Potassium** is abundantly present in the growing regions. It is essentially a constituent of the protoplasm but is absent from the nucleus and the plastids. Potassium is known to help synthesis of carbohydrates and proteins and also growth of the plant body; starch grains are not formed in its absence. In the absence of potassium the stem becomes slender and the leaves lose their colour and gradually wither.

(2) **Magnesium** is present in the chlorophyll to the extent of about 5.6% by weight and, therefore, in its absence chlorophyll is not formed, and the plant becomes stunted in growth. It is present to a considerable extent in the seeds of cereals and leguminous plants.

(3) **Calcium** occurs in the cell-wall, particularly in the middle lamella, as calcium pectate. Calcium neutralizes certain organic acids to form insoluble salts such as calcium oxalate; otherwise the acids would be injurious to the protoplasm. It promotes the growth of roots. Plants like lemon, orange, shaddock, etc., grow well in a soil rich in calcium (lime). Fruits in general, and stone-fruits in particular, require plenty of calcium for their normal development. Plants become stunted in growth in the absence of calcium, and are liable to be diseased.

(4) **Iron** is essential for the formation of chlorophyll although it is not present in the latter. It may be associated with the plastids. Iron is always present in the protoplasm and in the chromatin of the nucleus. In the absence of iron the leaves become chlorotic, i.e. pale-yellow or pale-green in colour.

(5–6) **Sulphur and Phosphorus.** Sulphur is present in the protoplasm and enters into the composition of all plant proteins. It is an important constituent of mustard oil. In its absence leaves become chlorotic and the stem slender. Phosphorus is always present in the nucleoprotein, a constituent of nucleus, and in lecithin, a constituent of protoplasm; it promotes nuclear and cell-divisions. Phosphorus aids in nutrition and hastens maturity and ripening of fruits, particularly of grains. It promotes the development of the root system and other underground organs.

(7) **Carbon** forms the main bulk—45% or even more—of the dry weight of the plant. It is the predominant constituent of all organic compounds which are, in fact, known as compounds of carbon. Carbon is absorbed from the atmosphere as carbon dioxide, Although carbon dioxide occurs in the air to the extent of only 0.03%, air is still the only source of all the carbon for the plant, as proved by water culture experiments. **Carbon Cycle.** It is to be noted that there is a regular circulation of carbon dioxide and oxygen between the green plant and the atmosphere, and two processes are connected with it: one is photosynthesis and the other is respiration. In photosynthesis *green* plants take in carbon dioxide from the atmosphere *during the daytime* to manufacture food, and they give off oxygen (by the breakdown of water in the process). There is thus a tendency of the atmosphere becoming poorer in carbon dioxide and richer in oxygen. In the reverse process, i.e. in respiration, *all* plants and animals take in oxygen from the atmosphere *at all times,* and by oxidation and decomposition of food they give off carbon dioxide. In the combustion of coal and wood also carbon dioxide is given out to the atmosphere. Thus the atmosphere has a tendency of becoming richer in carbon dioxide and poorer in oxygen. There is thus a regular circulation of carbon dioxide and oxygen between the green plant and the atmosphere, and by the two processes mentioned above the total volumes of these gases are kept constant in the air.

(8) **Nitrogen.** Although nitrogen occurs to the extent of about 78 parts in every 100 parts of air by volume, it is not as a rule utilized by plants in its free state. Nitrogen occurs in the dry substance of the plant to the extent of 1–3% only. Nevertheless, it is indispensable to the life of the plant, as it is an essential constituent of proteins, chlorophyll and protoplasm. Nitrogen is essential for growth, more particularly of leafy herbs like lettuce. In the absence of this element leaves become yellowish.

Nitrogen of the Soil. Nitrogen is present in the soil in the form of inorganic and organic compounds. The chief forms of *inorganic compounds* are the nitrates and nitrites of potassium and calcium as well as ammonia and ammonium salts; while the *organic compounds* are chiefly the proteins. Normally the ammonia and ammonium salts present in the soil are made available for the use of the green plants after conversion into nitrate by the action of certain micro-organisms—

the nitrifying bacteria—which live in the soil. The process is called **nitrification**. In this process the ammonia is oxidized into nitrate in two stages: (a) first, this is acted on by the nitrite-bacteria (*Nitrosomonas*) and oxidized into nitrite ($-NO_2$), and (b) second, the nitrite thus formed is then acted on by the nitrate-bacteria (*Nitrobacter*) and further oxidized into nitrate ($-No_3$). The nitrate thus produced is readily absorbed by green plants. In certain types of soils, however, ammonia is the chief form in which nitrogen is readily absorbed by plants. Most bacteria, some fungi and some algae can readily assimilate ammonia. In the absence of oxygen a portion of the nitrates, however, is disintegrated by denitrifying bacteria and fungi into free nitrogen which then escapes into the atmosphere (**denitrification**).

The chief forms of *organic compounds* of nitrogen are the various nitrogenous organic compounds, such as many amino-acids, amines, urea, etc., which are good sources of nitrogen for normal plant growth. Dead bodies of animals and plants containing them are decomposed by different putrefying bacteria and fungi present in the soil. In the first stage, *in the absence of oxygen,* the proteins contained in their bodies are reduced to amino-acids and then to ammonia (**ammonification**) by several groups of putrefying bacteria and fungi; and in the second stage, *in the presence of oxygen,* the ammonia undergoes nitrification, as stated above. The nitrate thus produced is readily absorbed by green plants.

Fixation of Atmospheric Nitrogen. The gaseous nitrogen of the air combines with other elements and is ultimately made available to the plants as compounds of nitrogen in the soil. The methods by which the free nitrogen of the air may be fixed are as follows: (1) discharge of electricity in the atmosphere, (2) activity of certain saprophytic bacteria, (3) activity of symbiotic bacteria, and (4) activity of blue-green algae.

1. **Nitrogen Fixation by Electric Discharge.** The free nitrogen of the air to some extent becomes available to the green plants by the discharge of electricity (lightning) during a thunderstorm. Under the influence of electricity nitrogen of the air combines with oxygen to form nitric oxide $-N_2 + O_2 = 2NO$ (nitric oxide). The latter at once unites with oxygen from the air and forms nitrogen peroxide—$2NO + O_2 = 2NO_2$ (Nitrogen peroxide). This nitrogen peroxide is finally washed down into the soil by rain as nitric acid (HNO_3) and nitrous acid (HNO_2)—$2NO_2 + H_2O = HNO_3 + HNO_2$. In the soil they combine with some metal like K or Ca and form corresponding salts—nitrate and nitrite. On an average the rain water brings down to the soil about 4 kilogrammes of nitrogen per year per hectare.

2. **Nitrogen Fixation by Saprophytic Bacteria of the soil.** Certain types of soil bacteria, particularly species of *Azotobacter* (aerobic) and *Clostridium* (anaerobic) have the power of fixing free nitrogen of the soil air in their own bodies in the form of amino-acids and finally building up proteins from them. After the death of these bacteria the proteins are released to the soil. In due course these are acted on by the nitrifying bacteria and finally transformed into nitrates which are then made use of by the green plants. The chemistry of nitrogen fixation representing different intermediate stages is not, however, definitely known. But it is certain that molecular nitrogen is reduced to ammonia (NH_3) which is rapidly converted into some form of amino-acid. Ammonia is otherwise toxic to plants.

3. **Nitrogen Fixation by Symbiotic Bacteria: Nodule Bacteria** of *Leguminosae*. Agriculturists have known for a very long time that leguminous plants such as pulses grown in a soil make it fertile and lead to an increase in the yield of cereals. It was later discovered that the roots of these plants possess some swellings, called **nodules** or **tubercles** (Fig. 2), which are infected with some types of nitrogen-fixing

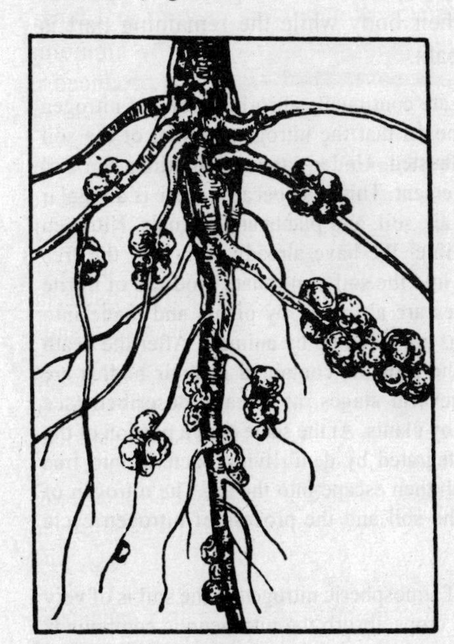

Figure 2. Nodules of a leguminous plant.

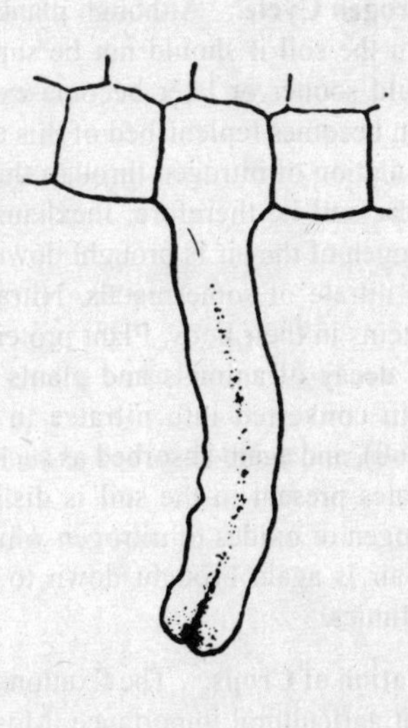

Figure 3. A root-hair infected with bacteria.

bacteria, particularly the different strains of *Rhizobium radicicola,* and these bacteria have the power of fixing the free nitrogen of the soil air in the nodules. Bacteria enter through the tip of the root-hair (Fig. 3) and pass into the root-cortex. Bacteria then multiply in number and colonize the cortex. By their activity and possibly also by some kind of secretion the cortical cells are stimulated, and they grow out here and there into small swellings or nodules of varying sizes (Fig. 2). Bacteria then fix in these nodules the nitrogen of the soil air in the form of some amino compounds. A portion of the amino compounds is absorbed into the plant body, another portion is excreted out of the nodules, and the remaining portion remains locked up in the nodules. Thus the soil becomes richer in nitrogen, more particularly so, if the nodule-bearing leguminous plants are ploughed into the soil. The leguminous plants supply the bacteria with carbohydrates, and the bacteria supply the former with nitrogenous food; so this is a case of **symbiosis**.

4. **Nitrogen fixation by Blue Green Algae.** Certain members of cyanophyceae like Nostoc, Anabaena which commonly grow in wet soil, specially in water logged rice fields have the power of fixing the free nitrogen from the air in their body. A part of the nitrogenous compounds is excreted into the soil from their body while the remaining part is released to the soil after their death.

Nitrogen Cycle. Although plants are continually absorbing salts of nitrogen from the soil it should not be supposed that the nitrogen content of the soil would sooner or later become exhausted. Under natural conditions the soil soon becomes replenished of this element. This is so because there is a regular circulation of nitrogen through the air, soil, and plants and animals. Nitrogen in the soil is, therefore, inexhaustible. We have already seen how the free nitrogen of the air is brought down into the soil as ultimate products of nitrite and nitrate of some metals. Nitrates are absorbed by plants and made into proteins in their body. Plant proteins are taken up by animals. After the death and decay of animals and plants the proteins contained in their bodies are again converted into nitrates in several stages, as already described (see p. 260), and again absorbed as such by plants. At the same time a portion of the nitrates present in the soil is disintegrated by denitrifying bacteria into free nitrogen or oxides of nitrogen which then escape into the air. The nitrogen of the air is again brought down to the soil and the process of nitrogen cycle continues.

Rotation of Crops. The fixation of atmospheric nitrogen in the soil is of very great agricultural importance. Most crops absorb the nitrogenous compounds from the soil and impoverish it. Leguminous plants, on the other hand, enrich it in nitrogen when their nodule-bearing roots are left in the soil. Thus leguminous crops such as pulses, *Sesbania cannabina* (B. DHAINCHA), cow pea (*Vigna sinensis*), etc., are grown in the field in rotation with the non-leguminous crops such as cereals (rice, wheat, maize, barley, oats, etc.) and millets. For the same reason certain leguminous plants—*Tephrosia* and *Derris*, for example—are grown in tea gardens as natural fertilizers (and also for shade). Root crops, such as turnip, radish, beet, etc., and tubers such as potato take plenty of potash, calcium and nitrogen from the soil.

Chapter 4

Absorption of Water and Mineral Salts

Roots and leaves are the main absorbing organs of plants. Roots absorb water and dissolved mineral or inorganic salts from the soil, while leaves take in gases—oxygen and carbon dioxide—from the atmosphere.

Water and Inorganic Salts from the Soil. Green plants absorb water and mineral or inorganic salts from the soil by the unicellular root-hairs which pass irregularly through the interstices of the soil particles and come in close contact with them. Absorption is also carried on by the tender growing regions of the roots. Surrounding each soil particle there is a film of water, thin or sometimes thick, loosely held to it by *capillary force,* with various mineral salts such as nitrates, chlorides, sulphates, phosphates, etc., dissolved in it. This *capillary water* is readily absorbed by the root-hairs. It is to be noted that water is absorbed in large quantities, always in excess of the requirements of the plant, by a process known as **osmosis** (see below), while the inorganic salts are absorbed in the form of **ions**. The two processes are independent of each other.

Experiment 2. Absorption of water. (a) An interesting experiment may be carried on in the following way. Take a cut branch of a lupin plant with white flowers in a glass cylinder filled with water coloured with eosin. Within a few minutes it will be seen that the white flowers turn pinkish—the colour of eosin—as a result of absorption of coloured water by the cut end of the branch. *Peperomia* plant may be similarly used and streaks of red noticed through the stem. (b) To demonstrate the **rate of absorption** proceed in the following way. Arrange the experiment as shown in Fig. 11A, and mark the level of water in the graduated tube. Note every few hours the gradual fall of the water level. At the end calculate the rate of absorption per unit of time. The experiment may be repeated under different conditions of light and temperature and the rates of absorption compared. It will be noted that strong light and high temperature enhance the rates of absorption.

Atmospheric Gases. Of the various gases present in the air it is only oxygen and carbon dioxide that are normally absorbed and utilized by the plant. Other gases may enter the plant body, but they are returned unused. Oxygen is absorbed and utilized by all the living cells of the plants for respiration; but carbon dioxide is absorbed by only the green cells for the manufacture of carbohydrates.

Composition of the Air. Of 100 parts of air by volume nitrogen occupies 78%, oxygen 21%, carbon dioxide 0.03%, and other gases such as hydrogen, ammonia, ozone, aqueous vapour, etc., occur in traces only.

Osmosis. There are certain membranes which allow a solvent (water, for example) to pass through them freely but resist the passage of a solute (salt or sugar in solution) so that only a minute quantity of the latter can pass through. On account of this property of selective transmission such membranes are said to be semipermeable or differentially permeable, e.g., parchment paper, fish or any animal-bladder, egg-membrane, etc. When weak and strong solutions are separated by such a membrane there is a net transfer of the solvent from the weaker solution to the stronger one. *This process of selective transmission of a solvent in preference to the solute through a semi-permeable membrane is termed* **osmosis**. Osmosis continues until the *hydrostatic* pressure due to the accumulated flow of the solvent has attained a value sufficient to stop further flow. This excess pressure is called the **osmotic pressure** of the stronger solution (see experiment 3). The greater the concentration of a solution the greater would be its osmotic pressure. A familiar example of osmosis is that raisins immersed in water are seen to swell up as a result of endosmosis. Similarly, grapes immersed in a strong solution of sugar or salt (say, 25% or 30%) are seen to shrink.

Importance of Osmosis in Plant Life. Root-hairs absorb water from the soil through the process of osmosis. All the cells of the plant body are saturated with water as a result of cell to cell osmosis. The cortex of the root generates root-pressure by this process. Parenchymatous cells surrounding xylem vessels absorb water from the latter by the same process. Similarly the mesophyll cells of the leaf draw water from the ends of veins generating a suction force. Osmosis gives rise to turgidity (see p. 265) which is responsible for some kinds of movements of plant organs.

Experiment 3. Process of osmosis (Fig. 4). Take a thistle-funnel with a long narrow stem, and tie a piece of animal bladder (fish bladder will do) or

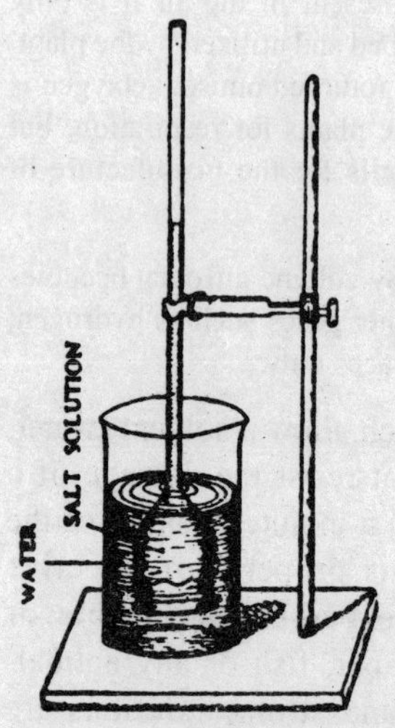

Figure 4. Experiment on process of osmosis.

parchment paper to the wide end of the funnel. Fill it with strong salt solution so that its level may stand a little above the neck of the funnel. Introduce it, stem upwards, into a beaker containing water. Mark the level of the solution in the stem. After an hour or so note that the level of the solution in the stem has gone up. This rise is due to the accumulation of water in the funnel as a result of a more rapid flow of the water into the thistle-funnel by osmosis (endosmosis) through the membrane. This rise is seen to continue until the level has gone sufficiently high up to exert a hydrostatic pressure on the membrane which then stops further net transfer of water by osmosis. This value of the hydrostatic pressure is equal to the **osmotic pressure** of the solution. At the same time a small quantity of salt also passes out through the membrane.

Parts played by Root-hairs (Fig. 5). In the case of root-hairs which contain some sugars and salts in solution, the cell-sap is stronger than the surrounding soil water. The two fluids (cell-sap and water) are separated by the cell-membrane (cellulose cell-wall + plasma membrane). As a consequence osmosis is set up. There is a flow of water from the soil into the root-hairs through the intervening cell-membrane (endosmosis). Osmosis, however, is not in this case a purely physical process. Although the cell-wall is permeable to both the water and the solutes, the plasma membrane is but differentially and selectively permeable, allowing the water to flow in, while checking the sugars and salts of the cell-sap from flowing out. This selective permeability is characteristic of the plasma membrane.

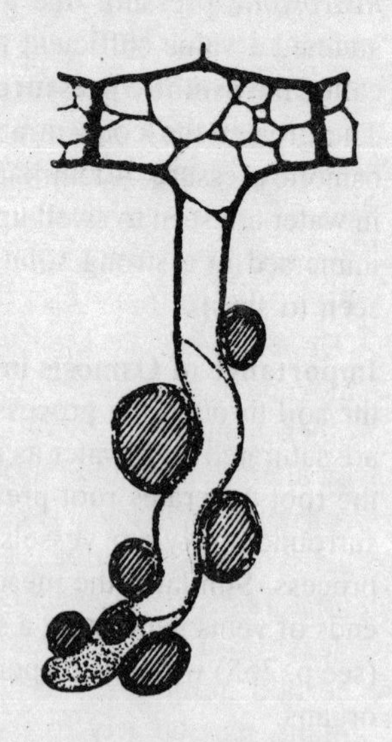

Figure 5. A root hair with soil particles adhering to it.

Turgidity. As the cell absorbs more and more water by osmosis, it increases in volume. Under this condition, the protoplasm is pushed outward against the cell-wall, and the latter also becomes much stretched. *The fully expanded condition of a cell with its wall in a state of tension due to excessive accumulation of water is called* **turgidity**. It will be noted that in a fully turgid cell two pressures are involved: outward and inward. The outward pressure exerted on the cell-wall by the fluid contents of the cell is called the **turgor pressure**, and the inward pressure exerted on the cell-contents by the stretched cell-wall is called the **wall pressure**. Normally these two pressures counterbalance each other and a state of equilibrium is maintained between them. Turgidity of a cell depends on three factors, viz. (1) formation of osmotically active substances inside the cell, (2) an adequate supply of water, and (3) a semi-permeable membrane.

Importance. A turgid condition is necessary for cell-to-cell osmosis. Turgidity is always the initial stage of growth. It is responsible for different kinds of movements of plant organs. Thus movements of the guard-cells of the stomata are due to changes in the turgidity of these cells; similarly, the rising and the falling of the leaf and leaflets of the sensitive plant (*Mimosa;* see Fig. 36), Indian telegraph plant (*Desmodium;* see Fig. 30), 'sleep' movement in leguminous and some other plants, etc., are brought about by alterations in the turgidity of the cells of the pulvinus. Turgidity of the cells of the root cortex is responsible for forcing the water into the xylem vessels. Turgidity also gives a certain amount of rigidity to the plant, particularly to the growing regions and the soft parts.

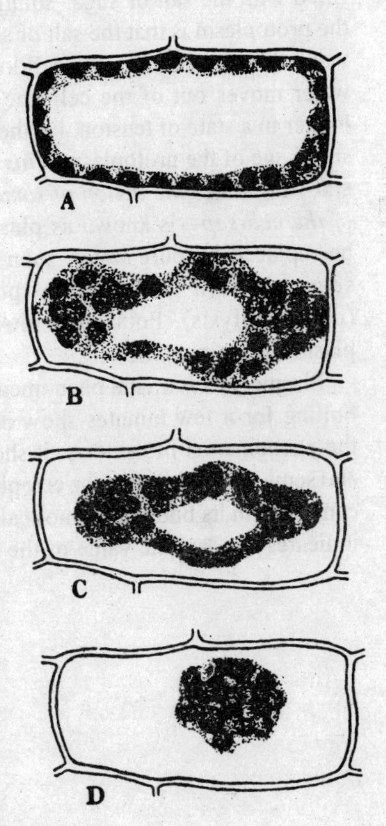

Figure 6. Plasmolysis in a cell of *Vallisneria* leaf under the action of 10% pottassium nitrate solution; *A,* a normal cell, *B–D,* stages in plasmolysis.

Plasmolysis (Fig. 6). If a section from a green leaf (e.g. *Vallisneria*), or an entire thin green leaf (e.g. *Hydrilla*), or a coloured petal, or a *Spirogyra* filament be placed in strong salt or sugar solution (say, 5 to 10% sucrose solution) and observed under the microscope, it will be seen that the cell as a whole contracts, and more obviously the protoplasm together with the nucleus and the plastids gradually shrinks away from the cell-wall and forms a rounded or irregular mass in the centre; while the space between the cell-wall and the protoplasmic mass becomes filled with the salt or sugar solution. The reason for such shrinkage of the protoplasm is that the salt or sugar solution being of greater osmotic value than the cell-sap, the cell loses water by outward osmosis. As the water moves out of the cell, the protoplasm and the cell-wall are no longer in a state of tension. Further loss of water evidently results in the shrinkage of the protoplasm. *This shrinkage of the protoplasm from the cell-wall under the action of some strong solution—stronger than that of the cell-sap*—is known as **plasmolysis**. If the salt or sugar solution be replaced by pure water, soon after plasmolysis, the protoplasm is seen to return to its normal position and the vacuole reappears (deplasmolysis). Potassium nitrate solution (10%) is a very good plasmolysing reagent.

Plasmolysis is a vital phenomenon since dead cells or those killed by boiling for a few minutes show no plasmolysis. The process explains the phenomenon of osmosis; it shows the permeability of the cell-wall and semi-permeability of the ectoplasm; it also shows that the protoplasm can retain in its body the osmotically active substances of the sap; and it indicates the osmotic value of the cell-sap.

Chapter 5

Conduction of Water and Mineral Salts

ROOT PRESSURE

The water with the mineral salts absorbed from the soil by the root-hairs gradually accumulates in the cortex. As a result the cortical cells become fully *turgid*. Under this condition, their elastic walls being much stretched exert pressure on the fluid contents and force out a quantity of them towards the xylem vessels, and the cortical cells become *flaccid*. They again absorb water and become turgid and the process continues. Thus an intermittent pumping action naturally gives rise to a considerable pressure. As a result of this pressure the water is forced into the xylem vessels through the passage cells of the endodermis, and the unthickened areas and pits that the vessels are provided with. Besides, the lignified walls of the vessels are also permeable to water. **Root pressure** *is thus explained as the pressure exerted on the liquid contents*

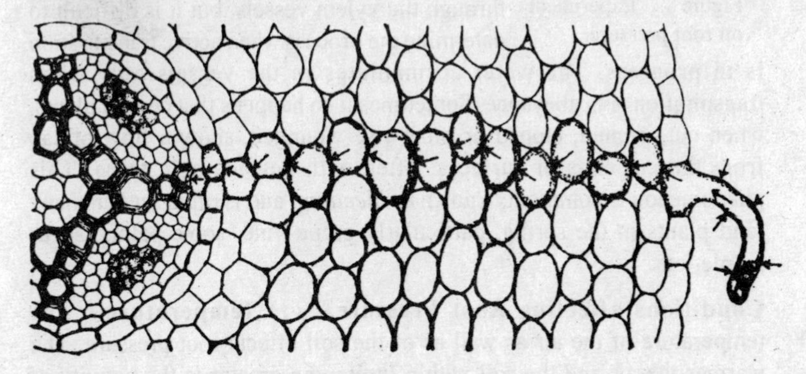

Figure 7. A root in transection showing the course of water from the root-hair to the xylem.

of the cortical cells of the root, under fully turgid condition, forcing a quantity of the liquid contents into the xylem vessels and through them upwards into the stem up to a certain height.

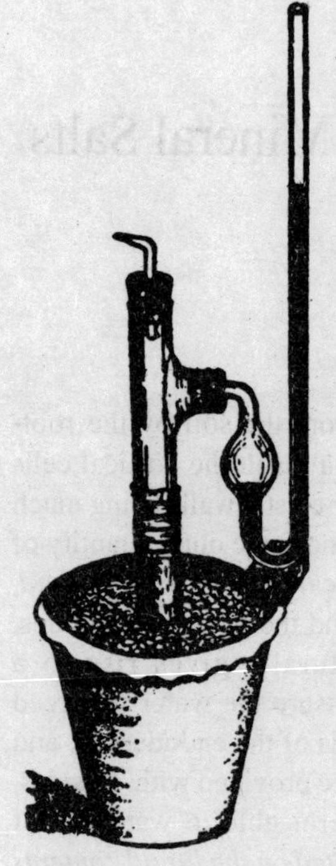

Figure 8. Experiment on root pressure.

Experiment 4. Root pressure (Fig. 8). Cut across the stem of a healthy plant (preferably a pot plant) a few cm. above the ground in the morning, and fix a T-tube to it by means of rubber tubing. Pour some water into the tube and freely water the soil. Fill a **manometer** (i.e. a U-tube with a long arm and a bulb) partially with mercury. Connect the manometer to the T-tube through a rubber cork. Insert a cork fitted with a narrow glass tube to the upper end of the T-tube. Make all the connections air-tight by applying melted paraffin-wax. Seal the bore of the narrow tube and note the level of mercury in the long arm of the manometer.

Observation. After a few hours note the rise of mercury-level in the long arm; also note the rise of water-level in the T-tube.

Inference. The rise of mercury is certainly due to accumulation of water in the T-tube and the pressure exerted by it. This phenomenon is evidently due to exudation of water from the cut surface of the stem. This experiment thus shows that the water is *forced* up through the stem by root pressure.

Root pressure is continually forcing up water through the xylem vessels, but it is difficult to determine the process when active transpiration is in progress. The water accumulates in the vessels only when transpiration is in abeyance. Sometimes it so happens that certain plants, when cut, pruned, tapped or otherwise wounded, show a flow of sap from the cut ends or surfaces, often with considerable force. This phenomenon is commonly known as *bleeding,* and is often seen in many land plants in the spring, particularly grape vine, some palms, sugar maple, etc.

Conditions affecting Root Pressure. (1) **Temperature.** The temperature of the air as well as of the soil affects root pressure. The warmer the air and the soil within limits, the greater is the activity of the root. (2) **Oxygen.** There must be an adequate supply of oxygen to the roots in the soil for respiration; otherwise their activity diminishes and may soon come to a standstill. (3) **Moisture in the soil.** A certain amount of moisture must be present in the soil. Within certain limits, the more the better. (4) **Salt in the soil.** Preponderance of salts, making the soil saline, greatly interferes with the absorption of water.

TRANSPIRATION

Plants absorb a large quantity of water from the soil by the root-hairs. Only a very small part (1–2%) of this water is retained in the plant body for the building-up processes, while the most part (98–99%) of it is lost in the form of water vapour. **Transpiration** *is the giving off of water vapour from the internal tissues of living plants through the aerial parts such as the leaves, green shoot, etc., under the influence of sun-light, regulated to some extent by the protoplasm.* It is not a simple process of evaporation since it is regulated by the vital activity of the protoplasm and some structural peculiarities of the transpiring organs (see p. 275). A detached leaf is seen to lose water much more rapidly than the one still attached to the plant, and this loss has been found to be 5 or 6 times greater. The total quantity of water that evaporates from a single plant is considerable. For example, a single sunflower plant transpires about 187.5 cc of water daily. Water vapour escapes into the atmosphere either through the stomata or through the thin cuticle. The former is called **stomatal transpiration**, and the latter **cuticular transpiration**. The stomatal transpiration is the rule amounting to 80–90%, and is evidently many times in excess of the cuticular transpiration. At night the stomata being closed, transpiration is checked. Lenticels (see Fig. II/58) are also concerned in the process of transpiration. This is called **lenticular transpiration**. In this case water vapour escapes through the loose mass of cells of the lenticel. **Transpiring Organs.** In dorsiventral leaves the lower surface with a larger number of stomata transpires water more vigorously than the upper; whereas in isobilateral leaves transpiration is more or less equal from the two surfaces containing equal number of stomata. The guard cells regulate transpiration by partially or completely opening the stoma or by closing it altogether.

Experiment 5. Transpiration: bell-jar experiment. Transpiration can be demonstrated in the following way. A pot plant with its soil-surface covered properly with a sheet of oil-paper is enclosed in a bell-jar and maintained at room temperature for some time. It is then seen that the inner wall of the bell-jar becomes bedewed with moisture.

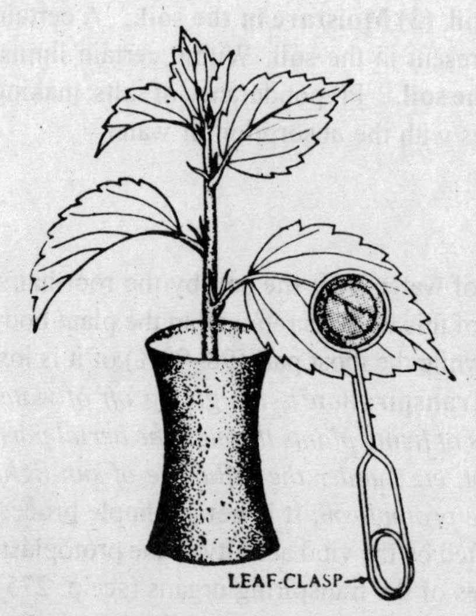

Figure 9. Unequal transpiration from the two surfaces of a leaf.

Experiment 6. Unequal transpiration from the two surface of a dorsiventral leaf (Fig. 9). Soak small pieces of filter paper or thin bloting paper in 5% solution of cobalt chloride (or cobalt nitrate) and dry them over a flame. The property of cobalt papers is that they are deep blue when dried, but in contact with moisture they turn pink. Place two dried cobalt papers, one on each surface of a thick, healthy leaf, as shown in the figure. Cover them completely with mica pieces or glass slides (or with a **leaf-clasp**, as shown in the figure), and clamp them properly to the leaf. Then quickly seal the sides with vaseline to prevent atmospheric moisture from coming in contact with the papers. It will be seen that the cobalt paper on the lower surface of the leaf turns pink sooner than the one on the upper surface. This change in colouration takes place within a few minutes. This evidently shows that the leaf transpires water more vigorously from the lower surface than from the upper. This is due to the occurrence of a large number of stomata on the lower surface, none or few being present on the upper.

Experiment 7. Measurement of the rate of transpiration current (Fig. 10). **Ganong's potometer** may be used for this purpose. The apparatus is filled with water, and a branch cut under water is fixed air-tight to the upper wide end of the apparatus through a cork. The distal end of the apparatus is dipped into water contained in a beaker. The water in the beaker may be coloured with eosin. As transpiration goes on, the coloured water is seen to enter the tube. Then remove the end of the tube from the beaker for a while and allow air to enter it. Dip it into water again. An air-bubble formed at the distal end of the tube is seen to rise and slowly travel through the horizontal arm of the potometer as a result of suction due to transpiration. Note the time that the bubble takes to cover the journey from one end of the graduation to the other and calculate (as the volume of the graduated tube is known) the rate of transpiration current. By opening the stopcock the bubble may be pushed back and the experiment re-started.

Experiment 8. Relation between transpiration and absorption (Fig.11). A wide-mouthed bottle with a graduated side-tube and a split India-rubber cork

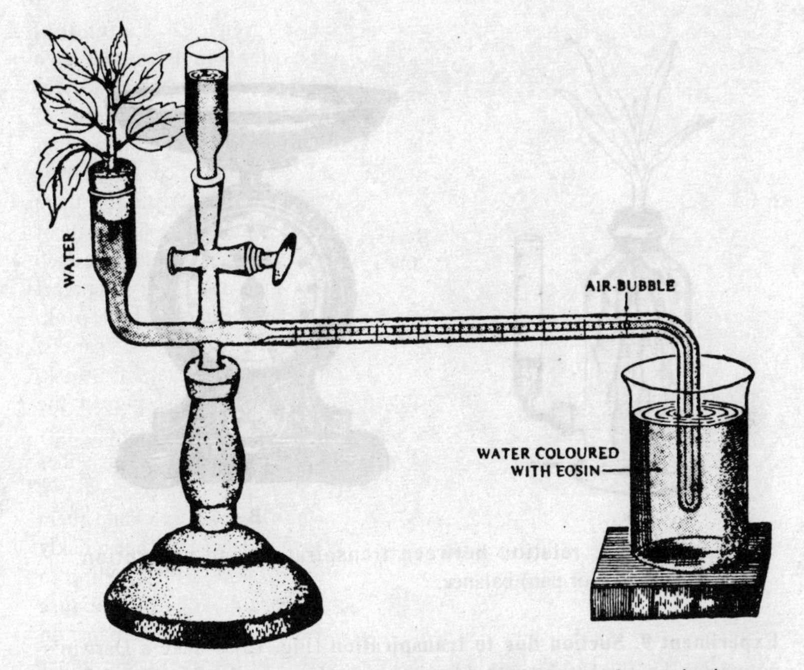

Figure 10. Ganong's potometer to demonstrate the rate of transpiration current.

are required for this experinient. A small rooted plant is introduced through the split cork into the bottle with is filled with water. The level of water is noted in the side-tubc, and 1 or 2 drops of oil poured into it to prevent evaporation of water from the exposed surface. The connexions are, of course, made air-tight. The whole apparatus is then weighed on a compression (or pan) balance (Fig. 11B) and the weight noted. It is seen after a time that the water-level has fallen, indicating the volume of water that has already been absorbed by the plant. The apparatus is then re-weighed. The difference in weight evidently shows the amount of water that has been transpired from the leaf-surfaces. If the experiment be continued for a period of 24 hours it will be seen that the volume of water (in c.c.) absorbed is slightly greater than the amount of water (in grams) lost by transpiration (1 c.c. of water = 1 gm.). In this way the relation between transpiration and absorption can be worked out for the various hours of the day and under diverse conditions.

Note. The experiment not only shows the relation between transpiration and absorption, but also separately proves 'absorption' and 'losss of water' by transpiration.

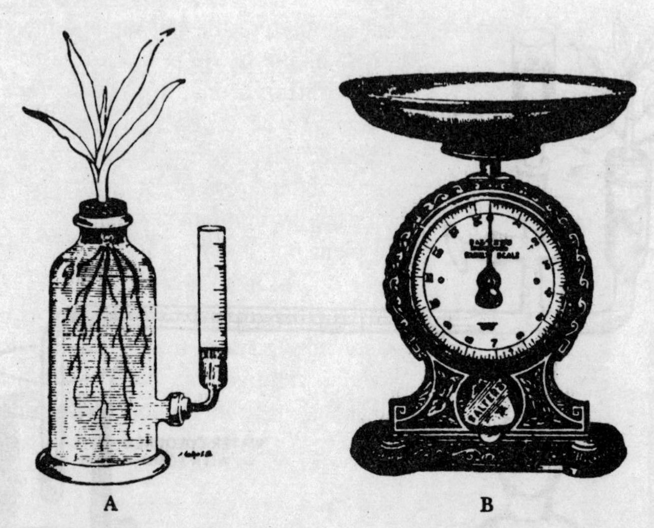

Figure 11. *A*, relation between transpiration and absorption; *B*, compression (or pan) balance.

Experiment 9. Suction due to transpiration (Fig. 12). Take a Darwin's potometer (i.e. the U-tube with a long arm, as shown in the figure) and fix to its lower end a long narrow glass tube. Completely fill the apparatus with water and insert a leafy shoot, with the cut end kept under water, into one of the arms of the potometer through a rubber cork. Close the other end with a cork. Make all the connexions air-tight by applying melted paraffin-wax. Dip the lower end of the tube into a beaker of mercury. As transpiration goes on water is absorbed, and within a few hours the mercury is seen to rise in the tube to some height. This rise of mercury indicates the suction exerted by transpiration. This suction force is known as transpiration pull.

Importance of Transpiration. Transpiration is of vital importance to the plant in many ways. (1) In the first place we find that roots are continually absorbing water from the soil, and this water is several times in excess of the immediate requirement of the plant; the excess is got rid of by transpiration. (2) There is a definite relation between transpiration and absorption. The greater the transpiration, the greater the rate of-absorption of water from the soil. (3) Absorption of water helps the intake of inorganic salts from the soil. It is, however, not a fact that the greater the transpiration, the greater the rate of absorption of inorganic salts from the soil. As a matter of fact, the intake of salts is independent of the quantity of water absorbed. (4) Transpiration secures

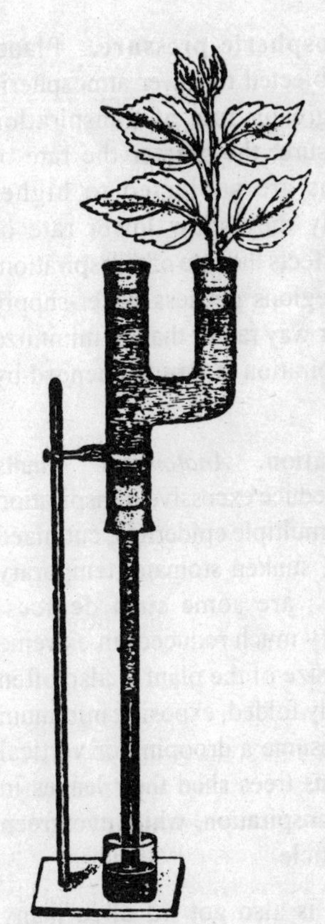

Figure 12. Suction due to transpiration with Darwin's potometer.

concentration of the cell-sap and thereby helps osmosis. (5) As a result of transpiration from the leaf-surface a suction force (see experiment 9) is generated which helps water to ascend to the top of lofty trees. (6) Transpiration also helps the distribution of water throughout the plant body. (7) As a result of transpiration, plants become cooler as a considerable amount of latent heat is lost in converting water from a liquid into a gaseous state. In the face of all these advantages the fact cannot be overlooked that excessive transpiration is often a real danger to plant life. Many plants are often seen to dry up and die when excessive transpiration takes place for a prolonged period.

Factors which affect Transpiration. (1) **Light.** Light is the most important factor. During the daytime stomata remain fully open and evaporation of water takes place normally through them. At night stomata remain closed and consequently transpiration is checked. During the daytime again heat-rays of the sun falling directly upon the leaves greatly enhance the rate of transpiration. (2) **Humidity of the Air.** There is an increase or decrease in the rate of transpiration according to whether the air is dry or moist. When the atmosphere is very dry it receives moisture very readily, but when it becomes very moist or saturated it can receive no more water vapour. Loss of water by transpiration is then very slight. (3) **Temperature of the Air.** The higher the temperature, the greater the transpiration; at high temperatures the water evaporates more freely than at low temperatures. When the two factors, viz. dryness of the air and high temperature combine, transpiration is markedly enhanced. (4) **Wind.** During high wind transpiration becomes very active because the water vapour is instantly removed and the area around the transpiring surface is not allowed to become saturated. (5) **Atmospheric pressure.** Plants growing in hills at higher altitudes are subjected to lower atmospheric pressure. Therefore, such plants show greater rate of transpiration because the lower the atmospheric pressure, the greater the rate of transpiration. Plants growing in plains are subjected to higher atmospheric pressure and comparatively they show lower rate of transpiration. Thus atmospheric pressure affects the rate of transpiration. (6) **Soil-water.** Plants growing in dry regions get less water-supply from the soil. As a result they have no other way rather than to minimize the rate of transpiration. The rate of transpiration is thus influenced by the soil-water.

Adaptations to reduce excessive Transpiration. *Anatomical.* Plants have developed many structural devices to reduce excessive transpiration which could be fatal to them. Thick cuticle, multiple epidermis, cutinized hairs and scales, a dense coating of hairs, sunken stomata, temporary closure of stomata, cork and bark, etc., are some such devices. *Morphological.* The leaf-area is often very much reduced; in extreme cases leaves are modified into spines. The size of the plant is also often reduced. Leaves may be rolled up or variously folded, exposing minimum surface for transpiration. They may also assume a drooping or vertical position to avoid direct sunlight. Deciduous trees shed their leaves in winter as a protection against excessive transpiration, while evergreen trees have their leaves well coated with cuticle.

Exudation of Water. The excess water is also got rid of in many herbaceous plants and undershrubs by a process commonly called *exudation* or *guttation.* In this process, as seen in rose, balsam, water lettuce, grape vine, many aroids, garden nasturtium, many grasses, etc., water escapes in liquid form at night and accumulates in drops at the ends of veins. These drops of water may be seen in the early morning. Exudation takes place in the absence of transpiration.

Transpiration and Exudation. (1) In transpiration water escapes in the form of vapour; while in exudation water escapes in liquid form. (2) In transpiration the water that escapes is pure; while in exudation the water that escapes contains minerals in solution. (3) in transpiration water escapes through the stomata and to some extent through the cuticle; while in exudation water escapes through the ends of veins and water stomata (or water pores). (4) Transpiration is regulated by the movement of guard cells; while exudation cannot be so regulated, the guard cells of water stomata having lost the power of movement. (5) Transpiration takes place during the daytime only when the stomata open;

while exudation takes place at night in the absence of transpiration. (6) By transpiration excess heat is removed from the plant body; while exudation has no such effect.

ASCENT OF SAP

The water absorbed from the soil by the root hairs slowly moves up through the plant body to the leaves and the growing regions of the stem and the branches, usually at the rate of 1–2 metres per hour. In the case of tall trees, e.g. *Eucalyptus,* some conifers, etc., which may be as high as 90 metres or more. The water column has to move up to that height against a considerable pressure and resistance. Two questions naturally arise in this connexion: what is the path of movement of sap and what are the factors responsible for the ascent of sap?

Path of Movement of Sap. The path of movement of sap may be determined in the following way. A small herbaceous plant (e.g. *Peperomia*) or a small branch of a plant (e.g. lupin) may be immersed in eosin solution. After a short time sections, cross and longitudinal, are prepared from it at different heights or the branch split open, and examined under the microscope. Sections or the split branch will show the presence of coloured solution only in the vessels and tracheids. Therefore, these are the elements through which movement of sap, or **transpiration current** as it is called, takes place.

Factors Responsible for the Ascent of Sap. Various theories have been advanced from time to time to explain the ascent of sap, but none has proved satisfactory yet. It is believed that root pressure forces up the water to a certain height and transpiration exerts a suction force on this column of water from above. In short, it may be said that root pressure gives a 'push' from below and transpiration exerts a 'pull' from above. In this respect transpiration is a more powerful factor. Probable theories regarding the ascent of sap are as follows:

(1) **Root Pressure.** Root pressure is regarded as one of the forces responsible for the ascent of sap. By alternate expansion and contraction of the cortical cells of the root a pumping force is no doubt generated, but this force, which is the root pressure, is only adequate to force up water in herbs, shrubs and low trees, and that too in the absence of transpiration. Root pressure can hardly generate 2 atmospheres of pressure (although 3–6 atmospheres have been recorded in special cases) and the maximum height to which a column of water may be raised by

this pressure is only 19 metres. The process is also slow and cannot keep pace with the water lost by transpiration. In many plants root pressure is absent or feeble at certain times of the year. Besides, water still rises through the stem if the roots are decapitated and the cut end of the stem dipped into water.

(2) **Cohesion Theory.** According to Dixon and Joly (1895) the water molecules cohere together and form a long continuous column from the root to the leaf with no air bubbles in it. This column does not break anywhere in its entire length even under a state of very high tension. Apparently this water-column behaves as a solid column. Then as transpiration takes place from the leaf-surface, a suction force or transpiration pull is generated, i.e. a pull is exerted on the water-column at its upper end (see experiment 9). As a result the whole water-column is bodily pulled up. This theory explains how the water can be lifted through the vessels to the height of the tallest trees (90 metres or more), and has been strongly suported by many later workers.

(3) **Imbibition Theory.** Sachs (1874) suggested that water moves along the walls of xylem vessels (and not through their cavities) as a result of imbibition of water by the solid particles of vessel-walls. But when the cavities of the vessels are artificially blocked with oil, air or gelatin the branches are seen to wilt, showing thereby that the amount of water absorbed by this process cannot at all keep pace with the amount of water lost by transpiration.

(4) **Capillarity.** The level of water inside the capillary tube is always higher than the level outside. The smaller the bore of the tube, the higher will be the rise of water in it. Xylem vessels may be regarded as capillary tubes. Therefore, it is also suggested that water may rise up through them. But from the known diameter of the vessels it is evident that the rise of water can hardly exceed a metre or so in them.

(5) **Vital Force Theory.** It is also believed by some that the activity of living cells, e.g. wood parenchyma and medullary ray cells surrounding xylem, is responsible for the rise of sap through the plant body. The role played by the living cells is like that of relay pumps. The living cells take up water from the vessels at a particular level and then force it again into the vessels at a higher level, and the sap thus rises. Strasburger (1891), however, refuted the idea of vital force by killing the living cells by the application of heat as well as by poisonous chemicals.

(6) **Pulsation Theory.** According to the late Sir J.C. Bose (1923), the ascent of sap is due to active *pulsation* of the internal layer of the

cortex abutting upon the endodermis. Conduction of water takes place through this layer even in the absence of root pressure and transpiration. Xylem vessels being dead and inactive no pulsation is exhibited by them, and these were regarded by him as only reservoirs of water. All the living cells exhibit pulsation to a greater or less extent, but the activity of the internal cortex is exceptionally great. Anatomical and experimental evidence, however, does not support this view.

Chapter 6

Manufacture of Food

Food of Plants. Food consists of certain organic substances which are more or less directly utilized by the living organisms for their nourishment. In this respect there is hardly any difference between the food of plants and that of animals. Such substances are **carbohydrates**, **proteins**, and **fats** and **oils**. Animals, non-green plants and non-green cells of plants have to depend directly or indirectly on the organic food prepared by the chloroplast-bearing cells of green plants. It is evident, therefore, that green plants hold a key position so far as the living world is concerned.

I. *CARBOHYDRATES*

Photosynthesis. *Photosynthesis* (*photo,* light; *synthesis,* building up) *consists in the building up of simple carbohydrates such as sugars in the green leaf by the chloroplasts in the presence of sunlight* (*as a source of energy*) *from carbon dioxide and water absorbed from the air and the soil respectively.* The process is accompanied by a liberation of oxygen (see experiment 10). The volume of oxygen liberated has been found to be equal to the volume of carbon dioxide absorbed. But it is to be noted that all the oxygen liberated in the process is released exclusively from water (H_2O) and not from carbon dioxide (CO_2), as first proved by Hill in 1937 and later by others by using radioactive oxygen, O^{18}, in water (H_2O^{18}). Oxygen escapes from the plant body through the stomata. This formation of carbohydrates, commonly called **carbon-assimilation**, is the monopoly of green plants only, chlorophyll being indispensable for the process. By this process not only are simple carbohydrates formed but also a considerable amount of *radiant* (*light*) *energy* absorbed from sunlight is stored up as *potential chemical energy* in the organic substances formed. It must be noted that the whole process of photosynthesis takes place in the body of the chloroplasts in green cells and, therefore, mainly in the leaf.

Mechanism of Photosynthesis. Photosynthesis involves two distinct phases: the initial phase of light reactions or photochemical reactions and the second phase of dark reactions. Photosynthesis is essentially an oxidation-reduction process in which hydrogen is transferred from water (oxidised) to carbon dioxide (reduced). The overall reaction may be represented as:

$$6CO_2 + 12H_2O \xrightarrow[\text{chlorophyll}]{\text{Light}} C_6H_{12}O_6 + 6H_2O + 6O_2$$

Light reactions. Light reactions take place in the lamellar region (see p. 169) of the chloroplasts (Park and Pou, 1961) and require the presence of sunlight (as a source of radiant energy). In this phase, chloroplasts absorb radiant energy and convert it into chemical energy (located in phosphate bonds, i.e. formation of ATP) and water molecules undergo ionization (i.e., photolysis of water). As a result, reducing agent $NADP.H_2$ is formed and molecular oxygen, evolving from water (Hill, 1937–39, Ruben, 1941), escapes to the atmosphere. It is to be noted that at this phase, CO_2 does not play any role. Further, ATP (i.e. chemical energy) is formed in two ways—cyclic set of photophosphorylation and non-cyclic set of photophosphorylation. In the latter the formation of reduced $NADP.H_2$ and evolution of oxygen due to photolysis of water also take place.

Cyclic set of photophosphorylation. When the light falls upon a cholophyll molecule, the latter becomes excited. This means when a photon (a quantum of light, which is a discrete packet of light energy) strikes the chlorophyll molecule, an electron is lifted into high energy level helping the chlorophyll molecule to gain an amount of energy equivalent to that of a photon. It is now known that the important light absorbing pigment is chlorophyll-a.

The light energy thus absorbed by chlorophyll-a molecules (also called antenna Chl.-a molecules) are finally transferred to the specific chlorophyll-a molecule (called reaction centre or P_{700}) which then only expells electrons(e^-). The expelled electron is picked up by an electron carrier known as ferredoxin (an iron containing co-enzyme present in the chloroplast). From ferredoxin electron passes through a series of other electron carriers such as cytochrome, Vitamin-K, etc. During this

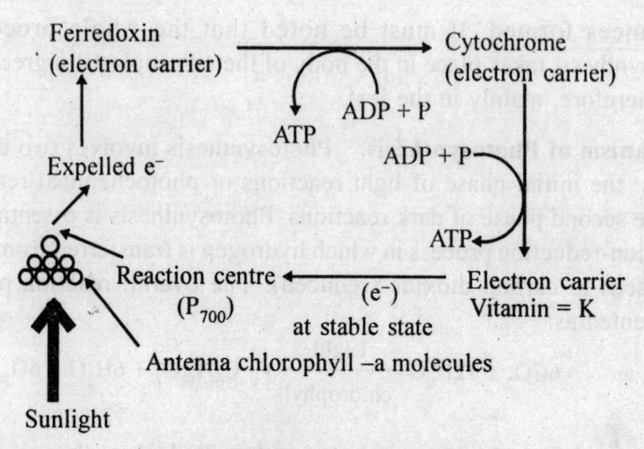

Cyclic Photophosphorylation

process, high energy of the electron is used for the addition of phosphate radical (P) to ADP by the phosphate bond to form ATP (ADP + P = ATP). Finally the electron comes back to its stable state and is returned to the chlorophyll (from which it was expelled). In this cycle only ATP molecules (usually 2) are formed and act as the source of chemical energy.

Non-Cyclic set of photophosphorylation. In this process radiant energy (light) is absorbed separately by two different pigment systems or Photosystems generally known as PS-I and PS-II (Emerson, 1950). The two systems are excited at different wave lengths of light. In PS-I, light is absorbed by Chl.-a molecules at a particular wave length (683 nm.). The excited chlorophyll-a molecules pass the absorbed light energy to P_{700} which then expells electrons. The latter is picked up by Ferredoxin. But here two such electrons are transferred by Ferredoxin to one molecule of oxidised NADP (present in the chloroplast). Thus NADP becomes negatively charged and develops an affinity to pick up two positively charged hydrogen ions (released as a result of ionization of water). Whereas in PS-II, light is absorbed primarily by another pigment (probably chlorophyll-b) in addition to Chl-a at another wave length (673 nm.). But here also it is only P_{700} Chl-a molecule that expells electron. The expelled electrons with high energy then pass through electron carriers Q, plastoquinone, cytochromes, plastocyanin. During this process 2 ATP molecules are formed. Electrons are returned finally at stable state to Chl-a molecule of PS-I. Now it is also known that 24 molecules of H_2O (four times more than $6CO_2$ required for the formation of one molecule of glucose in the dark reactions) undergo ionization

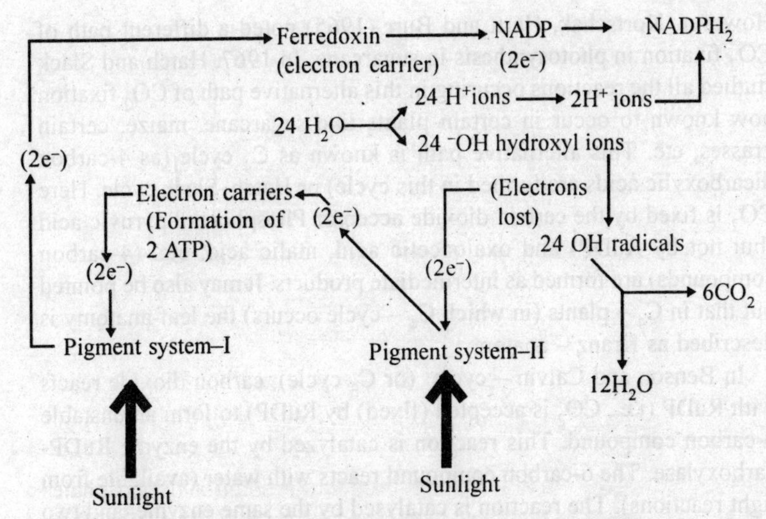

Non-cyclic Photophosphorylation

into 24⁻OH (hydroxyl) and 24 H^+ (hydrogen) ions in this non-cyclic set of reactions. One negatively charged NADP picks up two positively charged hydrogen ions and becomes NADP.H_2 (i.e. electrically neutral). In this way 12 NADP.H_2 are formed per 24 H_2O. The ⁻OH ions lose electrons and become 24 OH radicals. Such lost electrons replace the electrons expelled by pigment system-II. 24 OH radicals combine to form 12 H_2O and $6O_2$. The fact that oxygen is evolved only from water in photosynthesis has been experimentally verified by using radioactive isotope O^{18}, i.e. , H_2O^{18} (Ruben, 1941). Oxygen escapes to the atmosphere, and $12H_2O$ is utilized in dark reaction.

Dark Reactions. Dark reactions represent the second phase of photosynthesis. Recently with the help of chromatography (a technique used for the separation and identification of different components in a mixture) and by the use of radioactive isotope, C^{14} i.e. $C^{14}O_2$ (the technique is called **tracer technique**), much more has been known about the dark reactions. Dark reactions occur in the stroma region of chloroplast. The cycle of reactions taking place in this phase is called C_3 cycle (as compounds containing three carbon atoms are formed in successive stage of this cycle) or Benson and Calvin cycle (after the name of Benson and Calvin, 1950, who could detect the reactions in this cycle). In C_3 cycle, CO_2 is fixed by RuDP and the first formed stable intermediate product is 3-carbon compound Phosphoglyceric acid.

However, Kortschak, Hartt and Burr (1965) noted a different path of CO_2 fixation in photosynthesis in sugarcane. In 1967, Hatch and Slack studied all the reactions occuring in this alternative path of CO_2 fixation now known to occur in certain plants like sugarcane, maize, certain grasses, etc. This alternative path is known as C_4 cycle (as 4-carbon dicarboxylic acids are formed in this cycle) or Hatch–Slack cycle. Here CO_2 is fixed by the carbon dioxide acceptor Phosphoenol pyruvic acid (but not by RuDP) and oxaloacetic acid, malic acid, etc. (4-carbon compounds) are formed as intermediate products. It may also be pointed out that in C_4 – plants (in which C_4 – cycle occurs) the leaf-anatomy is described as Kranz – anatomy.

In Benson and Calvin – cycle (or C_3 cycle), carbon dioxide reacts with RuDP (i.e., CO_2 is accepted (fixed) by RuDP) to form an unstable 6-carbon compound. This reaction is catalyzed by the enzyme RuDP-carboxylase. The 6-carbon compound reacts with water (available from light reactions). The reaction is catalysed by the same enzyme, and two molecules of phosphoglyceric acid (a stable first intermediate product, 3-carbon compound) are formed which then are reduced to two molecules of phosphoglyceraldehyde (3-carbon compound) by $NADP.H_2$ under the action of enzyme triphosphate dehydrogenase, in the presence of ATP. Here per molecule of CO_2, 2 molecules of phosphoglyceric acid and from the latter 2 molecules of phosphoglyceraldehyde are formed. Or in other words, per six molecules of CO_2, 12 molecules of phosphoglyeric acid and 12 molecules of phosphoglceraldehyde are formed. Out of 12 molecules of phosphoglyceraldehyde, 2 molecules convert into fructose diphosphate which then through the following sequence finally form the sugar (glucose, a six carbon compound):

Fructose diphosphate → Fructose phosphate → Glucose phosphate → Glucose (sugar). Whereas the remaining 10 molecules of phosphoglyceraldehyde, through various reactions, regenerate Ribulose phosphate which then reacts with ATP and converts into RuDP. Various enzymes are involved in all these reactions. The Dark reactions, in brief, can be represented as follows:

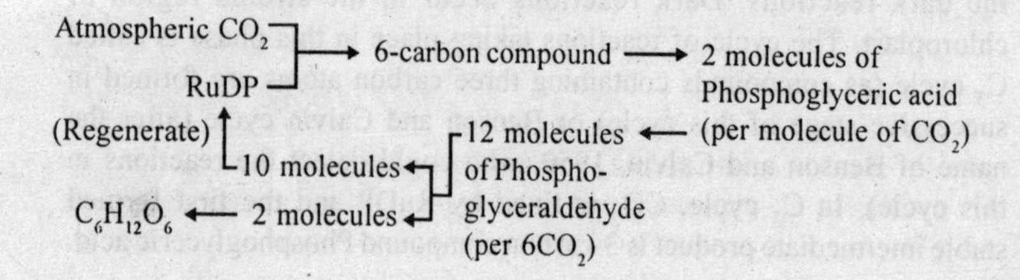

Oxidized = – H : NADP = Oxidized nicotinamide adenine dinucleotide phosphate
Reduced = + H : NADP.H_2 = Reduced NADP.H_2
ADP = Adenosine diphosphate
ATP = Adenosine triphosphate
RuDP = Ribulose diphosphate

Production of Oxygen and Starch in Photosynthesis. Production of oxygen or starch or both is used as a criterion for experimentally testing the process of photosynthesis. Of these two, oxygen is formed as a *by-product* as a result of the breaking-down of water (H_2O) at the initial stage of photosynthesis, while starch (always derived from sugar) is the *final-product* of the process. Oxygen escapes from the leaf through the stomata (see experiment 10), while starch accumulates in the leaf (see experiment 11).

Experiment 10. Photosynthesis: to show that oxygen is given off during photosynthesis (Fig. 13). Place some cut branches of *Hydrilla* (an aquatic plant) under cover of a funnel in a large beaker filled with water. Expose the apparatus to bright sunlight. Within a few minutes streams of gas bubbles will be seen to rise from the cut ends of the branches and collect in the test-tube by displacing the water. If the apparatus be removed to a dark or semi-dark room or covered with a black paper or cloth no bubbles are seeds to come out. Then remove the test-tube with the gas to pyrogallate solution. It rises and completely fills up the tube. Pyrogallate solution absorbs oxygen. The gas in the tube is, therefore, oxygen.

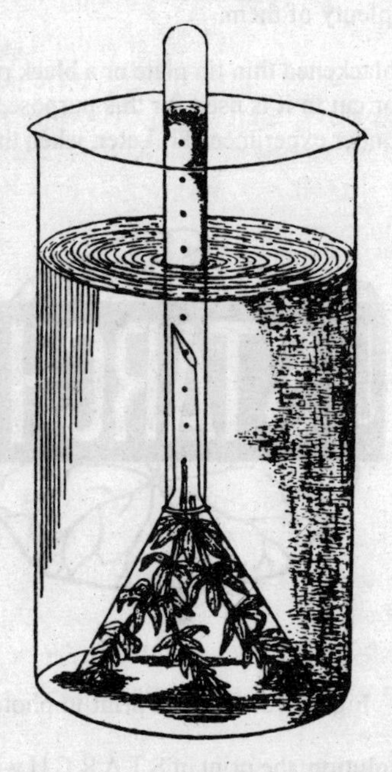

Figure 13. Evolution of oxygen bubbles in photosynthesis of a water plant (*Hydrilla*).

Experiment 11. Photosynthesis: to demonstrate that starch is formed in photosynthesis (Figs. 14–15). Select a healthy green leaf of a plant *in situ* and cover a portion of it on both sides with two uniform pieces of black paper, fixed in position with two paper clips or soft wooden clips either in the morning before the sun rises or the previous evening, so that the experiment is performed with a starch-free leaf. Now expose the plant to light for the whole day. Then collect

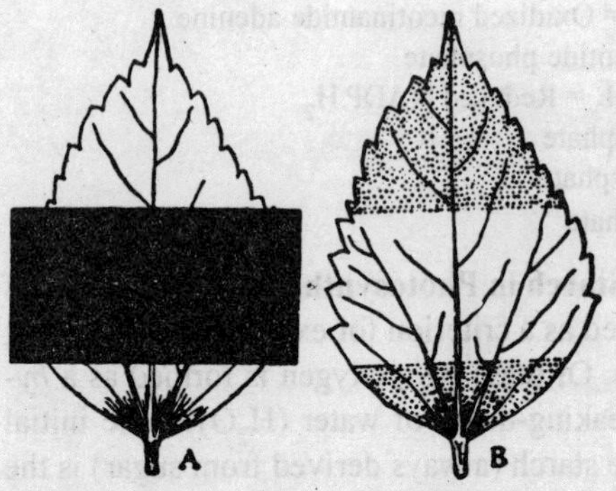

Figure 14. Formation of starch grains in photosynthesis of land plants. *A*, leaf partially covered with black paper; *B*, covered portion without starch grains; uncovered portions with plenty of them.

the leaf and decolorize it with methylated spirit. Dip it into iodine solution for a minute or so. Note that the exposed portions turn blue or black showing the presence of starch, while the screened portion turns yellowish brown, there being no starch formed in it; this yellowish brown colour is due to the action of iodine solution on protoplasm and cellulose.

A very interesting experiment known as the **starch print** (Fig. 15) may be carried out in the following way. A stencil (which may be a blackened thin tin plate or a black paper) with the letters S T A R C H punched or cut in it is used for this purpose, the procedure being the same as described under experiment 11. Later, when the leaf is decolorized and treated with iodine

Figure 15. Starch print in photosynthesis.

solution, the print of S T A R C H will stand out boldly in black on the bleached leaf owing to the formation of starch grains which have turned black by contact with iodine.

Instead of loose black paper or stencil a **light-screen**, as shown in Fig. 16, may be used to cover a portion of the leaf. The advantage of the light-screen is that it allows free ventilation, while it cuts off all light.

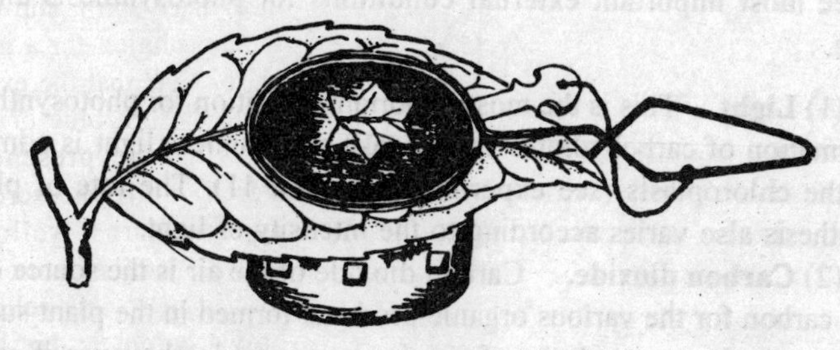

Figure 16. A light-screen.

Experiment 12. To show that plants cannot photosynthesis unless carbon dioxide is available: Moll's experiment (Fig. 17). Take a wide-mouthed bottle and a split cork of appropriate size. Pour a small quantity of dilute caustic potash solution into the bottle. Before sunrise cut a healthy green leaf, evidently starch-free, and place it—half inside the bottle and half outside—between the two halves of the split cork. Lay the bottle flat on a wooden tray, with the petiole dipped into a dish of water. Smear the edges of the split cork with vaseline to make the bottle air-tight. The tray with the bottle and the dish is then exposed to direct sunlight till the evening. Then remove the leaf, decolorize it with methylated spirit and dip it into iodine solution. It will be seen that the portion of the leaf lying outside the bottle turns black; while the portion inside the bottle turns yellowish. This evidently shows that no starch grains are formed when carbon dioxide is not available, all the carbon dioxide contained in the bottle having been absorbed by the caustic potash solution.

Figure 17. Moll's experiment on photosynthesis.

Experiment 13. To show that chlorophyll is essential for photosynthesis. Select a garden plant with variegated leaves, e.g. garden croton or garden tapioca or *Coleus*. Cut out a small branch from it and dip the cut end into water in a bottle. Keep it in a dark room for 1 or 2 days to free the leaves of starch grains. Then mark the green portions in 1 or 2 leaves, and expose the branch to bright sunlight for the whole day. In the evening collect the marked leaves, decolorize them with methylated spirit and dip them into iodine solution for a few minutes. Note that only the green portions of the leaf turn black indicating the presence of starch grains; while the non-green portions turn yellowish. It is thus obvious that without chlorophyll photosynthesis cannot take place.

Conditions necessary for Photosynthesis. Light intensity, temperature, and carbon dioxide concentration of the air are the three most important external conditions for photosynthesis and its rate.

(1) **Light.** This is the most important condition for photosynthesis. Formation of carbohydrates cannot take place unless light is admitted to the chloroplasts (see experiments 10 and 11). The rate of photosynthesis also varies according to the intensity of light.

(2) **Carbon dioxide.** Carbon dioxide of the air is the source of all the carbon for the various organic products formed in the plant such as sugar, starch, etc., and, therefore, the process is in abeyance if carbon dioxide is not available to the plant (see experiment 12). Under favourable conditions of light and temperature if carbon dioxide concentration rises from 0.03 per cent in the air to 0.1 per cent or even more, carbohydrate formation greatly increases. But higher concentration of CO_2 is harmful to plants.

(3) **Water.** Water is indispensable for photosynthesis because water and carbon dioxide undergo chemical changes leading to the formation of carbohydrates under the influence of chloroplasts and in the presence of sunlight. It is, however, a fact that less than 1 per cent of the water absorbed by the roots is utilized in photosynthesis.

(4) **Temperature.** Photosynthesis takes place within a wide range of temperature. It goes on even when the temperature is below the freezing point of water, but the maximum temperature lies at about 45°C. The optimum temperature, i.e. the most favourable temperature for photosynthesis, may be stated to be 35°C.

(5) **Chlorophyll.** This is essential for photosynthesis; chlorophyll absorbs light and initiates the process of photosynthesis. The plastid are powerless in this respect without the presence of chlorophyll. For the same reason non-green parts of plants cannot photosynthesize (see experiment 13). Fungi and saprophytic and parasitic phanerogams have altogether lost this power, being devoid of chlorophyll.

(6) **Potassium.** Potassium helps synthesis of carbohydrates and, therefore, in its absence starch grains are not formed. Potassium does not enter into the composition of carbohydrates but acts as a catalyst helping in their synthesis.

Conditions necessary for the Formation of Chlorophyll. A number of factors, both internal and external, are responsible for the formation of chlorophyll. In the absence of any of them chlorophyll synthesis is in abeyance.

(1) **Light.** Without light chlorophyll cannot develop; continued absence of light decomposes chlorophyll, and the plants become *etiolated,* i.e. pale, sickly and drawn out (see Fig. 27). Very strong light also decomposes chlorophyll in the leaf, particularly in shade-loving plants.

(2) **Temperature.** Chlorophyll develops within a wide range of temperature: very high temperature (45–48°C), however, decomposes chlorophyll.

(3) **Iron, Magnesium and Manganese.** In the absence of the salts of these metals chlorophyll is not formed, and seedlings assume a sickly yellow appearance. In this condition they are said to be **chlorotic**. Of these metals it is only magnesium that enters into the composition of chlorophyll.

(4) **Nitrogen.** Nitrogen enters into the composition of chlorophyll and, therefore, in the absence of nitrogen chlorophyll fails to develop.

(5) **Water.** Leaves, when they dry up in the absence of water, are seen to lose their green colour. Desiccation thus brings about decomposition of chlorophyll. In prolonged droughts the leaves of many plants turn brownish in colour.

(6) **Oxygen.** This is also necessary for the formation of chlorophyll. Etiolated seedlings fail to develop chlorophyll in the absence of oxygen, even when these are exposed to sunlight.

(7) **Carbohydrates.** Cane-sugar, grape-sugar, etc., are also necessary for the formation of chlorophyll. Etiolated leaves, without soluble carbohydrates in them, develop chlorophyll and turn green in colour when floated on sugar solution.

(8) **Heredity.** This is a powerful factor and determines the formation of chlorophyll in the offspring. Familiar examples are those with multi-coloured leaves, e.g. garden crotons, garden tapioca, garden amaranth, certain aroids (e.g. *Caladium*), *Coleus,* etc.

Chemistry of Chlorophyll. Chlorophyll is a mixture of four different pigments, as follows:

1 Chlorophyll *a,* $C_{55}H_{72}O_5N_4Mg$—a blue-black micro-crystalline solid.
2 Chlorophyll *b,* $C_{55}H_{70}O_6N_4Mg$—a green-black micro-crystalline solid.
3 Carotene, $C_{40}H_{56}$—an orange-red crystalline solid.
4 Xanthophyll, $C_{40}H_{56}O_2$—a yellow crystalline solid.

Photosynthesis and Chemosynthesis by Bacteria. *Photosynthetic bacteria* develop a pigment closely related to chlorophyll and utilize light as a source of energy. Such bacteria are **purple sulphur bacteria** and **green sulphur**

bacteria—both anaerobic. They decompose hydrogen sulphide (H_2S) into hydrogen (H) and sulphur (S). The hydrogen thus released is used to reduce CO_2, to carbohydrate in a series of dark reactions, as in green plants. This may indicate a primitive type of photosynthesis.

Chemosynthetic bacteria are colourless and aerobic; they do not utilize light energy. The chemical energy required for their synthetic processes is derived from the oxidation of certain inorganic compounds present in their habitat. The energy thus released by the oxidative process is used by such bacteria in their chemosynthetic work leading to the production of carbohydrate and other organic compounds. Common such bacteria are: sulphur bacteria, iron bacteria and nitrifying bacteria. **Sulphur bacteria** grow in sulphur springs and in stagnant water containing H_2S. They obtain necessary chemical energy for their synthetic processes by the oxidation of sulphur compounds present in water. Thus they oxidize H_2S to free sulphur and finally to H_2SO_4 and use the chemical energy thus released by this oxidative process to reduce CO_2 and synthesize carbohydrate and other organic compounds. The sulphur deposited in the bacterial cells in the process is again used for the above purpose. **Iron bacteria** grow in lakes and marshes and in water containing ferrous iron. They obtain their chemical energy for synthesis of organic compounds by oxidation of ferrous hydroxide to ferric hydroxide. Bog iron ore and deposits of iron oxides in lakes and marshes are due to their activity. **Nitrifying bacteria** live in the soil and are of two types: nitrite bacteria and nitrate bacteria (see p. 259).

II. *PROTEINS*

Nature of Proteins. These are very complex organic nitrogenous compounds formed in plants. Analyses of plant proteins show that carbon, hydrogen, oxygen, nitrogen, and sulphur and sometimes phosphorus enter into their composition, but we know little about their molecular structure. Protein-molecules are very large and extremely complex, usually consisting of thousands of atoms, and are composed of several chains of amino-acid molecules. Various kinds of proteins are found in plants, often abundantly in many cells. They are the chief constituents of protoplasm and all other living bodies. Amino-acids are the initial stages in the formation of proteins, and are also the degradation products of the latter.

Synthesis of Proteins. Proteins are normally formed from nitrates absorbed from the soil by the root-hairs. But the chemical reactions leading to the formation of these complex compounds are only imperfectly known. Protein synthesis mostly takes place in the meristematic tissues (root-tip and stem-tip) and storage tissues, and to some extent in most living cells. In this process, unlike photosynthesis, light is not absolutely necessary. The whole process of protein synthesis takes place in three successive stages.

(a) **Reduction of Nitrates.** Nitrogen is an essential constituent of all proteins. We know that the nitrate of the soil is the main source of supply of nitrogen to the higher plants. After it is absorbed from the soil into the plant body it is reduced to nitrite, and the latter further reduced to ammonia, as follows: —$NO_3 \rightarrow NO_2 \rightarrow NH_3$. This reduction of nitrate to ammonia takes place either in the root or in the leaf under the action of two separate enzvmcs—*nitrate reductase* for the first reaction, and *nitrite reductase* for the second reaction. Ammonia now holds a key position in the pathway of protein synthesis.

(b) **Synthesis of Amino-acids.** Amino-acids play major role in the synthesis of proteins. They are mainly formed in the root and in the leaf. All amino-acids contain at least one acidic or carboxyl group (–COOH) and one basic or amino group ($-NH_2$). Ammonia directly reacts with α-ketoglutaric acid, an important intermediate product of aerobic respiration. The result is the formation of the first amino-acid; i.e. *glutamic acid,* under the action of a specific enzyme. It may be noted that the amino-acid formation is correlated with root-respiration supplying necessary energy, and also with photosynthesis supplying carbon, hydrogen and oxygen (in weight over 80% of the amino-acids are non-nitrogenous). Glutamic acid now holds a central position from which several other amino-acids may be formed by transfer of its amino group ($-NH_2$) to the carboxyl group (–COOH) of any other keto acid under the action of specific enzymes. Thus the amino group of the glutamic acid may be transferred to oxalacetic acid and to pyruvic acid which in their turn produce two other amino-acids, viz. *aspartic acid* and *alanine* respectively. In the same way other amino-acids may be formed. There are 20 different amino-acids known to be constituents of plant proteins. The amino-acid *cystine* formed in all plants also contains sulphur. Amino-acids may travel from their seat of origin to distant tissues, particularly to the meristematic and storage tissues where protein synthesis mostly occurs. It may be noted that animals do not normally utilize ammonia and cannot, therefore, produce an amino-acid.

(c) **Synthesis of Proteins.** Protein molecules, as stated before, are very large and complex, and hundreds of them may occur in a single cell. Proteins are made of different amino-acids. A protein molecule may finally be formed by linkages of hundreds or thousands of amino-acid molecules arranged in long chains. Linkages of different amino-acids in specific sequences in the chains under the action of specific enzymes result in the synthesis of an infinite variety of proteins. The arrangement of particular amino-acids, often with repetition of one or more of them, in a specific sequence in the chain determines the particular kind of protein to be formed. It is evident, therefore, that the omission of a single amino-acid required in the chain directly checks the synthetic process. Emil Fischer first suggested (1899–1906) that proteins are formed by condensation of several amino-acids. By linkages of several amino-acids proteins are formed in the following stages: amino-acids → dipeptide → polypeptides → peptones → proteoses → proteins. It may be noted that protein synthesis and protein breakdown go on simultaneously in the living cells. The amino-acids, thus formed by breakdown, may again combine into proteins. All plant proteins contain sulphur, and more complex ones like nucleoproteins also contain phosphorus. Sulphur and phosphorus are obtained from the soil as sulphates and phosphates. Proteins cannot travel as such from one cell to another like amino-acids. It may further be noted that protein synthesis is mostly localized in the ribosomes which occur in plenty in the cytoplasm (see below). Some proteins may also be formed in the nucleus. Proteins contribute to the nourishment of plants and animals.

DNA and Protein Synthesis. Recent brilliant researches on nucleic acids (DNA and RNA; see p. 163) have established the fact that DNA (see Fig. II/6) is the controlling centre of the cell. It controls the synthesis of proteins. It does not, however, take a direct part in this synthetic process but its working copy is RNA, i.e. DNA works through RNA. as follows. DNA prepares a master plan for the specific kinds of proteins to be built up in the cells of a plant. It transmits a message *in coded form* to RNA (messenger RNA or mRNA) which then travels from the nucleus, through the pores present in the nuclear membrane to the surrounding cytoplasm and finally reaches a ribosomes. The mRNA comprises a series of nucleotides (see p. 163) arranged in triplets. Each such triplet comprising three nucleotides (represented by three code 'letter' e.g. ACA, CUG, etc.) is called a codon. The sequence of three nucleotides in a codon is called a code word or coded message. Ribosomes occur in the cytoplasm in groups of two, three, four or five ribosomes. These groups are called polyribosomes where protein synthesis occurs. mRNA releases the code to a ribosome and moves to another ribosome and then to another; soon, however, it gets degraded. A second type of RNA (transfer RNA or *tRNA*) or soluble RNA (or sRNA) carries a particular amino-acid from the cytoplasm to the ribosome. Several such RNAs occur in the cytoplasm, and thus several amino-acids may be brought to the ribosome. The sequence of nucleotides in triplet on the tRNA is called anticodon. It is now for the ribosomal RNA (a third type - rRNA) to do the rest of the work. It receives the code from mRNA and the required amino-acids from tRNA; it then deciphers the code, i.e., translates it into the language of physiology, and finally brings about linkage of the amino-acids in a specific order into a specific protein molecule. Or in other words, the decoding of codon is accomplished through anticodon of tRNA (carrying a specific amino acid) which is complementary to the codon of mRNA. tRNA is then released to the cytoplasm for further work. A specific enzyme is of course indispensible for each chemical reaction involved, and the energy required for the whole process is supplied by ATP. DNA may coin an infinite number of codes through its two nitrogen bases (purines and pyrimidine; see Fig. II/6) and thus produce an endless variety of proteins. Although DNA is the same in all plants it produces only specific kinds of proteins for each species of plants.

III. *FATS AND OILS*

Fats are usually synthesized in the microsomes or endoplasmic reticulum of the cells. Fats are formed by the combination of glycerol with fatty acids.

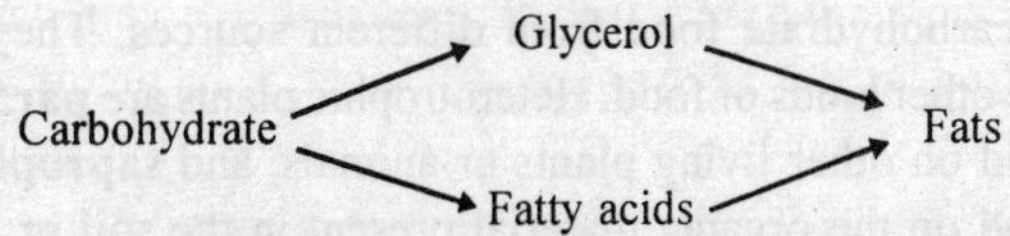

We know that both these compounds appear in the living cells as a result of breakdown of some carbohydrates (particularly glucose and fructose) during the anaerobic phase of respiration. Fatty acids and glycerol do not accumulate in the cells as such but soon after they are formed they combine into fats under the action of *lipase*. Fats in ordinary temperature convert into oils. It is also known that in a reverse reaction fats and oils may again be converted into glycerol and fatty acids by the action of the same enzyme, lipase. Fats and oils are insoluble in water and cannot, therefore, diffuse out of the cells in which they are formed. They are utilized as food, their energy value being more than double that of carbohydrates.

Chapter 7

Special Modes of Nutrition

Green plants are **autotrophic** (*autos,* self; *trophe,* food) or self-nourishing, that is, they are able to manufacture carbohydrates from raw or inorganic materials and thus nourish themselves. Non-green plants on the other hand are **heterotrophic** (*heteros,* different), that is, they cannot prepare carbohydrates and nourish themselves. They get their supply of carbohydrate food from different sources. They can, however, prepare other kinds of food. Heterotrophic plants are **parasites**, when they depend on other living plants or animals, and **saprophytes**, when they depend on the organic material present in the soil or in the dead bodies of plants and animals. Their nature and mode of nutrition have already been discussed (see pp. 9).

Carnivorous Plants. These plants are known to capture lower animals of various kinds, particularly insects. They digest the prey and absorb the nitrogenous products (proteins) from its body. Digestion is extracellular in all carnivorous plants. Being green in colour, they can manufacture their own carbohydrate food. Altogether over 450 species of carnivorous plants have till now been discovered representing 15 genera belonging to 6 families; of them over 30 species occur in India. According to the mode of catching the prey they may be classified into four groups.

(a) Plants with sensitive glandular hairs on the leaf-surface, secreting a sweet sticky fiuid, e.g. sundew (*Drosera*).

(b) Plants with special sensitive hairs—trigger hairs—on the leaf-surface, e.g. Venus' fly-trap (*Dionaea*) and water fly-trap (*Aldrovanda*).

(c) Plants with leaves modified into pitchers. e.g. pitcher plant (*Nepenthes*).

(d) Plants with leaf-segments modified into bladders, e.g. bladderwort (Utricularia)—floating or submerged aquatic herbs.

(1) **Sundew** (*Drosera;* Fig. 18)—100 sp.; only 3 sp. in India. They are small herbs, a few to many cm. in height Each leaf is covered on the upper surface with numerous glandular hairs known as the **tentacles**. Each gland secretes a sticky fluid which glitters in the sun like dew-drops and hence the name 'sundew'. When any insect, attracted by the glistening fluid, which is possibly mistaken for honey, alights on the leaf, it gets entangled in the sticky fluid, and the tentacles bend down on it from all sides and cover it. When it is suffocated to death the process of digestion begins. The glands secrete an enzyme, called *pepsin hydrochloric acid,* which acts on the insect and changes the proteins in its body into soluble and simple forms. The carbonaceous materials are rejected in the form of waste products. If the tentacles are poked with any hard object, they hardly show any movement. On the other hand, a bit of raw meat (evidently containing proteins) placed on the leaf induces movement. This shows that the glands are sensitive and reacts only to chemical stimulus.

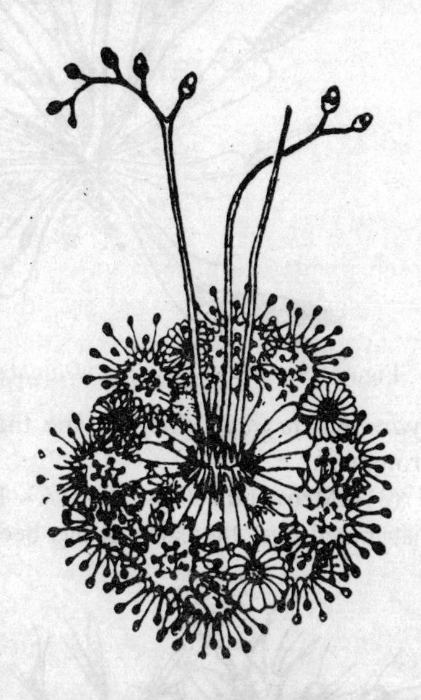

Figure 18. Sundew (*Drosera*).

(2) **Venus' Fly-trap** (*Dionaea;* Fig. 19)—1 sp. The plant is a native of the U.S.A. It is herbaceous in nature and grows in damp mossy places. Each half of the leaf-blade is provided with three long pointed hairs—trigger hairs—placed triangularly on the leaf-surface. The hairs are extremely sensitive from base to apex. The slightest touch to any of these hairs is sufficient to bring about a sudden closure of the leaf-blade, the mid-rib acting as the hinge. The upper surface of the leaf is thickly covered with reddish digestive glands. When the insect is caught, or any nitrogenous material such as meat, fish, etc., is placed on the leaf, it closes suddenly and the glands begin to secrete *pepsin*

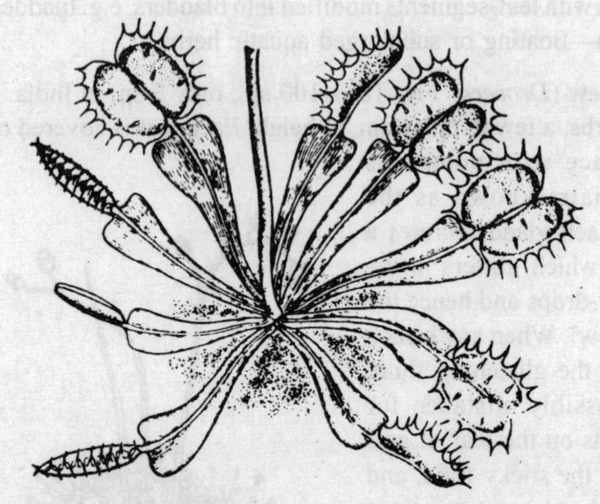

Figure 19. Venus' fly-trap (*Dionaca*).

hydrochloric acid. The enzyme then brings about the digestion of proteins.

(3) **Aldrovanda** (Figs. 20–21A—B)—1 sp. This plant is very widely distributed over the earth. It has been found in abundance in the salt-

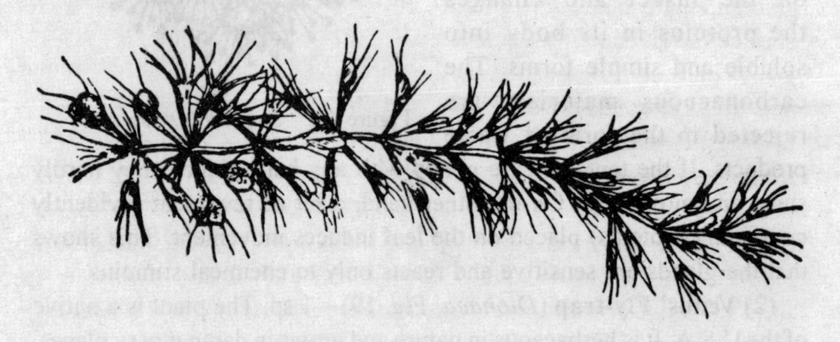

Figure 20. Water fly-trap (*Aldrovanda*).

lakes of the Sundarbans, the salt-marshes south of Calcutta, the freshwater 'jheels' of Bangladesh and in several tanks in Manipur. *Aldrovanda* may be regarded as a miniature *Dionaea* in some respects. It is, however, a rootless, free-floating plant with whorls of leaves. The mechanism for catching prey is practically the same as that of *Dionaea,* but instead of only six sensitive hairs there are a number of them here on either side of the mid-rib, and the leaf is protected by some bristles.

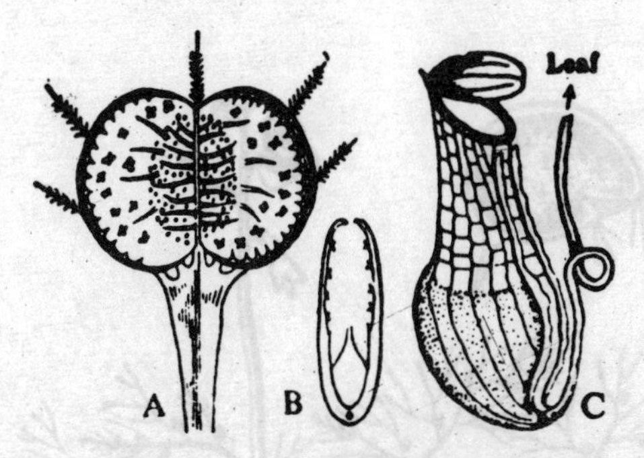

Figure 21. *A, Aldrovanda* (an entire open leaf); *B,* the same (a closed leaf in section); *C,* a pitcher of *Nepenthes*.

There are, of course, numerous digestive glands on the upper surface of the leaf, and the margins are beset with minute teeth pointing inwards.

(4) **Pitcher Plant** (*Nepenthes;* Figs. 21C)—67 sp; only 1 sp. (*N. khasiana*) in India (in the Garo Hills and the Jaintia Hills). Pitcher plants are climbing herbs or under-shrubs which often climb by means of the tendrillar stalk of the pitcher. Each pitcher may be as big as 20 cm., sometimes somewhat bigger. The mouth of the young pitcher remains closed by a lid which opens afterwards and stands more or less erect. Below the mouth the inside of the pitcher is covered with numerous smooth, sharp hairs, all pointing downwards. Lower down, the inner surface is studded with numerous, large, digestive glands. The pitcher is also partially filled with a fluid. Animals, as they enter, slip down the smooth surface, and get drowned in the fluid. After their death the process of digestion commences. The digestive agent secreted by the glands is in the nature of a *trypsin* which digests proteins into peptones and the peptones into amines. Amines are readily absorbed by the pitcher. Bits of egg-white, meat, etc., dropped into the pitcher, as was first found by Hooker, are seen to become dissolved and ultimately absorbed in the form of amines. Carbohydrate and other materials remain undigested in the pitcher as waste products.

(5) **Bladderwort** (*Utricularia;* Fig. 22)—120 sp; over 20 sp. in India. They are mostly floating or slight submerged, rootless aquatic herbs; there are a few terrestrial species also. The leaves are very much segmented, looking like roots except that they are green in colour. Some of these segments become transformed into *bladders*. Each bladder is about 3–5 mm in diameter and is provided with a trap-door entrance. The trap-door acts as a sort of valve opening only inwards when pushed from outside. Very small aquatic animals enter inside the bladder

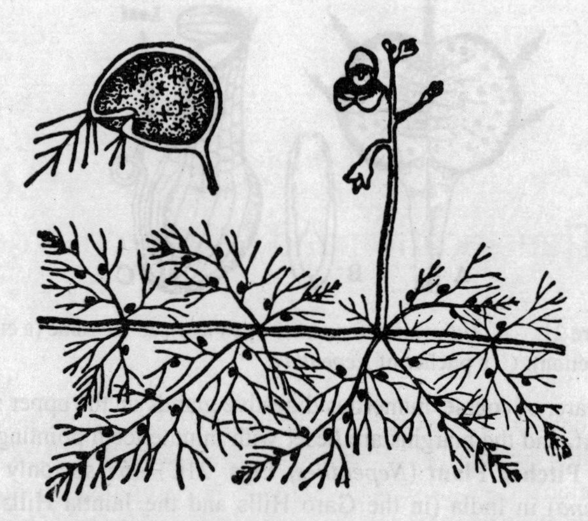

Figure 22. Bladderwort (*Utricularia*) with many small bladders; *top,* a bladder in section (magnified).

by pushing the trap-door. Once inside the trap (bladder) there is no escape for them. The inner surface of the bladder is dotted all over with numerous digestive glands. After the death of the animals the process of digestion begins.

Chapter 8

Translocation and Storage of Food

Translocation. Food materials are prepared mostly in the leaves. From there they are translocated to the storage organs which often lie at a considerable distance. For this purpose there are definite and distinct channels extending through the whole length of the plant body. These are the sieve-tubes and associated cells. Through them soluble-proteins, amines, amino-acids and sugars travel downward to the storage organs. Such materials can easily pass through the perforated sieve-plates. The protoplasmic threads extending through the pores in the sievc-plates may also help in this respect. In the storage organs these substances are converted into insoluble (complex) proteins and starch grains, and stored up as such.

Later during the period of active growth—formation of buds and flowers—the various forms of stored food are rendered soluble and, therefore, suitable for travelling. Now an upward movement of the soluble food materials takes place through the phloem and finally they are brought to the growing organs. At this time of active growth a part of the food also may move upward through the xylem (Fischer, 1915; Atkins, 1916; Dixon, 1923). The forces responisible for the downward or upward movement of food through the phloem are not known. That the phloem is definitely the principal channel for conduction of food can be proved by chemical analysis of its contents. Such analysis reveals the presence of soluble carbohydrates (sugars), proteins and other nitrogenous compounds in it.

Storage. Food is prepared in excess of the immediate need of plants. This surplus food exists in plants in two conditions—either *suitable for travelling* or *suitable for storage.* The travelling form is characterized by *solubility,* and the storage form by *insolubility* in the cell-sap.

Storage Tissues. Tissues meant for storage of food are mostly made of living parenchymatous cells with thin cellulose walls. If the walls are thick they are provided with many simple pits in them. All parts made of large-celled parenchyma always contain a certain amount of stored food. The cortex of roots is particularly rich in it, as also are the large pith of the monocotyledonous root and that of the dicotyledonous stem. There is also a quantity of food stored up in the endodermis, medullary rays and xylem parenchyma of the stem, and border parenchyma of the leaf.

Storage Organs. Food materials are stored up in the endosperm or in the thick cotyledons of the seed for the development and growth of the embryo. In the fleshy pericarp of the fruit there is a considerable amount of food stored up. Food is specially stored up in the fleshy roots such as the fusiform, napiform, conical and other roots, and in the underground modified stems such as the rhizome, tuber, corm, etc. All fleshy stems and leaves, as in Indian aloe (*Aloe vera*), American aloe (*Agave*), purslane (*Portulaca*), etc., and fleshy scales of onion always contain a store of food in them. The swollen stem-base of kohl-rabi also contains stored food.

Forms of Stored Food. The various forms in which the food materials are stored in these different organs and tissues may now be considered. The food materials are carbohydrates, proteins, and fats and oils (see also pp. 178). Among the carbohydrates **starch** is most abundant in almost all the storage organs; **glucose** accumulates as such in grapes to the extent of 12–15%, and **sucrose** in sugarcane and beet to the extent of 10–15%, and 10–20% respectively; **inulin** in the tuberous roots of *Dahlia;* **reserve cellulose** (see Fig. II/8) in the endosperm of date seed, vegetable ivory-palm seed, etc.; and **glycogen** in fungi. Among the nitrogenous materials various kinds of **proteins**, particularly **aleurone grains**, are found in both starchy and oily seeds but bigger aleurone grains occur in oily seeds; pulses are rich in proteins, while amino-compounds are scarce in storage organs. **Fats and oils** are found in almost all living cells; they are specially common, however, in seeds and fruits. In oily seeds very little carbohydrate is found.

Food Stored in the Seed. There is always a considerable amount of food stored up in the cotyledons and in the endosperm of the seed for the use of the embryo as it grows. Food materials occur there in insoluble forms and these are first digested, i.e. rendered soluble and chemically simpler under the action of specific enzymes (see next chapter), and then utilized by the embryo for its nutrition and growth. Common forms of such food materials are the following: (1) **Starch** is a very common form of carbohydrate stored up in the seed. Cereals such as rice, wheat, maize, oat, barley, etc., are particularly rich in starch. (2) **Reserve cellulose** is deposited as thickening matter in the cell-walls of the endosperm of many palm seeds, e.g. date-palm, betelnut-palm, nipa-palm, vegetable ivory-palm, etc. (3) **Oils** are deposited in most seeds to a greater or less extent. There is a special deposit of them in seeds like groundnut, gingelly, coconut, castor, safflower, etc. (4) **Proteins** also occur in all seeds in varying quantities. In seeds like pulses they occur in fairly high percentage. Soyabean contains 35% or more of proteins. Oily seeds also contain a high percentage of proteins, e.g. castor seed.

Chapter 9

Digestion and Assimilation of Food

Digestion. The stored food materials are generally insoluble in water or cell-sap and also indiffusible, but when translocation is necessary they are rendered soluble and diffusible by the action of enzymes. It is only in the soluble forms that food materials are absorbed by the protoplasm. *This rendering of insoluble and complex food substances into soluble and simpler forms suitable for translocation through the plant body, and assimilation by the protoplasm is collectively known as* **digestion**.

The process of digestion is chiefly intracellular, that is, it takes place inside the cell. Extracellular digestion occurs in a few cases, as in the digestion of proteins by carnivorous plants, parasites, fungi, etc. Digestion, like all other physiological functions, is performed by the protoplasm. For this purpose it secretes digestive agents known as **enzymes**.

Enzyme. Enzymes are organic (biological) catalysts, each being a certain kind of protein secreted by the living cells, mostly in the mitochondria, to bring about thousands of specific biochemical reactions in various metabolic processes such as photosynthesis, respiration, digestion of food, etc., without themselves undergoing any chemical change. Enzymes actually hasten and regulate the rate of a chemical action and, therefore, but for their activity life would have been very slow or even still. Evidently, enzymes are of utmost importance in plant and animal life. In digestion, the enzymes act upon different types of complex food substances and render them chemically simpler and suitable for assimilation. They are soluble in water and, when dry, form a white amorphous powder. A temperature of 70°C destroys the properties of most enzymes.

Properties of Enzymes. (1) The action of an enzyme is mostly specific, i.e., for a particular substance there is a particular enzyme; for instance, the enzyme that acts on starch will not act on protein or any other substance. This is expressed as the *lock and key* action. (2) The enzyme is never exhausted; a small quantity of it can act on an almost unlimited supply of the substance, provided that the products of digestion are removed from the seat of its activity. (3) The enzyme acts as a catalytic agent; this means that the presence of the enzyme induces some chemical action in the substance without itself undergoing any change. Thus the enzyme may be regarded as a *specific organic catalyst.*

Kinds of Enzymes and Nature of Digestion

- **Diastase** converts starch into dextrin and maltose.
- **Maltase** converts maltose into glucose.
- **Invertase** changes sucrose into glucose and fructose.
- **Cellulase** converts cellulose into glucose.
- **Cytase** converts hemicellulose into glucose.
- **Inulase** changes inulin into fructose.
- **Pepsin** changes proteins into peptones.
- **Trypsin** transforms proteins into amino-acids.
- **Erepsin** transforms peptones into amino-acids.
- **Lipase** breaks up fats into fatty acids and glycerine.

Assimilation. *Assimilation is the absorption of the simplest products of digestion by the protoplasm into its own body and conversion of these products into similar complex constituents of the protoplasm* (*the term assimilate means to make similar*). The various kinds of carbohydrates are converted into sugar, particularly glucose, and the various kinds of proteins converted into peptones and amino-acids. These simplest products of digestion travel to the growing regions where the protoplasm is very active. Here glucose is mostly broken down during respiration, releasing *energy*; while the digested products of proteins are assimilated by the protoplasm into its own body. We know that the protoplasm itself is a living substance composed of very complex proteins. The food proteins are, therefore, changed into complex protoplasmic proteins, i.e. into, 'live' proteins or, in other words, food passes from non-life into life, that is, protoplasm. This is the goal of nourishment. We do not know how this mysterious change takes place. We know only that the protoplasm has the power of bringing it about.

Chapter 10

Respiration

Respiration *is essentially a process of oxidation and decomposition of organic compounds, particularly simple carbohydrates such as glucose, in the living cells with the release of energy.* The most important feature of respiration is that by this oxidative process the *potential* energy stored in the organic compounds in living cells is released in a stepwise manner in the form of active or *kinetic* energy under the influence of a series of enzymes and is made available, partly at least, to the protoplasm for its manifold activities such as manufacture of food, growth, movement, reproduction, etc. Often, a considerable amount of energy escapes from the plant body in the form of heat, as seen in germinating seeds. It is, primarily glucose that undergoes oxidation, but sometimes in its absence other materials like fats, proteins, organic acids and even protoplasm under extreme conditions may also be oxidized. The main facts associated with respiration are: (1) consumption of atmospheric oxygen, (2) oxidation and decomposition of a portion of the stored food resulting in a loss of dry weight as seen in the seeds germinating in the dark, (3) liberation of carbon dioxide and a small quantity of water (the volume of CO_2 liberated being equal to the volume of O_2 consumed), and above all (4) release of energy by the breakdown of organic food. The overall chemical reaction may be stated thus: $C_6H_{12}O_6 + 6O_2 \rightarrow 6CO_2 + 6H_2O$ + Energy (sugar + oxygen → carbon dioxide + water + energy). This shows that for oxidation of one molecule of sugar, six molecules of oxygen are used and that six molecules each of CO_2 and H_2O are formed. By burning sugar at a high temperature CO_2 and H_2O are also formed, but in the living cells this process is carried on by a series of enzymes at a comparatively low temperature.

All the living cells of the plant, however deeply seated they may be, must respire day and night in order to live. In the process they are continually absorbing oxygen and giving out carbon dioxide. If the supply of air is cut off by growing the plant in an atmosphere devoid of oxygen, it soon dies. Growing organs, such as the floral and vegetative buds, the germinating seeds, and the stem-and root-tips, respire actively; while adult organs do so comparatively slowly. Entry of oxygen and exit of carbon dioxide normally take place through the stomata, and in shrubs and trees through lenticels also (see Fig. II/58). For diffusion of gases through the plant body, a network of intercellular spaces and air-cavities develops in it.

Aerobic and Anaerobic Respiration (*aer,* air; *an,* not; *bios,* life). Normally free oxygen is used in respiration resulting in complete oxidation of stored food and formation of carbon dioxide and water as end products; this is known as **aerobic respiration.** A considerable amount of energy is released by this process as represented by the equation: $C_6H_{12}O_6 + 6O_2 \rightarrow 6CO_2 + 6H_2O + 674$ kg.cal.(sugar + oxygen $\rightarrow$ carbon dioxide + water + 674 kg. cal. of energy). Under certain conditions, as in the absence of free oxygen, many tissues of higher plants, seeds in storage, fleshy fruits and succulent plants like cacti temporarily take to a kind of respiration, called **anaerobic respiration**, which results in incomplete oxidation of stored food and formation of carbon dioxide and ethyl alcohol; and sometimes also various organic acids such as malic, citric, oxalic, tartaric, etc. Very little energy is released by this process to maintain the activity of the protoplasm. This may be represented by the equation: $C_6H_{12}O_6 \rightarrow 2C_2H_5OH + 2CO_2 + 28$ kg. cal. (sugar ethyl alcohol + carbon dioxide + 28 kg. cal. of energy). It is otherwise known as *intramolecular respiration.* Anaerobic respiration may continue only for a limited period of time, at most a few days, after which death ensues. Certain bacteria and fungi normaly take to anaerobic respiration for release of energy.

Experiment 14. Respiration (Fig. 23). A flask with a bent bulb, called **respiroscope** (or an ordinary long-necked flask) with some germinating seeds in it is inverted over a beaker containing a good quantity of mercury. A small caustic potash stick is introduced into the flask. The respiroscope is fixed in a vertical position with a suitable stand and clamp. The enclosed air in the flask is completely cut off from the outside atmosphere. Now leave the apparatus in this position for some hours, preferably till the next day. It will then be seen that mercury has risen in the flask to the extent of *nearly* one-fifth the total volume of the flask. The rise of the mercury is evidently due to absorption of a certain volume of gas contained in the flask. Since caustic potash absorbs carbon dioxide, it may be safely inferred that the gas absorbed is carbon dioxide. This gas must have been exhaled by the germinating seeds during respiration.

Note. Instead of mercury, water may be used in the beaker, and the respiroscope with the germinating seeds inverted over it. A short test-tube with a small piece of caustic potash stick in it may be floated on water inside the respiroscope. Subsequently the rise of water will be one-fifth the total volume of the enclosed air. This indicates that the volume of carbon dioxide evolved is equal to the volume of oxygen absorbed since oxygen occupies one-fifth the total volume of air.

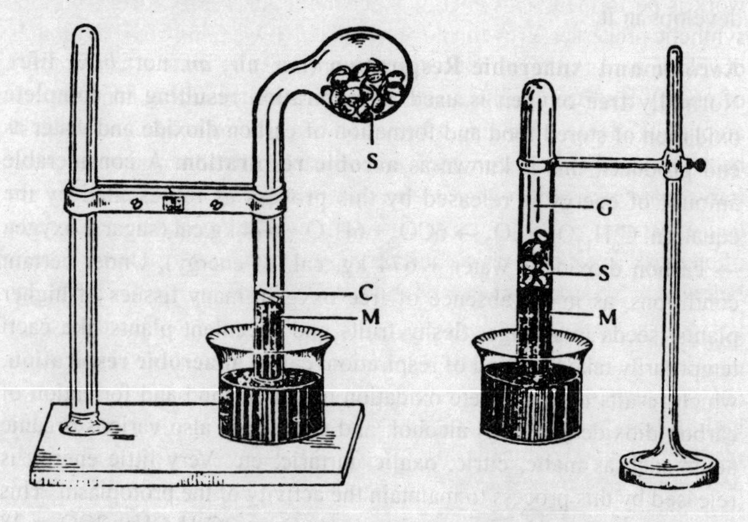

Figure 23. Experiments on Respiration. Aerobic respiration.
Figure 24. Anaerobic or intramolecular respiration. *S*, seeds; *C*, caustic potash stick; *M*, mercury; *G*, gas.

Experiment 15. Anaerobic respiration (Fig. 24). Completely fill a short narrow test-tube with mercury (*M*), close it with the thumb and invert it over mercury contained in a beaker. Keep the tube in a vertical position with a suitable stand. Take some germinating seeds, and remove the seed-coats from them to get rid of the enclosed air (oxygen). With the help of forceps hold the skinned seeds under the test-tube, and release them one after another so that the released seeds rise to the closed end of the tube. Introduce, in this way, five or six seeds. They are now free from oxygen. Prior to their introduction it is better to soak the seeds in distilled water. This keeps the seeds moist. Note on the following day that the mercury column has been pushed down, owing to the exhalation of a gas (*G*) by the seeds. Within one or two days nearly the whole of the mercury is seen to be pushed out of the tube. Introduce a small piece of caustic potash stick into the test-tube with the help of the forceps. It floats on mercury and, coming in contact with the gas, absorbs it quickly. The mercury rises again and fills up the test-tube. The gas evidently is carbon dioxide.

Respiration is a destructive process consisting of the decomposition of some of the food materials, more particularly the simple carbohydrates, and this decomposition is brought about by the action of specific enzymes secreted by the protoplasm. Nevertheless, it is highly beneficial to the life of the plant for the reason that respiration sets free *energy,* by which work is performed. This energy is absolutely necessary for the various synthetic processes, growth, movements, etc. If we think of the enormous development of a large tree we can at once realize what a vast amount of energy has been utilized in constructing that body. A considerable amount of energy, of course, escapes from the plant body in the form of heat. During vigorous respiration heat is generated. A thermometer thrust into a mass of germinating seeds will show a marked rise in temperature (see experiment 16). This production of heat is an easily detectable form of energy. Respiration results in a loss of dry weight of the plant. This is believed to be due to the escape of carbon dioxide.

Experiment 16. Heat generated in respiration (Fig. 25). Take two thermoflasks, and fill one of them (A) with germinating seeds and the other (B) with the same killed by boiling for a few minutes and then soaked in 5% formalin to prevent any fermentation in the flask. Insert a sensitive thermometer in each as shown in the figure and pack the mouth of the flask with cotton. It is better to place, half-immersed in the lump of seeds, a small test-tube containing a small piece of caustic potash stick. Wait for some time and note a remarkable rise in temperature in the case of the flask A containing germinating seeds; while flask B containing killed seeds shows no rise of temperature (the dotted line indicating the original temperature). This evidently proves that heat is evolved in respiration.

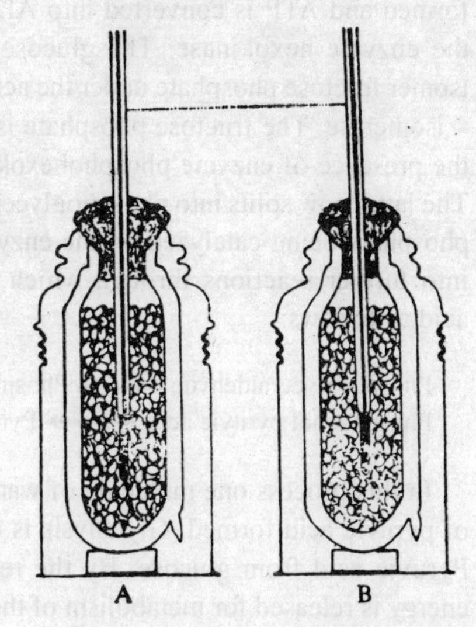

Figure 25. Experiment to show that heat is generated in respiration (see text).

Mechanism of Respiration. Chemical reactions in respiration (from the breakdown of glucose to the release of CO_2 and H_2O) are many and varied but proceed in a stepwise manner under the action of a group of complex enzymes. The whole process of respiration is, however, complete in two distinct phases:

Glycolysis, the initial anaerobic phase, i.e. in which oxygen is not required, and Krebs cycle or Tricarboxylic acid cycle (TCA-cycle) or citric acid cycle, the second phase in which oxygen is requried.

Glycolysis. Glycolysis, the initial phase of respiration, consists of a series of reactions resulting in the breakdown of one molecule of carbohydrate like glucose ($C_6H_{12}O_6$) to two molecules of pyruvic acid ($C_3H_4O_3$) in the absence of oxygen. Glycolysis is also known as Embden – Meyerhoff – Parnass (EMP) pathway. This phase is common to both aerobic and anaerobic respiration and it occurs in the cytoplasm of a cell.

In this phase, Glucose is first phosphorylated, i.e. a phosphate group is added to glucose by ATP (Adenosine triphosphate), an active energy rich phosphate compound which is formed in mitochondria and then available in cytoplasm (see pp. 169). As a result, glucose phosphate is formed and ATP is converted into ADP. This reaction is catalyzed by the enzyme hexokinase. The glucose phosphate now converts to its isomer fructose phosphate under the action of the enzyme phosphogluco – isomerase. The fructose phosphate is then phosphorylated by ATP in the presence of enzyme phosphohexokinase into fructose diphosphate. The latter now splits into phosphoglyceraldehyde and Dihydroxyacetone phosphate being catalyzed by the enzyme, aldolase. The former enters into further reactions through which finally it converts into pyruvic acid as follows:

Phosphoglyceraldehyde ⟶ Phosphoglyceric acid ⟶
Phosphoenol pyruvic acid ⟶ Pyruvic acid

In this process one molecule of water is also liberated per molecule of pyruvic acid formed. Glycolysis is completed with the formation of Pyruvic acid from glucose. By the reactions of this phase very little energy is released for metabolism of the cell. It is estimated that there is net yield of eight molecules of ATP from one molecule of glucose by this process. It is to be noted that Pyruvic acid holds a key position from where, either in the absence of oxygen Pyruvic acid is incompletely broken down into carbondioxide and alcohol (anaerobic respiration), or in presence of oxygen Pyruvic acid (first being changed to acetic acid and then to acetyl CoA) enters into Kreb's Cycle in which carbondioxide and water are liberated (aerobic respiration).

Krebs Cycle. Krebs cycle represents the second phase of respiration in which the availability of oxygen is a must (i.e., it is an aerobic phase) and this cycle occurs in the mitochondria. At first, by oxidation of one molecule of pyruvic acid (a 3-carbon acid) an active compound acetyl-coenzyme A (acetyl Co.A also called active acetate, a 2-carbon compound) is formed, during which a molecule of carbon dioxide is released. The acetyl Co.A then actually initiates the reactions of Kreb's cycle. Kreb's cycle was first worked out by H. A. Krebs, an English biochemist in 1943. In the first step of Krebs cycle the 'active' acetyl portion of acetyl Co.A reacts with Oxaloacetic acid (a 4-carbon acid) under the action of the enzyme citrogenase and citric acid (a 6-carbon acid) is formed and Co.A is released for the repetition of the process. Citric acid transforms to its isomer isocitric acid (6-carbon acid) by the action of enzyme aconitase through the intermediate product cis – aconitic acid (6-carbon acid). In the follow up cycle, the isocitric acid converts into α-Ketoglutaric acid (a five carbon acid). In this process, one molecule of carbon dioxide is evolved. It may be noted here that during amino acid synthesis, some of α-Ketoglutaric acid (see p. 289) is also utilized in the formation of amino acid (specially glutamic acid). This reaction continues and from α-Ketoglutaric acid, succinic acid, fumaric acid, malic acid and finally oxaloacetic acid (each a 4-carbon acid) are formed in successive steps. These reactions are catalyzed by enzyme dehydrogenases. One molecule of carbon dioxide is also released along with the formation of succinic acid. Further, 2-molecules of water are also finally formed in this cycle. Oxaloacetic acid, thus regenerated as the end product of Krebs-cycle, again reacts with acetyl coenzyme A to renew the cycle. Kerbs cycle involves 30 molecules of ATP (per oxination of one molecule of glucose or two molecules of Pyruvic acid). The outline of Krebs cycle is given on page 308.

Therefore, considering both glycolysis and Krebs cycle, in total 38 molecules of ATP, 6 molecules of water and carbon dioxide each are evolved while six molecules of oxygen are consumed per breakdown of one molecule of glucose (carbohydrate) in respiration.

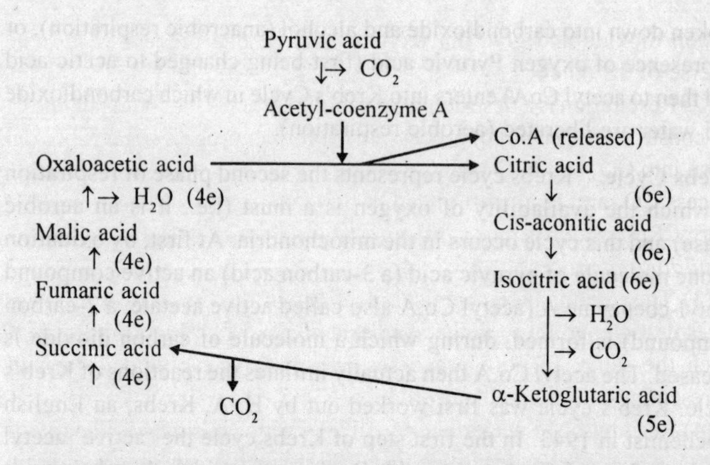

Conditions Affecting Respiration. (1) **Oxygen.** The presence of free oxygen is the most essential condition since respiration is mainly an oxidation process, particularly its aerobic phase. Thus a fall in the concentration of oxygen in the surrounding air markedly affects the process. Under this condition the process slows down and may even come to a standstill below 5% concentration. With the increase in concentration the rate of respiration increases but not far beyond the normal rate. (2) **Temperature.** This also markedly affects the rate of respiration. The minimum rate is reached at 0°C, and the maximum at 45°C or even at 40°C. Beyond these limits the protoplasm is injured. (3) **Light.** Its effect is only indirect; in bright sunlight, however, the respiratory activity is greater than in subdued light, possibly because under this condition the stomata remain wide open helping free exchange of gases. (4) **Water.** Protoplasm saturated with water respires more vigorously than in desicated condition, as in dry seeds. (5) **Vitality of Cells.** Respiration in young active cells is always more rapid than in old cells. Thus it is seen that vegetative buds, floral buds and germinating seeds respire more vigorously than the older parts of the plant. (6) **Carbon dioxide Concentration.** If CO_2 be allowed to accumulate as a result of respiration, the process slows down and may even altogether stop. (7) **Nutritive Materials.** Soluble carbohydrates like glucose affects respiration markedly since this is quickly broken down in the process releasing energy.

Respiration and Photosynthesis. (1) In respiration plants utilize oxygen and give out carbon dioxide, while in photosynthesis plants utilize carbon dioxide and give out oxygen; that is, one process is just the reverse of the other.

(2) Respiration is a destructive (catabolic) process, but photosynthesis is a constructive (anabolic) process. In the former process sugar is broken down into CO_2 and H_2O with the liberation of energy, while in the latter process CO_2 and H_2O are utilized to build up sugar with the storage of energy. Respiration is thus a *breaking-down* process, and photosynthesis a *building-up* process.

(3) The intermediate chemical reactions in the breakdown of sugar in respiration and those in the synthesis of sugar in photosynthesis are much the same. In both processes phosphoglyceric acid is formed representing an intermediate product.

(4) Respiration is performed by all the living cells of the plant at all times, i.e. it is independent of light and chlorophyll; while photosynthesis is performed only by the green cells, and that too, only in the presence of sunlight. Although photosynthesis persists only for a limited period, this process is much more vigorous than respiration.

(5) Respiration results in a loss of dry weight of the plant due to breaking-down of food materials and the formation of carbon dioxide which escapes from the plant body; but photosynthesis results in a gain in dry weight due to formation of sugar, starch, etc., which accumulate in the plant body.

Fermentation is the incomplete oxidation of sugars (particularly glucose) into alcohol and carbondioxide under the influence of several species of yeast and bacteria in the absence of oxygen. The change is due to the action of an enzyme, known as *zymase,* secreted by the micro-organisms, and not due to their direct action on sugar. Fermentation is most readily seen in date-palm juice, where sugar is broken up by unicellular yeast plants into alcohol and carbon dioxide, the frothing being due to the formation of this gas. This process is called alcoholic fermentation. The process is also analogous to anaerobic respiration and may be represented by an identical formula—$C_6H_{12}O_6$ + zymase → $2C_2H_5OH$ + $2CO_2$ + zymase (sugar + zymase → alcohol + carbon dioxide + zymase + energy). The release of energy in fermentation is due to re-arrangements of molecules in the compounds acted on. The process of fermentation is, therefore, also called *intramolecular respiration.* Little energy is released by this process, but it suffices for the activity of the anaerobic micro-organisms. There are other types of fermentation, e.g. acetic acid fermentation, or conversion of alcohol into acetic acid (vinegar) by acetic acid bacteria, lactic acid fermentation (souring of milk), or conversion of milk sugar into lactic acid by lactic acid bacteria, etc. Some of the fermentation processes leading to the production of alcohol, vinegar, manufacture of flavoured butter and cheese, conversion of hide into leather, retting of jute, etc., are of great commercial importance. It may be noted that the organic products formed by the fermenting micro-organisms are toxic to them. This checks their unlimited growth and multiplication in nature.

Chapter 11

Metabolism

Two series of chemical changes or processes are simultaneously going on in a plant cell, one leading to the construction or building-up of the protoplasm, and the other to its decomposition or breaking-down. These processes, which are constructive on the one hand and destructive on the other, are together known as **metabolism**. Metabolism takes place only in the living cells, and is one of the characteristic signs of life. The processes that lead to the construction of various food materials and other organic compounds and finally of protoplasm are together known as **anabolism**, and those processes leading to their destruction or breaking-down as **catabolism**. The main anabolic or constructive changes are: formation of sugars and other carbohydrates, formation of proteins, and formation of fats and oils. These changes or processes are regarded as anabolic because the protoplasm continually re-constructs itself with these nutritive substances. By anabolism, a considerable amount of *potential energy* is stored in those substances for future use of the protoplasm. The main catabolic or destructive processes are: digestion, respiration and fermentation. By these processes complex food substances are gradually broken down into simpler products, e.g. various carbohydrates into glucose, various proteins into amines and amino-acids, and fats and oils into fatty acids and glycerine. The potential energy already stored up in them is released by catabolism into *kinetic energy* for manifold activities of the protoplasm. Carbon dioxide and water are formed as a result of complete oxidation of glucose in aerobic respiration, and alcohols and organic acids as result of incomplete oxidation of glucose in anaerobic respiration or fermentation. Various secretary products such as enzymes, vitamins, hormones, cellulose, nectar, etc., and various waste products such as tannins, essential oils, gums, resins, etc., are the result of catabolic processes.

B. Physiology of Growth and Movements

Chapter 12

Growth

Growth is a vital phenomenon. The protoplasm assimilates the products of digestion and increases in bulk and weight. The cells divide and numerous new cells are formed; these increase in size and become fully turgid, and the plant grows as a whole. *Growth may be defined as a permanent and irreversible increase in size and form accompanied by an increase in weight*. Growth is usually very slow in plants, but it can be accurately measured with the help of an instrument, called the **arc indicator** or **lever auxanometer** (Fig. 26). There are, however, certain plants which show a remarkable rate of growth in length. Thus, young shoots of bamboo (Bambusa) and giant bamboo (Dendrocalamus) show an average growth of 46–60 cm a day; *Asparagus* shoots 30 cm. a day; some climbers, e.g. *Ipamoea nil* and some lianes, e.g. wood-rose (*I. tuberosa*) 20 cm. a day; inflorescence

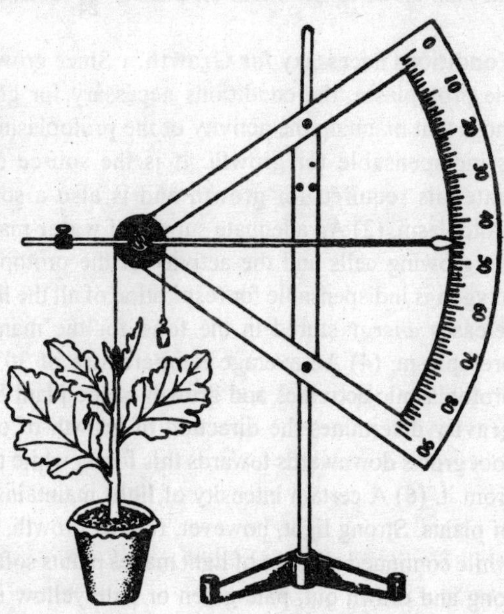

Figure 26. Arc indicator or lever auxanometer.

axis of *Agave* 15 cm a day; and tendrils of some *Cucurbita* 16 cm an hour.

Experiment 17. Growth in Length of the Shoot. The **arc indicator** is an instrument by means of which a small increase in length can be magnified many times. From this total known magnification recorded by the instrument the actual length attained by a plant within a certain specified time can easily be calculated.

The arc indicator consists of a movable lever or indicator fixed to a wheel and a graduated are fixed to a stand. A cord passes round the wheel. One end of the cord is tied round or gummed to the apex of the stem, and from the other end a small weight is suspended to keep the cord taut. As the stem increases in length the wheel rotates under the suspended weight and the indicator moves down the graduated arc. The growth in length of the plant is thus recorded by the instrument on a magnified scale. From the record thus obtained the actual increase in length of the stem is calculated; for instance, if the lever has traversed a distance of 45 cm in 24 hours, and the magnification is 90 times, the actual growth in the same period is $\frac{45}{90}$ cm, i.e., 0.5 cm or 5 mm and, therefore, in one hour the actual growth of the plant is $\frac{5}{24}$ mm., i.e. 0.2 mm.

Conditions necessary for Growth. Since growth is brought about by the protoplasm, the conditions necessary for growth are the same as those that maintain the activity of the protoplasm. (1) A supply of **food** is indispensable for growth. It is the source of necessary nutritive materials required for growth and is also a source of energy to the protoplasm. (2) An adequate supply of **water** maintains the turgidity of the growing cells and the activity of the protoplasm. (3) A supply of **oxygen** is indispensable for respiration of all the living cells. Respiration releases *energy* stored in the food for the manifold activities of the protoplasm. (4) An average **temperature** of 30°C is very suitable for protoplasmic activities and growth of the plant body. (5) The **force of gravity** determines the direction of growth of certain organs, e.g. the root grows downwards towards this force, while the stem upwards away from it. (6) A certain intensity of **light** maintains the healthy condition of plants. Strong light, however, retards growth, as during the daytime, while continued absence of light makes plants soft, weak, brittle, slender, long and drawn out, pale-green or pale-yellow in colour and sickly in appearance; such plants are said to be **etiolated** (Fig. 27). They seldom produce flowers. Moreover, stomata open and chloroplasts function only

Figure 27. Effect of light and darkness on the growth of seedlings. *Left,* gourd seedlings; *right,* gram seedlings. *A,* grown in light; *B,* grown in darkness.

in the presence of light manufacturing food materials which are required for growth.

Phases of Growth (Fig. 28). Growth does not take place throughout the whole length of the plant body, but it is localized in special regions called *meristems* which may be apical, lateral, or intercalary. The growth in length is due to gradual enlargement and elongation of the cells of the apical meristems (root-apex and stem-apex) and in dicotyledons and gymnosperms the growth in thickness is due to the activity of the lateral meristems, i.e., fascicular cambium, interfascicular cambium and cork-cambium. If the history of growth of any organ of a plant be followed three phases can be recognized in it.

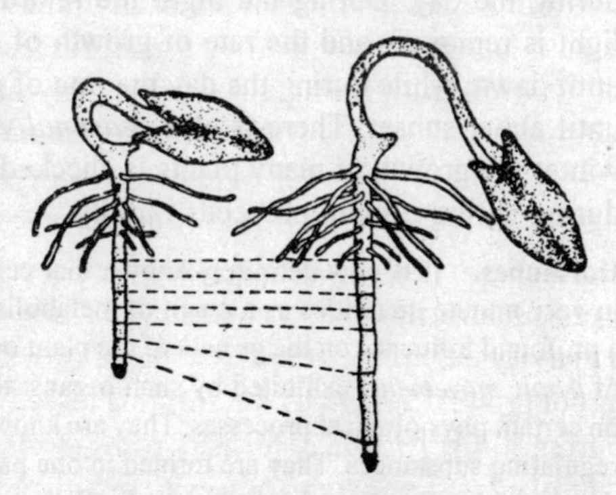

Figure 28. Phases of growth of root.

1. **The Formative Phase.** It is restricted to the apical meristem of, the root and the stem. The cells of this region are constantly dividing and multiplying in number. They are characterized by abundant protoplasm, a large nucleus and thin cellulose wall.

2. **The Phase of Elongation.** It lies immediately behind the formative phase. The cells no longer divide in this phase, but they increase in size; they begin to enlarge and elongate until they reach their maximum dimension. In the root this phase occupies a length of a few millimetres, and in the stem a few centimetres. In some of the climbers it may occupy a much longer space than this.

3. **The Phase of Maturation.** This phase lies further back. Here the cells have already reached their permanent size; the thickening of the cell-wall takes place in this phase.

Grand Period of Growth. Every organ of the plant body, in fact every cell that the organ is composed of, shows a variation in the rate of its growth. The growth is at first slow, then it accelerates until a maximum is attained, then it falls off rather quickly, and gradually slows down until it comes to a standstill. This growth of an organ or a cell or the plant as a whole extending over the whole period of life is called the **grand period of growth**. Within the grand period variations in growth occur owing to external and other causes. There is thus the *diurnal variation of growth*. Light inhibits growth, and too intense light even checks it altogether. Thus plants grow quicker during the night than during the day. During the night the retarding or inhibiting action of light is removed, and the rate of growth of a plant gradually increases until dawn; while during the day the rate of growth gradually decreases until about sunset. There is also *seasonal variation of growth;* during winter the growth of many plants is checked or becomes very slow, but during spring growth proceeds rapidly.

Hormones. It is now definitely known that certain organic products formed in very minute quantities as a result of metabolism inside the plant body have a profound influence on the *growth* of the plant organs and on the various kinds of *tropic movements* exhibited by such organs; they also have a marked effect on certain physiological processes. They are known as the **hormones** or growth-regulating substances. They are formed in one part of the plant body, chiefly in the apical meristem, and transported from there to another part to produce a particular physiological effect there. The presence of hormones was first demonstrated by experimental methods. It has now been possible to extract them from plants by appropriate chemical methods. At low concentration they stimulate growth, while at high concentration they retard growth. Of the various plant hormones (phytohormones) discovered till now auxins (auxin A and auxin B) and heteroauxin (indole-acetic acid—IAA, first isolated from human urine) are best known for their physiological actions, chemical composition and distribution in plants. Certain synthetic compounds such as indolylbutyric acid also act as hormones. Heteroauxin causes the formation of roots in stem-cuttings, leaf-cuttings (Fig. 29) and in grafting. Auxins are responsible for fruit development, seed germination, seedling growth and growth of plant organs; they also stimulate cell divisions in the meristematic tissue, and influence certain physiological processes; also, the role of homones in tropic responses has now been well established. Thus *phototropism* and *geotropism* are now explained on a hormonal basis. Hormones responsible for the development of the root, stem, leaf, flower, fruit, etc., have also been discovered.

Figure 29. Root formation in leaf-cutting of *Pogostemon,* treated with indolyl butyric acid.

Vitamins. Vitamins are a group of complex organic (biochemical) products of plants, essentially required by human beings for normal functioning of their body in various directions—nutrition, growth, development, reproduction and all other physiological processes of the body. As such vitamins have proved to be most valuable in preventing and curing deficiency diseases such as scurvy (livid spots on the skin and general debility), beri-beri, rickets, malnutrition, loss of appetite, poor physical growth, eye infection, nervous breakdown, etc., caused by the absence of vitamins in the food or their faulty absorption due to intestinal troubles. For over two centuries scurvy was a dreaded disease among the sailors, resulting in many deaths. About the year 1793 it was found that the use of orange or lemon juice dispelled scurvy from the navy. Evidently the juice contains something (now known as vitamin C) which cures the disease. It was only from the year 1906 that investigations on vitamins were made from the biological standpoint, and till today several vitamins have been discovered and their value established. Vitamins are required only in minute quantities for a particular effect and they are used up in the metabolic processes. They are mostly synthesized by plants and stored up in their different organs. Plants are, therefore, the main sources of vitamins for animals including human beings. It has now been possible to synthesize some of the vitamins, particularly vitamins A, C and D on a commercial scale. Some of the well-known vitamins are as follows:

Vitamin A is a growth-promoting vitamin, soluble in fats and oils, and fairly resistant to heat. It increases resistance to bacterial infections of the lungs and the intestines, prevents many eye-diseases—particularly night-blindness—and cures skin-diseases and nervous weakness. Carotene of green plants is the source of this vitamin, and animals can synthesize it in their body by taking food containing carotene of plants. Vitamin A is found in carrot, green leafy vegetables (spinach, lettuce, cabbage, etc.), cereals (particularly in their pericarp), pulses, many fruits (particularly yellow ones such as tomato, mango, orange, apple, papaw, etc.), fish-liver oils (e.g. cod-liver oil and halibut-liver oil), liver of mammals, milk, butter, egg-yolk, etc.

Vitamin B consists of a group of closely allied vitamins, commonly called *vitamin B complex;* they are soluble or sparingly soluble in water and more or less resistant to heat. Of these, vitamin B_1 (soluble in water and not very resistant to heat) prevents beri-beri (accumulation of water in the body resulting in serious diseases). Beri-beri was for a long time a dreaded disease in the rice-eating countries like India, Malaysia, China and Japan. Polished rice (evidently something removed from its pericarp, now known to be this vitamin) was found to be the cause of this disease which resulted in immense suffering and innumerable deaths. Other vitamins of this group are B_2, B_6, B_{12}, etc., each having its own function. Vitamin B complex is very widely distributed in plants. One or more of them are found in dry yeast, cereals, pulses, most vegetables, many fruits (e.g. orange, banana, apple, tomato, etc.), nuts, milk, cheese, egg, meat, fish, liver, etc.

Vitamin C (soluble in water and sensitive to heat, and therefore lost by cooking) prevents scurvy, mental depression, swelling and bleeding of gums, and degeneration of teeth. It is found in most fresh fruits (particularly orange, lemon, pummelo, tomato, pineapple, guava, papaw, etc.), green vegetables, sprouted pulses and cereals.

Vitamin D cannot stand strong light; otherwise it is sufficiently stable. Its deficiency causes rickets, softening of bones, dental caries, poor development of teeth, and inhibits proper absorption of calcium and phosphates. It is commonly found in dry yeast, butter, egg-yolk, fish and fish liver oils. Vitamin D can be produced in the human body by the action of ultraviolet rays (in sunlight or electricity) on the skin. It is formed from ergosterol when the latter is exposed to sunlight. Ergosterol, is widely distributed in plants and animals.

Vitamin E is resistant to heat and light but destroyed by ultraviolet rays. Its deficiency causes sterility in animals (not yet definitely proved in the case of human beings) and degeneration of muscles. It is found in green vegetables, germinating grains, wheat embryo, etc.

To summarize, it may be said that our daily diet should include at least some green vegetables, fruits like orange, banana, guava, papaw, tomato, etc. some of the vegetable oils, milk and milk products, eggs, animal liver, fish, meat, etc. as sources of vitamins, in addition to cereals and pulses.

Chapter 13

Movements

Movement is a sign of life. But most plants are fixed to the ground and cannot move bodily. Protoplasm, however, is sensitive to various external factors which act as stimuli, such as heat, light, gravity, certain chemicals, electricity, etc., and many plant organs or entire free organisms respond to them by some kind of movement. The capacity of plants or their particular organs to receive stimuli from outside and to respond to them is spoken of as *irritability*. Irritability expresses itself in some kind of movement and is a decided advantage to the plants in many respects. The various kinds of movements may be broadly classified into (I) movements of locomotion, and (II) movements of curvature.

(I) **Movements of Locomotion.** These are the movements exhibited by free unicellular or multicellular organs or entire organisms, or of protoplasm within the cell. They may be (A) spontaneous (or autonomic) and (B) induced (or paratonic).

A. **Spontaneous movement of locomotion** is the movement of certain minute free organs of plants or of entire free organisms *of their own accord,* i.e. without the influence of any external factor; it may be due to some internal cause not clearly understood. Common instances are: rotation and circulation of protoplasm, within the cells, ciliary and amoeboid movements of protoplasm, movements of all ciliate bodies, oscillating movement of *Oscillatoria* (see Fig. V/1), brisk movements of many unicellular algae like desmids and diatoms, etc.

B. **Induced movement of locomotion** is the movement of certain plant organs or of entire free organisms induced by some external factors which act as stimuli. The external factors may be in the nature of certain chemical substances, light and heat. The induced movement caused by any such factor is otherwise called *taxis* (pl. taxes) or *taxism*. Thus depending on the nature of the stimuli the taxic (or tactic) movements may be (1) **chemotaxis** when influenced by chemical substances, (2) **phototaxis** when influenced by light, and (3) **thermotaxis** when influenced by temperature.

(II) **Movements of Curvature.** These are the movements of certain organs of higher plants which, being fixed to the ground, cannot move. The movements shown by such organs are of different kinds, and broadly they may be mechanical or vital. The latter may again be spontaneous (or autonomic) and induced (or paratonic).

A. **Mechanical movement** is the movement of certain non-living organs of plants. Since this kind of movement has a definite relation with water, either quickly imbibing it or quickly losing it, this is otherwise known as: *hygroscopic movement*. This kind of movement is exhibited by the bursting of explosive fruits, e.g. *Ruellia* (see Fig. I/ 159), *Barleria, Phlox, Bauhinia vahlii* (see. Fig. I/160), etc., and by the bursting of moss and fern capsules. Besides, it is seen that the elaters of *Equisetum* spore coil round it when the air is moist and they uncoil and stand out stiffly when the air is dry (see Fig. V/86D–E).

B. **Spontaneous movements of curvature** exhibited by plant organs may be of the following two kinds:

(1) **Movement of variation** is the movement of *mature* organs due to *variation in turgidity* of the cells making up those organs. It is somewhat rapid, but instances are rather rare. The spontaneous movement of variation is, however, very remarkably exhibited by the *pulsation,* i.e. the rising and falling of the two lateral leaflets of the Indian telegraph plant (*Desmodium gyrans;* Fig. 30) during the daytime, the terminal leaflet remaining fixed in its position.

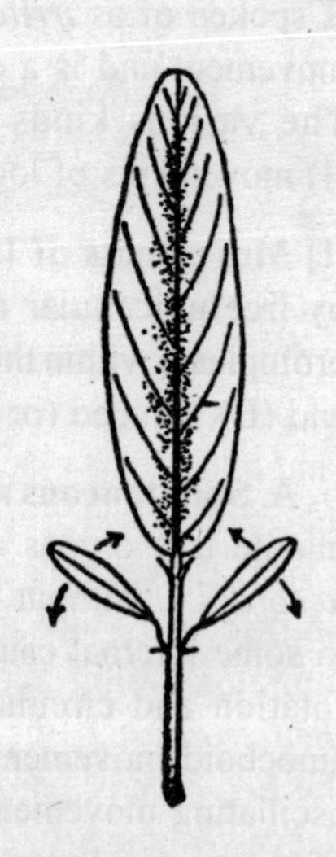

Figure 30. A leaf of telegraph plant showing spontaneous movement of its two lateral leaflets.

(2) **Movement of growth** is the movement of *growing* organs *due to unequal growth* on different sides of those organs. It is very slow. This kind of movement is seen in some trailers and creepers. In them the growth may be

restricted to one side of the stem for some time and then it passes on to the opposite side. In such an event the stem tip moves from one side to the other, i.e., it follows a zigzag course (**nutation**). More commonly the region of growth passes regularly around the stem (**circumnutation**), as in tendrils and twiners. Further, it is seen that young fern leaves at first remain closely coiled because of more rapid growth on the undersurface (**hyponasty**). Later, growth is more rapid on the upper surface (**epinasty**) and the leaves uncoil and straighten out. Opening of flowers is also a kind of growth movement.

C. **Induced movements of curvature** are those movements exhibited by certain plant organs under the influence of some external factors, particularly contact with a foreign body, light, gravity, and moisture, which act as stimuli. Such movements, as induced by external factors, may broadly be of two kinds: (a) tropisms and (b) nasties.

(a) **Tropisms** or tropic movements are the movement of plant organs influenced by external stimuli, particularly contact, light, gravity and moisture. Tropic movement is always directive, i.e. the organ concerned moves either towards the source of the stimulus or away from it. Depending on the nature of the stimuli the movements may be as follows:

(1) **Haptotropism** is the movement of an organ induced by contact with a foreign body. Twining stems and tendrils are good examples of haptotropism. In such cases the reaction is rather slow and, therefore, the contact must be of long duration to bring about the movement. When such organs come in contact with any support or any hard object the growth of the contact side is checked, while the opposite side continues to grow. The result is a mechanism for climbing. Some move clockwise and others anti-clockwise. If the direction be artificially altered, growth becomes arrested.

(2) **Heliotropism** or **phototropism** is the movement of plant organs in response to incidence of rays of light. Some organs are attracted by unilateral light and grow towards it; they are said to be *positively heliotropic,* e.g. the shoot; and others grow away from it and are said to be *negatively heliotropic,* e.g. the root. Dorsiventral organs such as leaves, runners, etc., grow at right angles to the direction of light so that their upper surface is exposed to light; such organs are said to be *diaheliotropic*. Positive heliotropism is seen markedly in potted plants, particularly seedlings, when these are grown in a closed box (heliotropic chamber; Fig. 31) with one open window on one side. They all tend to

grow towards the window, i.e. towards the source of light, and ultimately come out through it. The flower-stalk of groundnut (Fig. 32) grows towards light, but after pollination it becomes negatively heliotropic and positively geotropic like the root. The stalk bends down and quickly elongates, pushing the fertilized ovary into the ground where gradually the ovary ripens into a pod (fruit). Some species of *Trifolium* also exhibit the same phenomenon. In *Eucalyptus* it is seen that the edge of the leaf is turned towards intense light, and when the light is diffuse the leaf turns and its surface is exposed to it (the light).

Figure 31. Heliotropic chamber.

(3) **Geotropism** is the movement of plant organs in response to the force of gravity. Geotropism has a marked effect on the direction of growth of plant organs. The primary root is seen to grow towards the

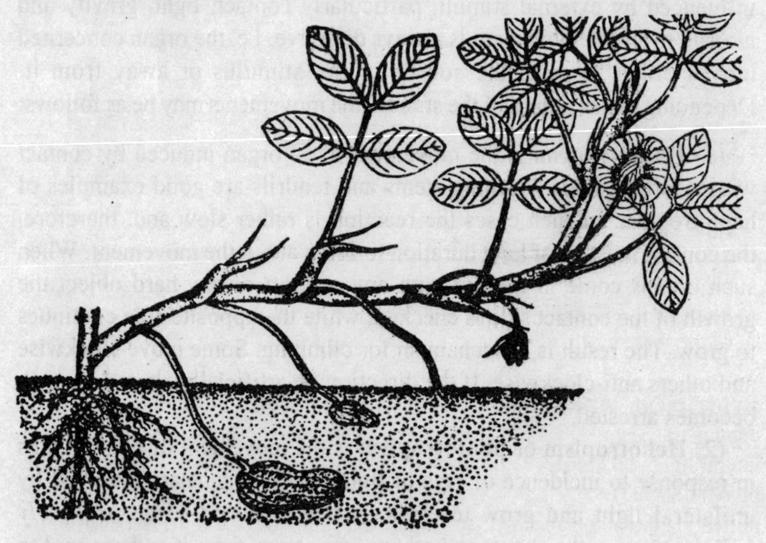

Figure 32. Groundnut showing that the flower-stalk is negatively heliotropic and positively geotropic after pollination of the flower.

centre of gravity, and the primary shoot away from it. The former is, therefore, said to be *positively geotropic,* and the latter *negatively geotropic.* The lateral roots and the branches usually grow at right angles to the force of gravity and are said to be *diageotropic.* That the direction of growth is determined by the stimulating action of the force of gravity is clearly seen in a seedling which has been placed in a horizontal position away from light. Both the stem and the root undergo curvature in their growing region behind the apex, passing through an angle of 90°; the root curves and grows vertically downwards, as does the stem upwards. It is the very tip of the root, for a distance of 1 to 2 mm in length, that is sensitive to this stimulus; but the actual bending takes place some distance behind the tip in the region of greatest growth. If the tip of the root be decapitated, no bending takes place. Besides, it is seen that the root of a germinating seed can, under the force of gravity, grow downwards even through mercury overcoming considerable pressure. Further, it has been found possible with the help of a **clinostat** (Fig. 33) to eliminate the effect of geotropic stimulus on the root and the shoot by introducing a centrifugal force (see Experiment 18).

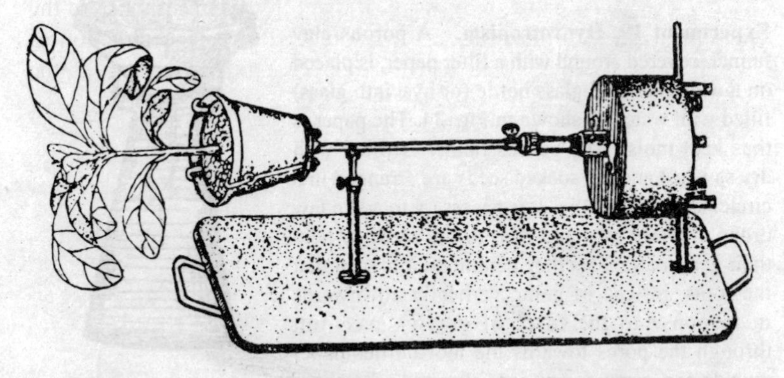

Figure 33. Clinostat in the horizontal position to eliminate the effect of the force of gravity.

Experiment 18. Geotropism. A clinostat (Fig. 33) may be used to demonstrate geotropism. A clinostat is an instrument by which the *effect* of lateral light and the force of gravity on an organ of a plant—root or stem—can be eliminated. It consists of a rod with a disc mounted on it, to which a small potted plant may be attached, and a clockwork mechanism for rotating the rod and the disc. The clinostat works slowly—its rotation being ordinarily ¼ to 4 turns per hour. A plant, preferably a pot plant, may be fixed in the clinostat in any position—vertical, horizontal or at an angle—and made to rotate by clockwork mechanism in the clinostat. When the plant is horizontal, the root and the stem grow horizontally, instead of the root curving downwards and the stem upwards. This is due to the fact that all sides of the growing axes are in turn directed downwards, upwards and sideways so that the force of gravity cannot act on any definite position. This results in the effect of the force being eliminated altogether. The root and the stem cannot, therefore, bend. If, however, the plant be fixed in a vertical position and the clinostat rotated, it is seen that the plant grows in a vertical direction—root downwards and the stem upwards.

(4) **Hydrotropism** is the movement of an organ in response to the stimulus of moisture. Roots are sensitive to variations in the amount of moisture. They show a tendency to grow towards the source of moisture, and are said to be *positively hydrotropic*. It is seen that roots of plants, growing in a hanging basket made of wire-netting and filled with moist sawdust, project downwards at first, coming out of the basket, under the influence of the force of moisture (moist sawdust of the basket), but turn back and pass again into the basket having formed loops.

Experiment 19. Hydrotropism. A porous clay funnel, covered around with a filter paper, is placed on a wide-mouthed glass bottle (or hyacinth glass) filled with water, as shown in Fig. 34. The paper is thus kept moist. The porous funnel is filled with dry sawdust and the soaked seeds are arranged in a circle, each near a pore. It is necessary to add a few drops of water now and then to the seeds to help their germination. As they germinate it is seen that the roots, instead of going vertically downwards in response to the force of gravity, pass out through the pores towards the moist filter paper outside and grow downwards alongside the paper into the bottle. Roots thus show movements towards moisture, or, in other words, they are positively hydrotropic.

Figure 34. Experiments on hydrotropism.

(b) **Nasties** or nastic movements are the movements of dorsiventral organs like leaves and petals, induced by external stimuli such as contact, light and temperature. But these movements are not directive, as are the tropisms, i.e. the direction of movement is not determined by the direction of the stimulus; in other words, whatever be the direction of the stimulus it equally affects all parts of the organs, and they always move in the same way and in the same direction. Two kinds of such movements are as follows.

(1) **Seismonasty** is the movement brought about by mechanical stimuli such as contact with a foreign body, poking with any hard object, drops

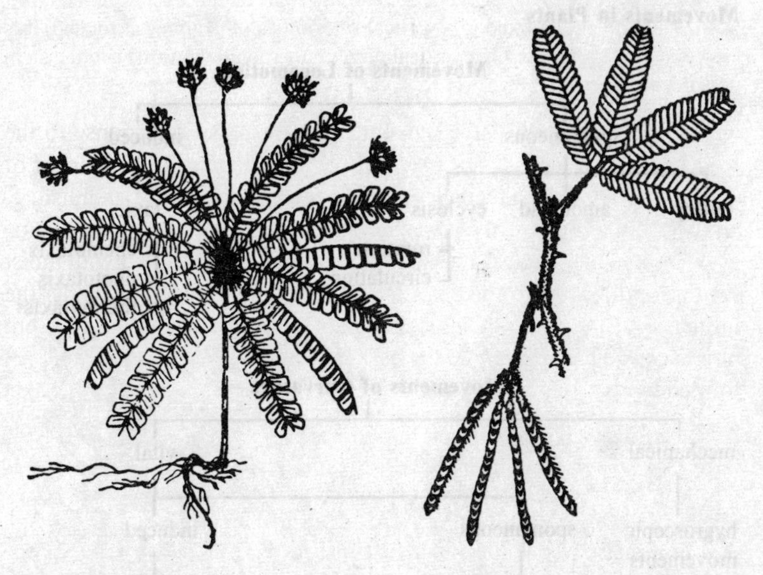

Figure 35. Sensitive wood-sorrel (*Biophylum sensitivum*).

Figure 36. Sensitive plant (*Mimosa pudica*).

of rain, a gust of wind, etc. Movements of the leaves (leaflets) of the sensitive plant (*Mimosa;* Fig. 36), the sensitive wood-sorrel (*Biophytum;* B. BAN-NARANGA; H. LAJALU—Fig. 35). *Neptunia* (B. PANI-LAJUK), carambola (*Averrhoa;* B. KAMRANGA; H. KAMRAKH), etc., are familiar examples. Leaflets of such plants close up when touched. The Venus' fly-trap (*Dionaea;* see Fig. 19) is another very interesting example.

(2) **Nyctinasty** is the movement induced by alternation of day and night, i.e. light and darkness. It is otherwise called **sleep movement**. Leaves and flowers, particularly the former, are markedly affected by nyctinasty. This kind of movement is most remarkably exhibited by the leguminous plants. Leaflets of these plants close up and often the leaf as a whole droops in the evening when the light fails, and they open up again when the light appears in the morning. A few other plants like goosefoot (*Chenopodium;* B. & H. BATHUA-SAK), carambola (*Averrhoa;* B. KAMRANGA; H. KAMRAKH), wood-sorrel (*Oxalis*), etc., also show the same phenomenon. Among the flowers showing nyctinasty mention may be made of *Gerbera* (a garden herb), *Portulaca* (wild or garden variety), etc.

Movements in Plants

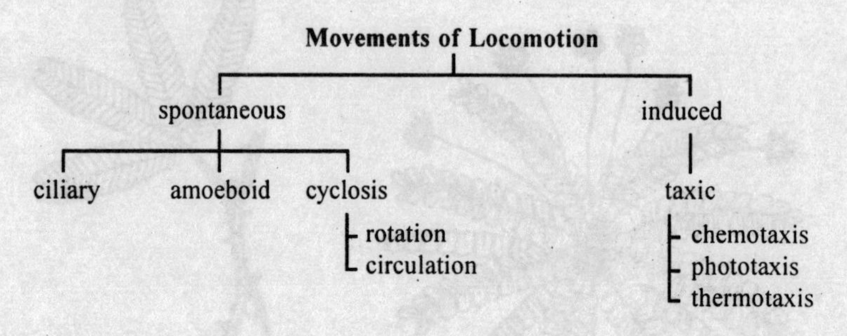
Movements of Locomotion
spontaneous
induced
ciliary
amoeboid
cyclosis
rotation
circulation
taxic
chemotaxis
phototaxis
thermotaxis

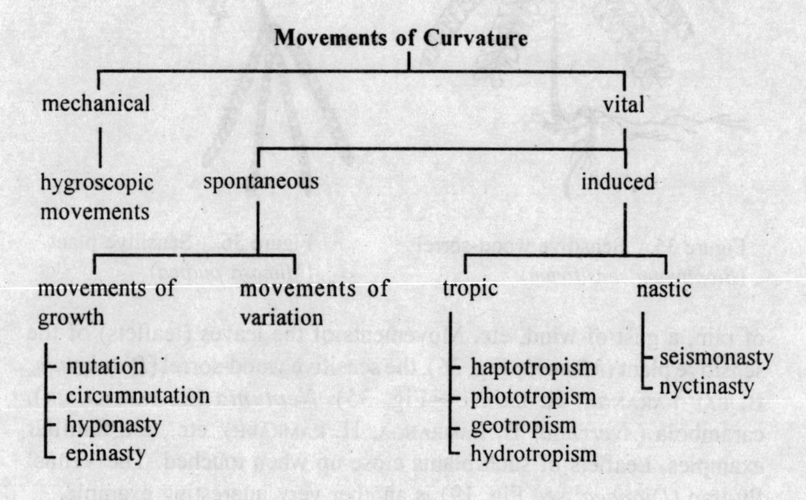
Movements of Curvature
mechanical
vital
hygroscopic movements
spontaneous
induced
movements of growth
movements of variation
tropic
nastic
nutation
circumnutation
hyponasty
epinasty
haptotropism
phototropism
geotropism
hydrotropism
seismonasty
nyctinasty

C. Physiology of Reproduction

Chapter 14

Reproduction

Since the life of an individual plant is limited in duration it has developed certain methods by which it can reproduce itself in order to maintain the continuity of the species and also to multiply in number. The following are the principal methods of reproduction: **vegetative**, **asexual** and **sexual**.

I. *VEGETATIVE REPRODUCTION*

A. *Natural Methods of Propagation.* In any of these methods a portion gets detached from the body of the mother plant, and gradually grows up into a new independent plant.

(1) **Budding.** In the case of yeast (Fig. 37A) one or more tiny outgrowths appear on one or more sides of the vegetative cell immersed in sugar solution. Soon these outgrowths get detached from the mother

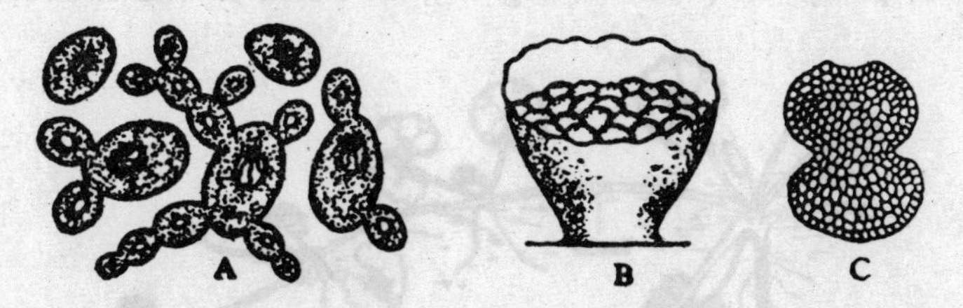

Figure 37. *A*, budding in yeast; *B*, gemma-cup and *C*, gemma in *Marchantia*.

cell and form new individuals. This method of outgrowth formation is known as budding. Often budding continues one after another so that chains and even sub-chains of cells are formed. The individual cells finally separate from one another and form new yeast plants.

(2) **Gemmae.** In *Marchantia* special multicellular bodies, known as gemmae (sing. gemma), develop in gemma-cups (Fig. 37B–C) on the thallus for the purpose of vegetative propagation (see Fig. V/57).

(3) **Leaf-tip.** There are certain ferns, commonly called walking ferns (e.g. *Adiantum caudatum*), which propagate vegetatively by their leaf-tips (Fig. 38). As the leaf bows down and touches the ground the tip strikes roots and forms a bud. This bud grows into a new independent fern plant. Ferns normally, however, reproduce vegetatively by their rhizome.

Figure 38. Walking fern (*Adiantum caudatum*).

(4) **Underground Stems.** Many flowering plants reproduce themselves by means of the rhizome, e.g. ginger; the tuber, e.g. potato; the bulb, e.g. onion; and the corn, e.g. *Gladiolus* and *Amorphophallus*. The buds produced on them gradually grow up into new plants.

(5) **Sub-aerial Stems.** The runner of wood-sorrel (*Oxalis*) and Indian pennywort (*Centella;* Fig. 39), the stolon of wild strawberry

Figure 39. Runner of Indian pennywort (*Centella*) showing vegetative mode of propagation.

(*Fragaria*), the offset of water lettuce (*Pistia*) and the sucker of *Chrysanthemum* are made use of by such plants for vegetative propagation.

(6) **Adventitious Buds.** In the sprout-leaf plant (*Bryophyllum pinnatum;* see Fig. I/45A) and in several species of *Kalanchoe* (Figs. 40–41) a series of adventitious (foliar) buds are produced on the leaf-margin, each at the end of a vein; these buds grow up into new plants.

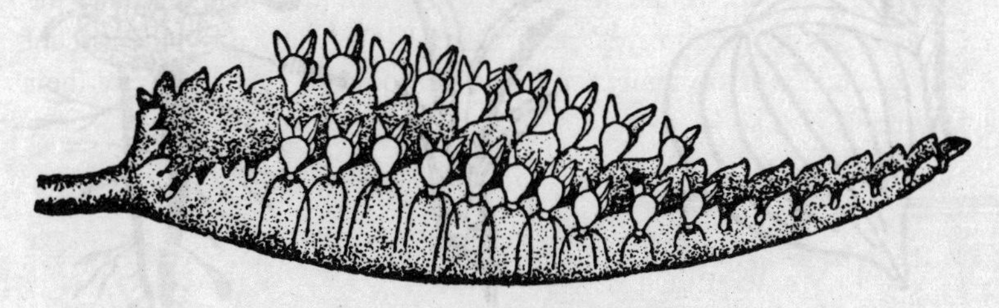

Figure 40. A leaf of *Kalanchoe sp.* with adventitious buds.

In *Kalanchoe verticillata,* however, such buds appear near the apex only (Fig. 41). In the elephant ear plant (*Begonia;* see Fig. I/45B) a few

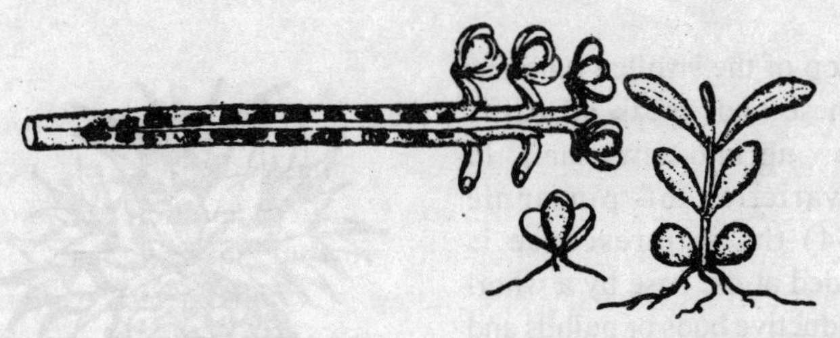

Figure 41. A leaf of *Kalanchoe verticillata* with adventitious buds.

adventitious buds are produced on the surface of the leaf from the veins and also from the petiole. Similarly the roots of some plants may produce adventitious (radical) buds for the same purpose, as in sweet potato.

(7) **Bulbils.** In *Globba bulbifera* (Fig. 42), American aloe (Fig. 43B) and garlic some of the lower flowers of the inflorescence become

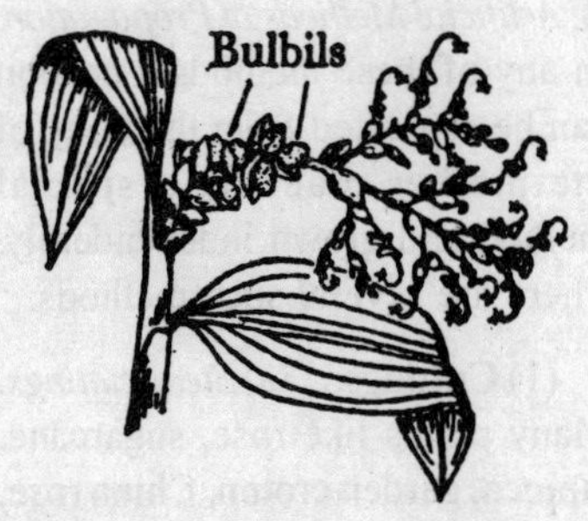

Figure 42. *Globba bulbifera.*

modified into small multicellular reproductive bodies called bulbils. These fall to the ground and grow up into new plants. Bulbils, big or small, are also produced in the leaf-axil of wild yam (*Dioscorea bulbifera;* B. GACHH-ALOO; H. ZAMINKHAND—Fig. 43A). In wood-sorrel (*Oxalis;* Fig. 43C) a large number of small buds (bulbils) may be seen

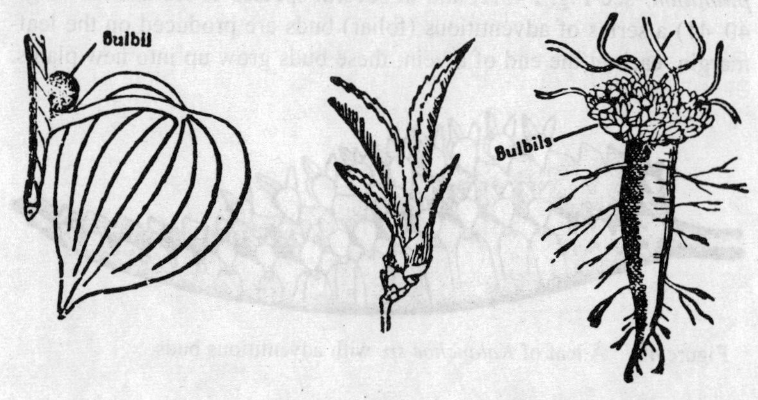

Figure 43. Bulbils. *A, Dioscorea bulbifera; B,* bulbil of American aloe (*Agave*); *C,* wood-sorrel (*Oxalis*).

on the top of the swollen tuberous root. These buds break off easily and grow up into new plants. In some varieties of pineapple (Fig. 44) the inflorescence is surrounded at the base by a whorl of reproductive buds or bulbils and also crowned by a few of them.

Figure 44. Pineapple with a crown and a whorl of bulbils.

B. *Artificial Methods of Propagation.* In any of these methods a portion can be separated from the body of the mother plant by a special method and grown independently. There are several such methods.

(1) **Cuttings.** (a) *Stem-cuttings.* Many plants like rose, sugarcane, tapioca, garden croton, China rose, drumstick (*Moringa*), *Duranta,*

Coleus (see Fig. I/32), etc., may be grown easily from stem-cuttings. When cuttings from such plants are put into moist soil they strike roots at the base and the buds present on them begin to grow up. Sometimes adventitious buds also develop and grow. (b) *Root-cuttings*. Sometimes, as in lemon, citron, ipecac (see Fig. I/39B), tamarind, etc., root-cuttings put into moist soil sprout forming roots and shoots.

(2) **Grafting.** Some of the common methods of grafting including layering and gootee adopted for the sake of fruits, e.g. mango, litchi, guava. sapodilla plum, lemon, etc., or for the sake of flowers, e.g, *Magnolia, Michelia, Ixora,* etc. (Figs. 45–51).

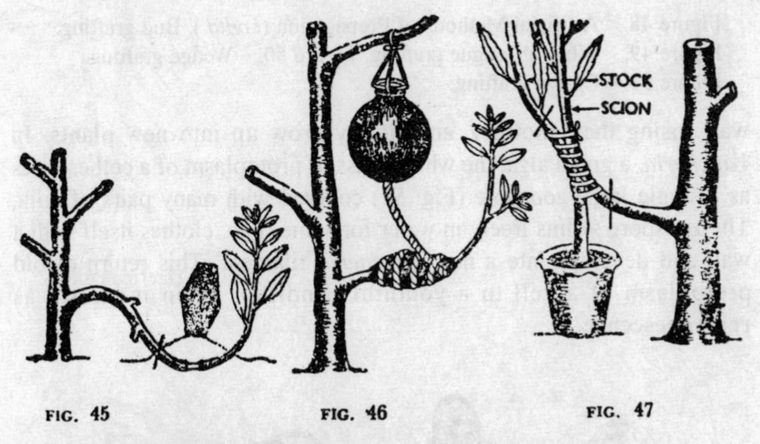

Figure 45. Artificial Methods of Propagation. Layering.
Figure 46. Gootee Figure 47. Inarching or approach grafting.

II. *ASEXUAL REPRODUCTION*

This takes place by means of special cells or asexual reproductive units, called **spores**, produced by the parent plant, which grow by themselves into new individuals, without two such cells fusing together, as in sexual reproduction. Spores are always unicellular and microscopic in size. They may be motile or non-motile.

(1) Motile spores are provided with one or more very slender, tail-like projections known as *cilia*. Such ciliate spores are called **zoospores**, as in many algae and fungi. Commonly they are formed in large numbers and are very minute in size. After escaping from the mother plant they swim briskly about in water for some time, clothe themselves with a

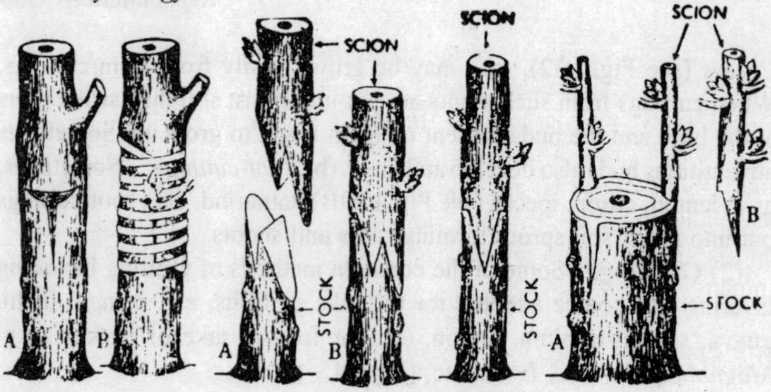

FIG. 48 FIG. 49 FIG. 50 FIG. 51

Figure 48. Artificial Methods of Propagation (*contd.*). Bud grafting.
Figure 49. Whip or tongue grafting. Figure 50. Wedge grafting.
Figure 51. Crown grafting.

wall losing their motility, and finally grow up into new plants. In *Vaucheria*, a green alga, the whole mass of protoplasm of a cell escapes as a single large zoospore (Fig. 52) covered with many pairs of cilia. The zoospore swims freely in water for some time, clothes itself with a wall and develops into a new *Vaucheria* filament. This return of old protoplasm of a cell to a youthful condition again is known as **rejuvenescence**.

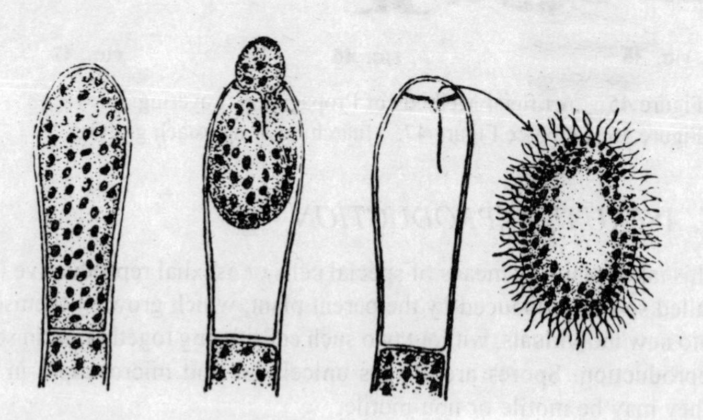

Figure 52. Rejuvenescence in *Vaucheria*.

(2) Non-motile spores commonly borne by terrestrial fungi are very light and dry, and provided with a tough coat. Such spores are well adapted for dispersal by wind and at the same time to meet the

ever-changing conditions of the atmosphere. They are of diverse kinds, and have received special names according to their mode of origin.

(3) True spores are always borne by a sporophyte. In mosses and ferns which show distinct alternation of generations reproduction takes place by both asexual and sexual methods. The sporophyte reproduces asexually by spores which on germination give rise to the gametophyte, and the gametophyte reproduces sexually by gametes which by fusion in pairs (male and female) give rise to the sporophyte again.

III. *SEXUAL REPRODUCTION*

This consists the fusion of two sexual reproductive units, called **gametes**. Gametes are always unicellular and microscopic in size. In sexual reproduction gametes of opposite sexes always fuse *in pairs*. The product of their fusion is a new cell called the **zygote**; the zygote develops into a new plant.

(1) **Conjugation.** In lower algae (e.g. *Spirogyra*) and lower fungi (e.g. *Mucor*) the pairing gametes are essentially *similar,* i.e. not differentiated into male and female, and are called *isogametes*. The union of two such similar gametes is known as **conjugation**, and the zygote thus formed is called the **zygospore**. The gametes may be *small, ciliate* and *motile* (often many in a single cell), or they be *large, non-ciliate* and *non-motile* (often one in a single cell).

(2) **Fertilization.** In all the higher forms of plant life, on the other hand, the pairing gametes are *dissimilar,* i.e. differentiated into male and female, and are called *hetero-gametes*. The union of two such dissimilar gametes is known as **fertilization**, and the zygote thus formed is called the **oospore**. In higher algae and fungi, mosses, ferns and allied plants male gametes are very minute, *motile, ciliate* and *active,* and are known as **antherozoids** or **spermatozoids**. The female gamete in them is *stationary, non-ciliate* and *passive,* larger in size, and is known as the **egg-cell, ovum** or **oosphere**. The corresponding male and female reproductive units in the 'flowering' plants are the **two gametes** of the pollen-tube and the **ovum** or **egg-cell** of the embryo-sac within the ovule.

IV. *SPECIAL MODE OF REPRODUCTION*

Parthenogenesis. The development of the zygote from the egg-cell without the act of fertilization, as seen in many lower plants, e.g. *Spirogyra, Mucor,* and in many ferns, is known as **parthenogenesis**. In some species of 'flowering' plants the embryo also may develop by parthenogenesis, i.e. without fertilization. The development of the fruit from the ovary without the act of fertilization is called **parthenocarpy**. Parthenocarpic fruits are almost always seedless. Examples are found in certain varieties of banana, pineapple, guava, grapes, apple, pear, papaw, etc. Sometimes mere spraying with certain chemicals (growth-promoting substances) like naphthalene-acetic acid results in the setting of fruits without fertilization (induced parthenocarpy).

Part IV

ECOLOGY

Preliminary Considerations
Ecological Groups
Ecosystem

Chapter 1

Preliminary Considerations

Ecology (*oikos,* home; *logos,* discourse or study) deals with the study of interrelationships between the living organisms (plants and animals) as they exist in their natural habitats, and the various factors of the environment surrounding them. Ecology, therefore, includes a detailed study of flora and fauna (together called **biota** or **biotic community**) of a particular region, and also the various conditions of the environment prevailing in that region. This interaction between the organisms and the surrounding environment is expressed as **ecological complex** or **ecosystem**. It includes an intensive study of all aspects of life from birth to death, as influenced by the environment. Evidently, this study relates to the various structural peculiarities (external and internal), functions such as metabolism, growth, reproduction, etc., survival capacity of the offspring, adaptations for distribution, migration, colony formation, etc., of the various species of plants and animals occurring in a particular region, including their identification and enumeration, and also the various environmental factors influencing these aspects of their life. A group of individuals of a particular species is called **population**. In a population, individuals interact and interbreed within a definite space. Whereas the assemblage of various plants or animals (or a group of populations) in a habitat is described as a **community**. Different plants and animals living together in a definite habitat forms a biotic community. Ecology is thus a vast subject, and a good preliminary knowledge of morphology, taxonomy, anatomy, physiology, climatology, soil science, geography, etc., is a prerequisite to the study of this branch of science. In addition to the above the interactions between plants and animals is also of considerable importance. We know that animals including human beings are directly or indirectly dependent upon plants, particularly for food and shelter; while the effects of animals and human communities on plants are also manifold (see below).

Further, in certain climatic regions of the earth it is seen that certain communities of plant life and animal life have become dominant and are distinct from the other life forms with which they remain associated. The dominant species constitute **climax** or major communities, and the associated species constitute minor communities. The climax communities have already established themselves in the region, while the minor communities are in their earlier stages of succession or are controlled by certain local conditions. Such a complex of communities (major and minor) of plants and animals of a region under identical climatic conditions constitutes a **biome**. Some of the important biomes of the world are various forests such as evergreen forests, deciduous forests, mangrove or littoral forests, coniferous forests, etc., grasslands, scrubs, deserts, Arctic tundra, etc. The above biomes (with the exception of the tundra) are well represented in India, while the coniferous forests are restricted to the hills, developing in successive stages according to altitude.

Interdependence of Plants and Animals. (1) **Food.** Green plants manufacture food, while animals including human beings depend on them, directly or indirectly for this basic need. (2) **Shelter.** Plants give shelter to a variety of forest dwellers, and also furnish building materials for human beings. (3) **Oxygen.** Green plants purify the atmosphere by absorbing carbon dioxide and giving out pure oxygen. (4) **Clothing.** Many plants, e.g. cotton and jute. furnish materials for clothing of mankind. Silk worms do likewise but they are reared on plants. (5) **Drugs.** Many plants (Rauwolfia, Ipecac, Cinchona, etc.) are sources of valuable drugs. (6) **Vitamins.** These are almost exclusive products of plants required by human beings, in fact by all animals, for their healthy growth. (7) **Pollination.** Many animals, particularly insects, often visit flowers in search of honey and bring about pollination which is a preliminary step towards the setting of seeds and fruits. (8) **Dispersal of Seeds.** Similarly, many animals, specially birds, act as useful agents for wide distribution of seeds and fruits. (9) **Parasitism.** Many plants and animals lead a parasitic life upon one another for their own existence. (10) **Disintegration of Soil.** Many bacteria, fungi and lower animals permanently dwelling in the soil disintegrate it and make it fertile for higher plants. (11) **Civilization.** Plants have materially contributed to the growth of civilization from time immemorial.

Environment. Environment, as stated before, includes all the factors that surround plants and animals and influence many aspects of their life. Living conditions, as known to all, are different in different regions of the earth, often widely divergent, e.g. temperate to tropic, salt-water seas to fresh-water rivers, lakes and ponds, plains to high altitude (extending up to the alpine region), arid region or desert (with little or no rainfall) to wet region (with plenty of rainfall), etc., and, therefore, depending on their power of adaptability distinct flora and fauna have developed in these regions. Transport from one such region to the other may often prove to be fatal to most of them. Environmental factors may be climatic, edaphic and biotic.

(1) **Climatic Factors.** These include all the conditions of the atmosphere such as temperature, light, water (rainfall), wind, humidity, etc., affecting primarily the shoot system of the plants.

(2) **Edaphic Factors.** These include all the conditions of the soil such as its chemical and physical nature, availability of water and air, temperature, acidity or alkalinity, etc., affecting primarily the root system of the plants.

(3) **Biotic Factors.** These include the action of soil bacteria, algae, protozoa, earthworms and burrowing animals which alter the soil, often making it loose and fertile; the competition with neighbouring plants for food, water and sunlight; parasitic fungi and bacteria, and also parasitic phanerogams; symbiosis; insects which help pollination and also damage plants; and the action of many higher animals including human beings; for example, grazing by herbivorous animals, dispersal of seeds and fruits by birds, woolly animals and even human beings, cultivation, deforestation, afforestation, etc.

Chapter 2

Ecological Groups

Although plants sometimes occur as isolated individuals, more commonly we find that they become adapted to the same environment and are associated together in groups. The groups may include different plant species, belonginig to different families, and differing in shape, size, form and relationship, but living under the same climatic and edaphic conditions. Some of the common groups are as follows.

1. **Hydrophytes.** These are plants that grow in water or in very wet places. They may be submerged, partly submerged, floating, or amphibious. Their structural adaptations are mainly due to high water content, deficient supply of oxygen, and weak light.

Adaptations. The main features of aquatic plants are the reduction of protective tissue (epidermis here is meant for absorption, and not for protection), supporting tissue (lack of sclerenchyma), conducting tissue (minimum development of vascular tissue) and absorbing tissue (roots mainly act as anchors, and root-hairs are lacking), and the special development of air-chambers for aeration of internal tissues.

The Root. The root system is on the whole feebly developed, and root-hairs and root-caps are absent. Some floating plants are rootless, e.g. bladderwort (*Utricularia;* see Fig. III/22), hornwort (*Ceratophyllum*), *Wolffia,* etc.; while others have a cluster of fibrous roots but no root-caps and root-hairs, e.g. water lettuce (*Pistia;* see Fig. I/52), water hyacinth (*Eichhornia;* see Fig. I/62B) and duckweed (*Lemna;* see Fig. I/31); in them, instead of the root-cap, an analogous structure called the root-pocket is formed (see p. 35). In those fixed to the ground under water, either submerged, e.g. *Vallisneria* (see Fig. I/139) and *Hydrilla,* or partly submerged with floating leaves, e.g. water lily and lotus, there is scanty development of roots.

The Stem. This may be in the form of a rhizome, small or large, or it may be long and slender, either branched or unbranched. The stem and the branches, particularly the latter, are soft and spongy, containing a large number of air-cavities filled with gases (oxygen and carbon dioxide for respiration and photosynthesis). They also help the plants

Figure 1. Giant water lily (*Victoria amazonica* = *V. regia*).

to keep in position under water or to float. There is minimum development of mechanical and vascular tissues. Xylem and phloem are reduced to a few narrow vessels and sieve-tubes respectively. The epidermis is without cuticle and is meant for absorption of water. There may be some chloroplasts in it. In some plants prickles are present for self-defence.

The Leaf. Leaves of submerged plants are thin and generally ribbon-shaped, finely dissected or linear, rarely broad. Cuticle and stomata are absent. The epidermis is thin and contains some chloroplasts so that it can utilize the weak light under water for photosynthesis. It is, however, primarily meant for absorption of water. The mesophyll is not differentiated into palisade and spongy tissues. Leaves of floating plants are well developed, and have a thick cuticle and a large number of stomata on the upper surface. Exchange of gases takes place through the upper surface, and absorption of water through the lower. Many air-cavities develop in them and also in the petiole for the purpose of aeration and necessary buoyancy. Amphibious plants subjected to alternate flooding and drying often show **heterophylly** (*heteros,* different; *phylla,* leaves), i.e. they bear different kinds of leaves on the same individual (see pp. 83).

Examples. (a) **Submerged:** *Vallisneria.* (see Fig. I/139), *Hydrilla, Naias, Potamogeton,* etc. (b) **Floating:** bladderwort (*Utricularia;* see Fig. III/22), hornwort (*Ceratophyllum*), duckweed (*Lemna;* see Fig. I/31), water lettuce (*Pistia;* see Fig. I/52), water hyacinth (*Eichhornia;* see Fig. I/62B), water chestnut (*Trapa;* see Fig. I/43), *Azolla, Salvinia,* etc, (c) **Partly submerged:** water lily (*Nymphaea*), lotus (*Nelumbo = Nelumbium*), *Euryale* (B. & H. MAKHNA), giant water lily (*Victoria amazonica = V. regia;* see Fig. 1). *Limananthemum, Ottelia,* etc. (d) **Amphibious** (showing heterophylly): water crowfoot (*Ranunculus*), water plantain (*Alisma*), arrowhead (*Sagittaria;* see Fig. I/91), *Cardanthera triflora* (see Fig. I/90A), *Limnophila heterophylla,* etc.

2. **Mesophytes.** These are plants that grow under average conditions of temperature and moisture, the soil in which they grow is neither saline nor is it waterlogged, and the temperature of the air is neither too high nor too low. Mesophytes are, therefore, intermediate between hydrophytes and xerophytes.

Adaptations. The root-system is well developed with the tap root and its branches in dicotyledons, and a cluster of fibrous roots in monocotyledons; root-hairs are luxuriantly produced for the absorption of water from the soil. The stem is solid (and not spongy, as in water plants), erect, and normally branched. Thorns on the stem are absent or few. All the different kinds of tissues, particularly the mechanical and conducting tissues, have reached their full development in the mesophytes. The aerial parts of plants such as the leaves and the branches are provided with a cuticle. In dorsiventral leaves the lower epidermis is provided with numerous stomata; there are few stomata or none at all on the upper surface. In erect leaves, as in monocotyledons, stomata are more or less equally distributed on both surfaces.

3. **Xerophytes.** These are plants that grow in deserts or in very dry places; they can withstand a prolonged period of drought uninjured. For this purpose they have certain peculiar adaptations. Dominant factors

in a desert or a very dry region are: scarcity of moisture in the soil and extreme atmospheric conditions, such as intense light, high temperature, strong wind and aridity of air.

Adaptations. In such conditions the xerophytic plants have to guard against excessive evaporation of water; this they do by reducing evaporating surfaces. They have also to adopt special mechanisms for absorbing moisture from the soil and sub-soil and retaining it.

The Root. Plants produce a long tap root which goes deep into the sub-soil in search of moisture; many of the desert plants which live for a short period produce a superficial root-system to absorb moisture from the surface-soil after a passing shower of rain. To retain this water roots often become very fleshy and contain plenty of mucilage, as in *Asparagus*.

The Stem. Stems of many plants become very thick and fleshy, as in Indian aloe (*Aloe*) and American aloe or century plant (*Agave*). Aqueous tissue develops in them for storing up water; this is further facilitated by the abundance of mucilage contained in them. Stems are provided, with thick cuticle to prevent loss of water by transpiration. In some plants, as in *Gnaphalium* and *Aerua,* there is a dense coating of hairs. In many cases the stem becomes reduced in size and is provided with prickles, as in *Euphorbia splendens*. Modification of the stem into phylloclade for storing water and food and at the same time performing functions of leaves is characteristic of many desert plants, e.g. cacti (see Fig. I/57A).

The Leaf. In some desert plants leaves are very fleshy, containing aqueous tissue and mucilage, as in Indian aloe; in others they are reduced in size minimizing their evaporating surfaces. Thus they may be divided into small segments, as in *Acacia,* or modified into spines, as in many cacti and spurges (*Euphorbia*), or sometimes reduced to small scales only, as in *Tamarix* and *Asparagus*. The cuticle develops strongly on the epidermis to check evaporation of water, as in American aloe (*Agave;* Fig. II/38). For the same purpose sometimes multiple (many-layered) epidermis develops, as in oleander (see Fig. II/39). Stomata are fewer in number—usually 10–15 per sq. mm, and remain sunken in grooves and occluded, as in *Agave* and *Nerium* (see Figs. II/38–9). Modification of the leaf into phyllode, turning its edge in a vertical direction in strong sunlight to minimize transpiration, is characteristic of Australian *Acacia* (see Fig. I/80). Under conditions of extreme dryness leaves of most xerophytic grasses and also of many other plants roll up, considerably reducing their evaporating surfaces. Many of the xerophytic herbs lie prostrate on the ground, completing their life-history within a short time, e.g. *Solanum surattense* (B. KANTIKARI; H. KATELI) and *Tribulus terrestris* (B. GOKHRIKANTA; H. GOKHRU); some are perennial in habit. Many xerophytes are elaborately armed with prickles and spines.

Examples. Many spurges (e.g. *Euphorbia splendens, E. tirucalli*—see Fig. I/57D, *E. neriifolia, E. royleana,* etc.), most cacti (e.g. *Opuntia, Cereus, Nopalia, Pereskia,* etc.), dagger plant (*Yucca;* see Fig. I/92), Indian aloe (*Aloe vera*), American aloe (*Agave*), globe thistle (*Echinops*), prickly or Mexican poppy (*Argemone*), *Asparagus, Acacia, Tribulus terrestris* (B. GOKHRIKANTA; H. GOKHRU), *Solanum surattense* (*S. xanthocapum;* B. KANTIKARI; H. KATELL), *Aerua, Fagonia* and some grasses.

4. **Halophtes.** These are special types of plants growing in saline soil or saline water with preponderance of salts in it. Hence such plants show some special characteristics (or adaptations). It is to be noted that plant cells of ordinary land plants and freshwater plants usually maintain a low osmotic pressure which is equal to or more often slightly higher than the osmotic pressure of the medium (soil or water) where the plants are growing. But when the soil medium or water medium is definitely saline, as in the sea, sea-coast and salt-lake, its osmotic pressure becomes very high. The osmotic balance between these plants and their habitat thus becomes much disturbed. Evidently this condition tends to extract water from the plant tissues, rather than help them to absorb water from such a medium. The saline water or soil thus behaves as a *physiologically dry* medium for such plants. There are, however, special plants whose tissues have a concentration of salts, and, therefore, an exceptionally high osmotic pressure—as high as 35–40 atmospheres or even much higher. It is only these plants that can adapt themselves to such a situation, and they are called **halophytes**. They show the following characteristics.

Xeromorphic Adaptations. The majority of halophytes show xeromorphic characteristics. Many of them have fleshy leaves; leaves sometimes altogether absent and the stems fleshy and jointed; stems sometimes fleshy and mucilaginous; epidermis often strongly cutinized, sometimes with hairs; spines and prickles in several cases; and internal structure often like that of xerophytes. Common examples of halophytes are sea-blite (*Suaeda maritima*), saltwort (*Salicornia brachiata*), *Salsola foetida, Acanthus ilicifolius* (B. HARGOZA; H. HARKUCH-KANTA),

Asteracantha longifolia (B. KULEKHARA; H. GOKULA-KANTA) *Solanum surattense* (B. KANTIKARI; H. KATELI, KATITA), Indian spinach (*Basella rubra*—cultivated or wild), prickly amaranth (*Amaranthus spinosus*), etc.

Special Adaptations. Halophytes growing on sea-coasts and estuaries, and also in salt-marshes and salt-lakes occasionally inundated by sea-tides, form a special type of vegetation known as the **mangrove** (Fig. 2). Mangrove plants produce a large number of **stilt roots** from

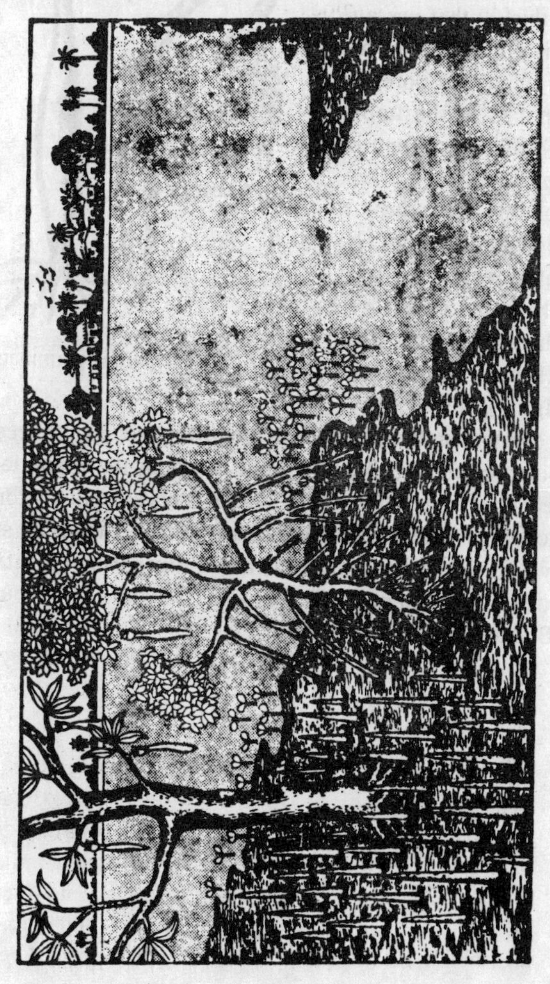

Figure 2. Mangrove plants showing (a) pneumatophores for respiration, (b) stilt roots for support, and (c) viviparous germination for survival.

the main stem and the branches. In several cases, in addition to the stilt roots, special roots called **respiratory roots** or **pneumatophores** are also produced in large numbers. They develop from underground roots, and projecting beyond the water level look like so many conical spikes distributed all round the trunk of the tree. In some places they grow so thickly that passage through them is difficult. They are provided with numerous pores or respiratory spaces in the upper part, through which exchange of gases for respiration takes place. Mangrove species also show a peculiar mode of germination. The seed germinates inside the fruit while it is still on the parent tree and is nourished by it. This kind of germination is known as **vivipary** (Fig. 3). The radicle elongates to a certain length and swells at the lower part. As the seedling drops, the root presses into the soft mud, keeping the plumule and cotyledons clear above the saline water. Lateral roots are quickly formed for proper anchorage. The advantage is that the fruit cannot be swept away by tidal waves. Typical mangrove plants are *Rhizophora* (B. KHAMO), *Ceriops* (B. GORAN), *Sonneratia* (B. KEORA), *Heritiera,* (B. SUNDRI), *Excoecaria* (B. GEO), etc.

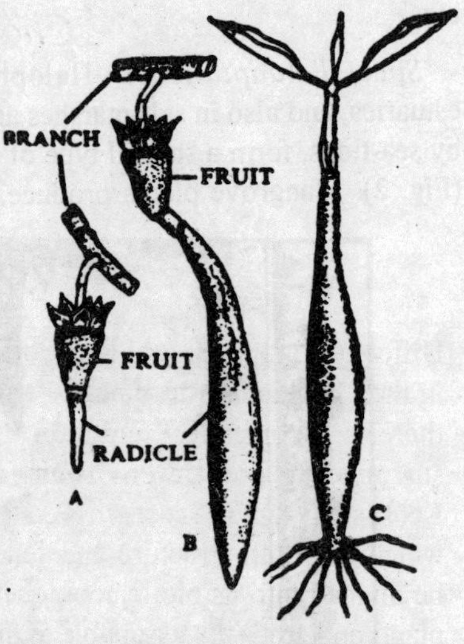

Figure 3. Viviparous germination.

Chapter 3

Ecosystem

Different communities of living organisms (i.e. biotic communities) exist in their respective surrounding environment (abiotic environment) and there is a constant interaction between a biotic community and its surrounding abiotic environment. Such a system is expressed as **Ecosystem** (A.G. Tansley, 1935). An ecosystem is an ecological unit, which has both structure and function. Structurally an ecosystem can be divided into its biotic components which include autotrophes (green plants and green algae capable of traping radiant energy and synthesizing food), and heterotrophes (non green plants and different categories of animals) and into abiotic components such as soil, water, air, solar radiants, gases, inorganic elements, organic compounds, etc. Biotic components may be classified as producers (green plants), consumers (Herbivores, Carnivores, Omnivores, etc.), decomposers and transformers (bacteria, fungi, etc.). Producers prepare food, herbivores (first level of consumers) depend on plants for food, carnivores (second level of consumers) depend on herbivores for food and Omnivores (third level of consumers) on both herbivores and carnivores. After the death of consumers as well as producers, their bodies are decomposed by bacteria and fungi (Decomposers) as a result of which organic substances present in them are converted into simpler forms and the latter are transformed into inorganic compounds and released to the soil, so that plants can re-absorb them. Whereas the study of functional aspect of an ecosystem consists of an analysis of the energy flowing in the system and through various trophic levels such as herbivores, carnivores, etc. (constituting food chains). Thus energy, while passing through the food chain gradually decreases and part of it is also lost in the process as heat. Finally, the energy captured by the producers is used up by the producers and a part of it is returned to the non-living world. The flow of energy, which is always unidirectional, and the nutrient cycles are the two most important functions of an ecosystem.

There are two major types of ecosystem namely aquatic ecosystem and terrestial ecosystem. Aquatic ecosystem may be further classified into Fresh water ecosystem (e.g. pond ecosystem, lake ecosystem, spring ecosystem and river ecosystem) and marine ecosystem. Whereas the terrestial ecosystem may be divided into forest ecosystem, grassland ecosystem, desert ecosystem and man-made ecosystem (e.g. crop fields, garden). The outline of a Pond ecosystem is described below:

Sun light, water, dissolved minerals, oxygen and carbon dioxide constitute the abiotic components and producers like algae, aquatic plants, consumers like aquatic herbivores, e.g. tadpole larvae, small fishes and other aquatic animals which consume algae and green plants, aquatic carnivores, e.g. big fishes, crabs, water-snakes which eat aquatic herbivores and third level of consumers like water loving birds, etc. which eat fish, etc. and decomposer, transformers (bacteria, fungi) all together constitute the biotic components of the pond system. There is a constant, interactions between the biotic and abiotic components of pond ecosystem involving unidirectional energy flow and nutrient cycles.

Part V

CRYPTOGAMS

Divisions and General Description
Algae
Microbes: Bacteria and Viruses
Fungi
Lichens
Bryophyta
Pteridophyta

Chapter 1

Divisions and General Description

Cryptogams are plants that do not bear flowers or seeds and hence are commonly known as 'flowerless' or 'seedless' plants. They are broadly classified as follows:

(1) **Thallophyta.** The plant body is a thallus, i.e. not differentiated into stem, leaf and roots. Thallophyta includes (a) **algae**, i.e. thallophytes containing chlorophyll and sometimes also other pigments, (b) **fungi**, i.e. thallophytes without chlorophyll, (c) **lichens**, i.e. plants consisting of specific fungi and algae living together symbiotically, and (d) **bacteria**, i.e. unicellular (prokaryotic), microscopic, non-green (without chlorophyll) organisms.

(2) **Bryophyta.** The plant body is thalloid (prostrate) or differentiated into stem, leaves but not into true roots (erect body); there is regular alternation of generations; the main body is always a gametophyte; the sporophyte always grows attached to the gametophyte as a dependent body. Bryophyta include (a) **Hepaticae (liverworts)** i.e. bryophytes with mostly thalloid plant body, e.g. *Riccia* and *Marchantia,* (b) **Anthocerotae**, i.e., plant body is simple and thalloid but sporophyte is complex and semi-independent on gametophyte, e.g. Anthoceros and (c) **Musci** or **mosses**, i.e. bryophytes with leafy stem e.g. Polytrichum, Sphagnum, etc.

(3) **Pteridophyta.** The plant body is differentiated into the stem, leaves and roots; there is regular, alternation of generations; the sporophyte and gametophyte are independent of each other; the main plant is always a sporophyte; vascular tissues are well developed (so Pteridophyta are also called *vascular cryptogams*). They include ferns and their allies.

Thallophyta are primitive plants and are regarded as lower cryptogams, while Bryophyta and Pteridophyta are advanced plants and are regarded as higher cryptogams.

Reproduction. Of the three methods of reproduction, viz., vegetative, asexual and sexual, a particular plant may adapt one or more methods. Vegetative reproduction takes place commonly by cell division or by fragmentation. Asexual reproduction takes place by fission or by spores of varied types in different groups of plants. Sexual reproduction takes place by the fusion of two similar gametes (isogametes), as in lower forms, or by the fusion of two dissimilar gametes (heterogametes) differentiated into male and female, as in higher forms. The former method is called **isogamy**, and the latter method **oogamy**.

Differences between Algae and Fungi. (1) Algae are green thallophytes containing the green colouring matter *chlorophyll*. In many algae the green colour may be masked by other colours; fungi, on the other hand, are non-green thallophytes having no chlorophyll in them. (2) Algae are *autotrophic* plants, i.e. they manufacture their own food with the help of chlorophyll contained in them; whereas fungi are *heterotrophic,* i.e. their modes of nutrition are diverse; they may get their food from decaying animal or vegetable matter, or from the tissue of a living plant or animal; accordingly they are either saprophytic or parasitic in habit. (3) The body of the algae is composed of a *true parenchymatous tissue;* while that of the fungi is composed of a *false tissue* or pseudoparenchyma which is an interwoven mass of fine delicate threads known as *hyphae.* (4) The cell-wall of an alga is composed of true cellulose, and that of a fungus of fungus-cellulose or chitin mixed with cellulose, callose, pectose, etc., in different proportions. (5) Algae live in water or in wet substrata; whereas fungi live as parasites on other plants or animals or as saprophytes on decaying animal or vegetable matter. (6) Reserve carbohydrate in algae is usually starch, but in fungi it is glycogen.

In structure both the groups may be unicellular, multicellular, filamentous or thalloid, and reproduction in them may take place vegetatively by cell division or by detachment of a portion of the mother plant, or asexually by fission or by spores, or sexually by gametes.

Alternation of Generations. The life-history of any of the higher cryptogams, e.g. liverworts, mosses and ferns, is complete in two stages or generations, alternating with each other. These two generations differ not only in their morphological characters but also in their modes of reproduction. One generation reproduces by the asexual method, i.e. by spores, and the other by the sexual method, i.e. by gametes. The former, therefore, is called the **sporophytic** or asexual generation, and the latter the **gametophytic** or sexual generation. To complete the life-history of a particular plant one generation gives rise to the other—the gametophyte to the sporophyte, and the sporophyte to the gametophyte, or in other words, the two generations regularly alternate with each other. This alternation of the gametophyte with the sporophyte and vice versa is spoken of as alternation of generations.

Cytological Evidence of Alternation of Generations. Alternation of generations can be traced on the basis of chromosome numbers. It is an established fact that the sporophyte bear's diploid or $2n$ chromosomes, while the gametophyte bears haploid or n chromosomes. The diploid number of the former is reduced (by meiosis) to the haploid number of the latter in the formation of *spores* (n). The spores gives rise to the gametophyte (n), evidently representing the beginning of the gametophytic generation. In due course the gametophyte bears gametes (n). Two gametes of opposite sexes (by fusion) give rise to the *zygote* ($2n$, i.e. $n + n$). The zygote grows into the sporophyte ($2n$); evidently representing the beginning of the sporophytic generation. The above stages are shown below with reference to fern and moss.

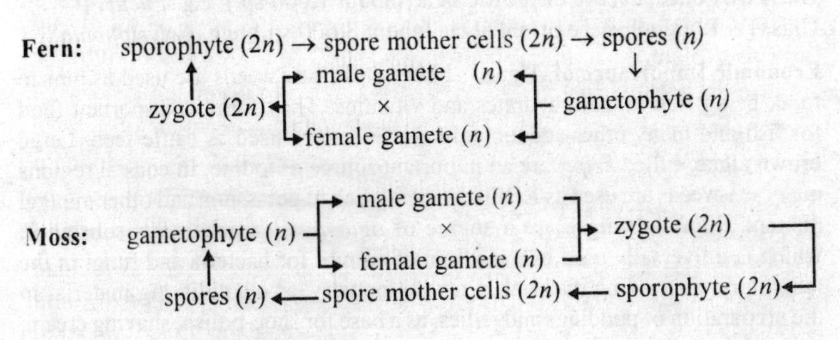

Chapter 2

Algae

Classification of Algae (20,000 sp.)

There are eleven classes of algae of which four main classes are:

Class I. Cyanophyceae or Myxophyceae or blue-green algae (1,500 sp.), e.g. *Oscillatoria* and *Nostoc*.

Class II. Chlorophyceae green algae (6,500 sp.), eg. *Chlamydomonas, Volvox, Ulothrix, Spirogyra* and *Vaucheria*.

Class III. Phaeophyceae or brown algae (about 1,000 sp.). e.g. *Fucus*.

Class IV. Rhodophyceae or red algae (about 3,000 sp.), e.g. *Polysiphonia*.

Economic Importance of Algae. Many of the sea-weeds are used as human food, being rich in carbohydrates and vitamins. They form an important food for fish and many other aquatic animals. Some are used as cattle feed. Large brown algae, called *kelps,* are an important source of iodine. In coastal regions many sea-weeds are used as fertilizers, being rich in potassium and other mineral matters. Some red algae are a source of *agar-agar,* a gelatinous substance, which is universally used as a medium of culture for bacteria and fungi in the laboratory, as a sizing material in textile industry, as a solidifying material in the preparation of puddings and jellies, as a base for shoe-polish, shaving cream, cosmetics, etc., and as a dyeing and printing material for textile goods. It is also used in medicine. Big deposits of diatoms in sea-beds, called *diatomaceous earth,* have a number of industrial uses as metal polish, tooth powder, heat insulators in boilers and furnaces, and as filters in refining sugar. Many green algae are a source of food for fish and many aquatic animals. They also purify the water by absorbing CO_2 and giving out O_2. Blue-green algae contribute to the fertility of the soil. Some of them are known to fix free nitrogen. But some of them are a nuisance to water reservoirs, sometimes polluting water, particularly during summer rains.

A. CYANOPHYCEAE OR BLUE-GREEN ALGAE

1. *OSCILATORIA* (100 sp.)

Occurrence. ***Oscillatoria*** (Fig. 1) is a dark blue-green filamentous alga. It commonly occurs floating in ditches, sewers, shallow pools of water and also on wet rocks and walls. Filaments of *Oscillatoria* are entangled in masses which float on water.

Structure. Each filament is slender, unbranched and cylindrical, consisting of a row of short cells (see Figure 1.A). The individual cells are the *Oscillatoria* plants, and the filament is regarded as a colony. All the cells of the filaments are alike except the end cells which is usually

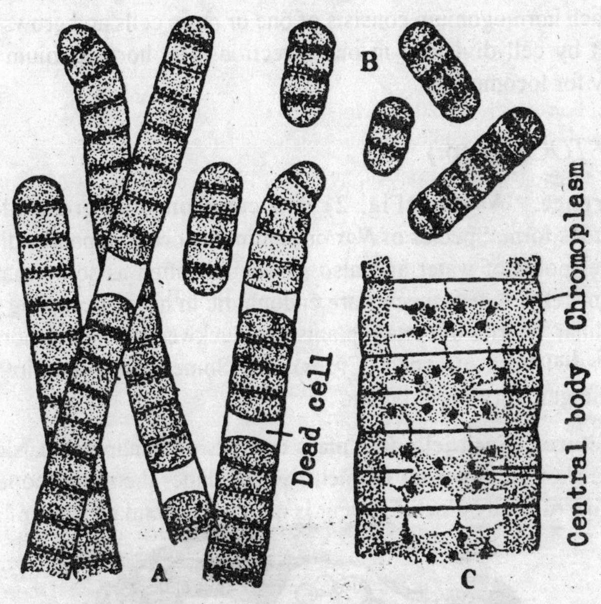

Figure 1. *Oscillatoria. A,* filaments; *B,* hormogonia; and *C,* a portion of the filament magnified.

convex, and there is no differentiation into the base and the apex. Here and there some dead and empty cells occur in some of the filaments. The protoplast of each cell is differentiated into two regions: a coloured peripheral zone—the **chromoplasm**—and an inner colourless zone—the **central body** (see Figure 1.C). The colour is due to the presence of cholorophyll and phycocyanin (a blue pigment) which diffuse through the chromoplasm. There is no plastid. True nucleus is also absent. The central body, however, is regarded as an incipient nucleus with only some chromatin but without nuclear membrane and nucleolus. Cell division takes place in one direction only. Each filament remains enveloped in a thin mucilaginous sheath. Under the microscope, a slow swaying or oscillating movement of the filaments with ends tossing from side to side may be distinctly seen. This is a characteristic feature of *Oscillatoria*. The filaments may sometimes exhibit a twisting or rotating motion.

Reproduction. In blue-green algae reproduction takes place vegetatively only. They do not bear any kind of ciliated body. Gametes and zoospores are also altogether absent. In *Oscillatoria* the filament breaks up into a number of fragments called **hormogonia** (see Figure 1.B). Each hormogonium consists of one or more cells and grows into a filament by cell divisions in one direction. The hormogonium has a capacity for locomotion.

2. *NOSTOC* (29 sp.)

Occurrence. ***Nostoc*** (Fig. 2) is a common blue-green alga of filamentous form. Species of *Nostoc* commonly occur in ponds, ditches, and other pools of water and also in wet soil often as somewhat firm masses of jelly. A few species are endophytic in habit, occurring in the intercellular cavities of certain plants like duckweed (*Lemna*), hornwort (*Anthoceros*), root of cycad (*Cycas*), etc. Some lead a symbiotic life with a fungus, forming a lichen.

Structure. Each jelly-like mass consists of innumerable, slender, long and short, interwoven filaments which under the microscope look like chains of beads. Each filament is unbranched and consists of a row

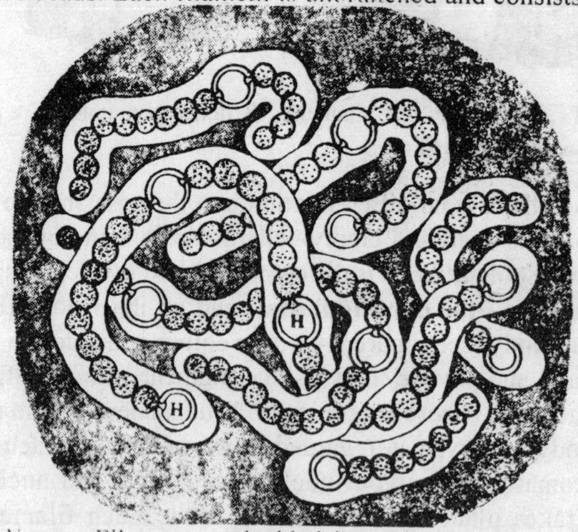

Figure 2. *Nostoc*. Filaments embedded in gelatinous matrix. Note the heterocysis (H) with the polar nodules.

of rounded or oval cells very much like a series of beads in a chain. There is often a gelatinous sheath covering each filament in addition to the general gelatinous matrix in which the tangled masses of *Nostoc* filaments remain embedded. Each cell is differentiated into two regions: an outer coloured region call the **chromoplasm**, and an inner colourless region called the **central body** (as in *Oscillatoria*) (see p. 362). The filament increases in length by cell divisions in one plane only. Some enlarged vegetative cells with thickened walls and transparent contents are seen to occur at frequent intervals and also at the ends; these are called **heterocysts**. A pore is present at each pole of the heterocyst, maintaining cytoplasmic connexions with the adjoining cells. There is, however, one pore in the terminal heterocyst. At a later stage the pore is closed by a button-like thickening of the wall, called the **polar nodule**.

Reproduction. *Nostoc* reproduces vegetatively by fragmentation of the filament, and sometimes asexually by resting cells (spores) called akinetes. **Fragmentation.** The filament breaks up at the junction of the heterocyst and the adjoining cell into a number of short fragments called **hormogonia**. Each hormogonium grows in length by cell divisions in one plane only. The heterocyst, as suggested by some, may otherwise be a food storage cell. **Akinetes.** Here and there certain vegetative cells of the filament may become enlarged and thick-walled, containing reserve food; these are resting cell (spores) called akinetes. Later they may germinate into *Nostoc* filaments.

B. CHLOROPHYCEAE OR GREEN ALGAE

3. *CHLAMYDOMONAS* (43 sp.)

Occurrence. ***Chlamydomonas*** is a unicellular green alga found in ponds, ditches and other pools of stagnant water. **Structure.** In shape the cell is usually spherical or oval, and is provided with a thin wall and two distinct long cilia (Fig. 3A). The protoplasm at the anterior end of the cell is clear, and contains two contractile vacuoles which are pulsating in nature, undergoing alternate expansion and contraction. These may be respiratory or excretory in function. There is a lateral orange or red pigment spot, commonly called the *eye spot*. This is sensitive to intensity of light. In the posterior region there is a single large cup-shaped chloroplast with a pyrenoid in it. The **pyrenoid** consists of a central protein body surrounded by numerous minute starch grains. There is a nucleus more or less centrally placed. By the lashing of the cilia the cells swim about briskly in water.

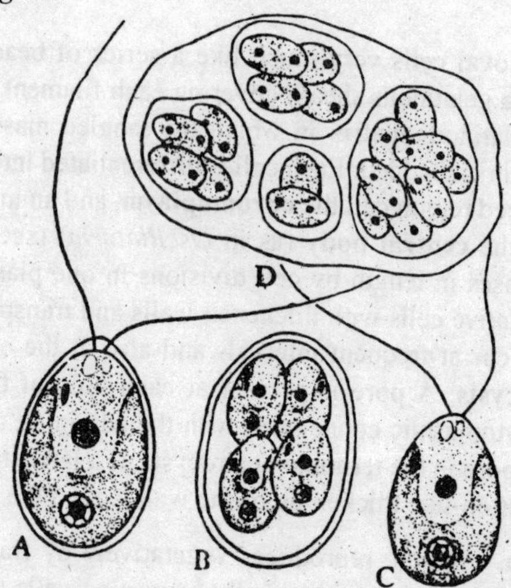

Figure 3. *Chlamydomonas*. *A*, a mature cell; *B*, four daughter cells formed by asexual method; *C*, a daughter cell after escape; *D*, palmelia stage.

Asexual Reproduction. This takes place by zoospores. In the formation of the zoospores the cilia of each cell are withdrawn, and the contents divide into 2, 4 or 8 daughter cells, seldom more (Fig. 3B). The cells grow, develop two cilia each, and become motile zoospores. The wall of the mother cell dissolves and the zoospores are set free (Fig. 3C), each acting as a chlamydomonas plant.

Palmella Stage. Under unfavourable conditions the daughter cells instead of forming zoospores divide repeatedly into numerous cells. Their walls become gelatinous, and the cells are held together in clusters in the gelatinous mass. This is known as the palmella stage (Fig. 3D). When the conditions are favourable the cells develop cilia, swim out of the gelatinous matrix, and become motile again. These cells then act as chlamydomonas cells.

Sexual Reproduction. This takes place by the fusion of motile ciliate gametes which are formed in the same way as the zoospores and are also like them but somewhat smaller in size and more numerous—16, 32 or 64, or even more (Fig. 4A–B). All gametes are similar and are called *isogametes* and their fusion is known ts *isogamy*. Gametes of different parents usually conjugate in pairs (Fig. 5A). A **zygospore**—the product of fusion of two similar gametes—is formed. The ciliate ends of the gametes conjugate first, and soon their complete fusion takes

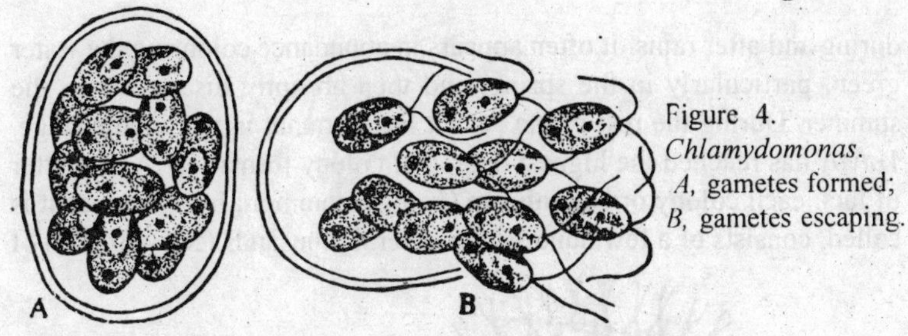

Figure 4. *Chlamydomonas*. *A*, gametes formed; *B*, gametes escaping.

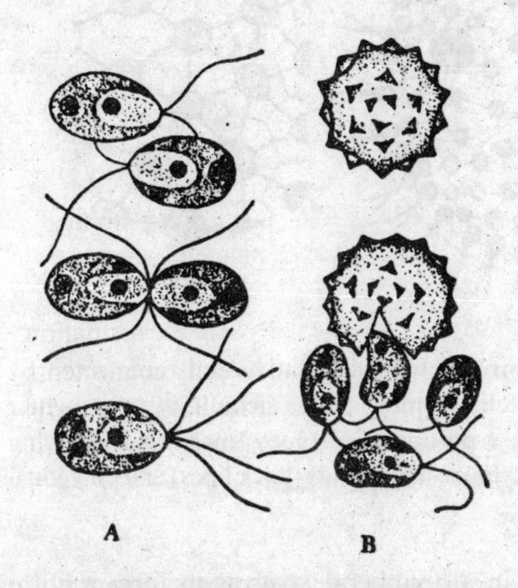

Figure 5. *Chlamydomonas*. *A*, stages in conjugation of gametes; *B*, (*top*) a resting zygote; (*bottom*) four daughter cells after escape from the zygote.

place. After this fusion the cilia are withdrawn and the zygospore clothes itself with a thick wall (Fig. 5B, *top*). It undergoes a period of rest, and then its contents divide and form 2 or 4 motile daughter cells (Fig. 5B, *bottom*). They grow in size, escape from the mother cell, and become individual motile *Chlamydomonas* cells.

Note. In *Chlamydomonas* gametes are mostly alike (isogametes); while there are cases showing slight differentiation of gametes (anisogametes). Further, similarity of gametes and zoospores is suggestive of the origin of sexual cells (gametes) from asexual cells (zoospores) by transformation of the latter into the former.

4. VOLVOX (over 12 sp.)

Occurrence and Structure. ***Volvox*** is a fresh-water, colony-forming, free-swimming, green alga occurring in ponds and other pools of water during and after rains. It often appears in abundance colouring the water green, particularly in the spring, and then abruptly disappears in the summer. During the rest of the year it lies dormant in the zygote stage. *Volvox* has reached the highest degree of colony formation. As a matter of fact, each colony or **coenobium** (*koinos,* common; *bios,* life), as it is called, consists of a few hundreds to several thousands (500–40,000) of

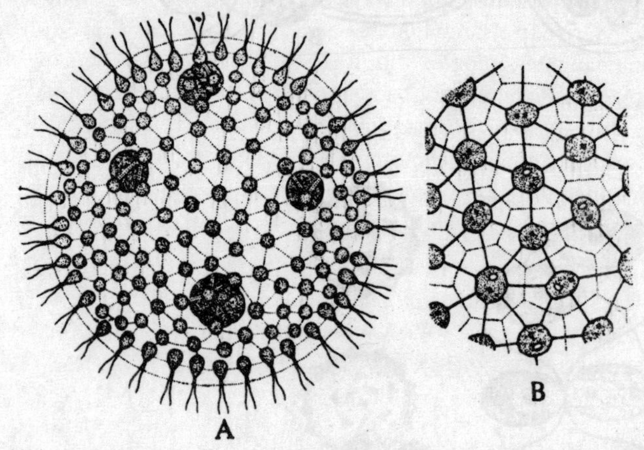

Figure 6. *Volvox. A,* a colony showing vegetative cells connected by cytoplasmic strands, four colony-forming cells (including two daughter colonies) and outer sheath: *B,* a portion of a colony (magnified) showing vegetative cells connected by cytoplasmic strands (thick lines) and polygonal sheaths (dotted lines).

cells which are so arranged in a peripheral layer as to form a hollow sphere (Fig. 6A) containing water or a dilute solution of a gelatinous material. Each cell (Fig. 6B) of the colony has a gelatinous sheath of its own, and at the same time the cells are held together in a colony by the sheaths secreted by the individual cells. The cells are connected by delicate but distinct strands of cytoplasm. The individual colonies, approximately 1 mm in diameter, sometimes up to 2 mm, freely swim about in water. Each *Volvox* cell is very much like that of *Chlamydomonas*.

A mature colony (Fig. 6A) shows two kinds of cells: numerous small vegetative cells and a few (5–20) large cells among the former. A vegetative cell has two cilia protruding outwards and vibrating, 2–5 contractile vacuoles, a central nucleus, a cup-shaped or plate-like chloroplast with one pyrenoid, and an eye-spot. The vegetative cells do not divide. The larger cells of the colony are the reproductive cells.

These cells may behave exclusively as asexual cells or as sexual cells. Normally they act as asexual cells in the beginning of the season, and as sexual cells at the close of the season.

Asexual Reproduction (Fig. 7). The above enlarged cells (called gonidia) of the mother-colony, after retracting their cilia and pushing back to the posterior side, divide and re-divide in the longitudinal plane and give rise to a large number of cells in one plane, thus forming new young daughter colonies within the mother-colony. When cell divisions cease the cells turn round, develop cilia and form again hollow spheres. These are seen to float and slowly revolve within the much enlarged hollow portion of the mother-colony. Soon they escape from their imprisoned state by a rupture of the membrane of the mother-colony or through a pore in it, and swim away as independent colonies.

Sexual Reproduction (Fig. 8). Sexual reproduction is oogamous in *Volvox*. In the monoecious species both types of gametes (male and female) are borne by the same colony (*homothallic*), while in the dioecious species these are borne by separate colonies (*heterothallic).*

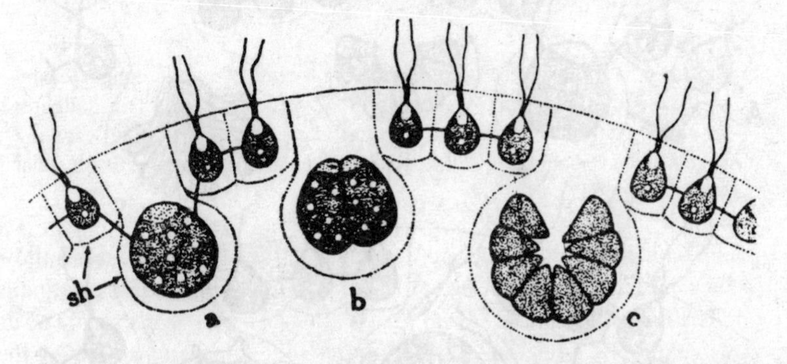

Figure 7. *Volvox:* Asexual Reproduction. Formation of a daughter colony within a mother-colony; *a,* an enlarged vegetative cell (gonidium); *b,* the same after first division; *c,* a young daughter-colony developed from it; *sh,* sheath. (*Redrawn After Fig. 12 in* Cryptogamic Botany, Vol. I *by G.M. Smith by permission of McGraw-Hill Book Company, Copyright 1938)*

The said gametes are borne by certain enlarged cells called gametangia (gamete-bearing cells) which lie in the posterior side of the colony. Some of these cells are antheridia or male reproductive organs, the protoplast of which divides many times and produces a cluster of minute biciliate male gametes called antherozoids or sperms (Fig. 8A); while other cells are oogonia or female reproductive organs, the protoplast of which forms a single large female gamete called egg or ovum (Fig. 8B). The egg is large, passive and non-motile; while the sperms are very minute, active and motile. The latter may be in a plate-like colony escaping from the mother-colony as a unit, or these may be arranged to form a hollow sphere. In the former case the unit as it approaches an egg breaks up into individual sperms, and in the latter case the sperms are liberated singly. The mode of fertilization is oogamous. The sperms swim and enter through the gelatinous sheath into the oogonium lying in the mother-colony, and one of them finally fuses with the egg (Fig. 8B). Thus fertilization is effected.

Zygote. After fertilization the zygote clothes itself with a thick spiny wall and turns orange-red (Fig. 8B). It is set free from the mother-colony only after the decay or disintegration of the latter. The zygote sinks to the bottom of the pool of water, and then after a period of rest it

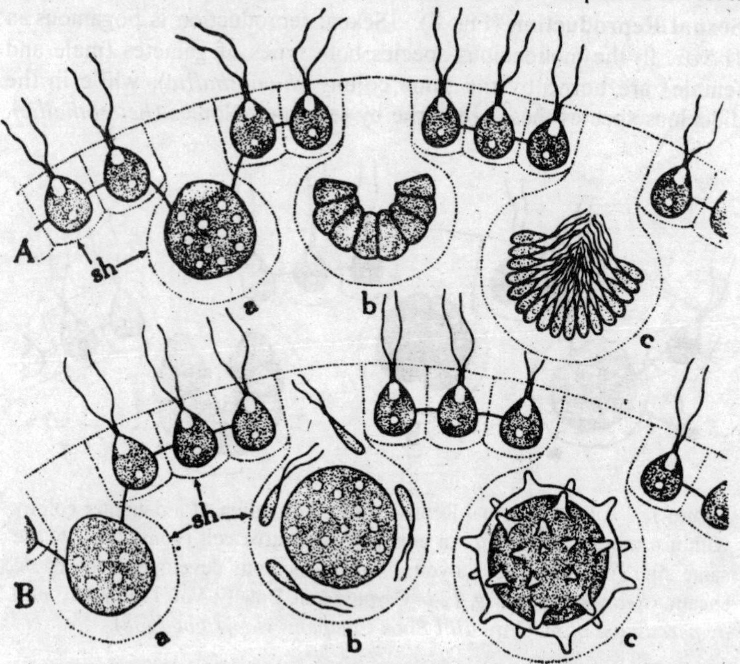

Figure 8. *Volvox:* Sexual Reproduction. *A,* formation of antheridium and antherozoids; *a,* an antheridium; *b,* the protoplast of it after first few divisions; *c,* antherozoids formed in a cluster; *sh,* sheath; *B,* formation of oogonium, egg and zygote; *a,* a young cogonium with an egg; *b,* the egg about to be fertilized (note the antherozoids surrounding it); *c,* zygote formed after fertilization; *sh,* sheath. (*Redrawn after Fig. 13 in* Cryptogamic Botany, Vol. I *by G.M. Smith by permission ot McGraw-Hill Book Company, Copyright 1938).*

germinates with the approach of the favourable season. The protoplast of the zygote undergoes reduction division prior to germination. In some species the protoplast of the zygote divides and forms a new colony directly, while in others it forms a single bi-ciliate zoospore which escapes by the rupture of the zygote wall and swims away. The free-swimming zoospore then divides and forms a new colony.

5. *ULOTHRIX* (30 sp.)

Occurrence. ***Ulothrix*** (Fig. 9) is a *green* filamentous alga occurring in fresh water in ponds, ditches, water-reservoirs, horse- or cow-troughs, slow streams, etc., particularly in the spring; a few species are marine.

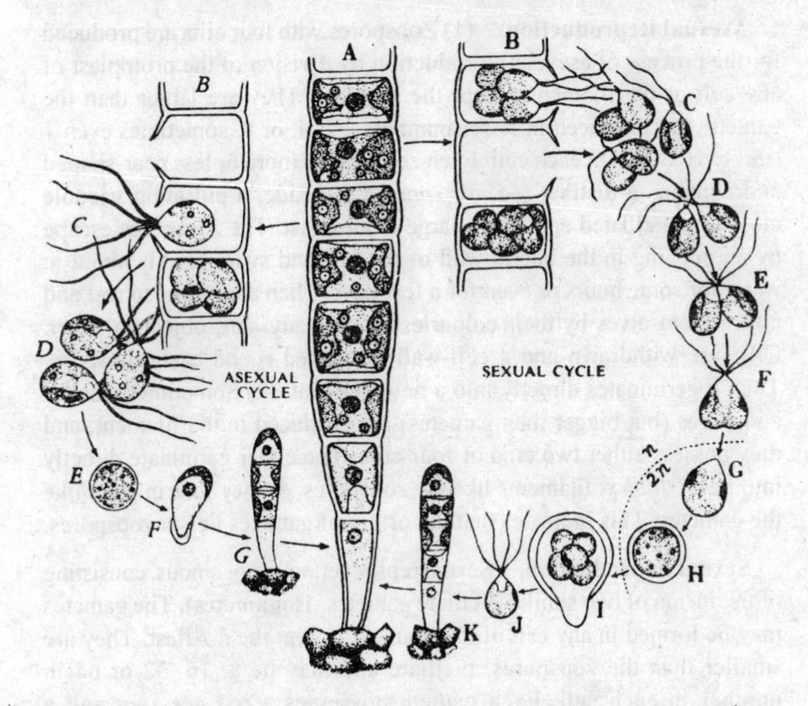

Figure 9. *Ulothrix.* Life-cyde: **sexual reproduction**—*A,* vegetative filament; *B,* formation of gametes; *C,* gametes swimming; *D–G,* stages in the conjugation of gametes; *H,* zygospore; *I,* the germ-plant with zoospores; *J,* a zoospore (quadriciliate); *K,* a young filament; **asexual reproductions**—*B,* a portion of the filament showing the formation of zoospores; *C,* a quadriciliate zoospore escaping; *D,* zoospores swimming; *E,* a zoospore rounded off; *F,* zoospore germinating; and *G,* a young filament.

It grows fixed to any hard object in water by the basal elongated colourless cell called the *holdfast*. The filament, if detached, may freely float on water.

Structure. The filament of *Ulothrix* is unbranched, cylindrical and multicellular, consisting of a single row of short cylindrical cells. Each cell of the filament contains a single nucleus and a peripheral band-like chloroplast with an entire or lobed margin. Usually there are many (sometimes one or few) **pyrenoids** lying embedded in the chloroplast. These are rounded protein bodies with a starchy envelope.

Reproduction takes place asexually by zoospores, sexually by gametes, and vegetatively by fragmentation of the filament.

Asexual Reproduction. (1) Zoospores with four cilia are produced for the process of asexual reproduction by division of the protoplast of any cell of the filament except the holdfast. They are larger than the gametes but produced in fewer numbers—2, 4, or 8, sometimes even 1 (rarely 16 or 32) in each cell. Each zoospore is more or less pear-shaped and contains a distinct red *eye spot* on one side, a pulsating vacuole close to the ciliated end, and a large chloroplast. The zoospores escape by an opening in the lateral wall of the cell and swim briskly about in water for some hours or even for a few days. Then they come to rest and attach themselves by their colourless end to any hard object in water. Cilia are withdrawn and a cell-wall is formed round each zoospore. Then it germinates directly into a new filament. (2) Sometimes smaller zoospores (but bigger than gametes) are produced in the filament, and they possess either two cilia or four cilia. The either germinate directly into new *Ulothrix* filaments like the zoospores, or they fuse in pairs like the gametes. This indicates that the origin of gametes lies in zoospores.

Sexual Reproduction. Sexual reproduction is isogamous, consisting of the fusion of two similar biciliate gametes (**isogametes**). The gametes may be formed in any cell of the filament except the holdfast. They are smaller than the zoospores, biciliate and may be 8, 16, 32 or 64 in number in each cell. Each gamete possesses a red *eye spot* and a chloroplast band. The gametes are set free from the cell in exactly the same way as the zoospores and they swim about in water with the help of their cilia for some time. Two gametes coming from two different filaments get entangled by their cilia and gradually a complete fusion (conjugation) of the two takes place laterally. Cilia are withdrawn towards the close of the process, and the fusion product still moves for a while but soon comes to rest. It rounds itself off and clothes itself with a thick cell-wall, and forms into a **zygospore**. The zygospore undergoes a period of rest till the next favourable season. It germinates then into a unicellular **germ plant** which produces 4 to 16 quadriciliate zoospores (or in some cases only non-ciliate spores). Each zoospore develops into a new plant.

Note. In *Ulothrix* we get a very early indicaton of sexual differentiation which becomes so pronounced in the higher plants. The behaviour of gametes or sexual cells and zoospores or asexual cells suggests that the former were originally derived from the latter. The gametes are similar in appearance, but not so in their behaviour. The passive one may be regarded as the egg-cell or female gamete, and the active one as the male gamete. *Ulothrix* thus shows the beginning of sexual differentiation.

Vegetative Reproduction. This takes place by **fragmentation** of the filament into short pieces, each consisting of a few cells. Each piece or fragment grows into a long filament by transverse divisions of cells and their enlargement.

6. *SPIROGYRA* (100 sp.)

Occurrence. *Spirogyra* (Fig. 10) is a *green* free-floating filamentous alga. It is found growing abundantly in ponds, ditches, springs, slow running streams, etc.

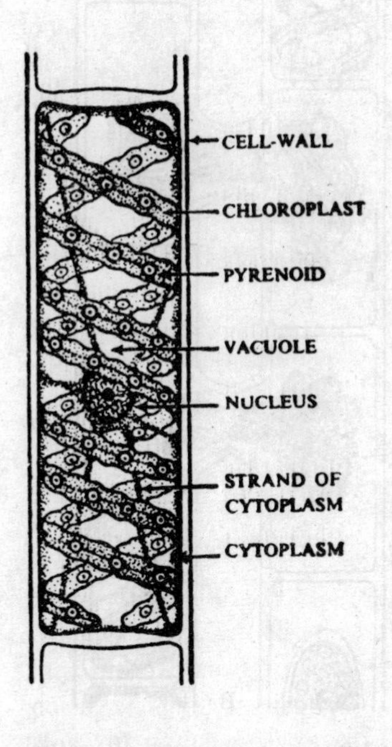

Figure 10. *Spirogyra*. A cell of the filament.

Structure. Each Spirogyra filament is unbranched and consists of a single row of cylindrical cells. The walls are made of cellulose and pectin. Pectin swells in water into a gelatinous sheath. The filament shows no differentiation into the base and the apex. Each cell has a lining layer of protoplasm, one or more (up to 14) *spiral bands* of chloroplasts with smooth, wavy or serrated margins, and a distinct nucleus situated somewhere in the middle. The *spiral* chloroplasts are the characteristic feature of *Spirogyra*. Each chloroplast includes in its body a number of nodular protoplasmic bodies known as **pyrenoids**, around which minute starch grains are deposited. If the filament happens to break up into pieces, they grow up into new filaments by cell divisions.

Reproduction. This takes place in *Spirogyra* by the sexual method only. It consists of the fusion of two similar gametes (isogametes). The fusion of two similar gametes is known as conjugation, which normally takes place between the cells of two filaments (scalariform or ladder-like conjugation, Fig. 11). Sometimes, however, conjugation takes place between two adjoining cells of the same filament (lateral conjugation; Fig. 13).

Scalariform Conjugation (Fig. 11). When two filaments come to lie in contact in the parallel direction they form tubular outgrowths from their opposite or corresponding cells. These tubular outgrowths, called **conjugation tubes**, give the whole structure the appearance of a

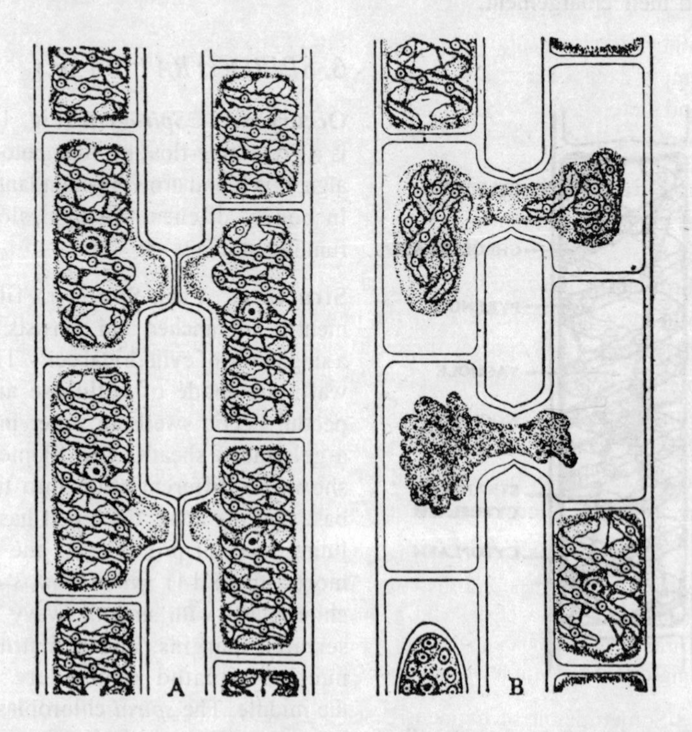

Figure 11. *Spirogyra.* Scalariform conjugation. *A–B* are stages in the process.

ladder (Fig. 11A) and hence the name scalariform or ladder-like conjugation. Their end- or partition-walls dissolve and an open conjugation tube is formed. In the meantime the protoplasmic contents of each cell loose water, contract and become rounded off in the centre. Each contracted mass of protoplast forms a **gamete**. All gametes are alike in appearance, but gametes of one filament (male) creep through the conjugation tubes into the corresponding cells of the adjoining filament (female) and fuse with the gametes of that filament. The fusion of two gametes results in the formation of a thick-walled **zygospore** (Fig. 12) which soon turns black or brownish-black. Zygospores are formed in a series in one filament (female), while the other filament (male) becomes practically empty except for a few vegetative cells here and there. Sometimes conjugation takes place between three filaments, the middle (female) one bearing the zygotes.

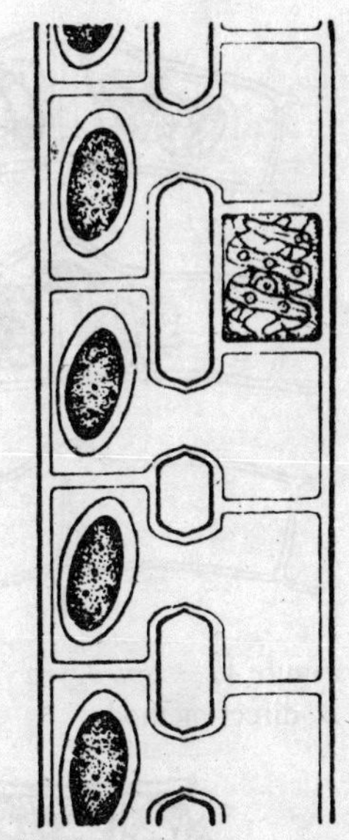

Figure 12. *Spirogyra*. Formation of zygospores after conjugation.

Lateral Conjugation (Fig. 13). This takes place between the cells of the same filament. (1) **Chain Type** (A). Commonly an outgrowth or conjugation tube is formed on one side of the partition wall, and through the passage thus formed, the gamete (male) of one cell passes into the gamete (female) of the neighbouring cell. (2) **Direct Method** (B). In certain species the male gamete pushes the partition wall and pierces it in the middle. Through the opening thus formed, the male gamete passes into the neighbouring cell and fuses with the female gamete. In lateral conjugation the gametes of alternate cells only move to the neighbouring cells, and thus later on the zygote-bearing cells are seen to alternate with the empty cells in the same filament.

Note. The zygospore is formed as a result of fusion of two gametes, each with n chromosomes, and, therefore, the zygospore nucleus has $2n$ chromosomes. It undergoes reduction division, giving rise to 4 nuclei, each with n chromosomes. Three of these nuclei degenerate so that the mature zygospore contains only a single nucleus with n chromosomes (see. Fig. 14E–H).

Sometimes it so happens that conjugation does not take place, and then a gametangium (a gamete bearing cell) may become directly

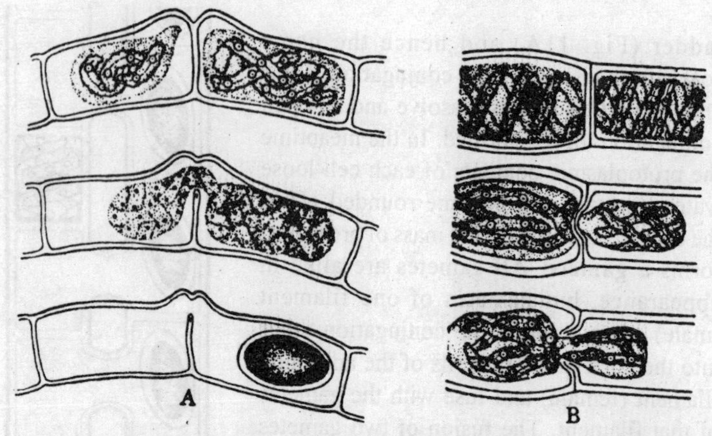

Figure 13. *Spirogyra*. Two methods of lateral conjugation. *A*, chain type; *B*, direct method.

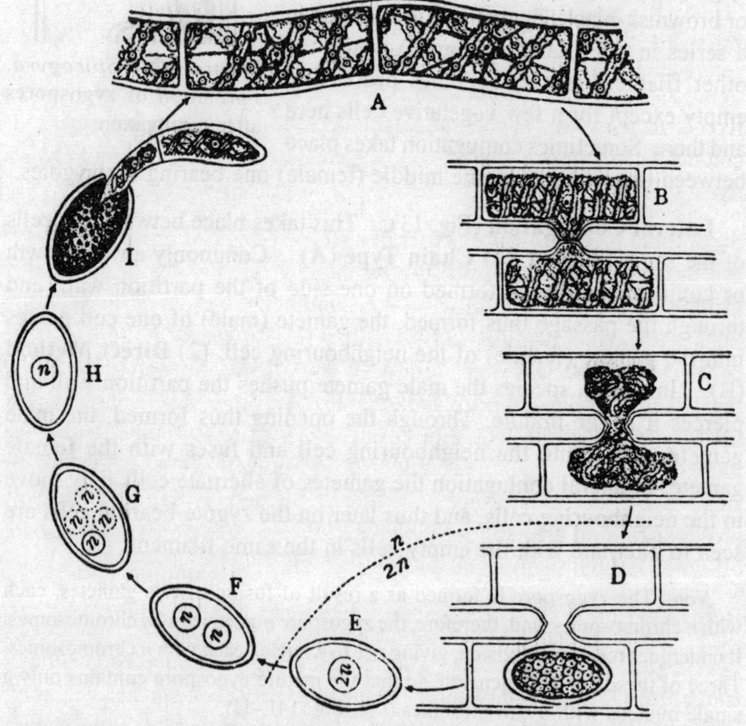

Figure 14. Life-cycle of *Spirogyra*. *A*, vegetative filament (portion); *B–C*, stages in conjugation; *D*, zygospore formed; *E–H*, reduction division and nuclear changes within the zygospore; and *I*, zygospore germinates.

converted into a zygospore-like body called the **azygospore** (or **parthenospore**). It germinates like the zygospore.

Figure 15. *Spirogyra.* Zygosporse germinating.

Germination of the Zygospore (Fig. 15). The zygospore is provided with a thick cellulose-wall. The filament decays and all the zygospores sink to the bottom of the pool of water. They undergo a period of rest till the next favourable season, and then they germinate. The thick wall of the zygospore bursts and the contents grow out into a short filament which escapes and floats to the surface of the water. Cells divide and the filament increases in length.

7. *VAUCHERIA* (35 sp.)

Occurrence. ***Vaucheria*** (Fig. 16) is a green freshwater alga, growing with other algae in ponds, ditches and also in the wet soil. It is not free-floating like *Spirogyra* but is mostly attached to a substratum by means of colourless rhizoids or 'holdfasts'. It is deep green in colour and always lives in a tangled mass.

Structure. The thallus consists of a single branched tubular filament. It is unseptate and contains numerous minute nuclei which lie embedded in the lining layer of cytoplasm surrounding, a large central vacuole. Such a structure is known as a **coenocyte**; *Vaucheria* is, therefore, a coenocyte. Septa, however, normally appear in connection with the reproductive organs. Injury also results in the production of septa, cutting off the injured parts which then develop into new plants. Filaments increase in length by apical growth. Chloroplasts are numerous, very small and discoidal in shape. They lie embedded in the lining layer of protoplasm, and are without pyrenoids. Protoplasm contains abundant oil-globules, but lacks in starch.

Reproduction. This takes place asexually as well as sexually.

Asexual Reproduction. This takes place by a large solitary **zoospore**. During its development the apex of the filament swells up, becomes club-shaped and is partitioned off from the rest of the filament by a septum. This club-shaped body is known as the **zoosporangium** (Fig. 16B). Its protoplasmic contents become rounded off forming a single zoospore. The wall of the zoosporangium ruptures at the apex, and the zoospore escapes through the terminal pore (Fig. 16C) and begins to rotate. The zoospore (Fig. 16D) is an oval body of large size. The

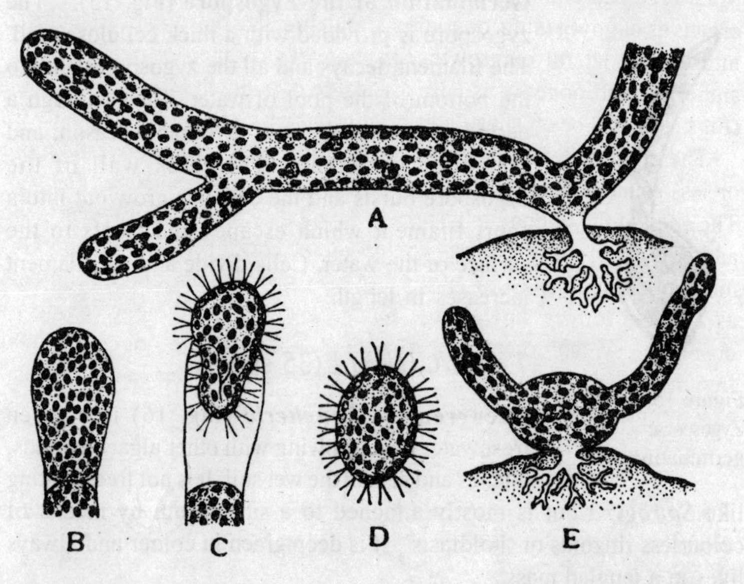

Figure 16. *Vaucheria. A,* a *Vaucheria* filament; *B,* formation of zoosporangium; *C,* zoospore escaping; *D,* free-swimming zoospore; *E,* zoospore germinating.

central part of it is occupied by a large vacuole, and in the surrounding zone of protoplasm there lie embedded numerous small chloroplasts, giving the zoospore an intensely deep green colour. The whole surface of the naked (without cell-wall) zoospore is covered with numerous short cilia arranged in pairs and under each pair there lies a nucleus. For this reason the zoospore is regarded as a compound one. The zoospores generally escape in the morning. They swim about freely in water for a while (half an hour or less) by the vibration of their cilia and soon come to rest. The cilia are immediately withdrawn and a cell-wall is developed round them. After coming to rest the zoospores germinate (Fig. 16E) almost immediately by the protrusion of one or more tube-like filaments, one of which, at least, produces a colourless branched rhizoid, and attaches the plant to the substratum. The protoplasm leaves the old cell and rejuvenates, i.e. it becomes young and active; this method of asexual reproduction is known as **rejuvenescence**.

Sexual Reproduction. This takes place by the method of **fertilization**, that is, by sharply differentiated male and female organs. Male organs are known as **antheridia** (sing. antheridium) and female organs as **oogonia** (sing. oogonium), and these are developed at scattered intervals as lateral outgrowths. In monoecious species of *Vaucheria,* antheridia and oogonia usually arise side by side on the same filament (Figs. 17B–C) or on short lateral branches of it (Fig. 17A).

The outgrowth that forms the **oogonium** swells out, assumes a more or less rounded form, and is cut off by a basal septum (Figs. 17A–C). The apex of the oogonium generally develops a *beak,* either towards the antheridium or away from it. The protoplasm of the oogonium contains much oil, numerous chloroplasts, but only one nucleus, and its contents as a whole form a single large female gamete, i.e. the egg (ovum or oosphere), which completely fills the oogonium. The oogonium is at first multinucleate, but before the partition wall is formed all the nuclei except one return to the main filament or they degenerate.

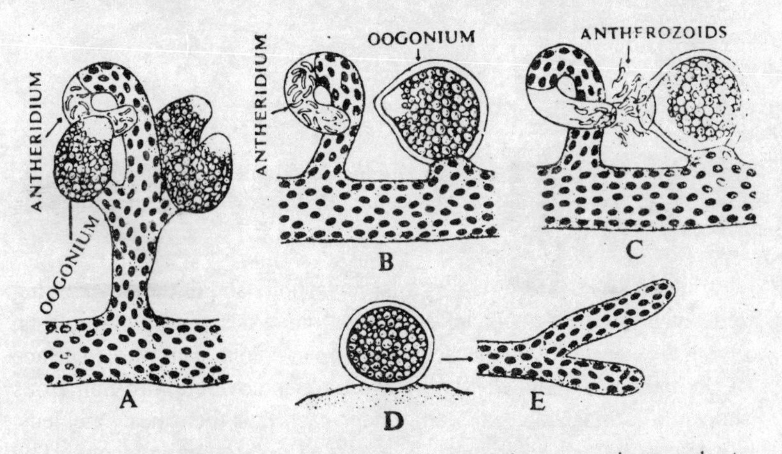

Figure 17. *Vaucheria. A,* an antheridium and three oogonia on a short lateral branch; *B,* the same borne directly by the filament; *C,* mature autheridium and oogonium; antherozoids discharged and ovum about to be fertilized; *D,* oospore; and *E,* a new filament developed from the oospore.

Each **antheridium** arises as a short tubular branch by the side of the oogonium, and simultaneously with it. The terminal portion of it is cut off by a septum and then it becomes the actual antheridium (Figs. 17A–C). As it matures it usually becomes much curved towards the oogonium. The protoplasm contains numerous chloroplasts and nuclei. Numerous male reproductive units or male gametes, known as **antherozoids**, are produced inside each antheridium. They are very minute in size and are bi-ciliated. The cilia point in opposite directions.

Fertilization. Self-fertilization is the rule, but in dioecious species cross-fertilization is apparent. The antheridium bursts at the apex and many antherozoids collect around the beak which opens at about the same time (Fig. 17C). Several antherozoids may enter the oogonium through the beak but only one of them fuses with the ovum, while the rest perish. Thus fertilization is effected. After fertilization the ovum becomes invested with a thick cell-wall, and is known as the **oospore** (Fig. 17D). The oospore undergoes a period of rest and then it germinatcs directly into a new *Vaucheria* filament. Reduction division takes place during the germination of the oospore.

Chapter 3

Microbes: Bacteria and Viruses

Bacteria. Anton von Leeuwenhoek (1652–1723) of Delft in Holland was the first to discover bacteria in 1676 with the help of the microscope considerably improved by himself (see also p. 157). Louis Pasteur (1822–1895) of France thoroughly established the science of bacteriology. He carried on extensive work on fermentation and decay, and the cause of hydrophobia. About the year 1876 Pasteur made known to the world the importance of bacteria. He was the first to prepare vaccine and use it for the cure of the disease. He saved many Russians from hydrophobia by the use of this vaccine, and the Tsar of Russia in honour of his marvellous discovery sent him a diamond cross and also a hundred thousand francs to build a laboratory in Paris—now called the Pasteur Institute. About the same year Robert Koch of Germany proved that anthrax disease, so common in cattle, was caused by a kind of bacteria. He also showed in 1882 that tuberculosis and Asiatic cholera were caused by bacteria.

Occurrence. Bacteria occur almost everywhere—in water, air and soil, and in foodstuffs, fruits and vegetables. Many float in the air; many are abundant in water; and many are specially abundant in the soil, particularly to a depth of half a metre, and also in sewage. A few thousands of them may occur in 1 c.c. of water, and a few millions in 1 gram of soil. Many live within and upon the bodies of living plants and animals. The intestines of all animals always contain a good number of different kinds of bacteria.

Structure. Bacteria are the smallest and the most primitive cellular organisms known to us, and number over 2,000 species. They are mostly single-celled—usually spherical, rod-like or branched. Their average size, particularly of the spherical ones, may be stated to be 0.5 to 2 μ (microns). The rodlike or filamentous forms may be as long as 10 μ or even longer.

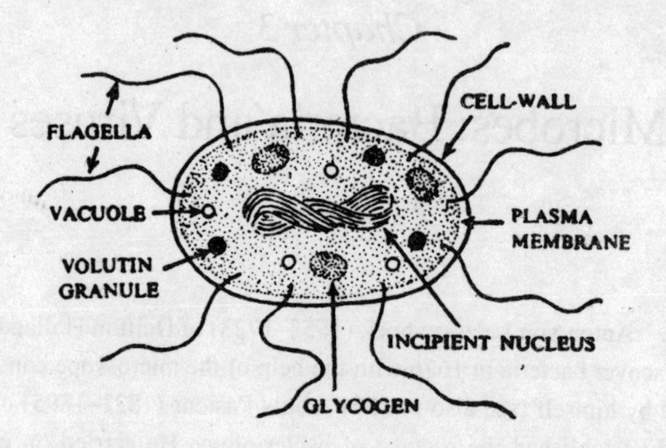

Figure 18. A bacterial cell as revealed by electron microscope.

Bacteria are only imperfectly seen under a compound microscope. The electron microscope, however, reveals the following structure in greater detail (Fig. 18). There is a distinct but complex **cell-wall** made of proteins and carbohydrates: *chitin* is often present but seldom any cellulose. Surrounding the cell-wall there is frequently a slime layer, often changed into a distinct sheath or **capsule**. The capsulated form is very resistant to adverse conditions and to treatments; such a form is commonly the cause of a disease. Several types of bacteria are provided with one or more slender, whiplike threads called **flagella** originating from the cytoplasm; such bacteria are motile. Internal to the cell-wall there is a thin **plasma membrane** formed by the cytoplasm. The **cytoplasm** spreads uniformly throughout the cell, and contains many small **vacuoles**, stored food granules such as volutin, glycogen and fats, sometimes sulphur also, and an **incipient nucleus**. Cytoplasm also includes ribosomes, mesosomes, etc. The active cell remains saturated with water, occurring to the extent of 90%. An organized nucleus, as found in higher plants, is absent in bacteria; nucleolus and nuclear membrane are absent. There is, however, a nuclear material (or chromatin) present in the bacterial cell in the form of 1 or 2 deeply, staining bodies. Chemically these bodies are composed of twisted and folded strands of DNA (deoxyribonucleic acid) which is the *genetic* material of the living cell. The DNA bodies divide prior to the division of the bacterial cell, and are distributed equally among the daughter cells.

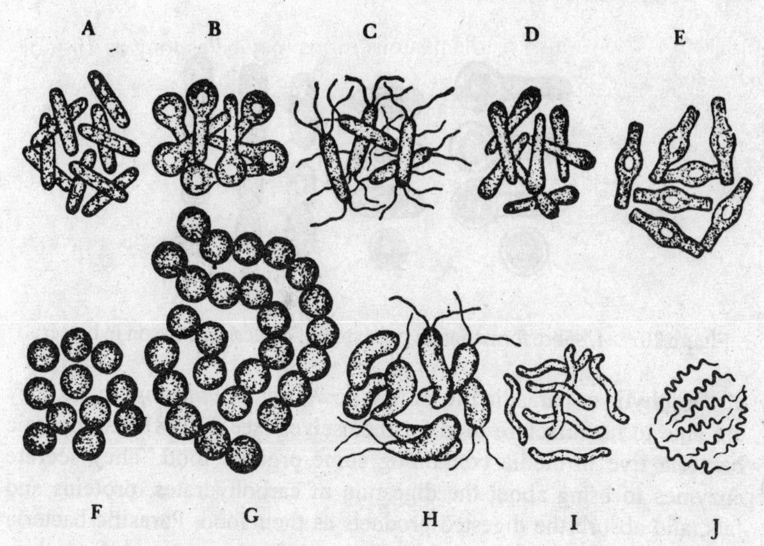

Figure 19. Bacteria. **Bacilli:** *A, Bacillus (=Mycobacterium) tuberculosis; B, B. (=Clostridium) tetani; C, B. typhosus; D, B. (=Corynebacterium) diphtheriae; E, B. anthracis,* **Cocci:** *F, Staphylococcus; G, Streptococcus,* **Comma:** *H, Vibrio cholerae,* **Spirilla:** *I, Spirillum (common in water); J, Spirochaete.*

Shapes of Bacteria. (1) **Bacilli** (sing. bacillus) are rod-shaped bacteria, e.g. *Bacillus* (=*Mycobacterium*) *tuberculosis, B.* (=*Clostridium*) *tetani, B. typhosus,* etc. (2) **Cocci** (sing. coccus) are spherical bacteria, e.g. *Staphylococcus, Streptococcus, Diplococcus, Micrococcus, Azotobacter,* etc. (3) **Spirilla** (sing. spirillum) are bacteria with the body spirally wound, e.g. *Spirillum, Spirochaete,* etc. (4) **Commas** are bacteria with the body slightly twisted like a comma, e.g. *Vibrio cholerae.*

Spore Formation (Fig. 20A). Some bacteria, particularly rod-shaped ones, form **endospores**, usually one in each bacterial cell; they are always 'resting' spores. The special advantage is that they can withstand very unfavourable conditions such as high temperature, freezing, extreme dryness, the presence of many poisonous chemicals, etc., for months or even several years. By this method bacteria, however, do not multiply in number.

Physiology of Bacteria. Bacteria are lacking in chlorophyll and thus are mostly unable to utilize carbondioxide for synthesis of organic compounds for their food. They are mostly *heterotrophic* in habit, leading

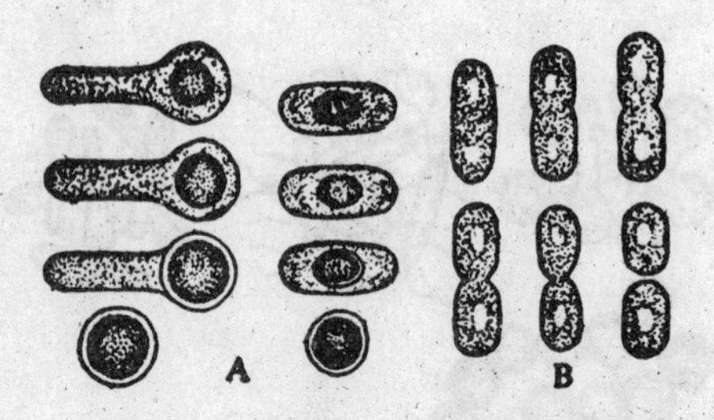

Figure 20. *A*, spore formation in two types of bacteria; *B*, fisson in bacteria.

a saprophytic or parasitic life. A few, however, are *autotrophic*, i.e. they are able to manufacture food for themselves (see p. 288). Saprophytic bacteria live in media containing some organic food. They secrete enzymes to bring about the digestion of carbohydrates, proteins and fats, and absorb the digested products as their food. Parasitic bacteria infect living plants and animals, and absorb food compounds from their body by the same process of enzyme-secretion and digestion.

Reproduction. Bacteria commonly multiply by **fission** (asexual method). Sexual reproduction in them is only imperfectly known.

Fission (Fig. 20B). Bacteria commonly divide by fission. A constriction appears around the middle of the cell and it becomes split up into two parts. These parts grow in size and form mature bacterial cells. By this methcd they may multiply rapidly. Hay bacillus (*Bacillus subtilis*), for instance, divides 2 to 3 times an hour under favourable conditions. At the minimum rate of division a single cell may give rise to over sixteen million (16,777,216) offspring at the end of twelve hours.

Sexual Reproduction. This is only imperfectly known. In 1947 Lederberg and Tatum observed a sort of conjugation between two strains of the common intestinal bacterium (*Escherichia coli*). They were of the view that during conjugation a part of the nuclear material (chromosome) of one strain is injected into the other. The new cell with materials from both the parent cells is more resistant to untoward external conditions than the latter. No multiplication of bacterial cells, however, takes placc by this method. In 1944 Avery, MacLeod and McCarty showed that certain bacteria can incorporate genetic material (DNA) from related strains into their genetic make up. This method is called transformation. Further in another method called transduction, the newly formed bacteriophages while leaving the cell of the infected bacterium pick up DNA of the same cell and implant them in other bacterial cells. These are all cases of genetic recombination. Therfore, true sexual reproduction leading to gametes and their zygote formation is not found in bacteria.

Harmful Effects of Bacteria. Many parasitic (or pathogenic) bacteria attack living plants, human beings and domestic animals, and cause various and often serious diseases in them, sometimes in epidemic form. They are always dreaded as an invisible enemy. Normally they infect the host through wounds or they may be breathed in or taken in with food, water and milk. After infection of the body they not only absorb the stored food and destroy the cells but also at the same time produce a toxin (poison). Some of the common disease-producing bacteria are: *Bacillus typhosus* causing typhoid fever, *B. anthracis* causing anthrax, *B.* (=*Clostridium*) *tetani* causing tetanus, *Clostridium botulinum* causing a dangerous type of food-poisoning (called **ptomaine** poisoning), *B.* (*Corynebacterium*) *diphtheriae* causing diphtheria, *B.* (=*Mycobacterium*) *tuberculosis* causing tuberculosis, *Mycobacterium leprae* causing leprosy, *B. dysenteriae* causing dysentery, *B.* (=*Diplococcus*) *pneumoniae* causing pneumonia, *Vibrio cholerae* causing cholera. Some species of *Streptococcus* (the blood-poisoning bacteria) are possibly the deadliest enemy of mankind. They have the remarkable power of dissolving the red corpuscles of the human blood, and are responsible for erysipelas and extremely dangerous kinds of blood-poisoning.

Parasitic bacteria also attack plants and cause various diseases such as canker of *Citrus,* wildfire of tobacco, black rot of cabbage, fire blight of apple and pear, ring disease of potato, etc. Canker of *Citrus* (orange and lemon) is a common disease caused by *Pseudomonas* (=*Xanthomonas*) *citri.* It appears as a dead area on the surface of the stem, leaves and fruits, with a depression in the centre, usually surrounded by a raised margin. It occurs in most of the *Citrus* orchards in India, sometimes taking a serious turn. The cankerous spots become corky and turn usually brownish in colour, affecting the shape, size, quality and appearance of the fruits. The disease may spread through the agency of wind, rain and insects, and also human beings. In plants, however, fungal diseases are far more common than bacterial diseases, while in animals the reverse is the case.

Many of the bacteria are also responsible for the decay (fermentation) of cooked food, meat, milk, vegetables, and fruits, etc., particularly in storage during summer months, often involving heavy loss.

Beneficial Effect of Bacteria. Although some bacteria (the disease-producing ones) are very harmful, it is a fact that a large number of them are very useful in various ways, particularly in agriculture and some industries. Many bacteria are nature's scavengers.

(1) **Agricultural.** (a) **Decay of Organic Substances.** But for the most useful work of many bacteria the dead bodies of plants and animals would remain unaltered, covering a vast area. Besides, organic compounds contained in such dead bodies would remain permanently locked up in them without any further use. Fortunately, bacteria act on these bodies and convert various organic compounds into simple forms such as nitrates, sulphates, phosphates, etc., for utilization by green plants again. (b) **Nitrification.** Proteins contained in the dead bodies of plants and animals are acted on by different kinds of bacteria and ultimately converted into nitrates which are then absorbed and utilized by the green plants (see p. 240). (c) **Nitrogen Fixation.** Fixation of free nitrogen of the air by many soil bacteria like *Azotobacter* and *Clostridium* directly in their own bodies, and *Rhizobium* (nodule bacteria) in association with the roots of leguminous plants is very important from an agricultural standpoint. (d) **Fertility of the Soil.** The fertility of the soil may largely be attributed to the activity of soil bacteria (and also other soil organisms). They bring about physical and chemical changes in the soil, particularly conversion of insoluble materials into soluble and suitable forms for absorption by green plants. Thus they make the soil fertile. In addition, the conversion of cowdung and animal excreta into manure, and the formation of humus or leaf-mould are due to bacterial activity.

(2) **Industrial.** From an industrial standpoint also many bacteria are most useful. Curing and ripening of tobacco leaves, fermentation of tea leaves, ripening of cheese, etc., for their characteristic flavours, retting of fibres as in jute and flax, manufacture of vinegar from alcohol by acetic acid bacteria, fermentation of sugar into alcohol by yeast and a few bacteria, curdling of milk by lactic acid bacteria, conversion of hide into leather, and such other cases of fermentation are specially important.

(3) **Medical.** (a) In the field of medicine valuable antibiotic drugs have been obtained from a number of bacteria and fungi. Thus, antibiotics like **penicillin** (isolated from *Penicillium notatum,* a mould), **streptomycin** (isolated from *Streptomyces griseus,* a bacterium), **chloromycetin** (isolated from *S. venezuelae*), **terramycin** (isolated from *S. rimosus*), etc., have been successfully used to control and cure some of the dreadful bacterial diseases such as pneumonia, diphtheria, wound infections, tuberculosis, typhoid, etc., saving the lives of millions of people throughout the world. (b) Bacteria always dwell in large numbers in the human system, particularly in the intestines. Some of the good ones often check the growth of the disease-producing bacteria and prevent diseases which may be caused by them. Thus, lactic acid bacteria in the

curd may prevent or cure dysentery. Certain bacteria are in some way connected, possibly by secretion of enzymes, with digestive activities in the intestinal tracts. In any case, some bacteria are considered essential for maintenance of normal health.

Viruses. Viruses (*virus,* poison) are the smallest and possibly the most primitive, acellular (they are not made of cells), non-protoplasmic, particle like bodies. They are very much smaller than the bacteria, and cannot be detected even under the most powerful microscope. Their presence is revealed only when they produce certain diseases in plants or animals. Meyer in 1886 first described the viral disease of tobacco and called it 'tobacco mosaic'. Later it was found that the sap, fresh or dried, of the diseased parts could infect healthy plants. By 1933–34 several viral diseases were discovered. Commonly such diseases are transmitted through insects or through contacts. They may also be seed-, leaf-, or stem-borne. All viruses are entirely parasitic and behave like living organisms (intracellular phase), while they are quite inactive and seem to be dead outside the host body (extracellular phase). Viruses can be isolated, purified and crystallized. In 1935 Stanley, an American microbiologist, first isolated tobacco mosaic virus (TMV) in the form of crystals. These crystals dissolved in water and rubbed on healthy tobacco leaves quickly produce disease symptoms. Since then many more have been obtained in such forms.

Recent investigations, using electron microscope and X-ray photography, have revealed some detailed facts about viruses. Virus particles have no cellular structure. They are of varying shapes and sizes ranging in lengths from 10mμ p to 200 mμ, sometimes up to 450 mμ (1 mμ = 1/1,000 μ) Further, a virus particle contains a core of nucleic acid, mostly DNA (or sometimes RNA) surrounded by a thin film of protein (protein shell). The protein shell is mainly protective in nature and is often very complex, while the DNA is a genetic (hereditary) material and is responsible for all biochemical activities.

There are three kinds of viruses infecting plants, animals and bacteria. A few hundred plant diseases caused by viruses have been recorded so far, e.g. mosaic diseases of tobacco, cabbage, cauliflower, groundnut, mustard, etc.; black ring spot of cabbage; leaf roll of tomato; chlorotic disease of apple, rose, etc.; leaf curl of cotton, bean, soyabean, *Zinnia,* etc.; spike disease of sandalwood; yellow disease of carrot, beet, marigold, peach, etc. Some human diseases like mumps, smallpox, chickenpox, measles, polio, yellow fever, influenza, common cold, cancer, etc., are supposed to be caused by viruses.

Bacteriophages. Viruses which attack bacteria and destroy their nuclear material are called bacteriophages (*phagein,* to eat). Such viruses have a tail and a head, surrounded completely by a contractile protein sheath. The head contains DNA. When the protein sheath contracts, the tail end penetrates into the bacterial cell. Then all of the virus DNA is injected into the latter, while the whole of the protein remains outside. The DNA of the virus now controls the biochemical activities of the bacterial cell and, peculiarly enough, induces the latter to produce more of virus DNA and protein. The result is the appearance, within about 20 minutes, of several new strains of virus particles, destruction of the bacterial genetic material and finally bursting of the bacterial cell (called lysis). No such effect is produced if the protein portion of the virus is injected into the bacterial cell. Evidently DNA is the genetic material (or gene).

Virus Reproduction. Outside the host body the virus is inactive. But inside it the virus DNA or RNA controls the biochemical activities of the infected cells which, peculiarly enough, now begin to make DNA or RNA and proteins characteristic of the invading virus. At the same time the virus particle replicates itself several times forming hundreds

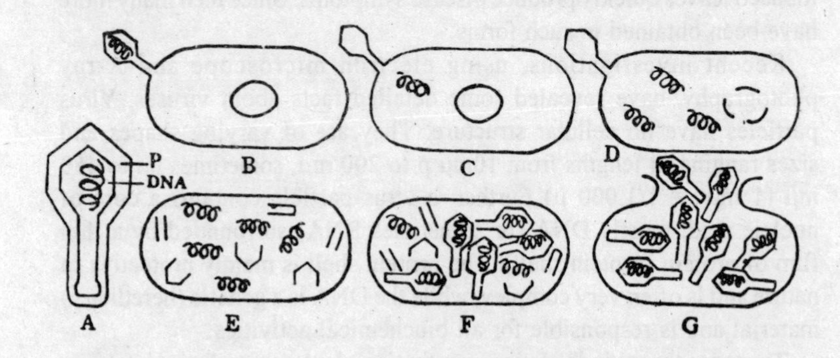

Figure 21. Bacteriophage. *A,* a virus particle showing its structure and composition (*P,* protein; *DNA,* deoxyribonucleic acid); *B–G,* stages showing how a bacteriophage infects a bacterial cell, destroys the bacterial chromosome (shown here diagrammatically as an oval body), replicates itself.

of particles like the original one with identical nucleic acids and proteins. The reproduction of virus has, however, been studied in greater details in connexion with the bacteriophage (see Fig. 21). Many types of viruses also undergo mutation producing new kinds of disease symptoms in host plants and animals.

Differences between Bacteria and Viruses

Bacteria	Viruses
1. Bacteria are very minute in size. They can only be seen under a compound microscope.	1. Viruses are smaller than bacteria. They reveal their identity only under an electron microscope.
2. Bacteria are living cellular organisms.	2. Viruses are non-cellular particles and only intracellularly (i.e. inside the living cells) they become active and living.
3. A bacterial cell is surrounded by a distinct cell wall.	3. The cell wall is absent in case of a virus particle and there is only a protein coat called capsid.
4. A bacterial cell contains protoplasm, nucleiod, vacuoles, ribosomes, mesosomes, etc. Genetic material consists of DNA.	4. A virus particle is made up of only nucleic acids and protein which represents the capsid. Genetic material is either DNA or RNA.
5. Bacteria can reproduce vegetatively, asexually and sexually.	5. Assembly of body parts is the only mode of reproduction, completed through penetration, replication, assembly and then the release of new particles.
6. Bacteria attain growth and show metabolic activities.	6. Viruses have no growth and metabolic activities.

Chapter 4

Fungi

Classification of Fungi (90,000 sp.)

Class I. Myxomycetes or slime fungi (400 sp.).

Class II. Phycomycetes or alga-like fungi (1,500 sp.), e.g. *Mucor, Rhizopus, Albugo* (=*Cystopus*), *Phytophthora,* etc. In them innumerable **spores** are formed *endogenously* in a case called **sporangium.**

Class III. Ascomycetes or sac-fungi (25,000 sp.), e.g. yeast, *Penicillium, Aspergillus,* etc. in them spores, called **ascospores,** usually 8 in number, are formed *endogenously* in a sac called **ascus.**

Class IV. Basidiomycetes or club-fungi (23,000 sp.), e.g. mushroom (*Agaricus*), rust (*Puccinia*), smut (*Ustilago*), etc. In them spores, called **basidiospores**, usually 4 in number, are formed *exogenously* on a club-shaped stalk called **basidium**.

Class V. Fungi Imperfecti (over 24,000 sp.), e.g. *Fusarium*. In them the life-history is imperfectly known.

Economic Importance of Fungi. The importance of fungi on human life, directly or indirectly, is immense. They have both harmful and beneficial effects in many ways. Like the bacteria many of them cause several, sometimes serious, diseases in man and domestic animals. Ringworm, ear and lung infections are common fungal diseases in man. Food-poisoning due to the infection of certain poisonous fungi is not uncommon. Ergot-poisoning (ergotism) now and then takes place in those countries (cold) where rye (a cereal) is grown for bread. Many fungi (often species of *Aspergillus* and *Penicillium*) spoil bread and other foodstuff. Many of such fungi grow on leather and leather goods, paper and books, linen and cotton clothes, rubber goods, wood, and even valuable optical lenses, and cause considerable damage and decay in them. Damage to foodgrains, vegetables and fruits in storage, often infected by moulds, particularly in a warm humid climate, results sometimes in a heavy loss. Many fungi often cause serious diseases of crop plants grown for food and industry. Vegetables, fruit trees and timber trees are often attacked by a variety of fungi causing damage and even destruction. On the other hand, many fungi have proved to be useful to mankind in many ways. Thus many soil fungi, like certain bacteria, act on dead bodies of plants and animals, decompose them and make the soil fertile. In the same way leaf compost (a valuable manure) is made in pits in the ground under their joint actions. Some industries dealing with the manufacture of alcohols, wines and other liquors, certain organic acids (e.g. citric, lactic, oxalic, glutonic, fumaric, etc.) and many other products have developed on the proper use of yeasts and certain other fungi (e.g. species of *Aspergillus* and *Rhizopus*). The universal use of yeasts in bread-making is too well-known. Besides, the yeasts are also a source of vitamins and enzymes. Proper ripening and flavouring of cheese depends on the use of certain species of *Penicillium*. Some valuable medicines like rye ergot (containing several powerful alkaloids), antibiotic like penicillin, some digestive enzymes, and also some other drugs are fungal products. Some fleshy fungi (mushrooms, for example) are widely used as food. There are, however, many other fleshy fungi (toadstool, for example) which are distinctly poisonous. Some wild animals like rabbits and squirrels feed upon a variety of fungi.

A. PHYCOMYCETES OR ALGA-LIKE FUNGI

1. *MUCOR* (50 sp.)

Occurrence. **Mucor** (Fig. 22) commonly called 'pin-mould,' is a saprophytic fungus. It grows on stale moist bread, rotten fruits, decaying vegetables, shed flowers, wet shoes, animal-dung and other organic media, spreading like a cobweb. It can be grown in the laboratory on a piece of moist bread kept under a bell-jar in a warm place for three or four days. [Another very common mould closely related to *Mucor* is *Rhizopus*; see p. 390].

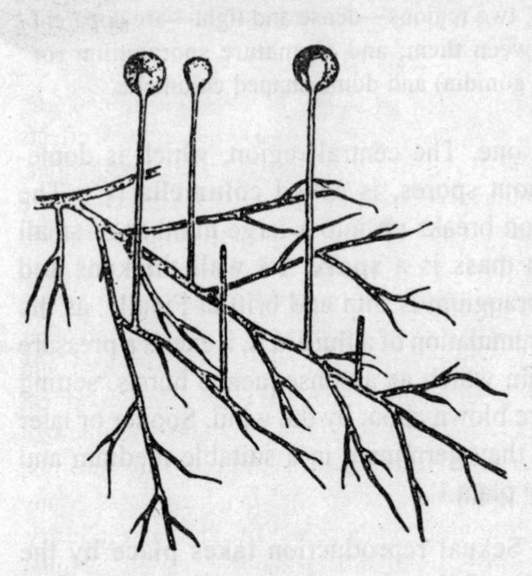

Figure 22. Mucor. Ramifying mycelia with some sporangia (or gonidangia).

Structure. The plant body is composed of a mass of white, delicate, cottony threads collectively known as the **mycelium** (Fig. 22). It is always very much branched, but is coenocytic, i.e. unseptate and multinucleate. Each individual thread of the mycelium is known as the **hypha** (pl. hyphae).

Reproduction. *Mucor* commonly reproduces asexually, and sometimes sexually.

Asexual Reproduction. This method of reproduction takes place by means of **spores** (or gonidia) which develop in a case, called **sporangium** (or gonidangium) under favourable conditions of moisture and temperature. It is seen that mycelia give off, here and there, numerous slender erect hyphae (**sporangiophores**), each ending in a spherical head—the **sporangium** (Fig. 23). The protoplasmic contents migrate to the spherical head (A) and become differentiated into two distinct regions—the outer and the inner. The outer region is dense and contains numerous nuclei, while the inner region is thin and vacuolate and contains few nuclei (B). A wall soon appears round the central regi

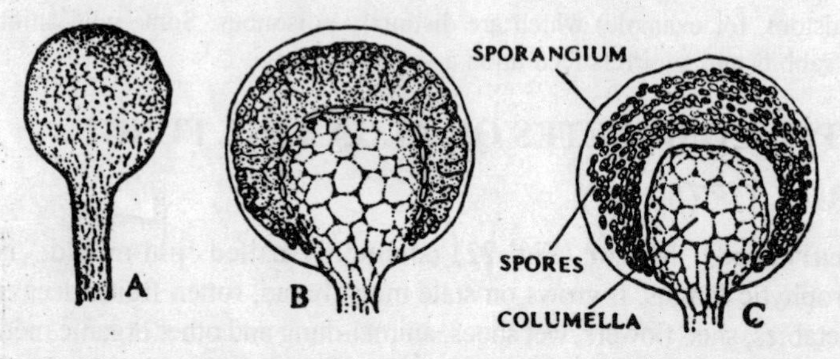

Figure 23. *Mucor.* Development of sporangium, spores and columella; *A,* the end of the hypha swells; *B,* two regions—dense and light—are apparent with a layer of vacuoles between them; and *C,* mature sporangium (or gonidangium) with spores (or gonidia) and dome-shaped columella.

separating it from the outer one. The central region, which is dome-shaped and sterile, i.e. without spores, is called **columella** (C). The protoplasm of the outer region breaks up into a large number of small multinucleate masses. Each mass is a **spore**. Its wall thickens and darkens. The wall of the sporangium is thin and brittle. Finally, as the columella swells owing to accumulation of a fluid in it, it exerts a pressure on the wall of the sporangium which as a consequence bursts, setting the spores free. The spores are blown about by the wind. Sooner or later under favourable conditions they germinate in a suitable medium and grow directly into the *Mucor* plant.

Sexual Reproduction. Sexual reproduction takes place by the method of **conjugation** (Fig. 24) only under certain conditions, particularly when the food supply becomes exhausted. Conjugation consists in the fusion of *two similar* **gametes**, i.e. isogametess (cf: *Spirogyra*). The process is as follows: When two hyphae borne by *two*

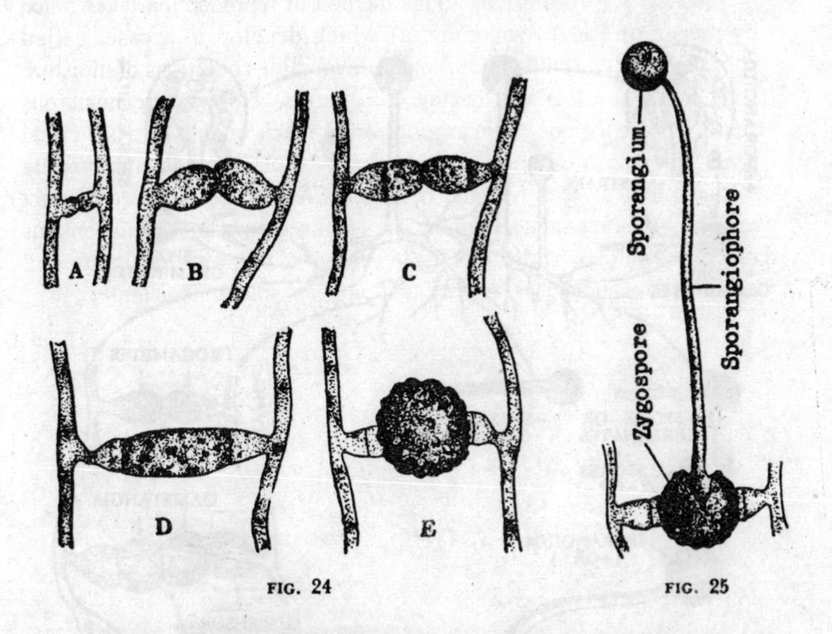

Figure 24. *Mucor*. Conjugation. *A–E* are stages in the process; note the thick-walled zygospore at *E*. Figure 25. Germination of zygospore.

different plants of opposite strains (called the + strain and the – strain) come close together, two short swollen protuberances, called the conjugation tubes, develop, forming a contact at their tips (A). As they elongate they push the parent hyphae apart front each other. Each tube enlarges and becomes club-shaped (B). Soon it is divided by a partition wall into a basal **suspensor** and a terminal **gametangium** (C). The protoplasmic contents of each gametangium constitute the **gamete**. The gametes, like the spores, are multinucleate. The two gametes are identical in all respects. The end- (or common-) walls of the two gametangia get dissolved, and the two gametes fuse together (D) and form a **zygospore** (E). The zygospore swells into a rounded body, and its wall thickens, turns black in colour and becomes warted. It contains an abundance of food, particularly fat globules.

Sometimes it so happens that conjugation does not take place, and then a gametangium may be converted into a zygospore-like body called the **azygospore** (or **parthenospore**). Germination of the azygospore has not been followed.

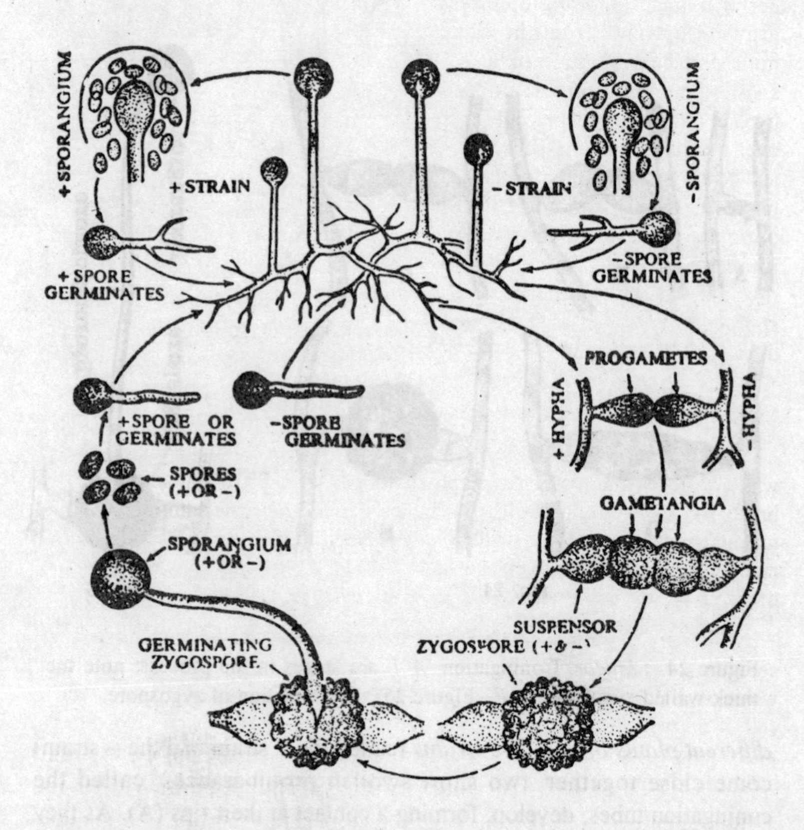

Figure 26. Life-cycle of *Mucor.*

Germination of Zygospore (Fig. 25). The zygospore undergoes a period of rest and then it germinates. The outer wall bursts and the inner wall grows out into a tube, called the **sporangiophore** or **promycelium**, which ends in a spherical sporangium. The sporangium contains numerous small **spores** but no columella. The spore germinates and gives rise to the *Mucor* plant.

Rhizopus (35 sp.). *R. nigricans* (Fig. 27) is a very common 'black mould' closely related to *Mucor*. It grows profusely in the same organic media and follows exactly the same life-history as *Mucor*. Morphological differences between the two, however, are as follows. *Rhizopus* consists of (a) clusters of much-branched absorptive hyphae (rhizoids) growing downwards into the food medium; (b) groups of aerial hyphae (sporangiophores) growing upwards from the same node and bearing sporangia; (c) swollen nodes; and (d) curved hyphae (stolons) growing outwards over the surface of the substratum and bearing rhizoids and sporangiophores from the same node in the opposite directions at frequent intervals. *Mucor* is lacking in rhizoids and stolons, and in it the sporangiophores grow singly and directly on the mycelia.

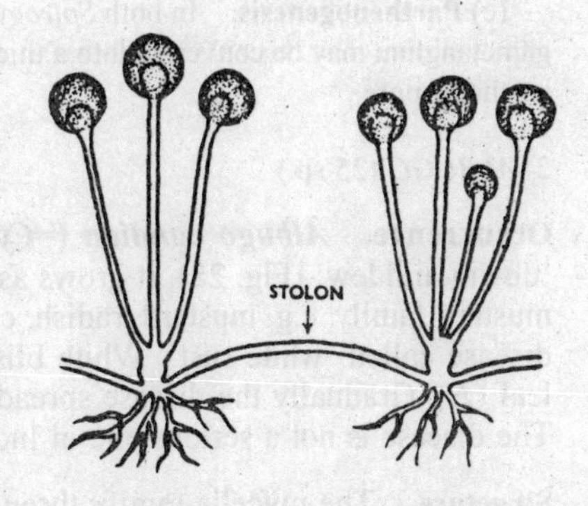

Figure 27. *Rhizopus*. Note the rhizoids, sporangiophores, swollen nodes and curved hyphae (stolons).

COMPARATIVE STUDY OF *SPIROGYRA* AND *MUCOR*

Habit and Habitat. *Spirogyra* is a green alga floating on stagnant water; while *Mucor* is a non-green saprophytic fungus growing in animal-dung, stale bread, wet shoes, rotting fruits, vegetables, etc. *Spirogyra* is autotrophic, i.e. it manufactures its own food; while *Mucor* is heterotrophic, i.e., it absorbs ready-made food from the medium on which it grows. Carbohydrate is in the nature of starch in *Spirogyra;* while it is glycogen in *Mucor*.

Structure. Each *Spirogyra* plant is a slender unbranched filament consisting of a row of cylindrical cells; while *Mucor* consists of a mycelium which is a network of white much-branched cottony threads (hyphae). Each cell of *Spirogyra* filament contains one or more spiral bands of chloroplasts with numerous pyrenoids in them, and one nucleus. In *Mucor* on the other hand each hypha is coenocytic, i.e. unseptate and multinucleate.

Reproduction. There is no regular vegetative reproduction in *Spirogyra* or *Mucor*. The normal method of reproduction is sexual in *Spirogyra* (asexual method being absent), and asexual in *Mucor* (sexual method being conditional).

(a) **Asexual Reproduction.** This is absent in *Spirogyra,* whereas in *Mucor* this is the commonest mode of reproduction. Thus the latter reproduces asexually by innumerable minute spores borne in sporangia, each at the end of an erect hypha. Each spore germinates into a *Mucor* plant.

(b) **Sexual Reproduction.** In *Spirogyra* sexual reproduction is commonly in the nature of scalariform conjugation; while in *Mucor* conjugation takes place between two hyphae of opposite strains (+ and –) only under certain conditions. In *Spirogyra* the zygote directly germinates into a new filament; while in *Mucor* it grows into a sporangiophore which ends in a sporangium with spores. The spores then germinate into *Mucor* mycelia.

(c) **Parthenogenesis.** In both *Spirogyra* and *Mucor,* if conjugation fails, a gametangium may be converted into a thick-walled spore called azygospore or parthenospore.

2 *ALBUGO* (25 sp.)

Occurrence. ***Albugo candida* (=*Cystopus candidus*)** is a common 'downy mildew' (Fig. 28). It grows as a parasite on many plants of the mustard family, e.g. mustard, radish, cabbage, turnip, etc., and causes a disease called 'white rust'. White blisters appear on the stem and the leaf (A). Gradually the disease spreads to the flowers and the ovaries. The disease is not a serious one in India.

Structure. The mycelia ramify through the intercellular spaces of the host plant and branch profusely. The hyphae are aseptate and multinucleate. Here and there they send globular or button-like haustoria (B) into the living cells of the host to absorb food from them.

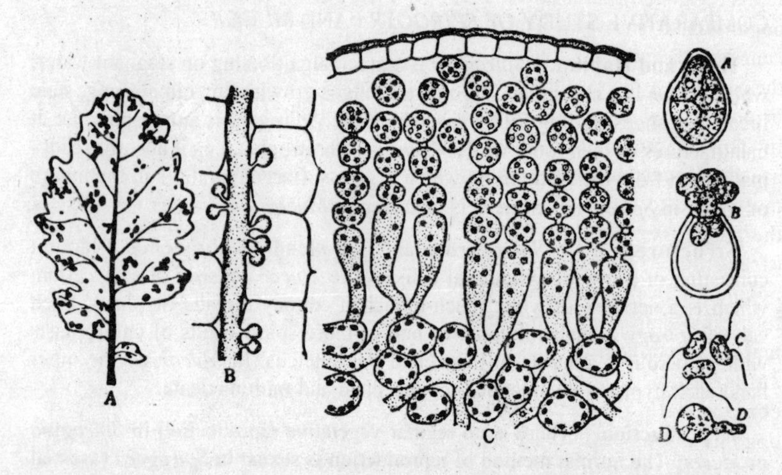

Figure 28. *Albugo. A,* an infected leaf of mustard; *B,* an intercellular hypha with button-like haustoria; *C,* an infected leaf in section showing chains of multinucleate sporangia under the epidermis (note the necks separating the sporangia); *D,* germination of a sporangium; *A,* sporangium dividing; *B,* zoospores escaping; *C,* biciliate zoospores swimming; and *D,* a zoospore germinating.

Reproduction. The fungus reproduces both asexually and sexually.

Asexual Reproduction (Fig. 28C–D). Hyphae grow luxuriantly at certain points below the epidermis of the host, and form clusters of club-shaped multinucleate hyphae (sporangiophores) which begin to cut off multinucleate

sporangia in chains at the tips. The sporangia are separated from one another by short necks made of gelatin. The epidermis soon gets ruptured and the sporangia appear on the surface as a white powdery mass. The sporangia are now blown by the wind to other plants. The contents of each sporangium (Fig. 28D, *A*) divide to form a few (4 or 8 or more) **zoospores**, each with two lateral cilia. The sporangium bursts and the zoospores escape (Fig. 28D, *B*). They swim in water for some time (Fig. 28D, *C*). Soon they lose their cilia, cover themselves with a wall and come to rest. Later they germinate by producing a **germ tube** (Fig. 28D, *D*) which enters the host plant through a stoma. Sometimes a sporangium germinates directly without forming zoospores.

Sexual Reproduction (Fig. 29). In the intercellular spaces of the host the hyphae form male and female organs called **gametangia**. The tip of a hypha swells and gives rise to a spherical multinucleate female gametangium called the **oogonium**. It shows two distinct zones: a dense central zone called the *ooplasm,* which is the **egg-cell** or oosphere with an **egg-nucleus** in it (other nuclei of this zone usually degenerate), and a lighter multinucleate outer zone called the *periplasm*. Similarly the tip of another hypha close to the oogonium swells and gives rise to a club-shaped multinucleate male gametangium called the **antheridium**. It soon comes in contact with the wall of the oogonium and produces a beak or **fertilization tube** which penetrates into the oosphere. One or more male nuclei are set free through this tube but only one of them fuses with the egg-nucleus. Thus fertilization is effected (*A*). The zygote (oospore) formed as a result o fertilization covers itself with a thick wall (*B*). The periplasm is used up in the process. The zygote is liberated only after the decay of the host tissue. Later it produces numerous (over 100) small zoospores which escape into a vesicle (zoosporaagium; *C*). Each zoospore develops two cilia laterally. The vesicle dissolves and the zoospores are set free to swim about in water (*D*). Finally they germinate under appropriate conditions by producing a **germ tube** which infects the host.

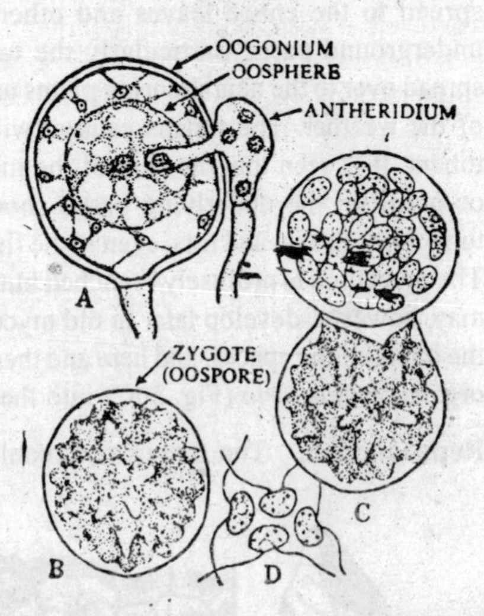

Figure 29. *Albugo*. *A*, fertilization; *B*, zygote (oospore); *C*, germination of zygote; zoospores escaping into a vesicle; and *D*, biciliate zoespores after escape.

3. *PHYTOPHTHORA* (20 sp.)

Occurrence. Species of **Phytophthora** are widely distributed, and in India several species occur as common parasites on potato,tomato, pepper, coconut-palm, palmyra-palm, tobacco, castor, taro (*Colocasia*), etc. *Phytophthora infestans* attacks potato plants and causes a serious disease known as 'late blight'. The disease takes a more serious turn in the hills than in the plains. In the hills the disease frequently appears in epidemic form, causing heavy damage to the crop. Black patches appearing on the leaves, mostly on their under-surface, indicate the diseased condition of the potato plants (Fig. 30A). The disease may spread to the entire leaves and other parts, extending down to the underground parts, particularly the tubers. The disease may rapidly spread over to the neighbouring plants under warm and humid conditions of the weather. The fungus causes 'wilting' of leaves and 'rotting' of tubers. It is seen that the skin of the tuber first turns slightly brownish or purplish; the underlying tissue soon softens and finally the entire tuber turns brown and rots, even in the field or later in storage. **Structure.** The mycelium is profusely branched but aseptate and cocenocytic. Septa may, however, develop later in old mycelia. The hyphae ramify through the intercellular spaces, and here and there send hooked or curled sucking organs or *haustoria* (Fig. 30C) into the living ceils.

Reproduction. The fungus commonly reproduces asexually. Sexual

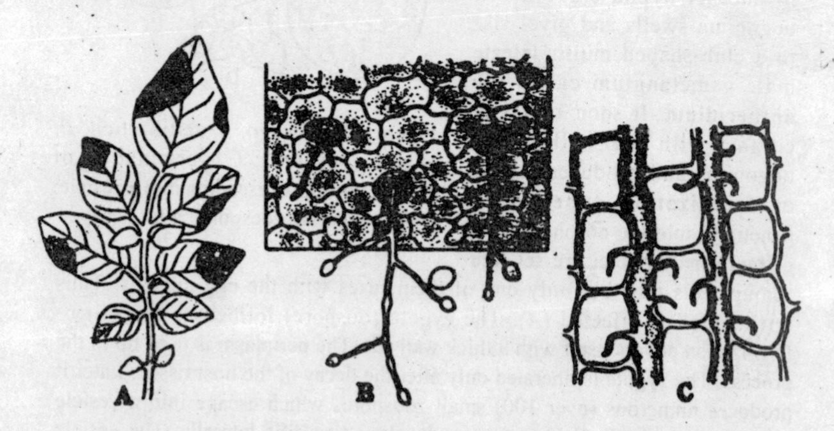

Figure 30. *Phytophthora. A,* a potato leaf showing the infected areas; *B,* sporangiophores protruding through the stomata and bearing sporangia; *C,* haustorial hyphae.

reproduction, though not common, was first observed in pure culture and later in infected tubers.

Asexual Reproduction. For this purpose sporangiophores grow out in small groups through the stomata on the lower surface of the leaf (Fig. 30B). On the tuber they, however, appear in large numbers. The **sparangiophore** is branched, and the branches bear several multinucleate lemon-shaped sporangia, each at the tip of a branch. Each branch, however, continues to grow pushing the sporangium aside, and again it forms a sporangium at its tip. The terminal sporangia thus become lateral as a result of zig-zag growth (a sympodium) of the sporangiophore. Further, a distinct nodular swelling of the branch is seen just above the sporangial base. When mature, the sporangia are dispersed by wind or washed away by rain. Commonly at high temperature the sporangium germinates directly on a potato plant by pushing out a *germ tube* when such sporangia are also described as conidia and the sporangiophores as conidiophores. At low temperature, however, the contents of the sporangium (Fig. 31A) divide into several uninucleate segments (Fig. 31B). Each segment then develops into a biciliate **zoospore**, with the two cilia attached laterally. The tip of the sporangium bursts and the zoospores escape (Fig. 31C). They swim for a while, then come to rest,

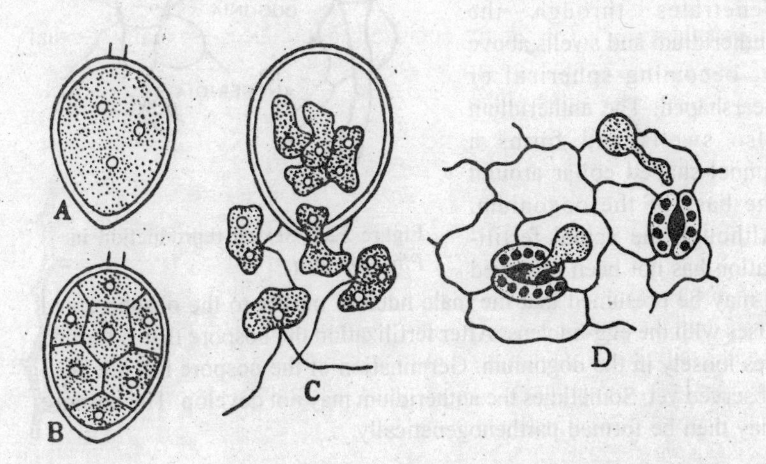

Figure 31. *Phytophthora*. Asexual reproduction; *A*, a sporangium; *B*, a sporangium becoming a zoosporangium; and *C*, zoospores escaping; *D*, zoospores germinating on the host plant and a germ tube penetrating through a stoma.

lose their cilia and form a wall round each. Under favourable conditions of temperature and moisture they germinate on the leaf by producing a *germ tube* which penetrates through a stoma into the tissue of the leaf (Fig. 31D). The disease may rapidly spread to many plants by this method. Sporangia and zoospores are short-lived, and the fungus hibernates in the tuber in the form of mycelia. Thus initial infection of the potato plant starts from this diseased tuber.

Sexual Reproduction. Antheridia (male) and oogonia (female) develop for this purpose from the two neighbouring hyphae (Fig. 32). The **oogonium** is spherical or pear-shaped, with a smooth reddish-brown wall. It contains a large oosphere or egg-cell, lying loose and free within it, surrounded by a scanty zone of protoplasm, called the *periplasm*. All the nuclei of the oogonium except one egg-nucleus of the oosphere degenerate. The **antheridium** is broadly club-shaped and develops before the oogonium. It contains many nuclei but finally all of them except one male nucleus degenerate. Peculiarly enough, the oogonium, as it grows, penetrates through the antheridium and swells above it, becoming spherical or pearshaped. The antheridium also swells and forms a funnel-shaped collar around the base of the oogonium. Although the actual fertilization has not been observed it may be presumed that the male nucleus moves to the oosphere and fuses with the egg-nucleus. After fertilization the oospore that is formed lies loosely in the oogonium. Germination of the oospore has not been observed yet. Sometimes the antheridium may not develop. The oospore may then be formed parthenogenetically.

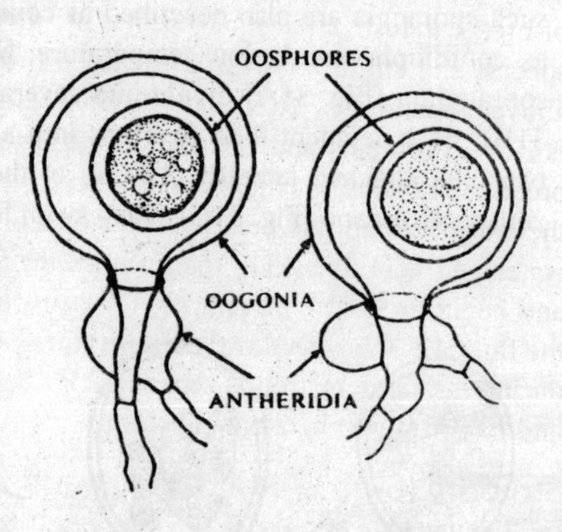

Figure 32. Sexual reproduction in *Phytophthora*.

Control of the Disease. (1) *Spraying.* The disease may be controlled by spraying the plants, when they are 15–20 cm high, with Bordeaux mixture which contains copper sulphate and lime. Spraying should be repeated every 10 or 15 days. (2) *Selection of Seed-tubers.* Tubers suspected to be diseased should be avoided. Seed-tubers obtained from non-infected areas are the best in this respect. (3) *Low Temperature.* Storage of seed-potatoes at a low temperature, 4–5° C, is also practised.

B. ASCOMYCETES OR SAC-FUNGI

4. *SACCHAROMYCES* (40 sp.)

Occurrence. **Yeast** (*Saccharomyces*) grows abundantly in sugar solution such as the juice of date-palm, grapes, etc.; it is also present in the soil of the vineyard and in the air. Yeast has the special property of changing sugar into alcohol and carbon dioxide. Because of this property yeast is regarded as very important economically since several industries have developed on this basis. Thus yeast is widely used in the making of bread, industrial alcohol, beer, wine, toddy, etc. Of course different species and strains of yeast and also different media are used for the above preparations. Yeast also prepares several vitamins in its body and is rich in proteins and fats. It has thus a medicinal value. In making bread the species *Saccharomyces cerevisiae* is likely employed all over the world. When this bread-yeast is added to the mixture of flour and water and kept at baking temperature the yeast cells multiply rapidly and begin to secrete an enzyme (*zymase*) which ferments the sugar of the flour, i.e. it breaks down the sugar into CO_2 and alcohol. CO_2 formed during the fermentation is retained in the bread making it soft, spongy and flavorous, while alcohol escapes.

Structure (Figs. 33–34). Yeast was first microscopically examined by Leeuwenhock in the year 1680. Its structure is simple. A single cell

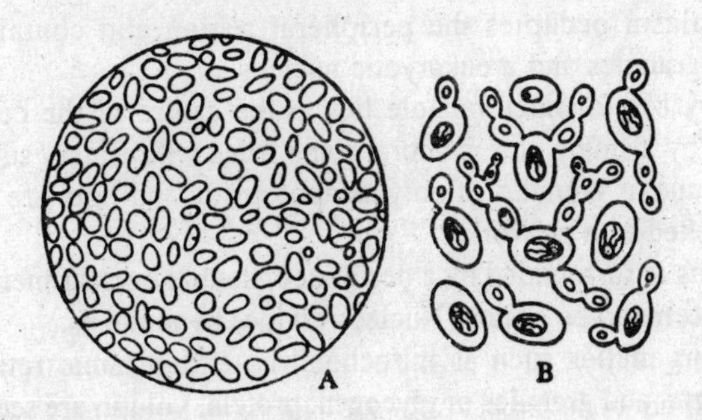

Figure 33. Yeast. *A*, yeast cells as seen under the microscope; *B*, budding.

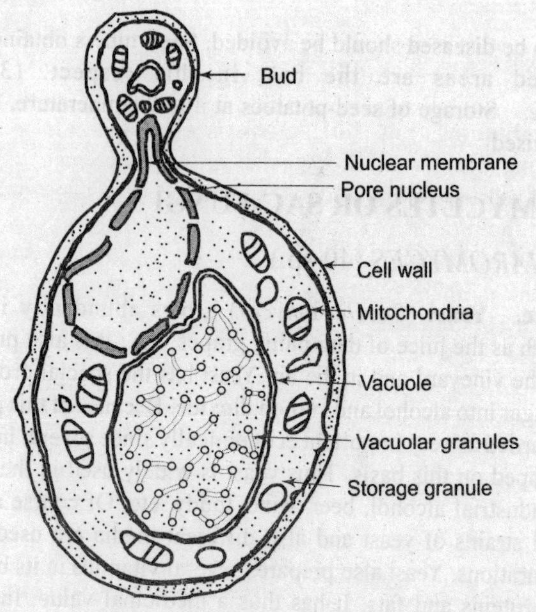

Figure 34. Unicellular somatic body of yeast (Sacchanomyces).

represents the whole body of the plant. It is very minute in size and looks like a pinhead under the microscope (33A). With the invention of electron micröscope, it has been possible now to study the structure of a Yeast cell in minute detail.

A Yeast cell (see Fig. 34) reveals the following structure, when examined under the electron microscope:

1. The cell is bounded by a distinct cell wall which in case of a matured cell shows prominent raised bud-scars.

2. Cytoplasm occupies the peripheral region and contains cell organelles, granules and a eukaryotic nucleus.

3. A very big, distinct vacuole lies in the centre of the cell. It is surrounded by a single unit membrane and filled with watery substance in which vacuolar granules of polymetaphosphate and lipid are seen to lie being linked in a network.

4. Nucleus is surrounded by a perforated, double nuclear membrane. Nucleolus seems to be absent. Nucleus divides by amitosis.

5. Cell organelles such as mitochondria, endoplasmic reticulum, ribosomes, etc. and granules of glycogen, protein, volutin are seen to be present.

Reproduction commonly takes place, by budding and sometimes by fission, and rarely sexually, as observed in a few species.

(1) **By Budding** (Fig. 33B). This is a common method in yeast cells growing in sugar solution. As they grow, two changes are noticed: budding of yeast cells, and alcoholic fermentation of sugar solution (see p. 313). In the process of budding each cell gives rise to one or more tiny outgrowths which gradually increase in size and are ultimately cut off from the mother cell; these then lead a separate existence. The budding may be repeated, resulting in the formation of one or more chains and even sub-chains (sometimes called *pseudomycelia*) of bead-like cells; these cells ultimately separate from one another into individual one-celled yeast plants.

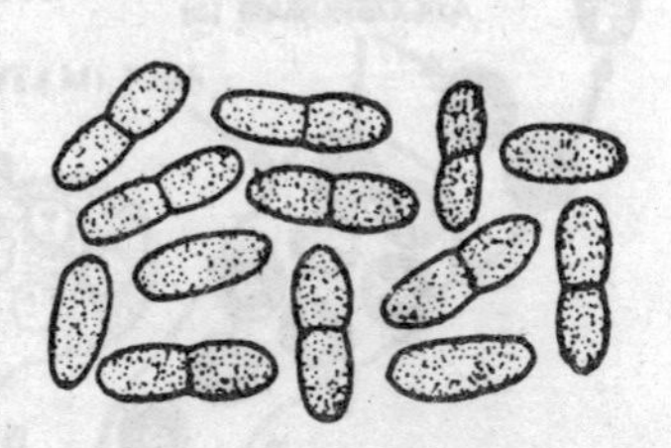

Figure 35. Fission in yeast.

(2) **By Fission** (Fig. 35). Some yeast cells, called fission yeasts, multiply by division. In this 'Process the mother cell at first elongates, and then its nucleus divides into two. The two nuclei move apart, and a transverse partition wall is formed somewhere in the middle of the mother cell, thus dividing it into two parts, each with a nucleus. The two parts then separate from each other along the partition wall, forming two independent yeast cells.

(3) **By Sexual Reproduction** (Fig. 36). Some species of yeast also reproduce by the sexual method (conjugation). In this connection it may be noted that the somatic (i.e. vegetative) cells of yeast may be diploid ($2n$) or haploid (n), while the ascospores are always haploid (n). The zygote is of course diploid ($2n$). Sexual reproduction may take place between two haploid somatic cells or between two ascospores, resulting in both the cases in a diploid cell (with $2n$ zygote). This diploid cell *by budding* gives rise to diploid somatic cells ($2n$), or it may behave as an ascus with 4 or 8 ascospores (n) formed in it *by meiosis*. The ascospores multiply by budding, or they may take to conjugation in pairs. Three patterns of sexual reproduction have so far been observed in different species of yeast (*Saccharomyces*), leading to three patterns of life-cycle, as follows.

(A) In certain species, as in *S. octosporus* (Fig. 36), the diploid ($2n$) phase is very short, while the haploid (n) phase is a prolonged one. In sexual reproduction two somatic cells (n) at first come in contact. At the point of contact they send

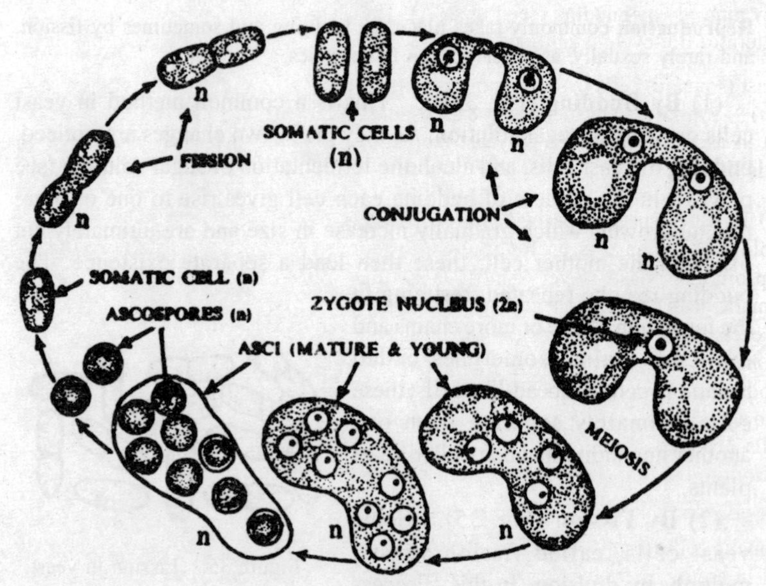

Figure 36. Sexual reproduction in yeast *Saccharomyces octosporus*.

out short, neck-like protuberances (conjugating tubes) which unite by their tips. The two nuclei then pass on to the conjugating tubes. The partition wall is dissolved, and the two nuclei fuse at the neck. The conjugating tubes widen and the contents of the two cells unite, resulting in a zygote ($2n$). The diploid cell thus formed grows and behaves as an **ascus**. The zygote nucleus now divides thrice, the first division being meiotic, and thus 8 nuclei are formed (each haploid or n). Each nucleus forms a wall round itself, enlarges and becomes an **ascospore** (n). Thus there are 8 ascospores in each ascus. The wall of the ascus breaks, and the ascospores are set free. Each ascospore enlarges and becomes a somatic cell (n) now. It divides by a transverse wall into two daughter cells (n). Sooner or later they may again take to conjugation with similar other haploid cells under suitable conditions. Commonly, however,

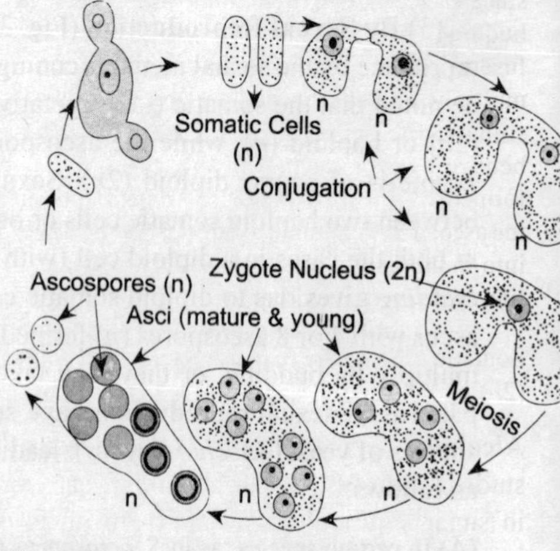

Figure 36A. Sexual reproduction and life cycle in *Saccharomyces octosporus*.

they multiply by budding (see Fig. 36A). The pattern of life-cycle is outlined below.

Yeast (somatic) cells (n) by fission in pairs ($n + n$) → zygote ($2n$) by meiosis → 8 ascospores (n) in ascus, by budding → yeast (somatic) cells (n).

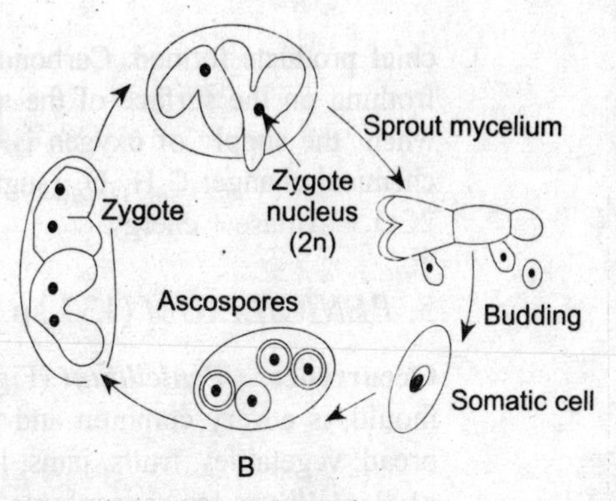

Figure 36B. Sexual reproduction and life cyclein *S. ludwigii*.

(B) In other species, as in *S. ludwigii,* the diploid ($2n$) phase is prolonged, while the haploid (n) phase is very short. Here the somatic cells ($2n$) enlarge and behave as asci. In each ascus 4 ascospores (n) are produced by *meiosis*. They fuse in pairs within the ascus, producing 2 zygotes ($2n$). Each zygote grows into a mycelium (sprout mycelium-$2n$), and its cells produce somatic (yeast) cells ($2n$) *by budding* (see Fig. 36B). The pattern of life-cycle is outlined below:

Somatic cells ($2n$) by meiosis → 4 ascospores (n) in ascus, by fusion in pairs → 2 zygotes ($2n$) → sprout mycelium ($2n$) by budding → somatic cells ($2n$).

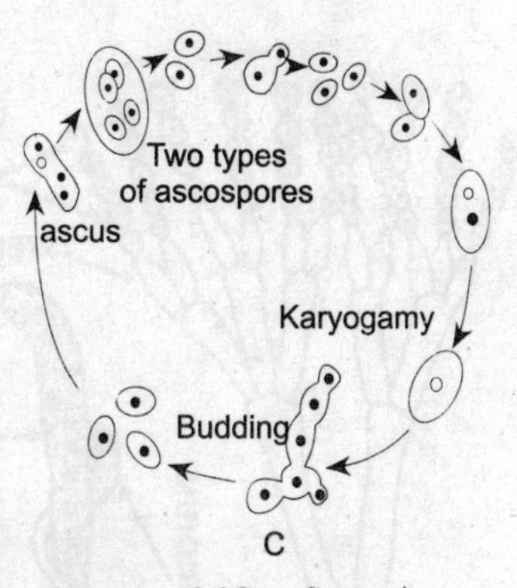

Figure 36C. Sexual reproduction and life cycle in *S. cerevisiae*.

(C) In still other species, as in bread yeast (*S. cerevisiae*) the two phases (haploid and diploid) are more or less equally important since each phase may continue indefinitely by budding. Two such haploid somatic cells may fuse to produce a zygote ($2n$). It continues to multiply by budding into innumerable large somatic (yeast) cells ($2n$). Later they may behave as asci, each with 4 ascospores (n) formed within it by meiosis. After liberation the ascospores continue to multiply by budding into innumerable small somatic (yeast) cells (n) (see Fig. 36C).The pattern of life-cycle is outlined below:

Somatic cells (n) by fusion in pairs → zygote ($2n$) by budding → somatic cells ($2n$) by meiosis → 4 ascospores (n) in ascus, by budding → somatic cells (n).

Alcoholic Fermentation. The process of fermentation was elaborately studied by Louis Pasteur about the year 1857. When the yeast cells grow in sugar solution, as in date-palm juice, palmyra-palm juice or grape juice, they set up fermentation (see pp. 309) in it by means of an enzyme (*zymase*). Sugar is decomposed, and alcohol and carbondioxide are the chief products formed. Carbondioxide escapes, and often gives rise to frothing on the surface of the solution. Fermentation takes place only when the supply of oxygen is cut off. Sugar undergoes the following chemical change: $C_6H_{12}O_6$ (sugar) + zymase $\rightarrow$ $2C_2H_5OH$ (alcohol) + $2CO_2$+ zymase + energy.

5. *PENICILLIUM* (137 sp.)

Occurrence. ***Penicillium*** (Fig. 37), commonly called blue or green mould, is a very common and widely distributed fungus, growing on bread, vegetables, fruits, jams, leather, leather shoes, etc. Most species of *Penicillium* are saprophytic in habit, while a few are parasitic on animals including human beings. Spores of this fungus are present almost everywhere in the soil and the air, and are often sources of contamination

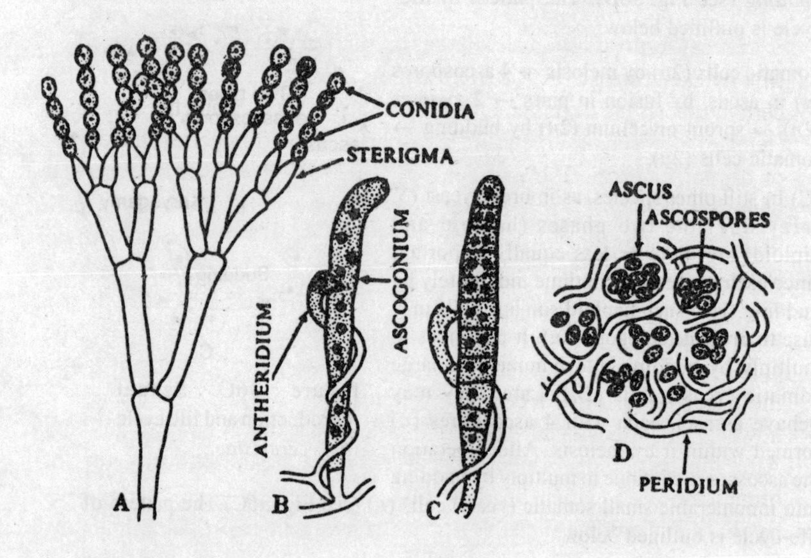

Figure 37. *Penicillium. A,* branched conidiophore; *B–C,* sex organs, *D,* cleistothecium.

of foodstuffs including fruits and vegetables. *Penicillium notatum* is the source of the world-famous **penicillin** (an antibiotic) which was first isolated from this fungus by the late Sir Alexander Fleming, a bacteriologist, in 1929.

Structure. The mycelium-consists of an interwoven mass of hyphae which spread over the surface of the substratum, penetrating here and there into it. The hyphae are branched, septate and multinucleate.

Reproduction. *Penicillium* freely reproduces asexually by conidia. It seldom reproduces sexually.

Asexual Reproduction. A number of hyphae stand erect from the vegetative mycelium; these are the **conidiophores** (Fig. 37A). They branch repeatedly near the apex, regularly or irregularly, in a broom-like fashion, and are septate. The slender ultimate branches known as the **sterigmata** (sing. sterigma) cut off chains of cells by the process of budding. These are the spores called **conidia**, and are formed in countless numbers. The branched condiophores including the sterigmata and the conidia are together called **penicillus** (a little brush). The conidia are easily dispersed by the wind, and under favourable conditions they germinate by producing a *germ tube.*

Sexual Reproduction. This has been observed only in a few species, first by Dangeard in 1907. Perfect stages, however, are not known yet. In certain species the development of the antheridia (male) and the ascogonia (female) has been observed (Fig. 37B); in others the antheridia are absent or do not function. The **ascogonium** is a long, straight and somewhat club-shaped body. It is at first uninucleatc but later it becomes multinucleate by repeated divisions of the nuclei. A slender hypha (antheridial hypha) arises from a separate vegetative hypha and grows twining round the ascogonium up to a certain height. Its swollen terminal cell is the **antheridium**. It is uninucleate, and its tip touches the ascogonium. The contact walls dissolve and the protoplast of the antheridium migrates into the ascogonium, however some scientists do not agree. Commonly, however, the ascogonial nuclei approach each other in pairs (Fig. 37C). After this pairing of the nuclei the ascogonium divides into a number of binucleate cells which produce numerous hyphae—the **ascogenous hyphae**. The pairing nuclei pass into the ascogenous hyphae which now become septate, with a pair of nuclei in each cell. The terminal cell of each such hypha produces a more or less globose ascus. The two nuclei fuse in the young ascus. In the meantinie a closed 'fruiting' body called **cleistothecium** (Fig. 37D) is formed from the surrounding vegetative hyphae, enclosing the asci. It has a protective wall called **peridium** which is pseudo-parenchymatous in nature. Meanwhile the zygote nucleus ($2n$) of the ascus divides thrice, the first division being meiotic, and thus usually 8 ascospores (n) are formed in each ascus. The asci dissolve away soon, leaving the ascospores free in the cleistothecium. After the decay of the peridium the ascospores are blown away by the wind.

6. *ASPERGILLUS* (78 sp.)

Occurrence. ***Aspergillus*** (Fig. 38), commonly called blue mould, is a very widely distributed fungus like *Penicillium,* and its spores are abundant in the air and in the soil. Species of *Aspergillius* commonly grow on almost all kinds of foodstuffs including butter, bread, fruits, vegetables and jams, and on leather goods, fabrics and books, sometimes causing considerable damage to them, particularly during the rainy

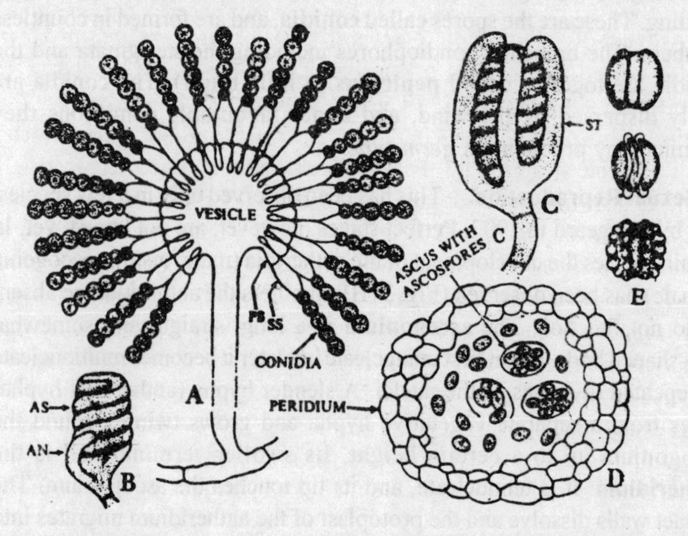

Figure 38. *Aspergillus. A,* conidiophore with primary sterigmata (*PS*), secondary sterigmata (*SS*), conidia, vesicle and foot cell (at the bottom); *B,* sex organs: antheridium (*AN*) and ascogonium (*AS*); *C,* sterile hyphae (*ST*) enclosing the ascogonium; *D,* cleistothecium; *E,* ascospores of different species.

season. Air-borne spores (conidia) are often the sources of contamination of foodstuffs and their decay. *A. niger,* a black mould, is one such spore. *Aspergillus* species are mostly saprophytic in habit, but a few (e.g. *A. fumigatus*) are parasitic on animals including human beings, causing diseases of the ear and the lungs.

Aspergillus can produce a large number of enzymes which enable them to grow on a variety of organic media. *Aspergillus* is economically an important fungus.

Some species (e.g. *A. oryzae*) are used industrially in the manufacture of alcohol from rice starch, and some species used in the manufacture of certain organic acids (e.g. citric, gluconic, etc.) on a commercial basis. Some enzyme preparations have also been made possible through their activity. Some species are sources of certain antibiotics.

Structure. The mycelium consists of an interwoven mass of hyphae which branch freely and spread through the substratum on its surface as well as into it. The hyphae are hyaline, septate, much-branched and multinucleate.

Reproduction. *Aspergillus* freely reproduces asexually by conidia. Only a fews species reproduce sexually.

Asexual Reproduction. Several hyphae stand erect from certain cells (called **foot cells**) of the vegetative hyphae; these are the **conidiophores** (Fig. 38A). They are long and erect but aseptate. Each conidiophore swells at the apex into a more or less spherical multinucleate head called the **vesicle**. Over the entire surface of the vesicle, growing from it, there are innumerable cells in 1 or 2 layers according to the species. The cells of the first layer are short and are called the *primary sterigmata,* while those of the second layer are long and bottle-shaped, and are known as the *secondary sterigmata.* If only one layer is present the cells become bottle-shaped. Each such sterigma produces spherical or oval spores, called **conidia**, in a chain in basipetal order. They are always formed in huge numbers. The fungal colony as a whole may then appear bluish greenish, blackish or brownish in colour according to the species. The mature conidia are blown away by the wind. They germinate in suitable medium by producing a *germ tube.*

Sexual Reproduction. In a few species of *Aspergillus* sexual reproduction has been observed but perfect stages not known. Antheridia (male) and ascogonia (female) may be formed in them for the above purpose (Fig. 38B). They develop from separate specialized vegetative hyphae lying close together, or from the same hypha at different levels. Both antheridium and ascogonium are unicellular but multinucleate. The ascogonial hypha, as it grows, becomes soon differentiated into three parts: a unicellular, tightly coiled structure—the **ascogonium**, a terminal cell (which is the receptive neck of the ascogonium)—the **trichogyne**, and a multicellular **stalk**. The antheridial hypha, growing by the side of the ascogonium, climbs it. Soon it cuts off a terminal cell—the **antheridium**. It further climbs and its tip reaches the trichogyne. The contents of the antheridium migrate through the trichogyne into the ascogonium where pairing of nuclei (male and female) takes place. Their actual fusion has not been observed. Commonly, however, the sister, nuclei of the ascogonium come together in pairs. The antheridium may also remain undeveloped or functionless, or it may not reach the trichogyne. After pairing of the nuclei within the ascogonium, whatever be the method, the latter becomes septate forming a number of binucleate cells. These cells now produce several **ascogencus hyphae** which branch within the 'fruiting' body (see below). They become septate, with binucleate cells, and produce asci at their tips. The **asci** are spherical, oval or pear-shaped, and binucleate. The two nuclei fuse in the young ascus, and the zygote nucleus ($2n$) divides thrice (the first division being meiotic) to form 8 nuclei. Each nucleus (n) clothes itself with a wall and becomes an **ascospore** (n). Thus there are 8 ascospores in each ascus. The ascospores are peculiar in shape being like pulley-wheels (Fig. 38E). In flat view, however, they are commonly spherical, oval or star-shaped.

In the meantime, soon after the formation of the sex organs, a number of sterile vegetative hyphae grow up the ascogonium (Fig. 38C) and enclose it as a sheath called the **peridium** which is pseudo-parenchymatous in nature. The peridium enclosing the asci appears as a small globose body; this is the 'fruiting' body of the fungus and being closed is otherwise. known as the **cleistothecium** (Fig. 38D). The asci within the cleistothecium soon dissolve away, leaving the ascospores free and scattered within it. The peridium decays, and the ascospores are blown away by the wind. They germinate under suitable conditions by producing a *germ tube.*

C. BASIDIOMYCETES OR CLUB-FUNGI

7. *PUCCINIA*

Puccinia, commonly called **rust,** is a destructive parasite attacking a number of economic plants, particularly cereals, and also other plants. Some species are *heteroecious* requiring two distinct host plants to complete their lifc-cycle; while others are *autoecious* requiring only one host plant. *P. graminis* is heteroecious, attacking wheat and barberry in rotation. It causes a serious disease in wheat, called the 'black rust', sometimes resulting in a heavy loss of crop. It also attacks barley, oat and rye.

Stages on Wheat Plant. (a) **Uredium and Uredospores** (Fig. 39). In late spring or early summer aeciospores (Fig. 42) are carried by wind from barberry to wheat plant. The aeciospore germinates on the wheat plant by producing a *germ tube* through a stoma (Fig. 39A). The infection spreads and soon reddish-brown streaks appear on the stem and the leaf (Fig. 39B). A section shows that the hyphae spread through the intercellular spaces, penetrating here and there into the living cells of the host. The hyphae are septate with binucleate cells (+ and –). On the surface spore-clusters called **uredia** (Fig. 39C) appear as reddish-brown streaks. The uredium consists of numerous slender hyphae, each ending in a rough-walled, reddish or brownish, binucleate (+ and –) spore called **uredospore**. This stage is known as the **'red rust'** of wheat. The spores, when mature, are blown about by the wind and they directly infect the

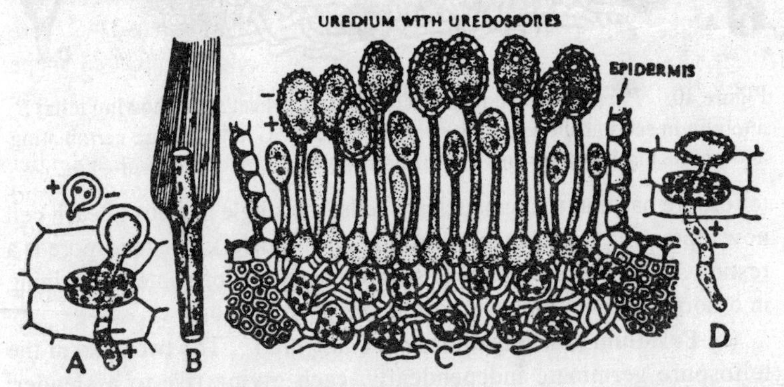

Figure 39. *Puccinia*. Uredium and uredospores. *A*, aeciospore germinating on wheat leaf through a stoma; *B*, wheat leaf with uredia; *C*, a uredium in section showing uredospores (binucleate, + and –); *D*, a uredospore germinating on wheat leaf.

wheat plants in the field. The spore germinate through a stoma by producing a *germ tube* (Fig. 39D). The disease may appear in an epidemic form, causing a heavy economic loss.

(b) **Telium and Teliospores** (Fig. 40). In late summer the mycelia still existing in the wheat plant grow and mass together below the epidermis, producing black spots or streaks on the stem (Fig. 40A) and the leaf. Each such spot or streak is called **telium** (Fig. 40B). The telium consists of a large number of slender hyphae, each ending in a black or dark-brown, elongated, heavy-walled, two-celled spore called the **teliospore**. This stage is called the **'black rust'** of wheat. A young teliospore has two binucleate cells. The two nuclei (one + and the other –) of each cell fuse, indicating a reduced form of sexual act. The mature

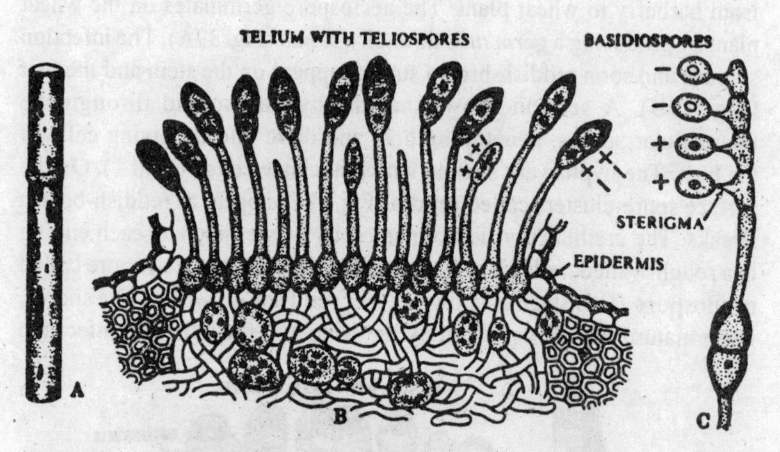

Figure 40. *Puccinia.* Telium and teliospore. *A,* wheat stem showing telia; *B,* a teliuin in section showing teliospores (see text); *C,* a teliospore germinating and producing basidium and four basidiospores (uninucleatc, + or –).

teliospore has thus two uninucleate cells (the single nucleus of each cell now represents the fusion product of + and – nuclei). The teliospore is a resting spore and can tide over severe conditions of winter, remaining in a dormant condition on wheat plants or in the soil.

(c) **Basidium and Basidiospores** (Fig. 40C). The two cells of the teliospore germinate independently, each giving rise to a slender, elongated hypha called the **basidium**. It has four terminal cells, each producing a short, slender stalk called the **sterigma**. The sterigma forms at its tip a spore called the **basidiospore**. The diploid nucleus (+ –) of the teliospore undergoes reduction division so that the basidial cells and the basidiospores become haploid. They are, of course, uninucleate but are of opposite strains (two + and two –). *Puccinia* is thus *heterothallic*. Basidiospores are blown about by the wind.

Stages on Barberry Plant. (d) **Spermogonium and Spermatia** (Fig. 41). The basidiospore (+ or –) germinates on the barberry leaf by producing a *germ tube* which penetrates into the leaf through the cuticle (Fig. 41A). Extensive mycelia formation takes place within a few days, and they mass together beneath the upper epidermis, forming yellowish or reddish raised spots called the **spermogonia** or **pyenia** (+ or –) on the leaf surface (Fig. 41B). As seen in a section (Fig. 41C) each is more or less flask-shaped and its inner wall is lined with numerous fine **paraphyses** at the upper part, projecting outwards through a pore called

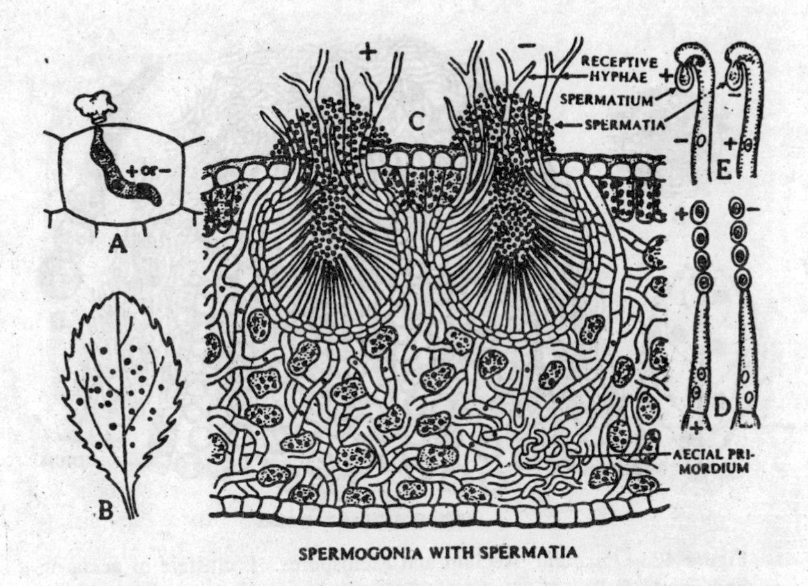

Figure 41. *Puccinia.* Spermogonia and spermatia. *A,* a basidiospore germinating on barberry leaf through the cuticle; *B,* infected leaf showing several spermogonia; *C,* infected leaf in section showing two spermogonia (+ or –); note the spermatia and receptive hyphae; *D,* spermatial hyphae cutting off spermatia (+ or –); *E,* spermatia and receptive hyphae of opposite strains uniting (nuclei, however, do not fuse).

the **ostiole**. The fertile hyphae cut off at their ends numerous uninucleate cells (Fig. 41D) called **spermatia** (+ or –). Besides, some special hyphae called the **receptive hyphae** also project outwards. The spermatia are exuded through the ostiole in a drop of sweet fluid. Now, insects attracted by this fluid visit the spermogonia and carry the spermatia from one spermogonium (say, +) to the rceptive hyphae of another spermogonium (say, –). The contents of the spermatium pass into the receptive hypha but no nuclear fusion takes place (Fig. 41E). The receptive hyphae, now with binucleate cells (one + and one –), grow and extend towards the lower surface of the leaf, forming the next stage (aecium).

(e) **Aecium and Aeciospores** (Fig. 42). The aecia appear on the lower surface of the leaf as cup-like blisters, commonly called the cluster-cups (Fig. 42A–B). To start with, the receptive hyphae with binucleate cells (+ and –) begin to cut off chains of large, orange or yellow, binucleate cells (spores) called the **aeciospores** (+ and –), and small sterile cells

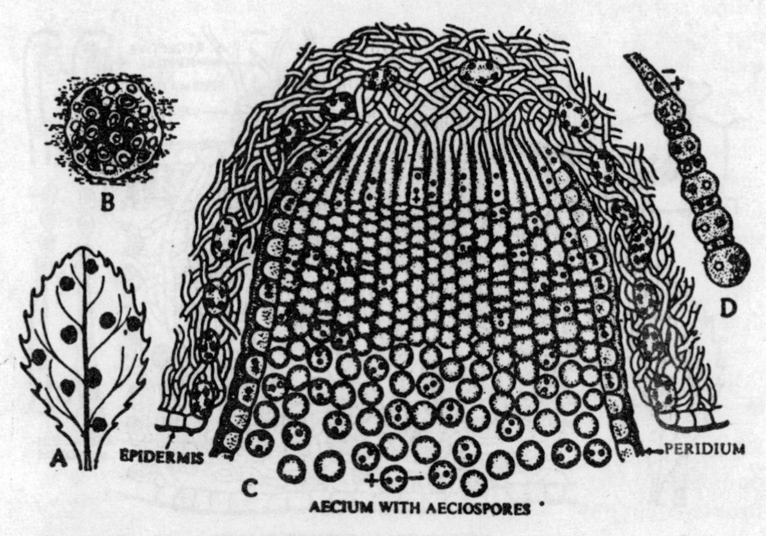

Figure 42. *Puccinia,* Aecium and aeciospores. *A,* clusters of aecia on barberry leaf; *B,* a cluster (magnified); *C,* an aecium in section showing aeciospores (binucleate, + and –) and peridium; *D,* a chain of aeciospores and sterile cells produced in an alternating manner.

(also binucleate, + and –) in an alternating manner (Fig. 42C–D). A protective layer called the **peridium** also develops surrounding the aecium. The peridium bursts and the binucleate aeciospores are blown about by the wind. If they happen to fall upon the wheat plant they infect it (Fig. 39A), and the life-cycle is repeated.

Control. No special method has been discovered yet to prevent or radically cure the disease. Certain methods, however, have been devised to control its intensity. (1) Eradication of barberry bush near a wheat field. (2) Evolving rust-resistant varieties by cross-breeding. (3) Elimination of cultivation of wheat in the hills during summer. (4) Selection of rust-resistant varieties.

8. *AGARICUS* (about 70 sp.)

Occurrence. *Agaricus,* commonly called mushroom, is a fleshy saprophytic fungus. It grows during the rainy season on damp rotten logs of wood, trunks of trees, decaying organic matter, and in damp soil rich in organic substances. *A. campestris* is such a common mushroom growing in open fields.

Edible and Poisonous Forms. There are about 200 species of fleshy fungi that are edible, many more are non-edible, and about 12 species distinctly poisonous. Most puff-balls and many species of *Agaricus* are edible, particularly when they are young. *A. campestris* is widely cultivated for food, particularly in U.S.A. Certain species of *Amanita* which resemble edible *Agaricus* are extremely poisonous however, they are usually distinguished from the latter by their possession of a cup-like structure (called **volva**) at the base, which is absent in *Agaricus*.

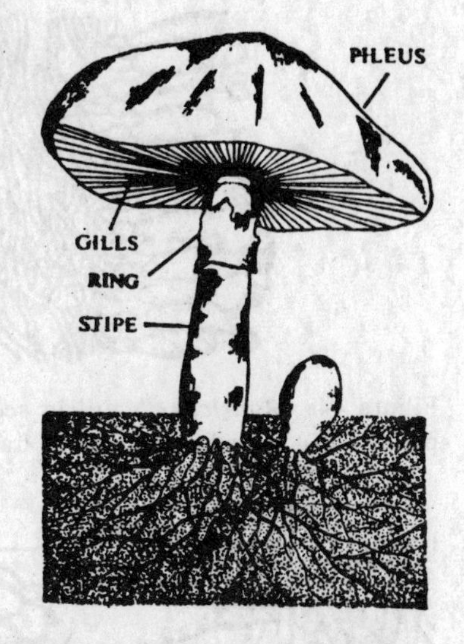

Figure 43. *Agaricus.* Two plants, young and old, with ramifying mycelia.

Structure (Fig. 43). The mycelium consists of a mass of much-branched hyphae which unite (anastomose) at their points of contact and form a network in the substratum. The hyphae are very slender, hyaline and septate, mainly consisting of binucleate cells (+ and –). These are secondary mycelia. At an early stage when a basidiospore (+ or –) germinates it produces a primary mycelium (+ or –) which consists of uninucleate cells. It is short-lived. Soon, however, two such mycelia of opposite strains (+ and –) come into contact and fuse. There is fusion of protoplasts only but not of nuclei. The new hypha thus formed is called the secondary mycelium, and its cells are typically binucleate (one + and one –). The secondary mycelium rapidly spreads in all directions through the substratum, perennates from year to year, and in season produces the main aerial fleshy body which is the **fructification** or fruit body of the fungus, otherwise called the **basidiocarp** (basidia-forming body).

Basidiocarp (Fig. 43). It consists of a fleshy stalk known as the **stipe** and an umbrella-likc head borne on its top, known as the **pileus** (a hat). The whole body of the fungus is composed of an interwoven mass of hyphae, looking in section like a tissue—a false tissue, known as pseudoparenchyma. When young, the fructification is spherical or oval in shape (button stage) and is completely enveloped by a thin membranous covering, called the **veil** or **velum**. With rapid growth of the pileus the velum gets ruptured while a part of it remains attached to

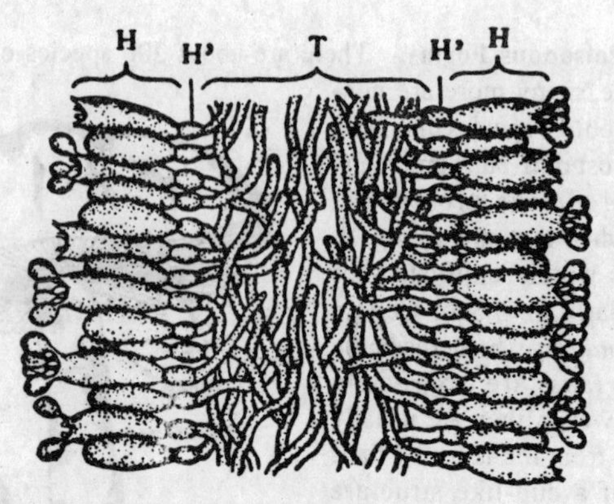

Figure 44. *Agaricus*. A gill in section. *H*, hymenium (basidia and paraphyses); *H'*, sub-hymenium; *T*, trama.

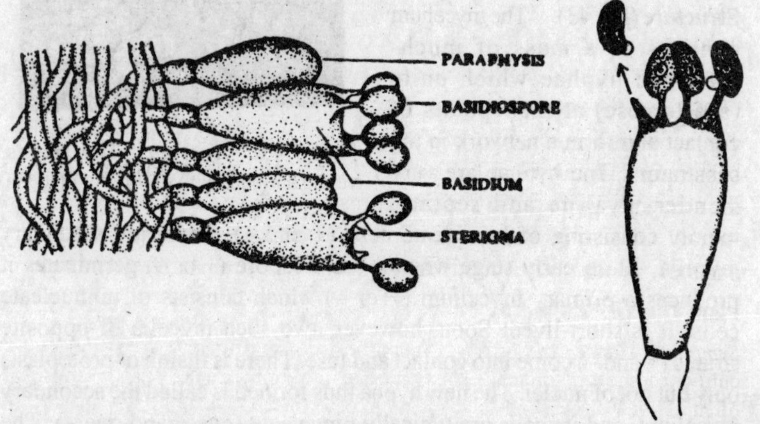

Figure 45. *Agaricus*. *Left*, a portion of the gill in section; *right*, a basidiospore shooting off.

the stipe in the form of a **ring** or **annulus**. Ultimately the pileus spreads in an umbrella-like fashion on the top of the stipe. From the under-surface of the pileus suspend a very large number of thin vertical plate-like structures, extending from the stipe to the margin of the pileus; these are known as the **gills** or **lamellae**. They vary in number from 300 to 600 for each fructification. Each gill bears innumerable spores (basidiospores) on both surfaces.

Gill (Figs. 44–45). A gill in section shows a central portion called **trama** which is an interwoven mass (false tissue) of long slender hyphae. The hyphal cells of the trama curve outwards on either side of the gill

and terminate in a layer of small rounded or oval cells—the **sub-hymenium**, and a layer of club-shaped cells—the **hymenium**. Many of the cells of the hymenium bear spores and are called **basidia;** while others are sterile and are called **paraphyses**. Each basidium bears four **basidiospores** (2 + and 2 –), each on a short slender stalk known as the **sterigma** (pl. sterigmata). In the cultivated mushroom only two basidiospores develop. The basidiospores, when mature, shoot off from the sterigma with a drop of water (Fig. 45 *right*). Later they germinate under favourable conditions, and the life-cyde is repeated.
Reproduction. There is no regular *asexual phase* in the life-cycle of *Agaricus*. Sometimes, however, it reproduces by a kind of 'resting' spores called **chlamydospores** which are certain enlarged thick-walled vegetative cells of a hypha. They germinate by producing a germ tube. *Sexual reproduction* takes place by the fusion of two primary mycelia (+ and –), as already described. Besides, a very short but regular *sexual phase* is represented by the complete fusion of two haploid nuclei of opposite strains (+ and –) in the young basidium. The diploid zygote thus formed (evidently + –) divides by meiosis to form 4 haploid nuclei (2 + and 2 –) in the basidium. The nuclei migrate singly, each into a basidiospore through the sterigma. Evidently 2 basidiospores are of + strain and the other 2 of – strain. On germination the + basidiospore produces + primary mycelium, and the – basidiospore produces – primary mycelium, as described before.

Plant Diseases caused by Fungi

Symptoms, Causes and Effect. Many parasitic fungi attack several field crops, cultivated and ornamental plants and even wild ones, and cause various and often serious diseases in them. Symptoms of diseases may appear in various forms such as leaf spots, curls and discolouration, rots, blights, rusts, smuts, mildews, wilts, hypertrophies, etc. The fungi plunder the food stored in the host plants, block the conducting tissues, destroy the affected cells and tissues, produce toxins (poisons) and finally cause their death. The annual loss in agricultural crops on this account alone is very heavy. A plant may suffer from more than one disease at a time. Some such diseases are as follows.

A. **Late Blight of Potato.** This is a serious disease, particularly common in the hills, and is caused by *Phytophthora infestans* (class Phycomycetes). In this disease black patches appear at first on the undersurface of the leaves and soon spread to the entire leaves and to all parts of the plant, including the tubers. If the weather is warm and humid, and the soil water-logged, the disease rapidly spreads to the neighbouring plants through the *multinucleate sporangia* of the fungus. The fungus finally causes 'wilting' of leaves and 'rotting' of tubers.

Control. (a) Spraying the young plants with Bordeaux mixture (a mixture of lime and copper sulphate). (b) Selection of seed tubers from non-infected areas. (c) Storage of seed tubers at a low temperature –4 to –5° C.

B. **Rusts.** Wheat suffers from a variety of rust diseases caused by species of *Puccinia* (class Basidiomycetes) such as black or stem rust by *P. graminis,* yellow rust by *P. glumarum,* and brown rust by *P. triticina.* All these rusts are common throughout India. *Puccinia graminis,* however, is a virulent type of parasite, and often causes a serious disease of the wheat plants both in the hills and the plains. The disease at first appears as reddish-brown spots and streaks on the stem, leaf-sheath and leaf due to the formation of *uredia with uredospores;* this is the 'red rust' stage of the disease. The uredospores, when mature, are blown about by the wind, and they may directly infect other healthy wheat plants. The disease may thus break out in an epidemic form, causing considerable damage to the crop. Later, dark spots appear on the stem, leaf-sheath and leaf due to the formation of *telia with teliospores;* this is the black or stem rust of wheat as it is the stem that is most severely affected. The final effect is the weakening of plants, reduction of grains in size and number, and their shrivelling up. The teliospores, however, do not infect wheat plants again. They produce basidiospores (+ and –) which attack barberry plants, their next hosts.

Control. (a) Cultivation of rust-resistant varieties. (b) Evolving rust-resistant varieties by cross-breeding. (c) Eradication of barberry bushes near a wheat field. (d) Elimination of cultivation of wheat in the hills during **summer**.

C. **Smuts.** These are sooty diseases caused by species of *Ustilago* (class Basidiomycetes). They are common and serious diseases of wheat, barley, maize, oats and sugarcane. The loose smut of wheat is caused by *Ustilago tritici.* The disease is widespread in the wheat-growing areas of India, as elsewhere. The fungus mainly attacks the flowers and often the whole inflorescence, and also the stem. A black sooty mass appears on the infected parts, particularly the 'ears', and all the grains are often totally destroyed, thus causing a heavy loss to the cultivators. It is seen that the entire spikelets become covered with a black powdery mass of spores (teleutospores), replacing the floral parts and the glumes. When the spores are blown away by the wind or washed away by the rain they spread over wide areas and infect flowers of healthy plants. The spores germinate on the stigma and the mycelium penetrates into the ovary. The mycelium remains dormant in the seed (grain), and as it germinates the mycelium ratifies through the intercellular spaces of the young plants and finally passes on to the flowers. In this way the disease is carried over to the next generation. The fungus evidently is seed-borne.

Control. (a) Varieties of wheat immune and resistant to smut should be cultivated. (b) Cross-breeding with types immune and resistant to smut is possibly the best method. (c) Hot water treatment of wheat grains and then drying them under strong sun may reduce the intensity of the disease..

D. **Mildews.** These diseases appear as whitish, yellowish or brownish spots on the leaves and also on other parts. There are two kinds of mildews: downy and powdery. The former are mainly caused by *Cystopus* (class Phycomycetes) and the latter mainly by *Erysiphe* (class Ascomycetes). (a) White rusts of crucifers (e.g. mustard, radish, cabbage, cauliflower, etc.) are caused by a downy mildew called *Cystopus condidus*. As a result of infection white or yellow blisters of variable shapes and sizes appear on the leaves (mainly), branches and even inflorescences. The fungus is endophytic, and if the attack be heavy the affected parts turn brown and dry up, soft parts disintegrate, and the flowers become deformed. The disease, however, appears in a mild form in India, and, therefore, no control measures are taken. (b) Powdery mildew of cereals (e.g. barley, oats, rye and wheat) and also several grasses is caused by *Erysiphe graminis*. The disease is common but not serious. The fungus is ectophytic; the mycelia and conidia form only superficial growths on the upper surface of leaves, stem and sometimes flowers, having a sort of powdery appearance which is white at first but turns reddish afterwards. The effect is that the plants become stunted in growth, and the leaves are shed or they become deformed and twisted. Since much damage is not done to any crop no control measures are taken.

It may also be noted that many moulds damage vegetables, fruits and food, particularly in storage; similarly they also damage fabrics, paper, books, leather, shoes, etc.

Control. *Prevention, Check and Cure.* Considering the heavy economic loss due to diseases the following methods have been devised to prevent them, to destroy the causative fungi and to keep them under check. (1) Spraying or dusting the affected parts with certain poisonous chemicals called fungicides, e.g. copper sulphate, sulphur, sulphur-lime, quick-lime, etc., or a mixture of them. (2) Fumigation (exposure to fumes) with sulphur dioxide gas. (3) Seed treatment—cautious application of hot water, formaldehyde or certain compounds of copper, sulphur or mercury, (4) Soil sterilization by burning wood or straw in the field or by application of steam or some poisonous chemicals. (5) Selection of disease-free seeds and plants. (6) Eradication and destruction of diseased plants. (7) Destruction of disease-carrying insects. (8) Breeding of disease-resistant varieties of plants. (9) Rotation of crops—growing some other crop in place of the existing one for one or more years.

Antibiotics. Antibiotics (*anti,* against; *bios,* life) are toxic chemical substances, possibly enzymes, secreted by several soil bacteria, mostly species of *Streptomyces,* and certain moulds, e.g. *Penicillium* and *Aspergillus,* which have a destructive effect on particular disease germs invading the human body and causing infectious diseases, often of a serious nature, e.g. pneumonia, typhoid, diphtheria, tuberculosis, cholera, erysipelas, etc. Antibiotics are the miracle drugs of modern times. They often act like magic bullets shooting down the germs which have invaded the human body. Within the last 30 years or so some 300 antibiotics have been discovered. Of these about 13 have an established therapeutic value. The first antibiotic was a 'chance' discovery. It is **penicillin**, discovered by late Sir Alexander Fleming, a bacteriologist, in 1928 from a blue-green mould of the soil, called *Penicillium notatum*. It has a powerful antibacterial action and is amazingly effective against a wide range of germ diseases like scarlet fever, rheumatic fever, sore throat, wound infections, erysipelas, abscesses, carbuncles, tonsilitis, tetanus, pneumonia, meningitis, etc. It came into general use from 1943–4 when mass production was well under way. Other antibiotics isolated from certain soil bacteria, particularly species of *Streptomyces*, came in fairly quick succession . Thus **streptomycin** was discovered by Waksman in 1944; it has proved to be very valuable against tuberculosis. Vigorous search for more antibiotics went on at this time at an almost incredible cost, and several thousands of soil samples were examined in this connexion. Soon another antibiotic called **chloromycetin** was discovered in 1947; it has proved to be a magic drug in the treatment of typhoid fever. Within the following few years some more antibiotics, **aureomycin**, **terramycin**, etc., have been discovered and put on the market for the treatment of one kind of bacterial disease or another. These wonder drugs have saved millions of human lives from death or from untold miseries, and that too within the shortest time possible. It is really a miracle that such drugs lay hidden in a spoonful of good earth for the relief of human beings.

Chapter 5

Lichens

General Description. ***Lichens*** numbering over 15,000 species form a large peculiar and interesting group of plants, being associations of specific fungi and algae, the former constituting the larger part of the lichen body. The associations of different fungi and algae give rise to distinct species of lichens. Commonly they occur as greyish-green or greenish-white incrustations of different dimensions on old walls, rocks, stems and branches of shrubs and trees, logs of wood and on ground. They have a variety of forms and colours: white, greyish-green, yellow, orange, brown, red and black. Many of them grow under extreme conditions of humidity and temperature, and may survive long periods of desiccation. They are very widely distributed over the earth, being specially common in tropical rain-forests as well as in cold, even very

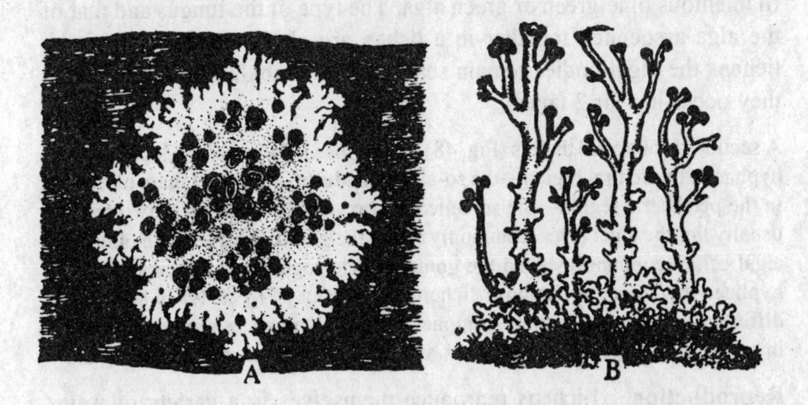

Figure 46. Lichens. *A*, a foliose lichen (*Parmelia*); *B*, a fruticose lichen (*Cladonia*).

cold regions. In lichens the fungi and the algae lead a symbiotic life, being of mutual help to each other. The fungi absorb water and mineral salts from the substratum, while the algae in their turn prepare necessary food for both. Lichens are thus very good examples of **symbiosis**. If separated from their associations they lead a precarious life, more particularly the fungi.

Classification. Depending on the nature of the fungi, the lichens have been classified into two groups: (1) **Ascolichens**, in which the fungi are members of Ascomycetes, reproducing by ascospores, and (2) **Basidiolichens**, in which the fungi are members of Basidiomycetes, reproducing by basidiospores; they are few in number.

Thallus. Lichen thalli take three different patterns of growth in different genera as follows: (a) **crustose lichens** forming hard granular crusts very tenaciously adhering to rocks and barks of trees, e.g. *Graphis* and *Lecanora;* (b) **foliose lichens** (Fig. 46A) forming flattened leaf-like thalli with lobed margin; they adhere to rocks, old walls, tree trunks and ground by delicate rhizoids (**rhizines**), e.g. *Parmelia* and *Physcia;* and (c) **fruticose lichens** forming much-branched shrub-like bodies and remaining attached to the substratum by their narrow basal portion only; such lichens may be standing erect, e.g. reindeer moss (*Cladonia;* Fig. 46B) or drooping from branches of trees, e.g. old man's beard (*Usnea;* Fig. 47A). The main framework of the thallus is made of an interwoven mass of hyphae of a fungus, commonly an ascomycete or in a few cases a basidiomycete, enclosing a certain unicellular or filamentous blue-green or green alga. The type of the fungus and that of the alga associated together in a lichen are always constant. In some lichens the algal bodies remain scattered in the thallus, while in others they occur in 1 or 2 layers.

A section through the thallus (Fig. 48) of a foliose lichen shows a loose mass of hyphae in the central region—the so-called **medulla**, a compact mass of hyphae in the peripheral region—the so-called **cortex**, and between these two regions usually lies the algal layer (commonly called the **gonidial layer**) with numerous algal cells. (commonly called the **gonidia**) held together in the meshes of the hyphae. In *Usnea,* a fruticose lichen, the thallus (Fig. 47B) is seen to be differentiated into a central compact core of hyphae, a region of loosely interwoven hyphae, an algal region, and externally another compact region,

Reproduction. Lichens reproduce themselves in a variety of ways: (A) vegetative, (B) asexual and (C) sexual. But it must be noted that reproduction is predominantly fungal in character.

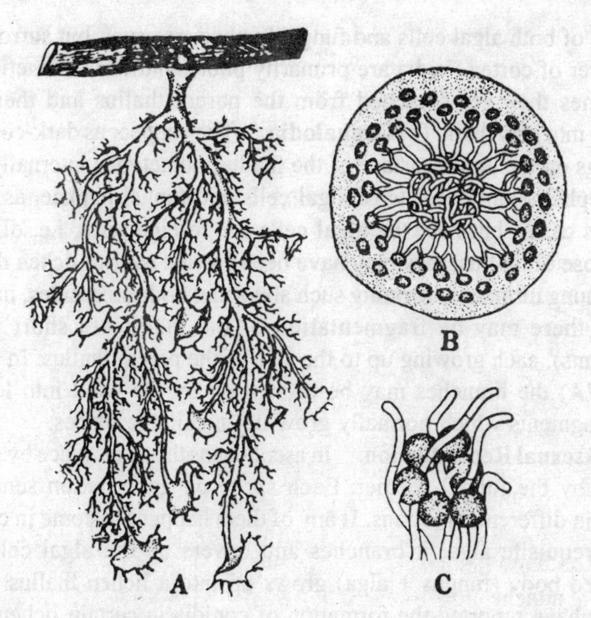

Figure 47. *A,* a fruticose lichen (*Usnea*); *B,* a section through the thallus of *Usnea; C,* a soredium.

(A) **Vegetative Reproduction.** This may take place by any of the following structures peculiar to lichens only. (a) **Soredia** (Fig. 47C). They are tiny granular bodies occurring in large numbers on the upper surface of the thallus as a greyish coating of powder. Each soredium consists of one to many algal cells wrapped up by fungal hyphae, as in *Usnea* and *Cladonia*. The soredia are blown about by the wind and they germinate directly into lichen thalli or sometimes they form new soredia. This is the commonest method of reproduction in lichens. (b) **Isidia.** In many lichens they are formed as minute outgrowths on the surface of the thallus. Each isidium

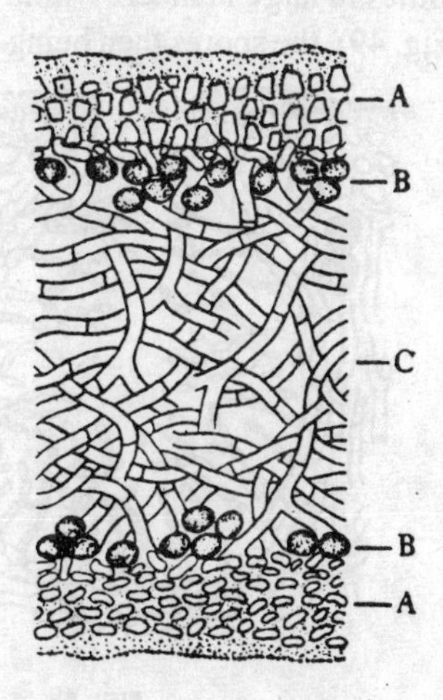

Figure 48. A section through the thallus of a foliose lichen. *A,* cortex; *B,* gonidial layer; and *C,* medulla.

consists of both algal cells and fungal hyphae, as usual, but surrounded by a layer of cortex. Isidia are primarily photosynthetic in function but sometimes they get detached from the parent thallus and then they develop into new thalli. (c) **Cephalodia.** They appear as dark-coloured swellings on the upper surface of the thallus, sometimes internally also. Each cephalodium consists of algal cells and fungal hyphae, as in the previous cases, but here the algal cells are foreign ones, i.e. different from those of thallus; they may have been carried over to lichen thallus, when young ultimately forming such abnormal bodies. Besides, in many lichens there may be fragmentation of the thalli into short pieces (fragments), each growing up to the size of the parent thallus. In *Usnea* (Fig. 47A) the branches may be broken up by the wind into long or short fragments which normally grow up in suitable places.

(B) **Asexual Reproduetion.** In ascolichens this takes place by **spores** formed by the fungal partner. Each spore on germination sends out hyphae in different directions. If any of them happens to come in contact with a requisite alga, it branches and covers up the algal cell. The combined body (fungus + alga) grows up into a lichen thallus. Some workers have reported the formation of conidia in certain lichens. But this is disputed. Many lichens, e.g. *Physcia,* produce small spore-like bodies in large numbers within a flask-shaped cavity, called **pyenidium** (Fig. 49), the spores then being called **pyenidiospores** (or pyenospores).

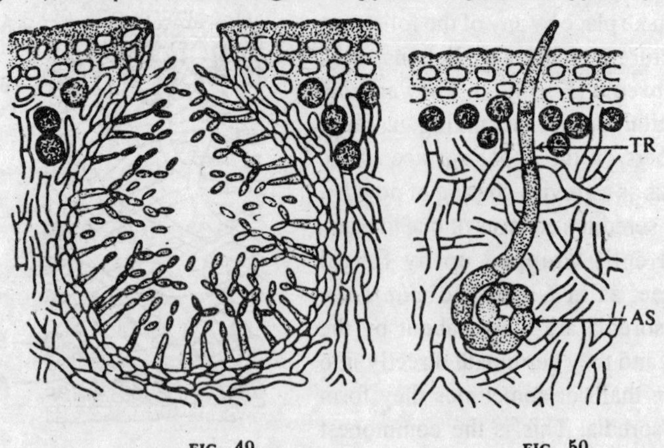

Figure 49. Lichen. A pyenidium (or spermogonium) of *Physcia* in longi-section showing pyenidiospores (or spermatia) formed in large numbers. Figure 50. A coiled ascogonium (*AS*) embedded in the thallus, with tube-like trichogyne (*TR*) protruding outwards.

These spores are known to germinate in certain species, producing a hypha. Coming in contact with an appropriate alga the combined body grows and forms into a lichen thallus. It may be noted that in certain other species, peculiarly enough, the pycnidia behave as male organs, then called **spermogonia**, and the pyenidiospores behave as male cells, then called **spermatia**. Basidiolichens, a few in number, e.g. *Cora* (an American genus), reproduce by basidiospores, very much like *Agaricus*.

(C) **Sexual Reproduction.** This takes place in certain asco-lichens, the fungus alone taking part in the process. Sexual reproduction results in the formation of a 'fruiting body', commonly a cup-shaped apothecium (Fig. 51A) or in some cases a flask-shaped perithecium, with numerous asci in it. Sex organs are differentiated into male and female, and they

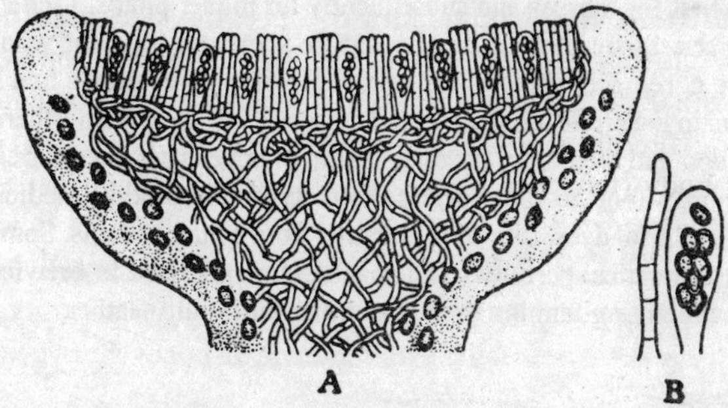

Figure 51. Lichen. *A*, a section through an apothecium; note the asci and the paraphyses; *B*, an ascus and a paraphysis.

develop in close proximity for facility of fertilization. The female organ is a multicellular stout filament of large cells and is known as the **carpogonium** (Fig. 50). It consists of a coiled basal portion, called the **ascogonium**, lying within the thallus, and a tube-like upper portion, called the **trichogyne**, usually protruding beyond the thallus. The male organ is a flask-shaped chamber (Fig. 49) known as the **spermogonium** (pyenidium), and the minute non-motile male cells formed within it known as the **spermatia** (cf. pyenidiospores). Several spermogonia (or pyenidia) may be formed in the thallus, and each is lined with a layer of short slender hyphae. Spermatia are formed from them by abstriction like condia. They occur in large numbers, and are liberated through the ostiole (apical opening) in slimy masses to float on the thallus.

Fertilization. This takes place when a spermatium comes in contact with the sticky protruding tip of a trichogyne. Its protoplast migrates into the trichogyne and apparently fuses with the ascogonium nucleus (though not actually observed yet). After fertilization several ascogenous hyphae begin to grow out from the basal portion of the ascogonium; they branch freely and are multicellular. The ascogenous hyphae may also develop parthenogenetically in some cases. Asci always develop at their ends. A mature ascus has 8 ascospores in it (Fig. 51B). A 'fruiting body'—apothecium (Fig. 51A) or perithecium—simultaneously develops with an inner palisade-like layer of paraphyses. The asci grow upwards into the 'fruiting body' in a palisade-like layer mixed up with the paraphyses. On liberation the ascospores germinate by producing hyphae, and those coming in contact with the right type of alga further grow and eventually produce lichen thalli.

Uses. Lichens growing on rocks disintegrate them to form soil, thus preparing the ground first for mosses and subsequently for higher plants. Lichens have a variety of uses. Some of them are a valuable source of food for wild animals and cattle, e.g. reindeer moss (*Cladonia;* Fig. 46B) which grows to a height of about 30 cm. in the arctic tundra; Iceland moss (*Cetraria*) in the northern regions is used as food and medicine. In some countries lichens are fried for cattle feed and even human food to some extent. Some types are used as medicines, and some yield beautiful dyes. Litmus is prepared from certain lichens. Some species are used in cosmetics, perfumes and soaps. Some are used in brewing liquor, and others containing tannins used in tanning hides into leather.

Chapter 6

Bryophyta

Classification of Bryophyta (23,725 sp.)

Class I Hepaticae or thalloid liverworts (8,450 sp.), e.g. *Riccia, Marchantia,* etc.

Class II Anthocerotae or horned livetworts (300 sp.), e.g. *Anthoceros.*

Class III Musci or mosses (14,975 sp.), e.g. *Funaria, Polytrichum, Barbula,* etc.

1. *RICCIA* (135 sp.)

Riccia (Fig. 52) is a rosette type of thalloid liverwort showing distinct dichotomous branching. The thallus is small and flat with a longitudinal groove on the upper surface along the mid-rib, and a number of slender unicellular hair-like structures called **rhizoids**, on the lower surface, serving as roots. Some scales may also be present. The plant grows during the rainy season as a green carpet on wet ground, old dam walls, old tree trunks and moist rocks, and dries up in winter.

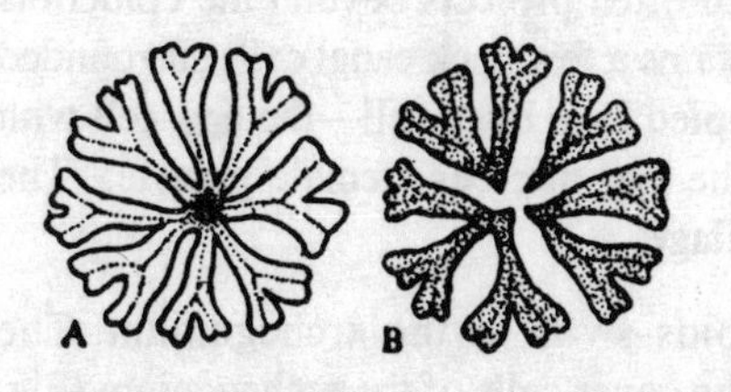

Figure 52. *A,* a *Riccia* plant; *B,* vegetative propagation.

Vegetative Reproduction. It may take place by the decay of the older portion of the thallus and the separation of the branches (Fig. 52B).

Gametophyte and Sexual Reproduction. *Riccia* plant is the **gametophyte**, i.e. it reproduces sexually by gametes. The two kinds of gametes—male and female–are borne in special structures known as the antheridia and the archegonia respectively (Fig. 53). Some species are *monoecious* and others *dioecious*. In the monoecious species antheridia and archegonia develop together in the median groove on the upper side of the thallus. Each **antheridium** (Fig. 53A) is more or less pear-shaped and consists

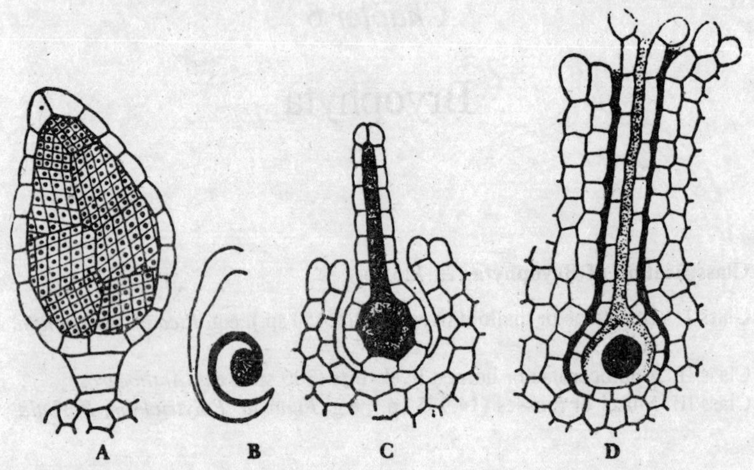

Figure 53. Riccia. *A,* an antheridium; *B,* an antherozoid; *C,* a young archegonium; and *D,* a mature archegonium.

of a short stalk, a wall and a compact mass of antherozoid mother cells. Each mother cell by a single division forms two cells, each of which becomes converted into a small twisted biciliate male gamete or **antherozoid** (Fig. 53B). Each **archegonium** (Fig. 53C–D) also lies sunken in the groove. It is a short-stalked, flask-shaped body with a swollen basal portion known as the **venter** and a narrow tubular upper portion known as the **neck** which often projects beyond the epidermis and turns purplish. The neck contains a few neck canal cells surrounded by a wall, and the venter is occupied by a large cell—the **egg-cell** with a distinct large nucleus in it—the **egg-nucleus** (female gamete). The canal cells degenerate into mucilage.

Fertilization. The antherozoids swim to the archegonium. The mucilage swells and forces out the cover cells of the archegonium (Fig. 53D). An open passage is thus formed, and the antherozoids enter through it into the archegonium. They pass down into the venter and one of them fuses with the egg-nucleus. After fertilization the ovum clothes itself with a wall and becomes the **oospore**.

Sporophyte and Asexual Reproduction. The oospore gives rise to the **sporophyte** which reproduces asexually by spores. The sporophyte is a simple spherical body called the **capsule** (Fig. 54). It consists of a spore-sac and a wall surrounding it, the latter made of a single layer. The capsule develops *in situ* within the venter of the archegonium. With

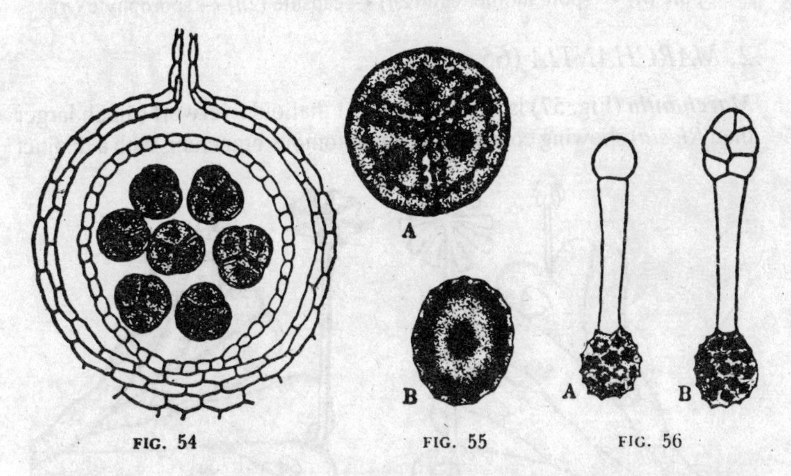

Figure 54. *Riccia.* Sporophyte (capsule) with spore tetrads within enlarged archegonium. Figure 55. Spores. *A,* spores in a tetrad; *B,* a single spore. Figure 56. *A–B,* early stages in the germination of spore.

the growth of the capsule the venter also grows and covers the capsule; this covering is called the **calyptra**. The spore-sac contains a loose mass of spore mother cells. Each mother cell undergoes reduction division and forms a *tetrad of spores* (Figs. 54 & 55A). Eventually by the rupture of the calyptra and the wall of the capsule the spores are set free. Each spore (Fig. 55B) is provided with a thick coat. The spore germinates at first into a short tube called the *germ tube* (Fig. 56) which gradually develops into *Riccia* thallus.

Alternation of Generations. The plant passes through two successive generations—gametophyte (haploid or *n*) and sporophyte (diploid or 2*n*)—to complete its life-history. The gametophyte begins with the spore and ends in the formation of the gametes—all haploid; while the sporophyte begins with the oospore and ends in the spore mother cells—all diploid. The gametophyte gives rise to the sporophyte through sexual reproduction, and the sporophyte to the gametophyte through asexual reproduction. Thus there is a regular alternation of generations in *Riccia*. Life-cycle showing alternation of generations is given below:

Riccia (gametophyte-n) → archegonium (n) → ovum (n) → oospore ($2n$)
↑ ↳ antheridium (n) → antherozoids (n) ↓
spore (n) ← spore mother cell ($2n$) ← capsule ($2n$) ← sporophyte (n)

2. *MARCHANTIA* (65 sp.)

Marchantia (Fig. 57) is a rosette type of thalloid liverwort (much larger than *Riccia*) showing conspicuous dichotomous branching with a distinct

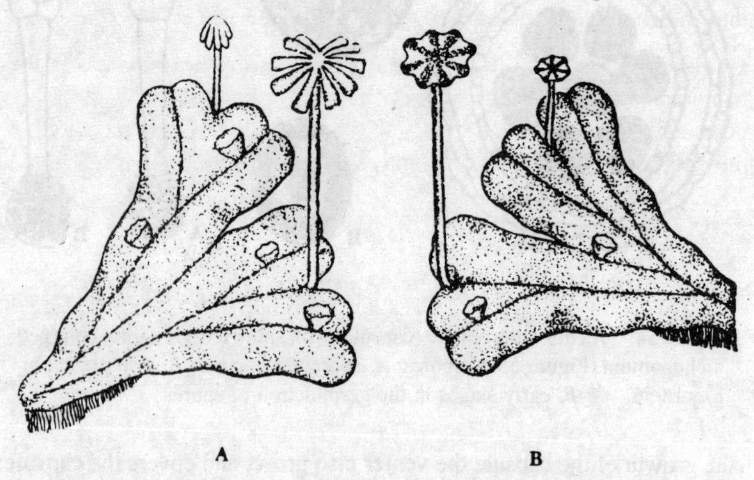

Figure 57. *Marchantia. A,* a female plant with archegoniophores and gemma cups; *B,* a male plant with antheridiophores and gemma-cups.

mid-rib. It grows on damp ground and old walls and spreads rapidly during the rainy season, forming a sort of green carpet. It grows luxuriantly in the cold climate of the hills. The plant dries up in winter. The thallus bears on its undersurface a number of unicellular **rhizoids** (hair-like structures functioning as roots), and also rows of scales. On the upper surface it bears a number of cup-like outgrowths, known as the gemma-cups, on the mid-rib (Fig. 57). *Marchantia* is dioecious. The male plant bears some special erect *male* reproductive branches (antheridiophores), each with a more or less circular disc or **receptacle** on the top (Fig. 57B). Similarly, the female plant bears special *female* branches (archegoniophores) with a star-shaped disc or **receptacle** with radiating rays or arms (Fig. 57A). The growing point of the thallus lies in its groove.

Vegetative Reproduction. It may take place (a) by the decay of the old basal portion of the thallus, thus separating a branch, or (b) by **gemmae** which develop in the **gemma-cup** (Fig. 58). Each gemma is a small, more or less circular, flattened structure with a conspicuous depression on each side. When the gemmae get detached from the gemma-cup, each grows out into a dichotomously branched green thallus.

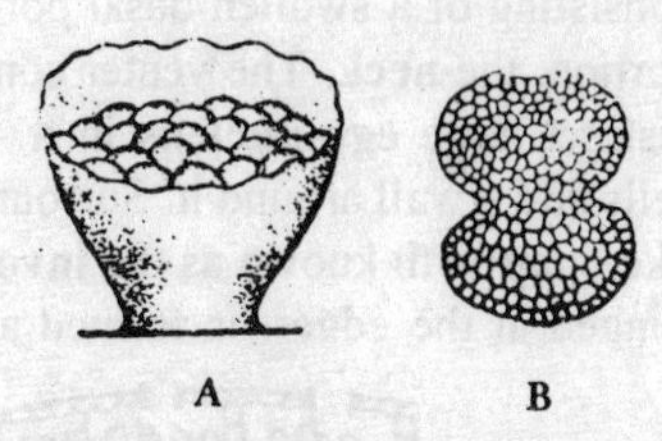

Figure 58. *Marchantia. A,* a gemma-cup with gemmae; *B,* a gemma.

Gametophyte and Sexual Reproduction. *Marchantia* plant is the gametophyte, i.e. it reproduces sexually by gametes. The male plant bears antheridia (or male organs) on the upper side of the receptacle of the antheridiophore (Fig. 59); and the female plant bears archegonia

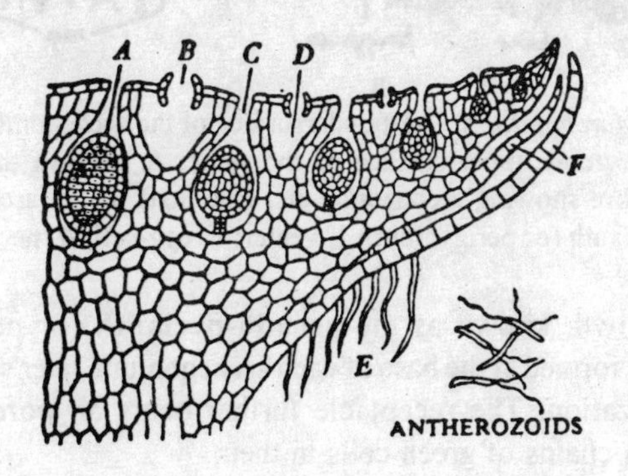

Figure 59. *Marchantia:* Section through the antheridiophore. *A,* antheridium; *B,* air-pore; *C,* ostiole; *D,* air-chamber; *E,* hairs, *F,* scales. Some antherozoids on the right.

(or female organs) on the lower side of the receptable of the archegoniophore (Fig. 60). The **antheridium** (Fig. 59A) is an ovoid body composed of a mass of antherozoid mother cells and surrounded by a wall. Each mother cell develops a spindle-shaped biciliate male gamete called the **antherozoid** or spermatozoid. The antherozoids escape through a narrow canal known as the **ostiole** (Fig. 59C). Besides, the receptacle has a number of air-pores (Fig. 59B) and air-chambers (Fig. 59D). The **archegonium** (Fig. 60B–C) is a flask-shaped body consisting of a swollen basal portion, the **venter** and a narrow tubular portion, the **neck**. The venter contains a large cell, the **egg-cell**, with a distinct large **egg-nucleus** in it. The neck contains a few neck canal cells and a wall around it. Surrounding a group of archegonia a curtain-like outgrowth known as the **involucre** (or perichaetium; Fig. 60A–B), fringed at the edges, is formed as a protective covering. Also, a cup-

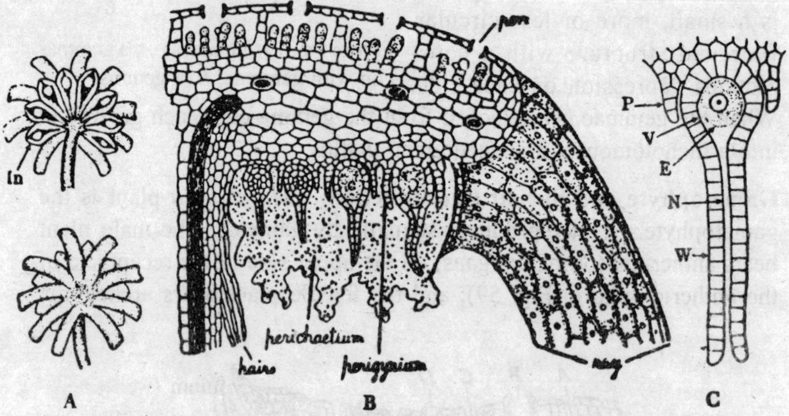

Figure 60. *Marchantia. A,* (*top*) undersurface of the archegoniophore; *In,* involucre; (*bottom*) upper surface of the same; *B,* section through the archegoniophore showing archegonia, etc, (see text); *C,* an archegonium; *P,* pseudo-perianth (or perigynium); *V,* venter; *E,* egg-cell; *N,* neck; *W,* wall.

shaped outgrowth known as the **pseudo-perianth** (or perigynium; Fig. 60B–C) is formed at the base of each archegonium, later surrounding it after fertilization. The receptacle further bears air-pores and air-chambers with chains of green cells in them.

Fertilization. After the antheridium bursts, the ciliate antherozoids escape through the ostiole and swim to the archegonium through the medium of dew or rain-water. Many of them enter into the venter through the neck. But only one of them fuses with the egg-nucleus. Fertilization is thus effected. After fertilization the egg-cell or ovum develops a wall round itself and becomes the **oospore**.

Sporophyte and Asexual Reproduction (Figs. 61–2). The oospore germinates *in situ* and gives rise to the sporophyte which reproduces asexually by spores. The sporophyte is a complex body and is known as the **sporogonium**. It consists of a **foot**, a short stalk called **seta**, and a **capsule**. The capsule consists of a single-layered wall, and a mass of

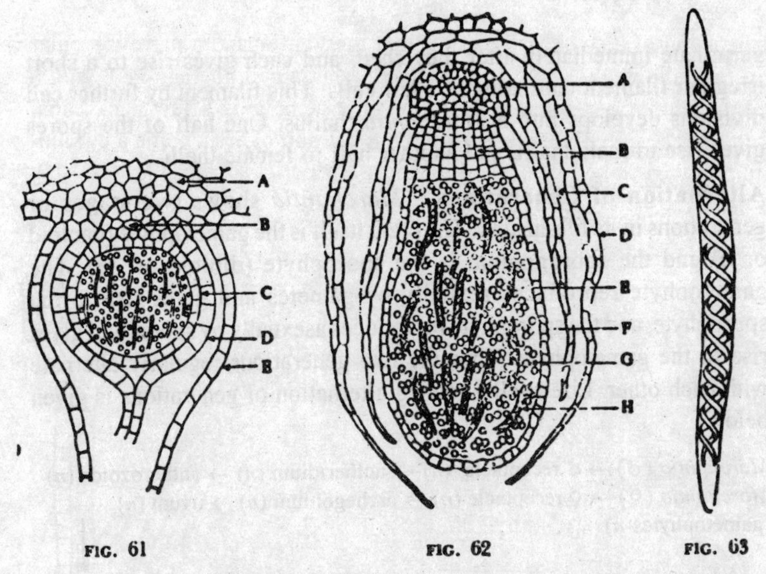

Figure 61. *Marchantia.* A young sporogonium; *A,* tissue of the gametophyte; *B,* foot; *C,* capsule (wall); *D,* archegonium (wall); and *E,* perigynium or pseudo-perianth. Figure 62. A mature sporogonium; *A,* foot; *B,* seta; *C,* remnant of venter (calyptra); *D,* perigynium or pseudo-perianth; *E,* capsule; *F,* wall of the capsule; *G,* spore; and *H,* elater. Figure 63. An elater (enlarged).

small cells. Some of these cells grow up into elongated, spindle-shaped, spirally thickened structures called **elaters** (Figs. 62–3), while others form **spore mother cells**. Each spore mother cell undergoes reduction division and forms *four* **spores** in a tetrad. Other parts of the archegonium also grow. Thus the wall of the venter grows and forms the **calyptra** which surrounds the capsule (Figs 62C); the neck withers and disappears. The **perigynium** (Figs. 61D and 62E) grows rapidly and ultimately surrounds the sporogonium. Finally the capsule dehisces rather irregularly, and the spores are discharged (Fig. 64). Under humid conditions the elaters undergo a twisting movement and push the spores out of the capsule. The spores

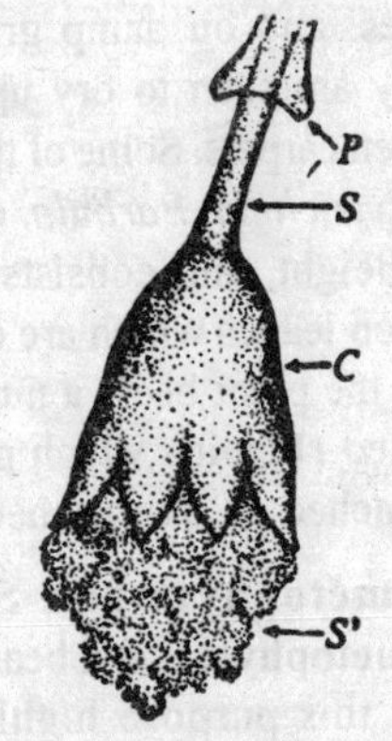

Figure 64. *Marchantia.* Sporogonium dehiscing and discharging spores; *P,* perigynium; *S,* seta; *C,* capsule; and S', spores.

germinate immediately after they shed, and each gives rise to a short irregular filament consisting of a few cells. This filament by further cell divisions develops into a *Marchantia* thallus. One half of the spores gives rise to male thalli, and another half to female thalli.

Alternation of Generations. *Marchantia* shows two stages or generations in its life-history. The plant itself is the gametophyte (haploid or n) and the sporogonium is the sporophyte (diploid or $2n$). The gametophyte reproduces sexually by gametes and gives rise to the sporophyte, and the sporophyte reproduces asexually by spores and gives rise to the gametophyte. Thus the two generations regularly alternate with each other. Life-cycle showing alternation of generations is given below.

Marchantia (♂) → ♂ receptacle (n) → antheridium (n) → antherozoids (n)
Marchantia (♀) → ♀ receptacle (n) → archegonium (n) → ovum (n) –
(gametophytes-n)
↑ ↑ ×↓
spores (n) ← spore mother cells ($2n$) ← capsule ($2n$) ← sporogonium (sporophyte-$2n$) ← oospore ($2n$)

3. *MOSS*

Mosses (Fig. 65) occur most commonly on old damp walls, trunks of trees, and on damp ground during the rainy season, while in winter they are seen to dry up. They form green patches or soft velvet-like, green carpets. Some of the common Indian mosses are species of *Funaria, Polytrichum, Barbula,* etc. A moss plant is small, usually 2–3, cm or so in height, and consists of a short axis with spirally arranged minute green leaves which are crowded towards the apex. True roots are absent but the plant bears a number of slender multicellular branching threads called **rhizoids** which perform the functions of roots. The axis may be branched or unbranched.

Gametophyte and Sexual Reproduction. Moss plant is the **gametophyte**, i.e. it bears gametes and reproduces by the sexual method. For this purpose highly differentiated male and female organs are developed, either together at the apex of the same shoot, or separately on two, often intermixed with some multicellular hair-like structures called **paraphyses**. The male organ is known as the **antheridium** (Figs. 66–7) and the female organ as the **archegonium** (Fig. 68A–B). The **antheridium** (Figs. 66–7) is a multicellular, short-stalked, club-shaped body filled with numerous small cells, known as the antherozoid mother

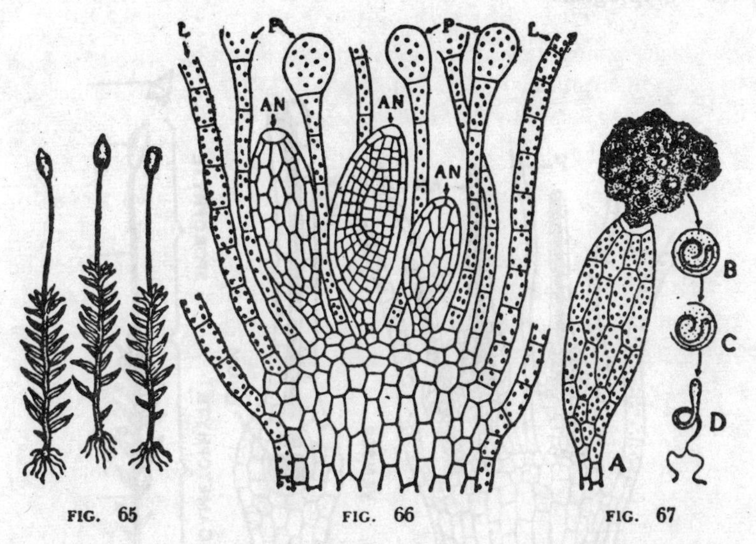

Figure 65. Moss. Three moss plants. Figure 66. Apex of a moss shoot in section showing antheridia (*AN*), paraphyses (*P*) and leaves (*L*). Figure 67. *A*, a mature antheridium discharging antherozoid mother cells; *B*, an antherozoid mother cell; *C*, wall of the mother cell dissolving; and *D*, a biciliate antherozoid.

cells. The antheridium bursts at the apex and the mother cells are liberated through it in a mass of mucilage (Fig. 67). The mucilaginous walls of the mother cells get dissolved in water and the antherozoids or male gametes are set free. They are very minute in size, spirally coiled and biciliate; after liberation they swim in water that collects at the apex of the moss plant after rain. The **archegonium** (Fig. 68A–B) is a multicellular, flask-shaped body. It consists of a short multicellular stalk, a lower swollen portion—the **venter** (belly), and an upper tube-like portion–the **neck**. The venter contains a large **ovum** or **egg-cell** with a distinct **egg** (egg-nucleus or female gamete) in it and slightly higher up a small ventral canal cell. The neck is long and straight and contains many small neck canal cells which together with the ventral canal cell soon degenerate into mucilage before fertilization.

Fertilization is effected through the medium of rain-water or dew that collects on the moss plants. When the archegonium matures it secretes mucilage with cane-sugar. This attracts a swarm of antherozoids which enter through the neck canal and pass down into the venter; one of them fuses with the egg-nucleus and the rest die. After fertilization the zygote clothes itself with a wall and is then known as the **oospore**.

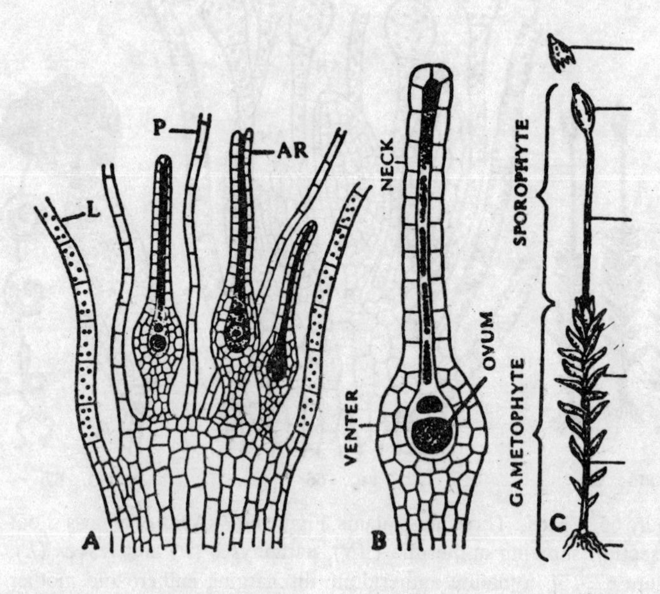

Figure 68. Moss. *A*, apex of a moss shoot in section showing three archegonia (*AR*), three paraphyses (*P*), and two leaves (*L*); *B*, an archegonium; *C*, a moss plant showing the sporophyte growing on the gametophyte.

Sporophyte and Asexual Reproduction. The oospore grows *in situ* and gives rise to the sporophyte on the moss plant (Fig. 68C). The sporophyte reproduces asexually by spores. It is a very complex structure known as the **sporogonium**. It consists of **foot**, **seta** (slender stalk) and **capsule** (case containing spores)—Figs. 69–71. The sporogonium grows as a semi-parasite on the moss plant. Although it draws most of its food from the moss plant it can manufacture its own food to some extent.

The **capsule** is a complex body with differentiated parts (Fig. 72). It is covered by a sort of loose cap known as the **calyptra** (Figs. 69–70) which is soon blown away by the wind. A longitudinal section through the capsule (Fig. 72) shows the following parts. (1) **Operculum** is the circular cup-shaped lid on the top. (2) **Annulus** is the ring-like layer of thickened cells at the base of the operculum. The capsule bursts at the annulus, and the operculum is thrown out. (3) **Peristome** is one or two rings of tooth-like projections at the rim of the capsule (Figs. 71–2). By their movements (opening and closing) the teeth help eject the spores

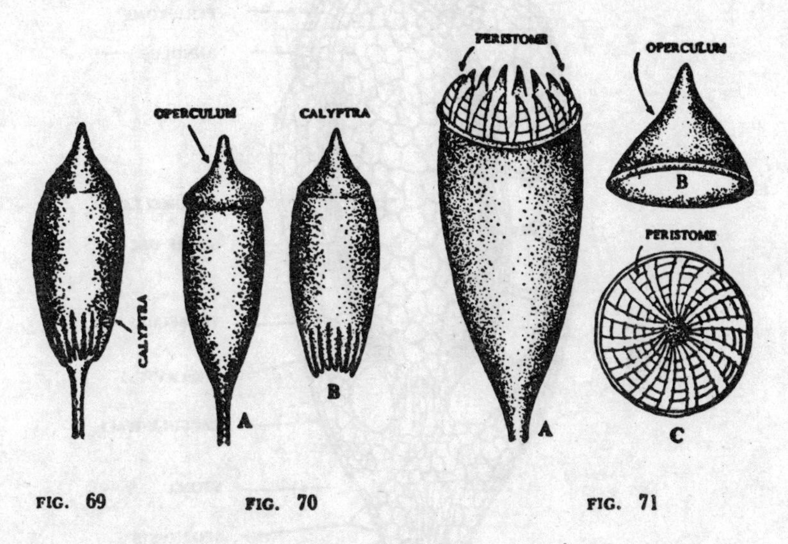

Figure 69. Moss Capsule. A capsule covered by calyptra. Figure 70. *A,* a capsule without calyptra; *B,* detached calyptra. Figure 71. *A,* a capsule showing peristome—open; *B,* operculum; *C,* peristome—closed (top view).

from the spore sac. (4) **Columella** is the solid central column, containing water and food. (5) **Spore-sac** is the hollow cylindrical sac surrounding the columella and bearing numerous spores. They are always formed in tetrads by reduction division of spore mother cells. (6) **Air-cavity** is the hollow cylindrical cavity surrounding the spore-sac, with delicate strands of cells (*trabeculae*) running across it. (7) **Capsule wall** with epidermis as external layer. (8) **Apophysis** is the solid basal portion of the capsule, having chloroplasts in many cells and stomata in the epidermis.

Germination of the Spore. After dehiscence of the capsule the spores are scattered by the wind, and they germinate under favourable conditions. The spore grows into a green, much-branched filament known as the **protonema** (Fig. 73). It produces here and there some slender rhizoids, and a number of small lateral buds which grow up into new moss plants. Thus the life-cyde of moss is completed.

Alternation of Generations (Fig. 74). The moss plant shows in its life-history two generations which regularly alternate with each other. The moss plant itself is the gametophyte (haploid or *n*) and it reproduces

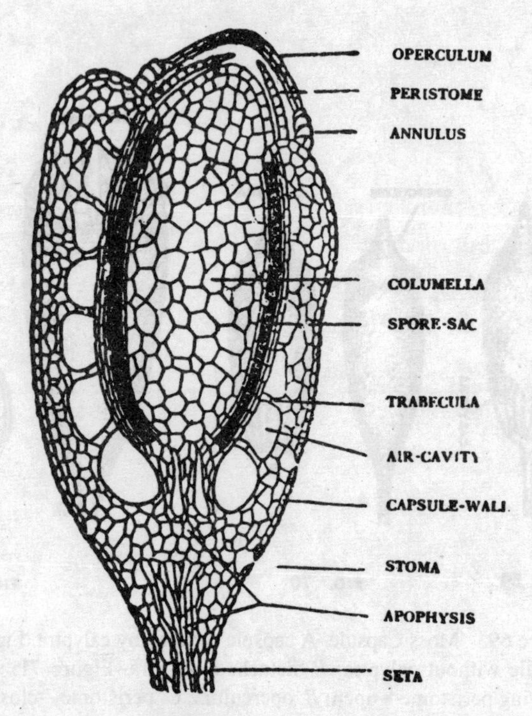

Figure 72. Moss capsule in longitudinal section.

sexually by the fusion of gametes (antherozoid and ovum, each *n*) in pairs to give rise to the sporophyte (diploid or 2*n*); while the sporogonium is the sporophyte and it reproduces asexually by spores to give rise to the gametophyte. Thus the two generations regularly alternate with each other. Life-cycle of moss showing alternation of generations is given below.

Figure 73. Protonema of moss (note of buds and rhizoids).

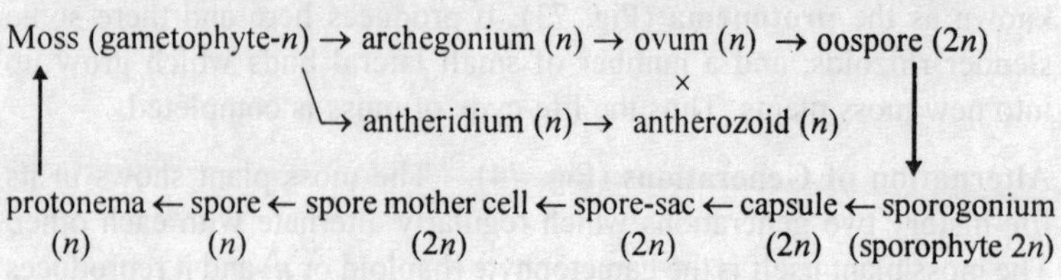

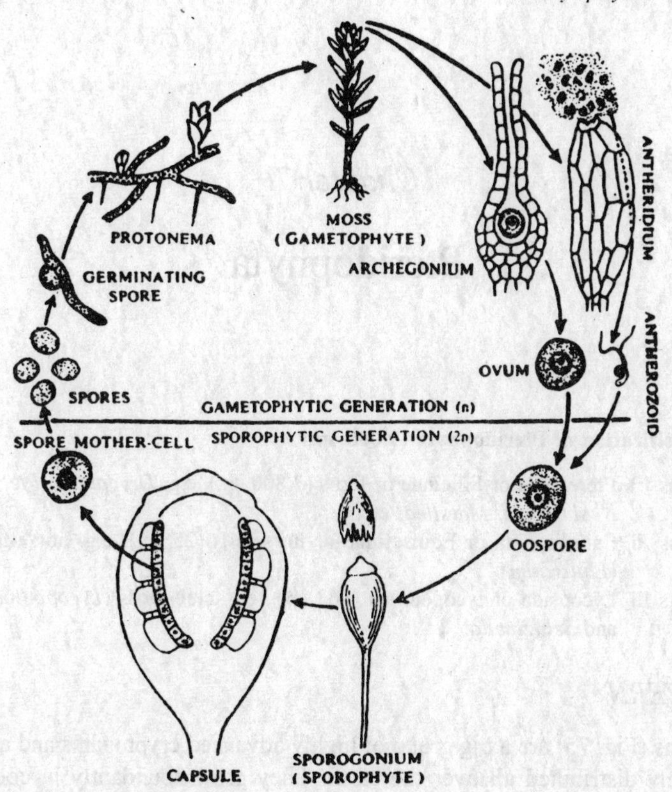

Figure 74. Life-cycle of moss showing alternation of generations.

Differences between Bryophyta and Pteridophyta

Bryophyta	Pteridophyta
1. Plant body is an independent gametophyte (haploid).	1. Plant body is an independent sporophyte (diploid).
2. Plant body is differentiated into stem, leaves and rhizoids (true roots are absent), though in certain bryophytes the plant body is thallus.	2. Plant body is differentiated into stem, leaves and true roots.
3. Vascular tissues are lacking.	3. Vascular tissues are present.
4. Plant body (gametophyte) reproduces sexually and gives rise to a sporophyte.	4. Plant body (Sporophyte) reproduces asexually and gives rise to a gametophyte.
5. Sporophyte is dependent on gametophyte (for fixation and nutrition).	5. Both sporophyte and gametophyte are independent bodies.

Chapter 7

Pteridophyta

Classification of Pteridophyta (9,000 sp.)

Class I Pteropsida or Filicinae or ferns (7,800 sp.), e.g. *Dryopteris, Pteris, Nephrolepis, Marsilea,* etc.

Class II Sphenopsida or Equisetinae or horsetails (25 sp.), e.g. horse-tail (*Equisetum*).

Class III Lycopsida or lycopodinae (963 sp.), e.g. club-moss (*Lycopodium*) and *Selaginella*.

1. *FERN*

Ferns (Fig. 75) are a big group of highly advanced cryptogams and are widely distributed all over the earth. They grow abundantly in cool, shady, moist places, both in the hills and in the plains. The stem is mostly a rhizome, but sometimes it is erect and aerial, as in **tree ferns**. Roots are adventitious (fibrous) growing profusely from the rhizome. Leaves are usually pinnately compound and *circinate* (rolled from the apex downwards) when young (Fig. 75B), the leaflets being known as the pinnae (sing. pinna). The stem and the petiole are covered with numerous brownish scales known as the **ramenta**.

Internal Structure of the Fern Stem (or **Petiole**; Fig. 76). (1) **Epidermis**—a single layer of cells with the outer walls thickened and cutinized. (2) **Sclerenchyma**—a few layers of sclerenchyma occur below the epidermis. (3) **Ground tissue** is made up of a continuous mass of polygonal parenchymatous cells. (4) **Endodermis**—a single layer of narrow cells surrounding each stele; this layer is often much thickened, particularly on the inner side. (5) **Steles**—in the young stem or petiole the stele is more or less horseshoe-shaped, but in the older part the stele is broken up usually into two or three smaller steles. Each stele consists of (a) **pericycle**, (b) **phloem** and (c) **xylem**. Pericycle surrounds the stele as a single layer (sometimes a double layer, particularly at

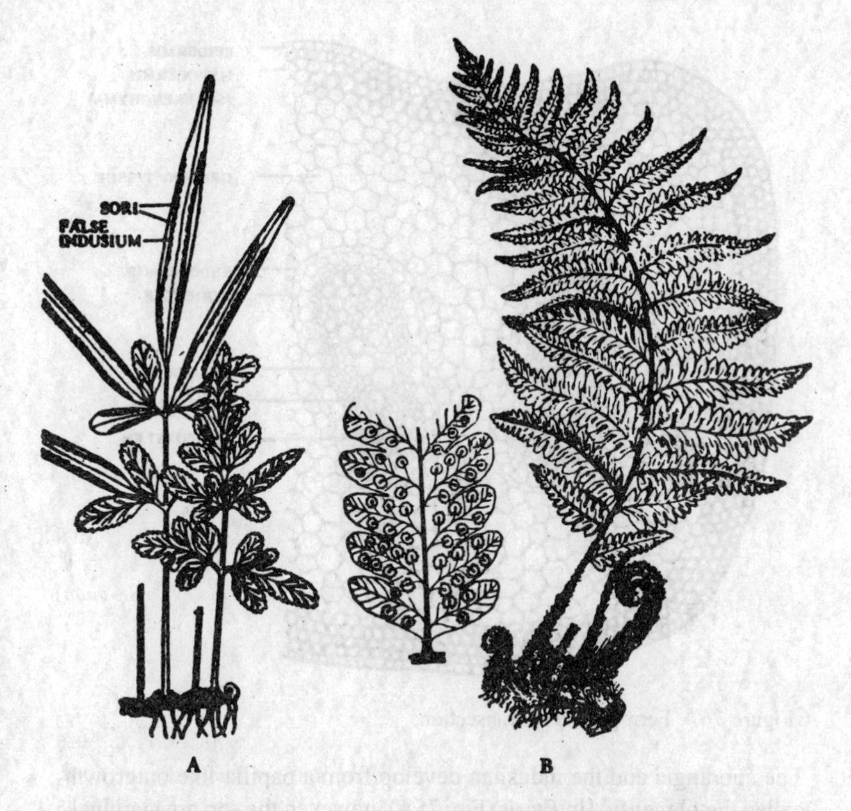

Figure 75. Fern Plants. *A, Pteris* with continuous linear sori and false indusium; *B, Dryopteris* with sori on veins of the pinna and reniform indusium.

the sides) and contains starch grains. Phloem surrounds the central xylem, bundle being concentric, and consists of sieve tubes and phloem parenchyma. Xylem lies in the centre surrounded by phloem. It consists usually of two groups of protoxylem at the two ends, and metaxylem in the middle. Protoxylem is made of spiral tracheids and metaxylem of scalariform tracheids.

Sporophyte and Asexual Reproduction. The fern plant (Fig. 75) is the sporophyte, i.e. it bears spores and reproduces by the asexual method. On the undersurface of the leaf or the **sporophyll** (as the spore-bearing leaf is called) a number of dark brown structures, pale green when young, may be seen; these are called **sori** (sing. sorus). They develop on the veins, and in *Dryopteris* (Fig. 75B) they are arranged in two rows in each leaflet or pinna of the leaf. Each sorus (Fig. 77) is a group of **sporangia** covered over by a kidney-shaped shield called the **indusium**.

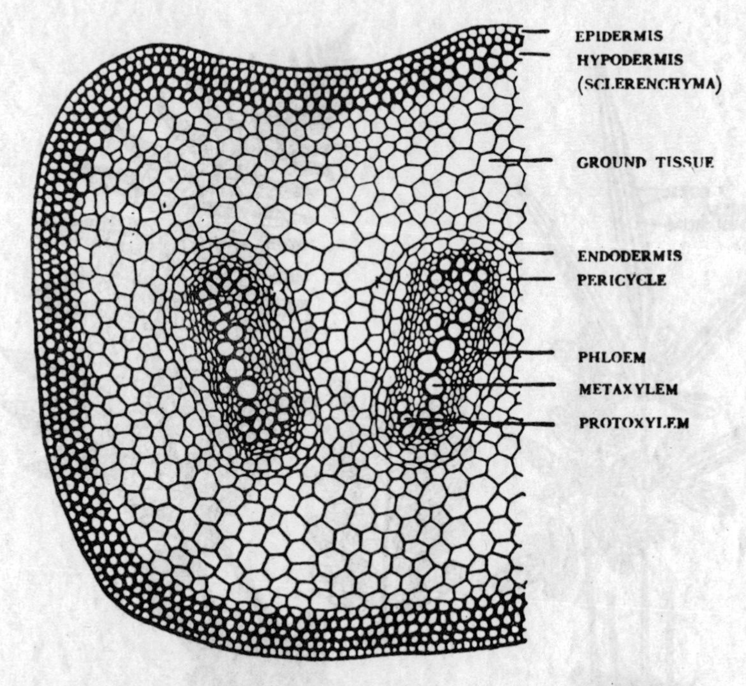

Figure 76. Fern petiole in transection.

The sporangia and the indusium develop from a papilla-like outgrowth called the **placenta**. [In *Pteris* (Fig. 75A), however, the sori are marginal, linear and continuous, covered over by the reflexed margin of the pinna; this type of indusium is called *false indusium*]. Each sporangium (Fig. 78) consists of a long slender multicellular **stalk** and a biconvex **capsule**. The capsule is filled with a mass of **spores**. They are always formed *in tetrads* by reduction division of spore mother cells. The capsule wall is thin but it has a specially thickened and cutinized band or ring running round its margin. This ring is called the **annulus**. The annulus has an unthickened portion known as the **stomium**. When the spores mature the capsule bursts at the stomium. The annulus bends back exposing the spores, and then suddenly it returns to its original position, ejecting the spores with a jerk (Fig. 78B). The spore germinates and gives rise to the gametophyte.

Gametophyte and Sexual Reproduction. The gametophyte in fern is a very small (more or less 8 mm across) green flat heart-shaped body known as the **prothallus** (Fig. 79). It bears gametes and reproduces

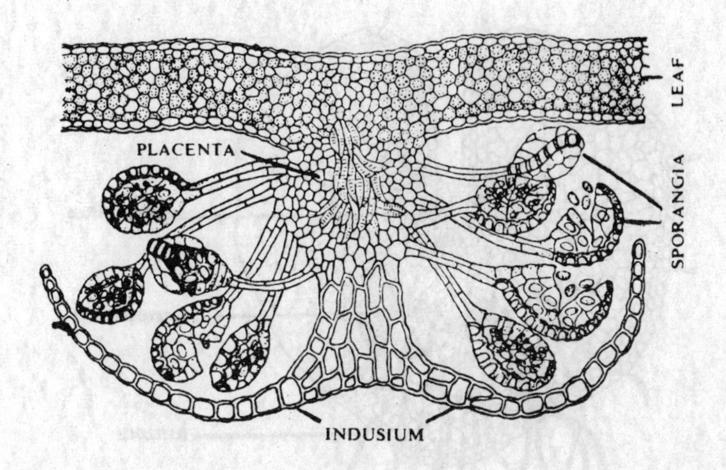

Figure 77. Fern . A sorus in section.

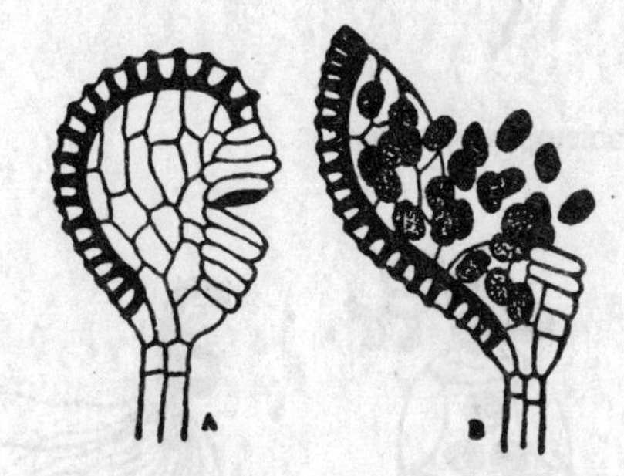

Figure 78. Fern. Sporangium (capsule and stalk); *A*, capsule just open at the stomium; *B*, the same after bursting, with the annulus bending back.

sexually by them. For this purpose the prothallus bears on its undersurface groups of highly differentiated reproductive organs called **antheridia** (male) and **archegonia** (female); it also bears many slender unicellular hair-like structures called **rhizoids** which function as roots. The **antheridium** (Fig. 80) is a spherical or oval body with many antherozoid mother cells in it. Each mother cell produces a single twisted and multiciliate **antherozoid** (male gamete). The **archegonium** (Fig. 81) is a flask-shaped body. The swollen basal portion of it is known as the **venter**, and the slender tube-like upper portion as the **neck**. The neck is short and curved in fern, and consists of a narrow neck canal cell with usually two nuclei in it, and a wall made of four vertical rows of cells. The venter lies embedded, partly at least, in the prothallus, and encloses a single large **egg-cell** with a distinct egg-nucleus or female gamete in it and slightly higher up a small ventral canal cell. The neck canal cell

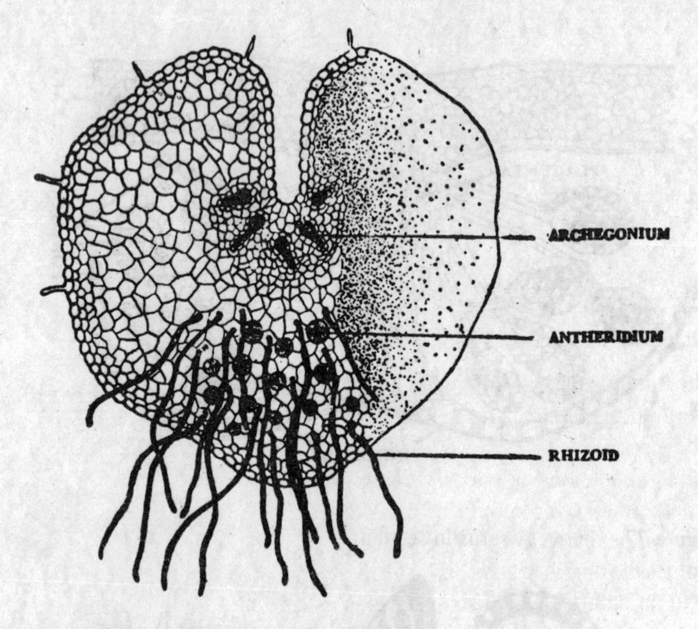

Figure 79. Prothalius (gametophyte) of fern.

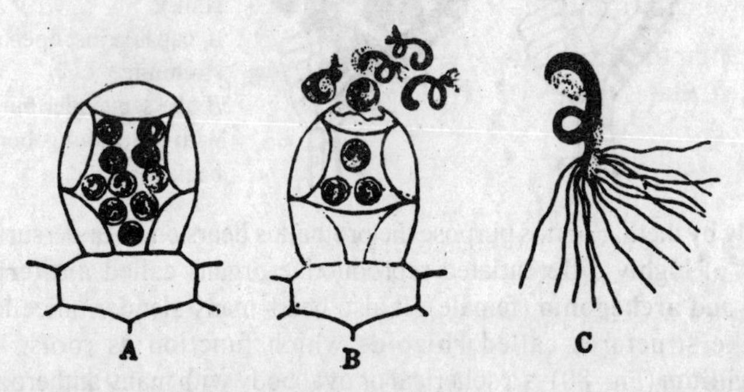

Figure 80. Fern. Antheridium. *A,* a young one with antherozoid mother cells; *B,* a mature one after bursting, antherozoids escaping; *C,* a multiciliate antherozoid.

and ventral canal cell disintegrate into mucilage which forces open the lid of the archegonium before fertilization.

Fertilization. After the antheridium matures and bursts, the antherozoids are liberated (Fig. 80). As the archegonium matures (Fig. 81B) it secretes mucilage and malic acid to attract the antherozoids. They swim to the

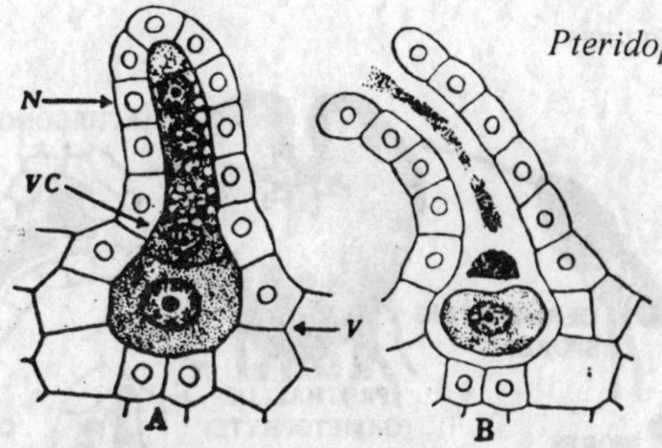

Figure 81. Fern. Archegonia. *A,* a young one; *N,* neck (wall and neck canal cell with two nuclei); *VC,* ventral canal cell; *V,* venter with an egg-cell and an egg-nucleus; *B,* a mature one ready for fertilization.

archegonium in large numbers, enter into it through the neck and pass down into the venter. They vibrate around the ovum for a while, and one of them soon fuses with the egg-nucleus. Fertilization is thus effected. The rest of the antherozoids die out. The fertilized ovum clothes itself with a cell-wall and becomes the **oospore**. The oospore gives rise to an embryo which soon develops into a young sporophyte (Fig. 82). The prothallus decays and the young sporophyte grows into a fern plant.

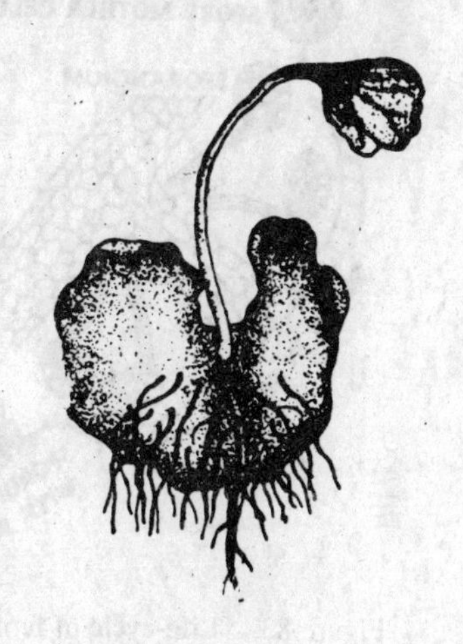

Figure 82. Prothallus of fern with young sporophyte.

Alternation of Generations (Fig. 83). The fern plant passes through two stages or generations to complete its life-history. The plant itself is the sporophyte (diploid or $2n$), and the prothallus the gametophyte (haploid or n). The sporophyte or the fern plant reproduces asexually by spores and gives rise to the gametophyte, i.e. the prothallus. The prothallus reproduces sexually by fusion of gametes (antherozoid and ovum) in pairs and gives rise to the sporophyte, i.e. the fern plant. Thus the two generations regularly alternate with each other. Life-cycle of fern showing alternation of generations is given below.

Fern → sporophyll ($2n$) → sorus ($2n$) → sporangium ($2n$) → spore mother cell ($2n$) ↓
(sporophyte–$2n$)
↑
embryo ($2n$) ← oospore ($2n$) ← ovum (n) × antherozoid (n) ← archegonium (n) / antheridium (n) ← prothallus ← spore (n)
(gametophyte-n)

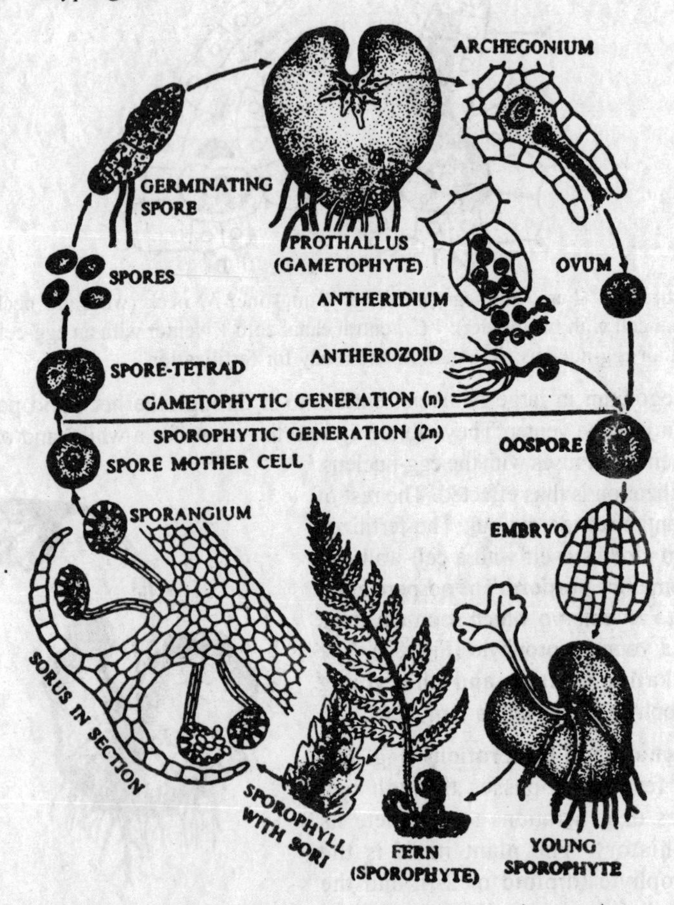

Figure 83. Life-cycle of fern showing alternation of generations.

COMPARATIVE STUDY OF MOSS AND FERN

Habit and Habitat. Moss plants are small, usually 2–3 cm, sometimes much more in height, growing in clusters from protonemal buds forming a green soft cushion on damp ground or damp old walls; while fern plants are much bigger in size, usually 25–40 cm, growing close together in cool shady moist places.

Structure. The structure of the moss plant is simple consisting of a short axis with spirally arranged minute leaves and a number of rhizoids at the base of the axis; while the fern plant is much more complicated in structure consisting of a rhizome with scales or ramenta, several adventitious roots and usually large well-developed leaves, often pinnately divided.

Vegetative Reproduction. Moss sometimes reproduces vegetatively by splitting of protonemal branches or by resting buds on the protonema; while fern reproduces by its rhizome.

Alternation of Generations. Both moss and fern are higher cryptogams showing a regular alternation of generations in their life-history. Moss plant is the gametophyte which is the dominant phase in its life-cycle; while the sporogonium is the sporophyte which is dependent on the gametophyte as a semi-parasite. The order is reversed in the case of fern. The fern plant is the sporophyte which is the dominant phase in its life-cycle; while the prothallus is the gametophyte which although an independent body is very much reduced in size and is inconspicuous. Thus from moss to fern a reduction of gametophyte and an advance of sporophyte are evident. In both the cases the sporophyte reproduces asexually by spores and gives rise to the gametophyte and the latter reproduces sexually by gametes (antherozoid and egg-nucleus) and gives rise to the sporophyte.

2. *EQUISETUM* (25 sp.)

Equisetum (Fig. 84), commonly called horsetail, is the only genus of the family *Equisetaceae*. It is widely distributed over the earth, being specially abundant in marshy places. Species of *Equisetum* are mostly short slender much-branched perennial herbs, hardly exceeding a metre in height. Such common Indian species are *E, arvense. E. debile* etc. The last one commonly found in Assam and Meghalaya, however, grows to a length of 3–4 metres. *E. giganteum,* a South American species, climbs neighbouring trees to a height of 12 metres. *Equisetum* plants consist of a long creeping underground rhizome and two kinds of erect aerial shoots—much branched and unbranched (simple), arising from the rhizome. Like the leaves, the branches also appear in whorls. The rhizome is provided with nodes and internodes and a whorl of scale-like leaves at each

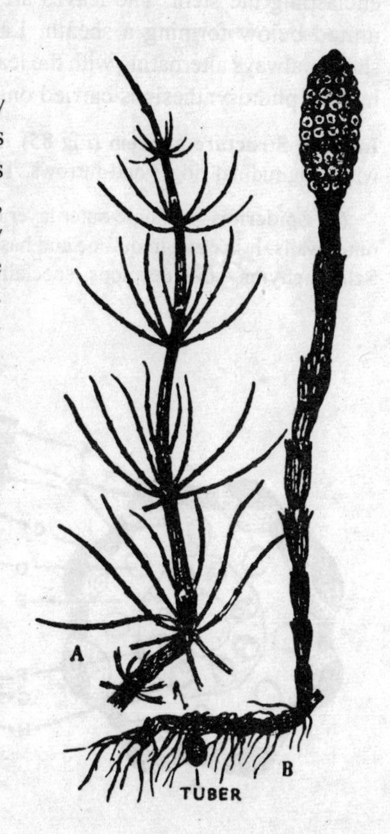

Figure 84. *Equisetum. A,* vegetative shoot with whorls of branches, *B,* a fertile shoot with a spike.

node, and bears several slender adventitious roots and often some tuber-like bodies which serve for food storage and vegetative reproduction. The branched shoots (Fig. 84A) are green in colour and usually *sterile,* being only vegetative in function, while the unbranched shoot (Fig. 84B) is non-green and *fertile,* being reproductive in function. The fertile shoot ends in a *cone,* and it dries up soon after the production of spores. All the shoots, fertile and sterile, are distinctly ribbed with ridges and furrows, and jointed with distinct nodes and internodes. Their surface is harsh to touch because of the presence of silica in their epidermal layer. Besides, there is a whorl of scale-like leaves at each node, partly enclasping the stem. The leaves are free and pointed at their tips but united below forming a sheath. Lateral branches grow piercing the sheath, always alternating with the leaves. The latter being much reduced in size, photosynthesis is carried on by the green shoots.

Internal Structure of Stem (Fig 85). It has distinct nodes and internodes, with longitudinal ridges and furrows. The internal structure is as follows:

(1) **Epidermis**—a single outer layer of cells with a deposit of silica in their outer walls. It is wavy in outline and has stomata in two rows in the furrow. (2) **Sclerenchyma**—this develops, specially in the ridges, below the epidermis,

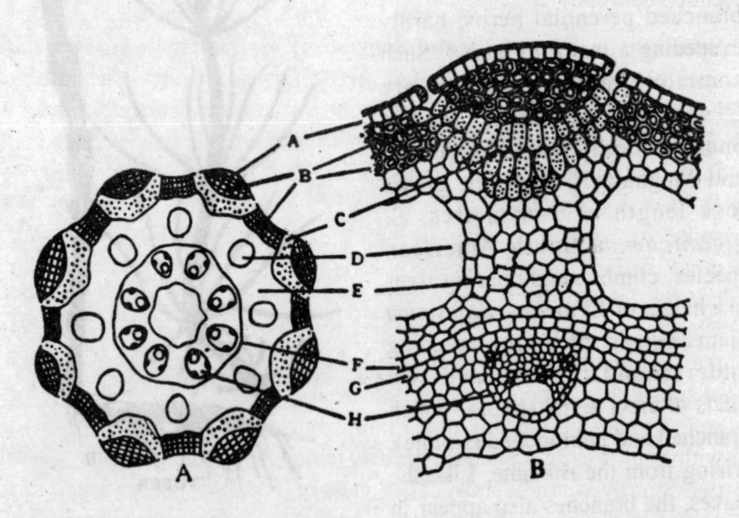

Figure 85. *Equisetum. A,* stem in transection, *B,* a sector of the section magnified). *A,* epidermis; *B,* hypodermis (sclerenchyma); *C,* outer cortex with chloroplasts (note the stomata); *D,* air-cavities; *E,* general cortex; *F,* endodermis; *G,* pericycle; *H,* vascular bundles (see text; note the pith and the

interrupted in the furrows by the underlying cortex. (3) **Cortex**—it is many-layered, and in the middle of it large air canals, each corresponding to the furrow, are formed. Outer layers of the cortex contain chloroplasts; leaves being scaly, carbon-assimilation is carried on by the cortex of the stem. The assimilating tissue extends up to the epidermis in the furrow where the stomata lie. (4) **Endodermis**—cortex is demarcated from the stele by a definite layer of endodermis. (5) **Pericycle**—the layer lying internal to the endodermis is the pericycle. (6) **Vascular bundles**—these are closed, collateral, and arranged in a ring. Each vascular bundle is very feebly developed with xylem on the inner side and a small patch of phloem on the outer; xylem consists of annular and spiral tracheids (protoxylem) with a water-containing cavity (*carinal cavity*) in it, and scalariform and reticulate tracheids (metaxylem) on the outer side laterally. (7) **Pith cavity**—there is a big air-cavity corresponding to the pith.

Life-cycle. The life-cycle of *Equisetum* is complete in two stages—sporophytic and gametophytic. The plant itself is the sporophyte and this is followed by another structure, called prothallus, which is the gametophyte (Fig. 87A).

Sporophyte. The *Equisetum* plant (Fig. 84) is the sporophyte, i.e. it reproduces asexually by spores which are borne by specialized leaves called sporophylls.

Sporophylls, Sporangia and Spores. The sporophylls are very much specialized in structure and take the form of somewhat flattened, hexagonal or circular discs, each supported on a short stalk, and are aggregated together in whorls at the apex of each fertile shoot in the form of a *cone,* called the **sporangiferous spike** or **strobilus** (Figs 84B & 86A). The lowest whorl is sterile and forms a **ring** at the base of the spike. In *Equisetum,* as in all higher plants, the reproductive region is quite distinct from the vegetative region. Each sporophyll (Fig. 86B) has the form of a stalked peltate disc. It bears on the undersurface a whorl of **sporangia** (5–10) which contain numerous small **spores**. Like the fern plant, *Equisetum* is also *homosporous,* bearing only one kind of spores. Each spore contains numerous minute chloroplasts and a large central nucleus. In addition to intine and exine, the spore is provided with a third layer, called **perinium**, which, when mature, ruptures into two spirally-wound bands, called **elaters** (Fig. 86D–E), attached to the spores at their centre; the elaters appear as four distinct appendages. They are extremely hygroscopic when the air is dry, they unwind and stand out stiffly from the spore, and when the air is moist they roll up spirally round it. *Functions of Elaters.* The elaters expand and help the dehiscence of the sporangium. The spores become entangled by the elaters and are carried away in clusters by air-currents; this helps the germination of spores close together for facility of fertilization.

Gametophyte. The prothallus (Fig. 87A) is the gametophyte, i.e. it bears gametes—male (antherozoid) and female (ovum), and reproduces by the sexual method. Spores remain alive only for a few days. Under favourable conditions their germination begins within 10 or 12 hours after they are set free from the sporangium. Spores can be easily grown in culture in the laboratory. On germination they give rise to prothalli which are small in size, dull brownish-green in colour and much-branched (lobed). The prothalli in most species are usually 3–6 mm in diameter; in *E. debile,* however, they may be as big as 3 cm in diameter. Prothalli, when they grow under favourable conditions, are normally monoecious bearing both antheridia and archegonia. However growing crowded together in the field or in culture they may be dioecious also, the smaller ones being male and the bigger ones female. It has been seen that the latter bear antherdia with age. This imperfect dioecism may be due to unfavourable conditions of growth. Prothalli usually live long, sometimes exceeding a period of two years. The prothallus is the

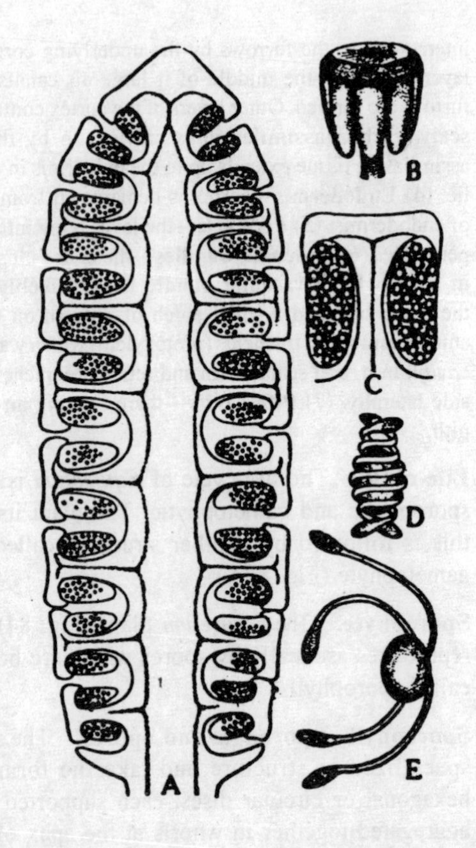

Figure 86. *Equisetum. A,* a spike in longi-section showing sporophylls, sporangia and spores; *B,* a sporophyll with a whorl of sporangia; *C,* the same in longi-section; *D,* a spore with elaters coiled; and *E,* the same with elaters uncoiled.

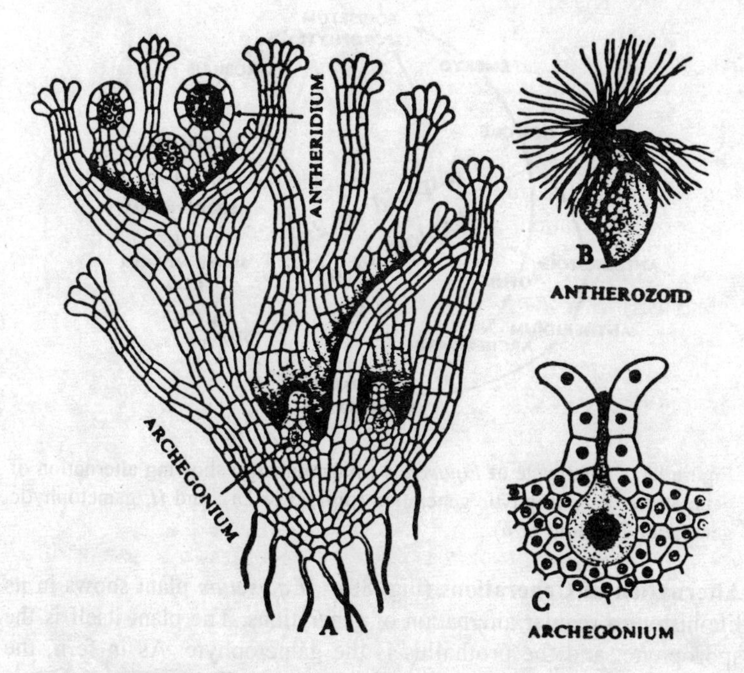

Figure 87. *Equisetum. A,* prothallus (monoccious); *B,* an antherozoid; *C,* a mature archegonium.

gametophyte, normally it bears both antheridia and archegonia. The sexual organs begin to appear in the gametophyte within 30–40 days of its growth, archegonia first and antheridia later. **Antheridia** develop at the apex of a branch (lobe) of the prothallus or on the margin of it. Each antheridium is more or less spherical in shape and contains numerous antherozoid mother-cells (usually 256). In each mother-cell one **antherozoid** or male gamete is produced. It is a large spirally coiled and multiciliated body (Fig. 87B). **Archegonia** (Fig. 87C) always develop in the axial region of the prothallus and in the axil of a branch of it. Each archegonium is flask-shaped with a swollen *venter* and a narrow *neck,* and contains an ovum (female gamete), a ventral canal cell and a neck canal cell.

Fertilization. The method of fertilization is the same as that of ferns. After fertilization the **oospore** gives rise to an **embryo** which develops into a branching rhizome. This then produces erect aerial shoots and a number of adventitious roots.

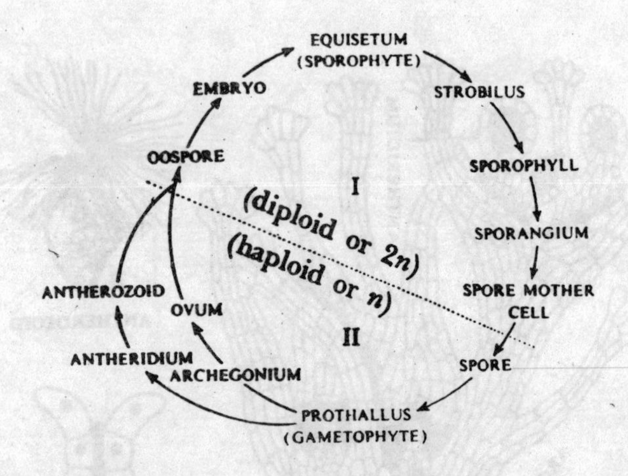

Figure 88. Life-cycle of *Equisetum* (diagrammatic) showing alternation of generations. *I,* sporophytic generation (diploid or $2n$); and *H,* gametophytic generation (haploid or n).

Alternation of Generations (Fig. 88). *Equisetum* plant shows in its life-history a regular alternation of generations. The plant itself is the sporophyte, and the prothallus is the gametophyte. As in fern, the sporophyte or *Equisetum* plant reproduces asexually by spores and gives rise to the gametophyte or prothallus, and the latter reproduces sexually by gametes—antherozoid and ovum, and gives rise to the sporophyte. Thus the two generations regularly alternate with each other. The sporophytic generation begins with the oospore and ends in the spore mother-cells because in all these stages $2n$ chromosomes are met with; while the gametophytic generation begins with the spores and ends in the gametes (antherozoid and ovum) because in all these stages there are only n chromosomes.

3. *SELAGINELLA* (700 sp.)

Selaginella (Fig 89) grows in damp places in the hills and in the plains. It is a slender much-branched plant creeping on the wall or on the ground. The slender stem bears four rows of leaves—two rows of small leaves on the upper surface and two rows of large leaves on the two sides. A scaly structure, called **ligule**, develops on the upper (ventral) surface of each leaf above its base. A long slender root-like organ is given off from the stem at the point of bifurcation; this is known as the **rhizophore**

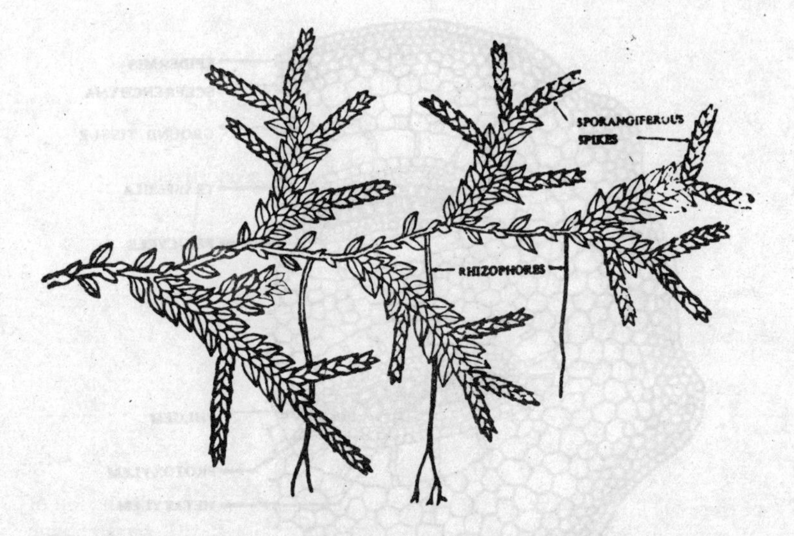

Figure 89. *Selaginella.* A portion of a plant showing four rows of leaves, a number of spikes and three rhizophores

(root-bearer). In some species the rhizophore bears small fibrous roots at the tip. It has no root-cap and grows exogenously like the stem; but it resembles a true root in its internal structure and bearing no leaves. It is thus regarded as an intermediate structure between the stem and the root. Many describe it as an organ *'sui generis'*, i.e. an organs growing in its own way.

Internal Structure of the Stem (Fig. 90). (1) **Epidermis**—a single layer with a cuticle. (2) **Sclerenchyma**—a few layers of sclerenchyma occur below the epidermis. (3) **Ground tissue**—a continuous mass of thin-walled, polygonal cells. (4) **Steles**—usually 2 or 3; each stele is surrounded by an air-space which is formed as a result of breaking down of some of the inner layers of the cortex, and remains suspended in it (air-space) by delicate strands of cells, called **trabeculae** (sing. trabecula). The stele, when young, is surrounded by a single-layered endodermis; later on, the cells of the endodermis separate laterally and elongate considerably in the radial direction. These long radiating cells formed as a result of stretching of the endodermal cells are the trabeculae; in the mature stele they act as bridges across the air-space. Each stele which is concentric in nature consists of (a) pericycle, (b) phloem, and (c) xylem. Internal to the air-space there is a layer, sometimes two, of rather large but thin-walled cells—the **pericycle**. **Phloem** surrounds the central spindle-shaped **xylem**. Protoxylem lies at the two ends and metaxylem in the middle.

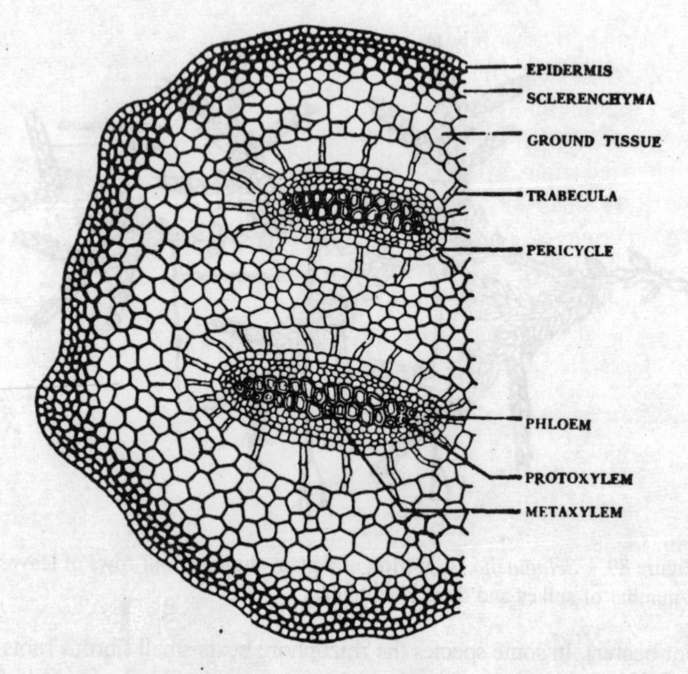

Figure 90. *Selaginella* stem in transection.

Sporophyte and Asexual Reproduction. *Selaginella* plant is the sporophyte. It is heterosporous and bears *two* kinds of spores—microspores and megaspores—and reproduces asexually by them. They are separately borne by micro-sporangium formed by microsporphyll, and megasporangium developed by megasporophyll. Both kinds of sporophylls may occur together in the same cone, or they may be borne in two separate cones either on the same plant (monoecious) or on two separate plants (dioecious). All the sporophylls are nearly of equal size and spirally arranged, usually in four rows, forming a 4-angled *cone,* called the **sporangiferous spike** (Figs. 89 and 91) at the apex of the reproductive shoot. The sporophylls are similar to the vegetative leaves in appearance, but are smaller in size. Each **megasporophyll** bears in its axil a single megasporangium with usually 16 megaspore mother cells. But only one of them divides, while others get disorganized. It undergoes reduction division and forms a tetrad of spores (Fig. 92A). Thus the megasporangium contains *four large* **megaspores**. A considerable amount of food material, chiefly oil, is stored up in the megaspore. The **microsporphyll** similarly bears in its axil a microsporangium with usually 16 microspore mother cells. All of them undergo reduction division and give rise to 64 small **microspores** (Fig. 93A) in groups of four (tetrads). The sporangia consist of a short stout stalk and a capsule with a distinct thick wall. The megasporangia are somewhat larger than the microsporangia.

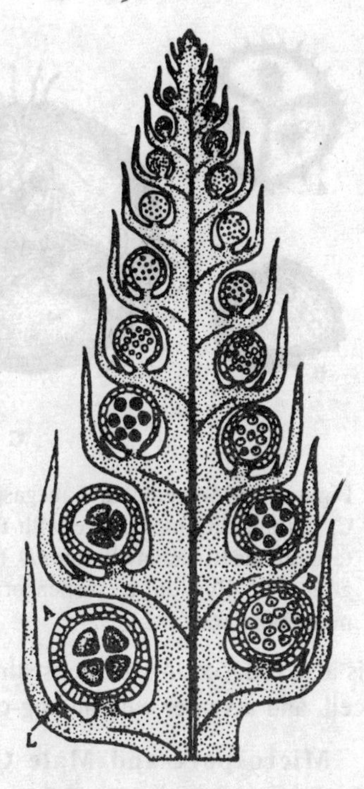

Figure 91. *Selaginella.* A spike in longisection. *A*, megasphorophyll with megasporangium and megaspores; *B*, microsporophyll with microsporangium and microspores; *L*, ligule.

Gametophytes and Sexual Reproduction. The two prothalli (male and female) are the gametophytes, i.e. they bear male gametes or antherozoids and female gamete or ovum, and reproduce sexually by them.

Megaspore and Female Gametophyte (Fig. 92C–D). The megaspore-nucleus divides repeatedly by the process of free cell formation, and gives rise to a mass of tissue within it. This is the female prothallus, i.e. the female gametophyte. *The megaspore begins to grow before it is set free from the megasporangium,* but the formation of the female gametophyte is completed after the spore has fallen to the ground. At an early stage of gametophyte formation a cavity appears at one end of it and is filled with reserve food, chiefly oil. This cavity subsequently becomes filled with cells. Further development of the gametophyte exerts pressure on the spore-wall which ruptures by a tri-radiate fissure, and the gametophyte becomes partially exposed. *The gametophyte is partially endosporous,* i.e. it is *not an independent structure* like that of fern, being enclosed by the spore-coat, and nourished by the food stored in the spore. It is also a much reduced structure compared to that of fern. A number of **archegonia** and some groups of rhizoids develop in the exposed green portion of the prothallus, while food is stored up in the inner non-green portion of it. Each archegonium

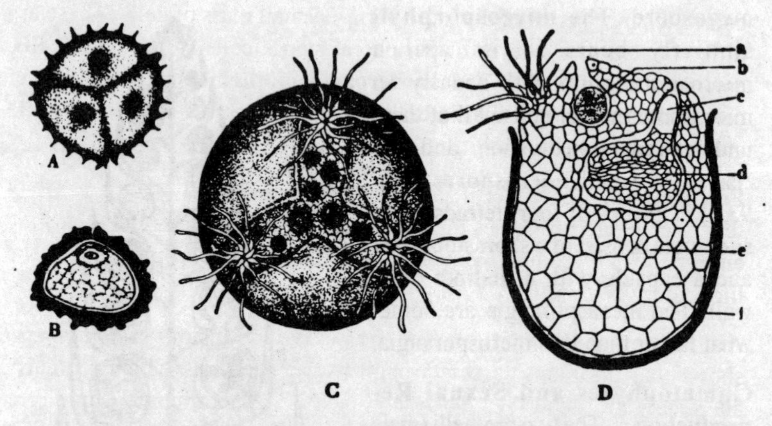

Figure 92. *Selaginella. A,* megaspores in a tetrad; *B,* a megaspore in section; *C,* germinating megaspore with the female prothallus partially protruding through the tri-radiate fissure; *D,* the same in section; *a,* rhizoids; *b,* oospore after first division; *c,* suspensor; *d,* embryo, *e,* prothallus tissue; and *f,* megaspore wall.

is also reduced in size, consisting of a short neck with one neck canal-cell, and a venter with an **egg-cell** and a ventral canal-cell.

Microspore and Male Gametophyte (Fig. 93C–D). The microspore germlinates and gives rise to the male prothallus, i.e. the male gametophyte. It begins to divide after it is set free from the microsporangium. A small cell is cut off at one end of the microspore; this is the **prothallus cell** representing an extremely reduced male gametophyte. The rest of the microspore forms a single **antheridial**

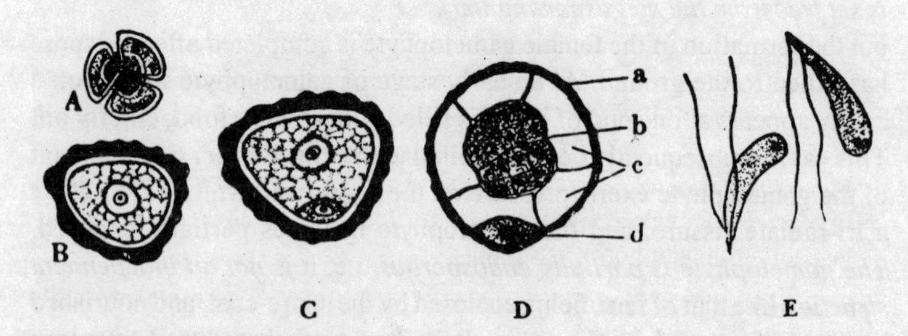

Figure 93. *Selaginella. A,* microspores in a tetrad; *B,* a microspore in section; *C,* germinating microspore with prothallus cell; *D,* male prothallus in section; *a,* microspore wall; *b,* antherozoid mother cells; *c,* jacket cells; *d,* prothallus cell; *E,* two antherozoids.

cell which by a series of divisions forms a small mass of cells. These are differentiated into a layer of peripheral sterile cells—the **jacket cells**, enclosing about 128 to 256 **antherozoid mother cells**. Each mother cell encloses a single biciliate slightly twisted **antherozoid** or male gamete (Fig. 93E).

Fertilization. After fertilization, which is essentially the same as in fern, the egg-cell becomes the **oospore**; the latter divides and forms the **embryo** which gradually develops into the *Selaginella* plant. Thus the life-history is completed.

Alternation of Generations (Fig. 94). The life-history of *Selaginella* shows that the plant passes through two generations which regularly alternate with each other. The plant itself is the sporophyte (diploid or 2*n*), and the two prothalli—male and female—are the gametophytes (haploid or *n*). The sporophyte reproduces asexually by two kinds of spores—microspore and megaspore—which give rise to the male

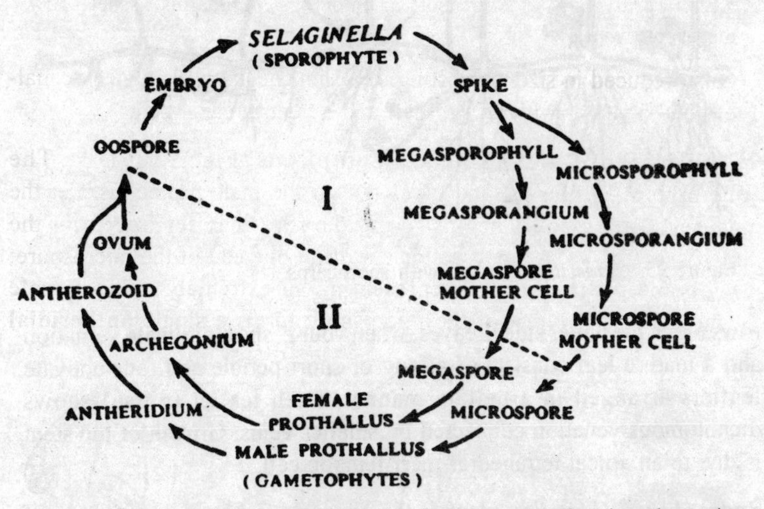

Figure 94. Life-cycle of *Selaginella* (diagrammatic) showing alternation of generations. I, sporophytic generation (diploid or 2*n*), and *II*, gametophytic generation (haploid or *n*).

prothallus and the female prothallus respectively. The male prothallus bears antherozoids in the antheridium, and the female prothallus bears the ovum in the archegonium, and the two prothalli (gametophytes) reproduce sexually by these two gametes, giving rise to the sporophyte again. Thus the two generations regularly alternate with each other.

4. *MARSILEA* (60 sp).

Marsilea (Fig. 95) is a slender prostrate herb, growing rooted to the mud-bottom at the edge of a tank or a ditch, and is very widely distributed throughout the world. *Marsilea quadrifolia* and *M. minuta* are two common species in India. The plant body consists of a slender prostrate dichotomously branched rhizome with distinct nodes and internodes, commonly rooting at the nodes and giving off leaves alternately in two

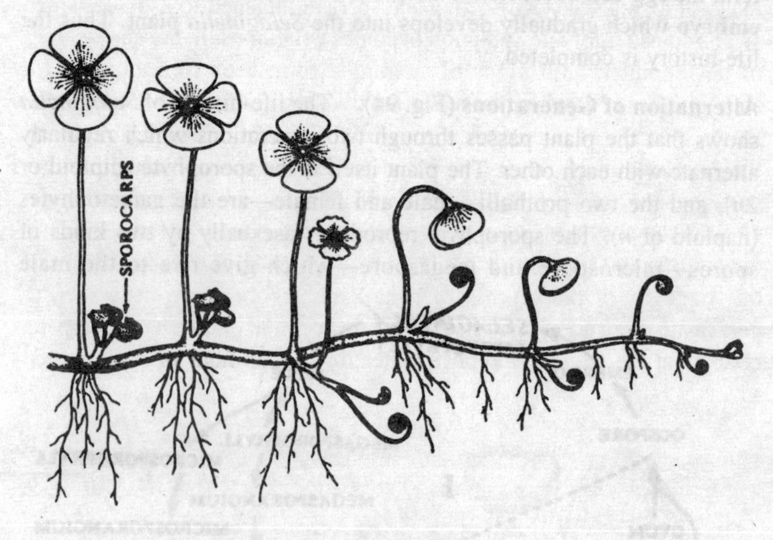

Figure 95. *Marsilea.* A plant with sporocarps.

rows along the upper side. Leaves, when young, show circinate vernation and a mature leaf consists of a long or short petiole and four obovate leaflets arranged in a peltate manner. Each leaflet (pinna) shows dichotomous venation connected by smaller veins. Growth of the stem is due to an apical tetrahedral meristematic cell.

Sporophyte. *Marsilea* plant is the sporophyte bearing two kinds of spores, i.e. it is heterosporous. Special structures, called **sporocarps**, are seen to grow, in small groups of 2–5, sometimes singly from the base of the petiole or a little above it as a segment of it (interpreted as modified fertile leaf-segment or entire leaf). Sporocarps develop only when the water recedes and the soil tends to dry up, and each is provided with a long or short stalk and has a very hard outer covering, when mature. In structure it is more or less bean-shaped (Figs. 95–6) and is about 8 × 6 mm in size. Each half of the sporocarp has distinct forked venation alternating with that of the other half. The sporocarp (Fig. 96B) contains within it usually 14–20 **sori**, each in a cavity, arranged in two rows on a receptacle; the sori develop in basipetal order. Each sorus is covered by a thin delicate layer known as the **indusium**. The sporangia at the apex of the receptacle are megasporangia and those lower down are microsporangia. The sporangium wall in each case is made of a single layer of cells. The sori are attached to a tissue which swells considerably in water and becomes gelatinous. In the early stages of development both kinds of sporangia form 8 or 16 *sporocytes* or spore mother cells which on reduction division produce 32 or 64 spores. But in the case of microsporangium all the microspores are functional, though very minute in size; while in the case of megasporangium only one megaspore grows larger in size and is functional; others degenerate.

With the development of the sporangia a stony layer is formed on the outer surface of the sporocarp. The resisting power of the sporocarp and the longevity of the spores are remarkable; spores have been seen to germinate even after many years of desiccation. If the sporocarp be cracked at the edge and kept in water for half an hour or so it is seen

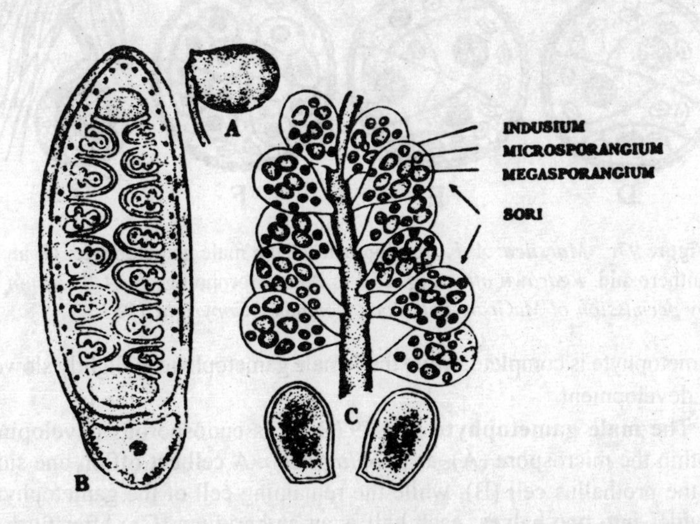

Figure 96. *Marsilea. A,* a sporocarp; *B,* the same in longi-section showing young sori in two rows, each sorus with megasporangium (terminal), microsporangia (lateral) and indusium (outer covering); *C,* part of the gelatinous ring carrying the sori; two empty valves of the sporocarp lying below.

that the gelatinized inner wall of the sporocarp pushes out of it in the form of a long gelatinous ring **(sorophore)** with the sori attached to it in an alternating manner (Fig. 96C). In nature, however, the sporocarp takes at least 2 or 3 years for the decay of its stony layer. Thereafter the spores are liberated on the conversion of the indusium and the sporangium walls into mucilage. If the gelatinous ring be left in water the development of the male and female gametophytes may be seen on the following day or the day after.

Gametophytes. The spores germinate very quickly, the microspore giving rise to the male gametophyte, and the megaspore to the female gametophyte. Within 12–20 hours the development of the male

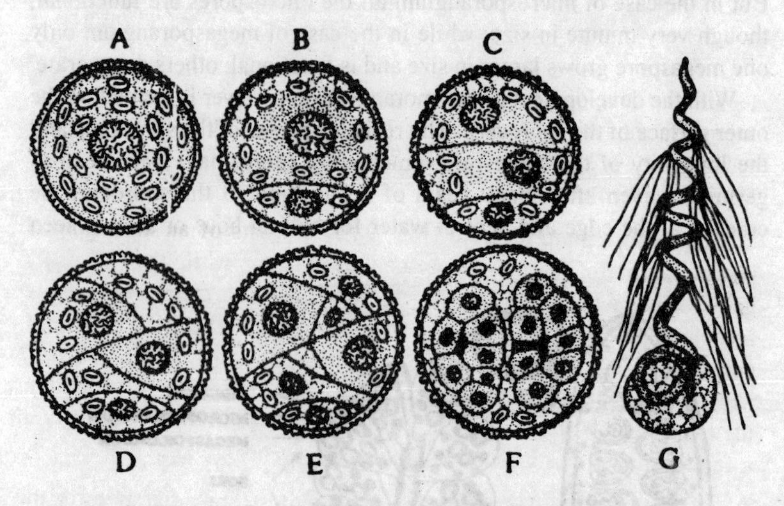

Figure 97. *Marsilea. A–F,* development of the male gametophyte; *G,* an antherozoid. *Redrawn after Fig. 252* in Plant Morphology *by A.W. Haupt by permission of McGraw-Hill Book Company. Copyright 1953.*

gametophyte is complete, while the female gametophyte is slightly slower in development.

The **male gametophyte** (Fig. 97B–F) is endosporous developing within the microspore (A), as in *Selaginella.* A cell cut off on one side is the prothallus cell (B), while the remaining cell of the gametophyte divides into two halves; each half is an antheridium (C). After further divisions two primary spermagenous cells are formed surrounded by a jacket layer (D–E). Each cell then produces 16 *antherozoid mother cells* (F). Each mother cell is metamorphosed into an antherozoid (G). This is much coiled, corkscrew-like and multiciliate, with a mucilagious vesicle containing some food.

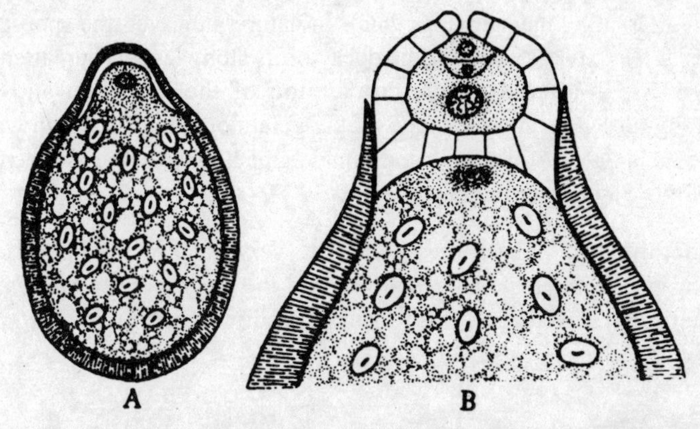

Figure 98. *Marsilea. A,* a megaspore in longi-section; note the nucleus and several starch grains; *B,* female gametophyte with an archegonium projecting out of the spore-coat; note the large egg-cell with the egg-nucleus, ventral cannal cell, neck canal cell and the sterile jacket.

The **female gametophyte** (Fig. 98B) consists of an archegonium protruding out of the megaspore-coat at its apex, while the rest of the gametophyte without any cellular differentiation is a food reservoir. The archegonium consists of (a) a venter with a comparatively large egg-cell with an egg-nucleus in it and a small ventral canal cell, (b) short neck with a neck canal cell, and (c) a sterile jacket of cells. Surrounding the female gametophyte there is a broad gelatinous envelope, which has a funnel-shaped portion converging to the archegonium.

Fertilization. Fertilization takes place almost immediately after the gametophytes are formed. Innumerable antherozoids swarm around the archegonium, many of them swim into the gelatinous envelope on the archegonium side, and some of them pass down the neck. Finally, however, one antherozoid fuses with the egg-nucleus.

Embryo. After fertilization the embryo is rapidly formed. The oospore divides and re-divides and the cells are arranged in four segments. The two segments, inner and outer, on one side give rise to the stem and the first leaf (cotyledon), while the remaining two segments, inner and outer, on the other side give rise to the foot and the root. The adjoining cells of the gametophyte form a cap-like structure or **calyptra** around the developing embryo. The embryo grows rapidly by bursting the calyptra.

Part VI

GYMNOSPERMS

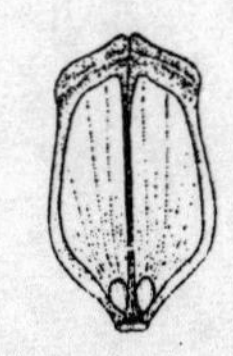

Introduction
Cycas (20 sp.)
Pinus (90 sp.)

Chapter 1

Introduction

Gymnosperms. (*gymnos,* naked; *sperma,* seed) are naked-seeded plants i.e. those in which the seeds are not enclosed within the fruit but are directly borne by the *open carpel* called megasporophyll (i.e. not closed to form the ovary, as in angiosperms). They form an intermediate group between the cryptogams and the angiosperms, being related to the higher forms of cryptogams on the one hand and to the lower angiosperms on the other. Gymnosperms, unlike angiosperms, are an ancient group of plants. They have been classified into 8 orders, of which 4 have become extinct, and the remaining 4 have living representatives, numbering about 700 species. The two common orders with living representatives are: (1) **Cyeadales** (or cycads) with 1 family, i.e. *Cycadaceae,* 9 genera and 100 species; *Cycas* is the typical genus; and (2) **Coniferales** (or conifers with 6 families, 41 genera and over 500 species; *Abietaceae* (=*Pinaceae*) is the largest family of this order with 9 genera and 230 species; Pinus is the typical genus of this family.

Differences between Angiosperms and Gymnosperms

Angiosperms	**Gymnosperms**
1. Branches of a plant are usually of one type and unlimited growth. Leaves are deciduous (annual).	1. Branches of a plant are of two types: of limited and unlimited growth. Leaves are evergreen.
2. Internally, vessels are present in the xylem tissue and companion cells in the phloem tissue.	2. Xylem mainly contains tracheids and phloem does not contain companion cells (except in Gnetum).
3. Flowers are either bisexual or unisexual and they are never arranged in strobili.	3. Flowers are always unisexual and they are arranged in strobili (except female flowers in cycas).
4. Ovules are protected within the ovary, i.e. closed seeded plants.	4. Ovules are freely exposed, i.e. naked seeded plants.
5. Ovules do not contain archegonia.	5. Ovules contain archegonia (except in Gnetum).

6. Endosperm develops after fertilization and it is a triploid ($3n$) tissue. Further, during the formation of male gametophyte, prothallial cells are never formed.

6. Endosperm develops before fertilization and it is a haploid tissue, and during the formation of male gametophyte prothallial cells are formed.

Chapter 2

Cycas (20 sp.)

Occurrence. The genus *Cycas* is represented in India by a few species, of which *C. circinalis* (of Malabar Coast), and *C. revoluta* (a Japanese sp.) and *C. rumphii* (of Malacca) are widely cultivated in Indian gardens; the first two species yield a kind of sago. Besides, *C. pectinata* grows in the low hills of Assam, and *C. beddomei* in the forests of Eastern Peninsula.

Figure 1. A female plant of *Cycas circinalis*

Morphology. *Cycad* (*Cycas;* Fig. 1) is a lower gymnosperm. It consists of an unbranched erect stout and palm-like stem with a crown of fern-like pinnately compound leaves arranged spirally round the apex. The plant bears a long primary (tap) root.

Cycads are dioecious, i.e. male and female flowers are borne by two separate plants. The male flowers are aggregated in a cone (Fig. 2) borne at the apex of the stem. The male cone consists of a collection of **stamens**

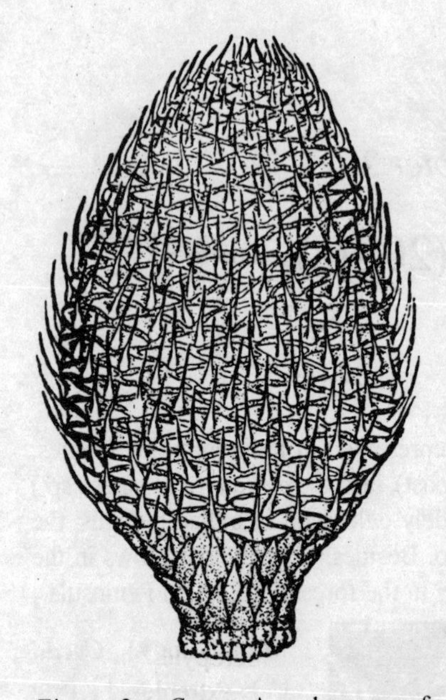

Figure 2. *Cycas.* A male cone of *Cycas pectinata* with stamens (micro-sporophylls).

or **micro-sporophylls** which are arranged spirally round the axis. Each stamen (Fig. 3C) is in the form of a scale, narrowed below and broadened above. It bears on its undersurface several **pollen-sacs** or microsporangia grouped in sori. There are usually 2 to 6 pollen-sacs in each sorus. In each pollen-sac there are numerous **pollen grains** or microspores. Before the pollen grain sheds from the pollen-sac its nucleus divides once, producing an extremely reduced male **prothallus cell** (Fig. 5A) on one side and a large **antheridial cell** on the other side. The latter divides again and produces a **generative cell** and a **tube cell**. Subsequent changes take place after pollination. All the stages are, however, diagrammatically shown below.

Pollen grain → prothallus cell
Pollen grain → antheridial cell → generative cell → stalk cell
antheridial cell ↘ tube cell
generative cell → body cell → spermatozoids (two)

In *Cycas* there is no female cone; the plant bears near its apex a rosette of **carpels** or **megasporophylls** (Figs. 1 & 3A–B) which do not form a cone but are arranged alternating with the leaves. They are usually 15–30 cm long, flattened or bent over like a hood, and often dilated above. In many species they are covered all over with soft brownish hairs. The margin of the carpel may be entire, crenate or pectinate (pinnately divided). Carpels are open, bearing usually 2–3 pairs of ovules, sometimes more, on their two margins. The ovules grow considerably, even before fertilization, and are commonly oval and fairly large.

The **ovule** in longitudinal section (Fig. 4A) shows: (a) a thick integument consisting of three layers, (b) a micropyle, (c) a pollen chamber, (d) a nucellus fused with the integument, (e) a **female prothallus** (often called the endosperm) which grows quickly after

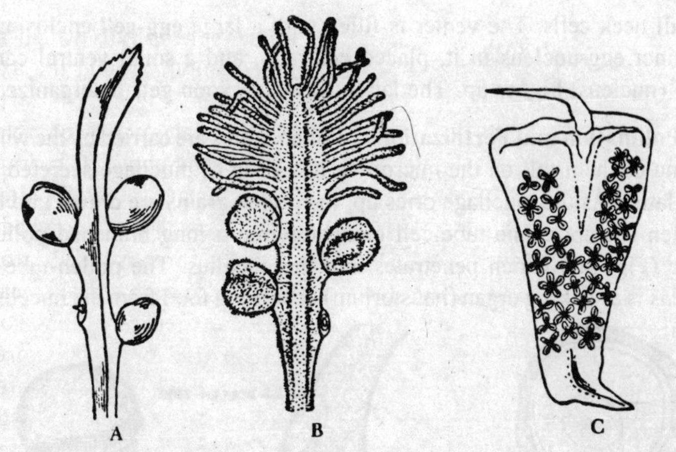

Figure 3. *Cycas.* Carpels and Stamen. *A,* a carpel of *Cycas circinalis; B,* a carpel of *Cycas revoluta; C,* a stamen of *Cycas pectinata* with numerous pollen-sacs.

fertilization and forms the major part of the seed, (f) a few archegonia (2–8) borne by the female prothallus towards the micropyle, and (*g*), an archegonial chamber. Each archegonium (Fig. 4B) is extremely reduced in size and consists of a swollen **venter** and a very short **neck** with two

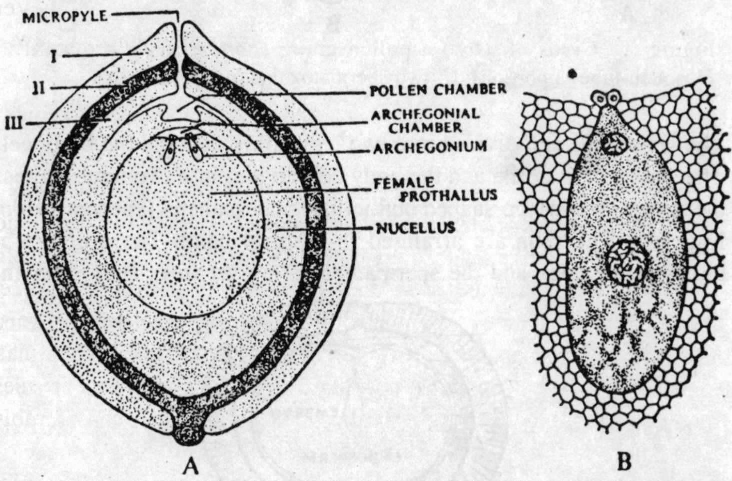

Figure 4. *Cycas. A,* an ovule in longitudinal section; *I, II* and *III,* outer, middle (stony) and inner layers of the integument; *B,* an archegonium; note the short neck with two neck cells, a large venter with egg-cell and egg-nucleus, and higher up a small ventral canal nucleus.

small neck cells. The venter is filled with a large **egg-cell** enclosing a distinct egg-nucleus in it, placed centrally, and a small ventral canal cell (nucleus) higher up. The latter, however, soon gets disorganized.

Pollination and Fertilization. Pollen grains are carried by the wind. Some of them fall on the micropyle in a drop of mucilage secreted by the latter. As the mucilage dries up, the pollen grains are drawn into the pollen-chamber. The tube-cell elongates into a long branched pollen-tube (Fig. 5B) which penetrates into the nucellus. The pollen-tube of *Cycas* is a sucking organ (haustorium) absorbing food from the nucellus.

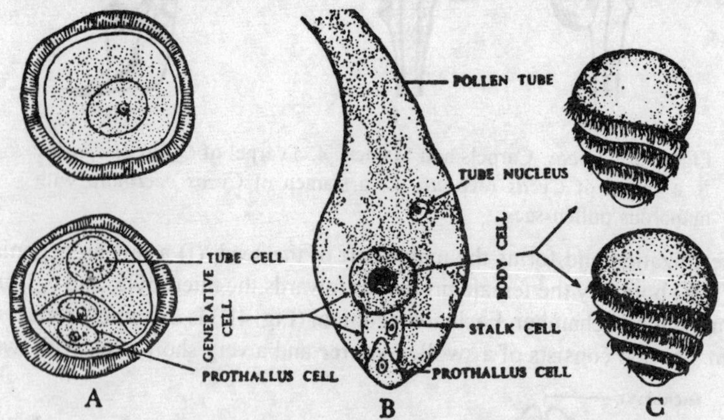

Figure 5. *Cycas. A,* (*top*) a pollen grain; (*bottom*) male prothallus; *E,* pollen-tube (a portion); *C,* two spermatozoids (or sperm cells).

The generative cell divides into two—the **stalk cell** and the **body cell**. The stalk cell is sterile and the body cell divides into two large (in fact, the largest known) top-shaped multiciliate male gametes (**spermatozoids** Fig. 5C). The cilia are arranged in a spiral band. The pollen-tube bursts at the apex and the spermatozoids are set free. They enter the

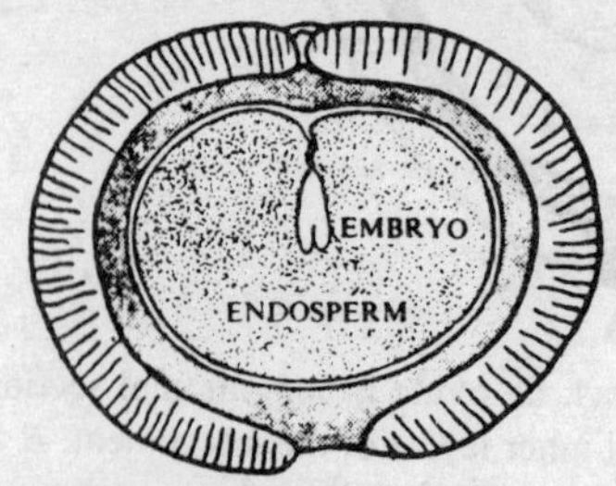

Figure 6. A *Cycas* seed in section.

archegonium and one of them fuses with the egg-nucleus. Fertilization is thus effected.

Seed. The fertilized egg-cell grows into an embryo, and the ovule as a whole into a seed. The mature seed bears only one embryo with two cotyledons lying embedded in the endosperm which again is surrounded by the integument or seed coat. The latter consists of an outer fleshy layer and an inner stony layer. The endosperm stores a considerable quantity of food for the embryo to be utilized at the time of germination.

Chapter 3

Pinus (90 sp.)

Occurrence. ***Pine*** (*Pinus;* Figs. 7–8) grows in rich abundance in the temperate regions of eastern and western Himalayas at an altitude of 1,200 to 3,300 metres. Common species are: Khasi pine (*P. khasya*), long-needle pine or CHIR pine (*P. longifolia*), and blue pine (*P. excelsa*)—the latter two growing in both eastern and western Himalayas. *P. sylvestris,* a British sp., is cultivated in hill stations.

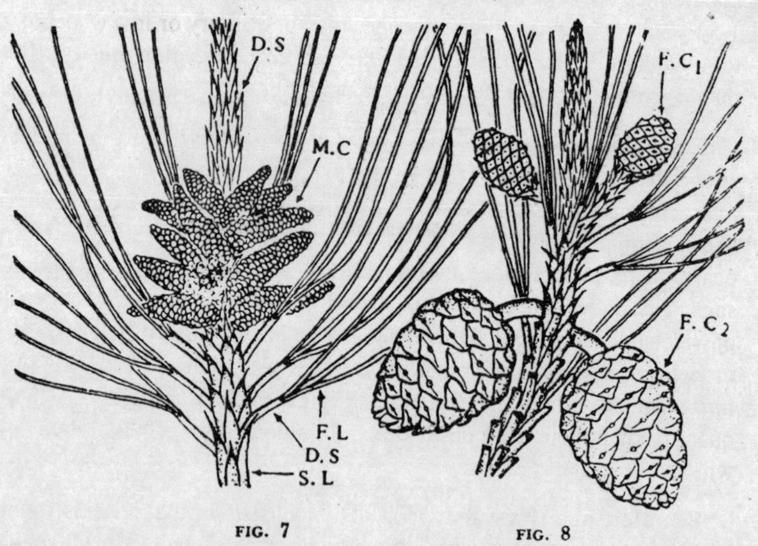

Figure 7. *Pinus.* A male shoot of unlimited growth. *M.C,* male cones; *D.S,* dwarf shoots of limited growth; *F.L,* foliage leaves or needles; *S.L,* scale leaves. Figure 8. A female shoot of unlimited growth. $F.C_1$, young female cone of the current year; $F.C_2$, maturing female cone of the previous year. For mature cone of the third year see Fig. 15.

Morphology. Pine is a tall erect evergreen tree, sometimes growing to a height of 45 m. with a basal girth of 3 m. The plant has a well-developed tap root and numerous aerial branches with green needle-like foliage leaves. The stem is rugged and covered with scale bark peeling off here and there in strips. Branches are of two kinds: long (of unlimited growth) usually in whorls, and dwarf (of limited growth; Fig. 7). Leaves are also of two kinds: long green needle-like foliage leaves (commonly called needles) borne only on dwarf branches (or foliar spurs, as they are called) and small brown scaly leaves borne on both kinds of branches. The number of leaves in a cluster varies from 1 to 5, 3 in *P. khasya* and *P. longifolia,* 5 in *P. excelsa,* 2 in *P. sylvestris,* and 4 or 1 in certain species.

Reproduction. *Pinus* is the sporophyte. It bears two kinds of cones (Figs. 7–8)—male and female—on separate branches of the same plant (monoecious). The male cone is made of **microsporophylls** or **stamens**, and the female cone of **megasporophylls** or **carpels**. Cones always develop on the shoots of the current year a little below the apex. Several male cones appear in clusters, each in the axil of scale leaves, and are 1.5 to 2.5 cm in length. Female cones may be solitary or in a whorl of 2 to 4, each in the axil of scale leaves. Male cones develop much earlier than female cones. Flowers have no perianth.

Male Cone (Figs. 7 & 9A). This consists of a number of **microsporophylls** or **stamens** arranged spirally round the axis (Fig. 9A). Each microsporophyll (Fig. 9 B–D) is differentiated into a stalk (filament) and a terminal leafy expansion (anther), and its tip is bent upwards. It bears on its undersurface two pouch-like **microsporangia** or **pollen-sacs**. The pollen-sac contains numerous microspore mother cells, each of which undergoes reduction division to produce a tetrad of **microspores** or **pollen grains**. Each pollen grain has two coats—*exine* (outer) and *intine* (inner). The exine forms two wings, one on each side. A huge quantity of pollen is produced for pollination to be brought about by wind.

Female Cone (Figs. 8 & 15). This consists of a short axis round which a number of small, thin, dry, brownish scales, called **bract scales** or **carpellary scales** (each corresponding to a carpel or megasporophyll; Fig. 10) are spirally arranged. They are inconspicuous in the mature cone. On the upper surface of each bract scale there is another stouter and bigger scale, woody in nature and somewhat triangular in shape,

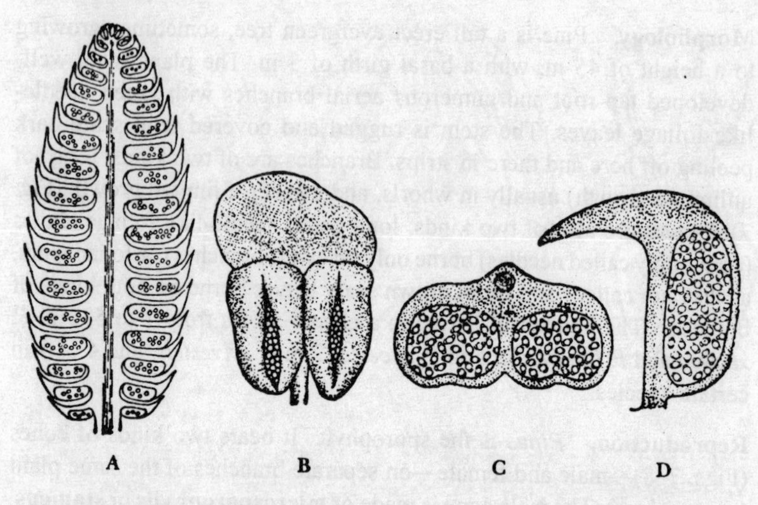

Figure 9. *Pinus. A,* a male cone in longi-section; *B,* microsporophyll with two microsporangia (pollen-sacs); *C,* the same in transection; *D,* the same in longi-section. Note the microspores (pollen grains) with wings.

known as the **ovuliferous scale** (Fig. 10). It bears two sessile **ovules** on its upper surface at the base. Each ovule (Fig. 11) is orthotropous, and consists of a central mass of tissue—the **nucellus** or **megasporangium**, surrounded by a single **integument** made of three layers. The integument leaves a rather wide gap known as the **micropyle** which is turned downwards. Within the nucellus a megaspore mother cell becomes apparent soon. It undergoes reduction division to produce four **megaspores** in a linear tetrad. Only one megaspore is functional, however; the other three degenerate.

Male Gametophyte (Fig. 13A). The microspore or pollen grain begins to divide before it is set free from the pollen-sac and gives rise to an extremely reduced **male prothallus** (male gametophyte) within it. The prothallus consists of (a) 2 or 3 small cells (**prothus cells**) at one end, which soon become disorganized, and (b) an **antheridial cell** which is the remaining large cell of the prothallus. The antheridial cell divides and forms a **generative cell** and a **tube cell**. The pollen grain sheds at this stage and subsequent changes take place on the nucellus after pollination.

Female Gametophyte (Fig. 11). The megaspore germinates *in situ* and gives rise to the **female prothallus** (female gametophyte) within the nucellus. The megaspore digests a big portion of the nucellus and

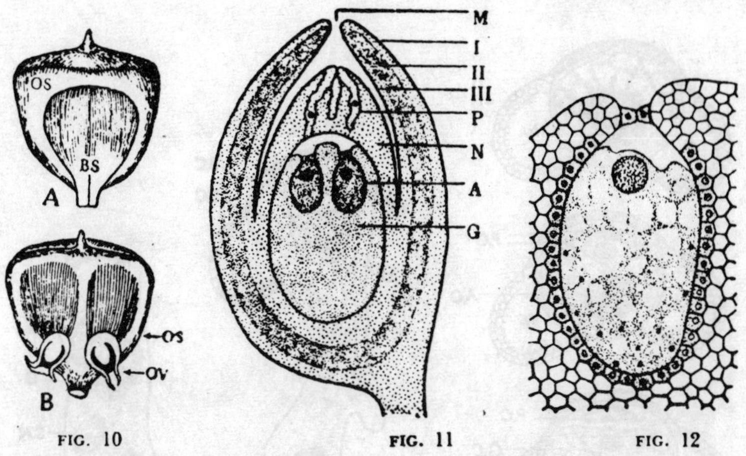

Figure 10. *Pinus*. Megasporophyll. *A*, lower surface; *B*, upper surface. *BS*, bract scale; *OS*, ovuliferous scale; *OV*, ovule with one integument. Figure 11. Ovule in longi-section. *M*, micropyle; *I*, *II* and *III* are outer, middle (stony) and inner layers of the integument; *P*, pollen-tube; *N*, nucellus; *A*, archegonium; *G*, female gametophyte (endosperm). Figure 12. An archegonium; note the neck with two neck cells, and a large venter with egg-cell and a distinct egg-nucleus in it.

enlarges considerably. It undergoes free nuclear division, finally forming a solid mass of tissue—thc female gametophyte (otherwise designated as the endosperm) within the nucellus, being completely *endosporous*. The endosperm grows quickly only after fertilization and surrounds the embryo in the seed (Fig. 17). At the micropylar end of the prothallus, lying embedded in it, develop 2 to 5 **archegonia**. Each archegonium (Fig. 12) consists of a swollen venter and a short neck. The venter encloses a large **egg-cell** with a distinct **egg-nucleus** in it, almost filling the cavity, and a ventral canal cell which, however, soon gets disorganized. The neck consists of 2 or more usually 8, neck cells but no neck canal cell. The archegonial wall is made of a layer of small but distinct cells called the *jacket cells*. The archegonia, however, mature slowly and become ready for fertilization only in the followIng year.

Pollination and Fertilization. ***Pollination*** takes place through the agency of the wind, usually in May/June, soon after the female cone emerges from the bud. The pollen sacs burst and the winged pollen grains are blown about by the wind like a yellow cloud of dust. Some of them happen to fall on the young female cone. Evidently there is a considerable wastage of pollen. The pollen grains pass between the two

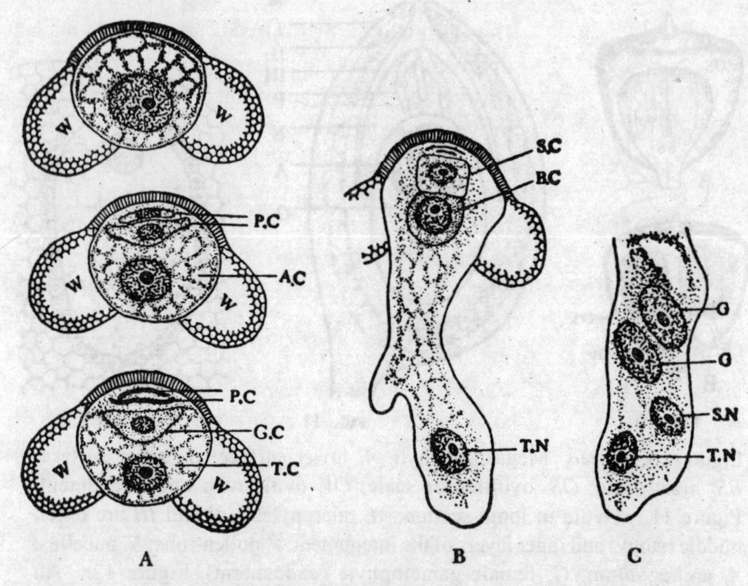

Figure 13. *Pinus. A,* pollen and male gametophyte; *B,* pollen-tube; *C,* lower portion of the same. *W,* wing; *P.C,* prothallus cells; *A.C,* antheridial cell; *G.C,* generative cell; *T.C,* tube cell; *S.C,* stalk cell; *B.C,* body cell; *T.N,* tube nucleus; *G,* male gamete; *S.N,* stalk nucleus.

slightly opened scales and are deposited at their base. A quantity of mucilage secreted at this time at the base of the micropyle is now drawn into the nucellus of the ovule together with the pollen grains. The latter then lodge somewhere at the apex of the nucellus for about a year. After pollination the scales close up, and so does the wide gaping integument. **Fertilization** (first discovered in 1883–84) takes place in the following year at about the same time when the archegonia mature. The outer coat (exine) of the pollen grain bursts and the inner coat (intine) grows out into a slender **pollen-tube** (Fig. 13B). The tube pushes forward through the nucellus and finally reaches the neck of the archegonium (Fig. 11). The tube nucleus passes into the pollen-tube. The generative cell divides and forms a **stalk cell** and a **body cell**. Both of them now migrate into the pollen-tube. The stalk cell is sterile, while the body cell is fertile. The latter divides and produces two passive (non-ciliate) **male gametes** (Fig. 13C). The pollen-tube bursts at the apex and the two gametes are liberated. One of them moves towards the cgg-nucleus, and soon fuses with it. Thus fertilization is effected. The other male gamete, the stalk cell and the tube nucleus become disorganized.

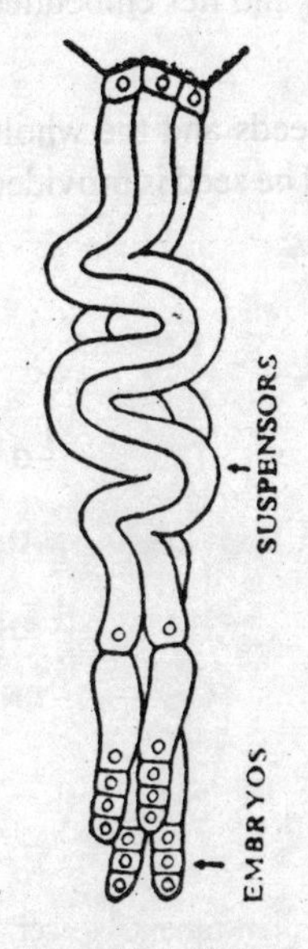

Figure 14. Embryos in *Pinus*.

Development of the Embryo. The fusion-nucleus or oospore undergoes two successive divisions within the egg-cell, and 4 nuclei are formed. They move to the bottom of the egg-cell and divide again into 8 nuclei. Walls now appear between them. Further divisions result in the formation of four tiers of four cells each; this 16-celled stage is called the **pro-embryo**. The lowest tier of four cells is the 'embryonic tier'. The embryonal cells further divide and give rise to *four potential* **embryos** (Fig. 14). The embryos are thrust into the endosperm by very long and tortuous **suspensors**. Besides, as there are a few archegonia, in the ovule, some more embryos may be formed. This **polyembrony** is characteristic of pines and other conifers. Of the many embryos thus formed the strongest one only survives and matures, while the rest degenerate. A mature seed has thus only one embryo (Fig. 17). The latter consists of an axis with a radicle, a hypocotyl and

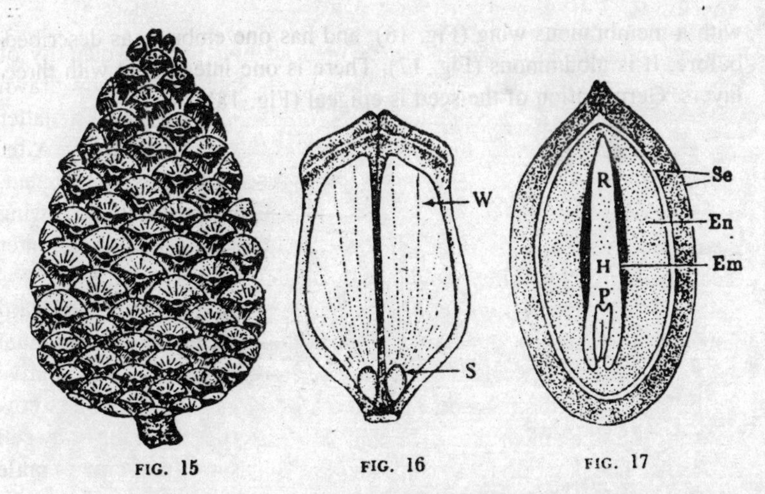

Figure 15. *Pinus.* A mature female cone. Figure 16. A mature megasporophyll with two seeds (winged); *W,* wing; *S,* seed. Figure 17. A seed in longi-section; *Em,* embryo (*P,* plumule with many cotyledons; *H,* hypocotyl; *R,* radicle); *En,* endosperm; *Se,* seed coat (with three layers).

a tiny plumule with a number of cotyledons (2 to 15) and lies embedded in the endosperm.

Seed. After fertilization the ovules develop into seeds and the whole female flower into a dry, hard, brown **cone** (Fig. 15). The seed is provided

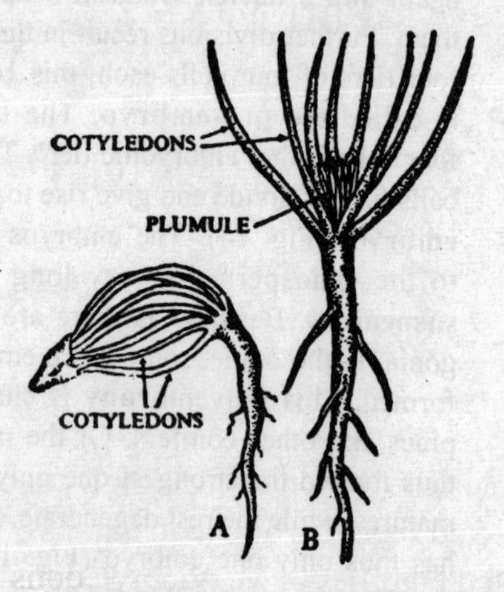

Figure 18. Germinating seed (*A*) and seedling (*B*) of *Pinus*.

with a membranous wing (Fig. 16), and has one embryo, as described before. It is albuminous (Fig. 17). There is one integument with three layers. Germination of the seed is epigeal (Fig. 18).

Part VII

ANGIOSPERMS

Principles and Systems of Classification
Selected Families of Dicotyledons
Selected Families of Monocotyledons

Chapter 1

Principles and Systems of Classification

Systematic Botany or Taxonomy. It deals with the description, identification and naming of plants, and their classification into different groups according to their resemblances and differences mainly in their morphological characteristics. So far as angiosperms or higher 'flowering' plants are concerned, it has been estimated that over 199,000 species (dicotyledons—159,000 and monocotyledons—40,000) are already known to us, and many more are still being discovered and recorded. Thus plants are not only numerous but they are also of varied types, and it is not possible to study them unless they are arranged in some orderly system. The object of systematic botany or taxonomy is to describe, name and classify plants in such a manner that their relationship with regard to their descent from a common ancestry may be easily brought out. The ultimate object of classification is to arrange plants in such a way as to give us an idea about the sequence of their evolution from simpler. earlier and more primitive types to more complex, more recent and more advanced types in different periods of the earth's history.

UNITS OF CLASSIFICATION

Species. A species is a group of individuals (plants or animals) of one and the same kind. Evidently they resemble one another in almost all important morphological characteristics—both vegetative and reproductive—so closely that they may be regarded as having been derived from the same parents. Further, all individuals of a species have normally the same number of chromosomes in their cells—2*n* or *diploid* in the vegetative cells, and *n* or *haploid* in the reproductive cells. Thus all pea plants constitute a species. Similarly all banyan plants, peepul plants, and mango plants constitute different and distinct species. Occasionally, owing to variations in climatic or edaphic conditions, individuals of a species may show a certain amount of variation in form, size, colour and other minor characteristics. Such plants are said to form **varieties**. A species may consist of one or more varieties or none at all.

Genus. A genus is a collection of species which bear a close resemblance to one another in the morphological characters of the floral or reproductive parts. For example, banyan, peepul and fig are different species because they differ from one another in their vegetative characters such as the habit of the plant, the shape, size and surface of the leaf, etc. But these three species are allied because they resemble one another in their reproductive characters, namely, inflorescence, flower, fruit and seed. Therefore, banyan, peepul and fig come under the same genus, and that is *Ficus*.

Binomial Nomenclature. As mentioned in introduction, this is the scientific method of naming species of plants or animals in two parts: the first refers to the genus and the second to the species. This system of naming plants or animals with a *binomial* was first introduced by Linnaeus in 1735 and the rules for its final adoption were drawn up by the International Botanical Congress held at Vienna in the year 1905. The name of the author who first described a species is also written in an abbreviated form after the name of the species, e.g. *Mangifera indica* Linn. Here Linn. refers to the author, Linnaeus; who first described the plant.

Family. A family is a group of genera which show general structural resemblances with one another, mainly in their floral organs. Thus in the genera *Gossypium* (cotton), *Hibiscus* (China rose), *Abelmoschus* (lady's finger), *Thespesia* (Portia tree), *Sida* (B. BERELA; H. BARIARA), *Malva* (mallow), *Althaea* (hollyhock), etc., we find free lateral stipules, epicalyx, twisted aestivation of corolla, monadelphous stamens, unilocular anthers, axile placentation, etc. So all the above mentioned genera belong to the same family, and that is *Malvaceae*.

SYSTEMS OF CLASSIFICATION

Theophrastus (370–285 B.C.) was the first to propose a system of classification basing on the habit of plants. He classified the plants, according to their heights, into trees, shrubs, under shrubs and herbs. But later this system was discarded and another new system, in which much importance was given on the characters of sex organs (stamens and carpels) of a flower, was adopted. A very well-known such a system of classification of plants was proposed by Carolus Linnaeus (1707–1778). This system is known as sexual or artificial system which is very helpful in identifying plants. In time, this system was also replaced by another system of classification called natural system of classification. Such a system was proposed by George Bentham (1800–1884) and Joseph Dalton Hooker (1817–1911) basing on gross morphology of plant structures, i.e. characters of both vegetative and reproductive structures. This system indicates the natural relationships existing between the plants. This system is still followed in many places. Recently another system called phylogenetic systems of classification, proposed by Engler (1886), Hutchinson (1942) and others, has come into the lime light. In this system, plants are classified basing on their true phylogenetic relationships and thereby indicating the sequence of evolution among the plants. Even the modern trend now is to classify the plants based on all the evidences that could be made available from different branches of studies like serum diagnosis, cytology, anatomy, biochemistry, plant geography, etc. besides morphology, in order to establish a pure phylogenetic relationship between the plants.

In the **artificial system** only one or at most a few characters are selected arbitrarily and plants are arranged into groups according to such characters; as a result, closely related plants are often placed in different groups, while quite unrelated plants are often placed in the same group because of the presence or absence of a particular character. This system enables us to determine readily the names of plants but does not indicate the natural relationship that exists among the individuals forming a group. It is like the manner of arrangement of words in a dictionary in which, except for the alphabetical order, adjacent words do not necessarily have any agreement with one another.

Linnaean System (1735). The best-known artificial system is the one compiled by Linnaeus and published by him in 1735. Linnaeus classified plants according to the characteristics of their reproductive organs, viz. stamens and carpels. According to this system plants are mainly divided into 24 classes: 23 of phanerogams and one of cryptogams. Phanerogams were further sub-divided into groups and sub-groups according to the following characteristics: unisexual or bisexual flowers, monoecious or dioecious plants, number of stamens, adhesion or cohesion of stamens, number and length of stamens, number of carpels, apocarpous or syncarpous pistil, etc.

In the **natural system** all the important characteristics are taken into consideration, and plants are classified according to their related characteristics. Thus according to their similarities and differences, mostly in their important morphological characteristics, plants are first classified into a few big groups. These are further divided and sub-divided into smaller and smaller groups until the smallest division is reached and that is a species. All modern systems of classification are natural and they supersede the artificial ones by the fact that they give us a true idea about the natural relationship existing between different plants and also of the sequence of their evolution from simpler to more complex types during different periods in the earth's history.

According to the natural system the plant kingdom has been divided into two *divisions,* viz. **eryptogams** or 'flowerless' plants (see Part V), and **phanerogams** or 'flowering' plants. Phanerogams have again been divided into two *sub-divisions,* viz. **gymnosperms** or naked-seeded plants (see Part VI), and **angiosperms** or closed-seeded plants. Angiosperms have further been divided into two *classes,* viz. **dicotyledons** and **monocotyledons** (see p. 481). The classes have been divided into sub-classes, series and *orders;* orders into *families;* families again into *genera* and *species;* and sometimes species into varieties.

Bentham and Hooker's System (1862–83). The natural system that is in practice in India is that of Bentham and Hooker which was published by them during the above period. According to these authors the dicotyledons have been divided into three *sub-classes,* as follows:

1. **Polypetalae.** Both calyx and corolla present; petals free; stamens and carpets also usually present; the former often indefinite and the latter apocarpous or syncarpous. Within the sub-class progress is indicated through polysepalous calyx to gamosepalous calyx, through indefinite number of stamens to definite number, and through hypogyny, perigyny and epigyny.

2. **Gamopetalae.** Both calyx and corolla present; the latter gamopetalous; stamens almost always definite and epipetalous; carpels usually two but sometimes more, free or united; ovary inferior or superior. This sub-class is also called Corolliflorae.

3. **Monochlamydeae.** Flowers incomplete; either calyx or corolla absent, or sometimes both the whorls absent; flowers generally unisexual. It usually includes the families which do not fall under the above two sub-classes.

According to Bentham and Hooker the monocotyledons are divided into seven *series*. A much simpler classification has been put forward by Vines in England. According to this author the monocotyledons are divided into three *sub-classes,* as follows:

1. **Petaloideae.** The perianth is usually petaloid.
2. **Spadiciflorae.** The inflorescence is a spadix, and is enclosed in one or more spathes.
3. **Glumiflorae.** The flower is enclosed in special bracts, called *glumes* (see p. 91).

Following the above scheme of classification any plant may be referred to its systematic position. Let us take BANI cotton.

Division	...	...	...	Phanerogam
Sub-division	...	...	...	Angiosperm
Class	...	...	...	Dicotyledon
Sub-class	...	...	...	Polypetalae
Series	...	...	...	Thalamiflorae
Order	...	...	...	Malvales
Family	...	...	...	*Malvaceae*
Genus	...	...	...	*Gossypium*
Species	...	...	...	*indicum*

A plant is always denominated by the *generic* and *specific* name, with the name of the author at the end. Thus BANI cotton is *Gossypium indicum* Linn.

Differences between Dicotyledons and Monocotyledons

	Dicotyledons	*Monocotyledons*
1. Embryo	with 2 cotyledons	with 1 cotyledon.
2. Root	tap root	fibrous roots
3. Venation	reticulate, with free ending of veinlets	parallel, with no free ending of veinlets
4. Flower	mostly pentamerous	trimerous
5. Vascular bundles	in stems collateral and open, arranged in a ring; in roots radial, xylem bundles usually 2 to 6	in stems collateral and closed, scattered; in roots radial, xylem bundles usually many, rarely few (5–8)
6. Secondary growth	present in both stem and root	absent (with but few exceptions)

Floral Diagram. The number of parts of a flower, their general structure, arrangement, aestivation, adhesion, cohesion and position with respect to the mother axis may be represented by a diagram known as the **floral diagram**. The floral diagram is the ground plan of a flower. In the diagram the calyx lies outermost, the corolla internal to the calyx, the androecium in the middle, and the gynoecium in the centre. Adhesion and cohesion of members of floral whorls may also be shown by connecting the respective parts with lines; as, for example, Fig. 1A shows that there are altogether ten stamens, of which nine are united into one bundle (cohesion) and the remaining one is free; while Fig. 35

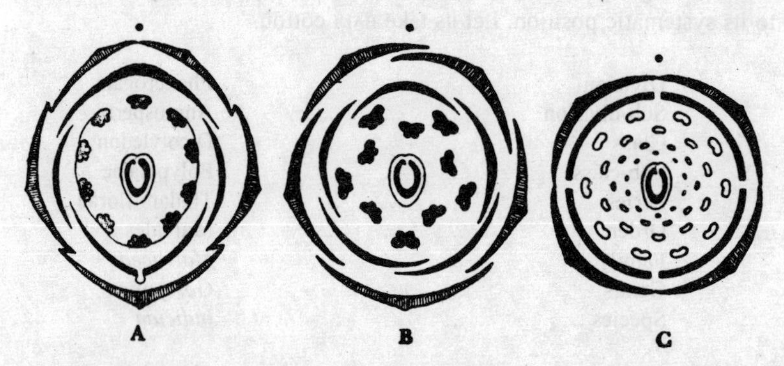

Figure 1. Floral Diagrams. *A, Papilionaceae; B, Caesalpinieae; C, Mimoseae.*

shows that petals and stamens are united (adhesion). The black dot on the top represents the position of the mother axis (not the pedicel) which bears the flower. The axis lies behind the flower and, therefore, the side of the flower nearest the axis is called the *posteior* side, and the other side away from the axis the *anterior* side. The floral characteristics of a species may be well-represented by a floral diagram, while to represent a genus or a family more than one diagram may be necessary.

Floral Formula. The different whorls of a flower, their number, cohesion, adhesion and their relative position may be represented by a formula known as the **floral formula**. In the floral formula K stands for calyx, C for corolla, P for perianth, A for androecium, and G for gynoecium. The figures following the letters K, C, P, A and G indicate the number of parts of those whorls. Cohesion of a whorl is shown by enclosing the figure within brackets, and adhesion is shown by a line drawn on the top of the two whorls concerned. In the case of the gynoecium the position of the ovary is shown by a line drawn above or below G or the figure. If the ovary is superior the line should be below it, and if it is inferior the line should be on the top. Thus all the parts of a flower may be represented in a general way by the floral formula; the floral characters of a family may also be represented by one or more formulae, as noted below. Besides, certain symbols are used to represent some particular features of flowers; e.g., ♂ male, ♀ female, ⚥ bisexual, ⊕ regular, % zygomorphic, ∞ indefinite etc.

Ranunculaceae : $\oplus$ ⚥ $K_5 C_5 A_\infty \underline{G}_\infty$

Cruciferae : $\oplus$ ⚥ $K_{2+2} C_4 A_{2+4} \underline{G}_{(2)}$

Malvaceae : $\oplus$ ⚥ $K_{(5)} \overline{C_{(5)} A_{(\infty)}} \underline{G}_{(5-\infty)}$

Solanaceae : $\oplus$ ⚥ $K_{(5)} \overline{C_{(5)} A_5} \underline{G}_{(2)}$

Labiatae : % ⚥ $K_{(5)} \overline{C_{(5)} A_4} \underline{G}_{(2)}$

Liliaceae : $\oplus$ ⚥ $P_{3+3} A_{3+3} \underline{G}_{(3)}$

Chapter 2

Selected Families of Dicotyledons

Family 1—*Ranunculaceae* (over 1,200 sp.—157 sp. in India)

Habit: mostly perennial herbs or climbing shrubs. **Leaves:** simple, often palmately divided, sometimes compound, alternate (rarely opposite), often both radical and cauline, usually with sheathing base. **Inflorescence:** typically cymose (racemose in larkspur and aconite). **Flowers:** mostly regular except in larkspur and aconite, bisexual and hypogynous; sepals and petals in whorls; stamens and carpels typically spiral on the elongated thalamus. **Calyx:** sepals usually 5, sometimes more, free. **Corolla:** petals 5 or more, free, sometimes absent, often with nectaries, imbricate; perianth leaves (i.e. calyx and corolla not distinguishable) free and petaloid. **Androecium:** stamens numerous, free, spiral. **Gynoecium:** carpels usually numerous, sometimes few, free (apocarpous), spiral with one or more ovules in each. **Fruit:** an etaerio of achenes or follicles, rarely a berry or capsule. **Seeds:** albuminous. *Floral formula*—$\oplus ⚥ K_5C_5A_{\infty}\underline{G}_{\infty}$.

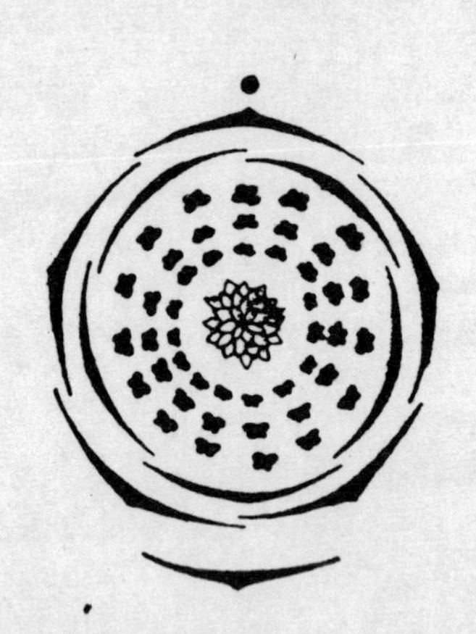

Figure 2. Floral diagram of *Ranunculaceae.*

Examples. Useful plants: monk's hood or aconite (*Aconitum ferox;* B. KATBISH; H. BISH)—medicinal, tuberous roots containing a very poisonous alkaloid, black cumin (*Nigella sativa;* B. KALAJIRA; H. KALOUNJI)—seeds used as a condiment; **ornamental:** larkspur (*Delphinium*), wind flower

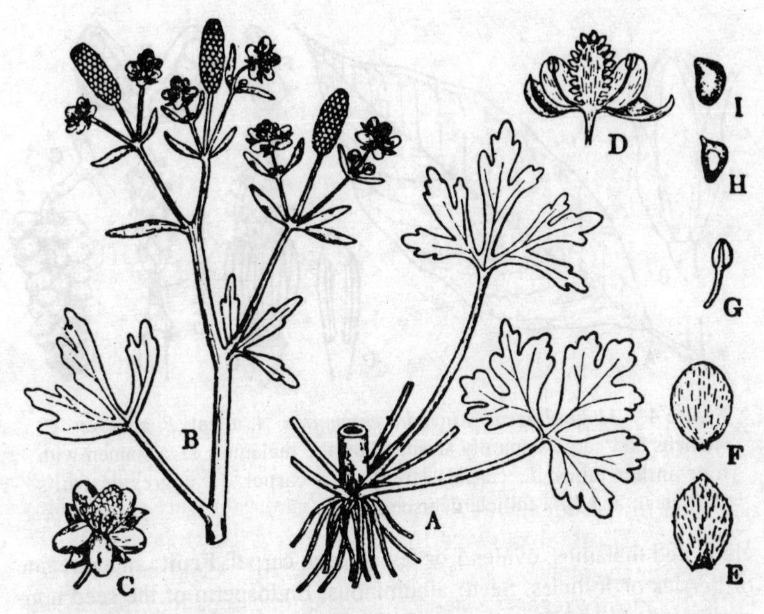

Figure 3. *Ranunculaceae. Ranunculus sceleratus. A,* basal portion of the plant with leaves and roots; *B,* upper portion of the same with inflorescence; *C,* a flower; *D,* flower cut longitudinally; *E,* a sepal; *F,* a petal; *G,* a stamen; *H,* a carpel; and *I,* a fruit (achene).

(*Anemone*)—a small tuberous plant with woolly achenes for wind-dispersal, virgin's bower (*Clematis*)—a climbing shrub, buttercup (*Ranunculus*), etc.; **other common plants:** some species of *Ranunculus,* e.g. Indian buttercup (*Ranunculus sceleratus;* Fig. 3) usually growing on river- and marsh-banks, water crowfoot (*R. aquatilis*) growing in water and showing heterophylly, etc., traveller's joy (*Naravelia*)—a climbing shrub, etc.

Family 2—*Magnoliaceae* (250 sp.—30 sp. in India)

Habit: shrubs and trees. **Leaves:** simple, alternate, often with large stipules covering young leaves. **Flowers:** solitary, terminal or axillary, often large and showy, aromatic; they are regular, bisexual and hypogynous. **Perianth leaves:** all alike petaloid, deciduous; either cyclic being arranged in whorls of 3 (trimerous) or acyclic (spiral); sometimes the outer whorl sepaloid. **Androecium:** stamens numerous, free; filament short or absent; anther-lobes linear, 4; with prolonged connective. **Gynoecium:** carpels numerous, free, arranged spirally around the

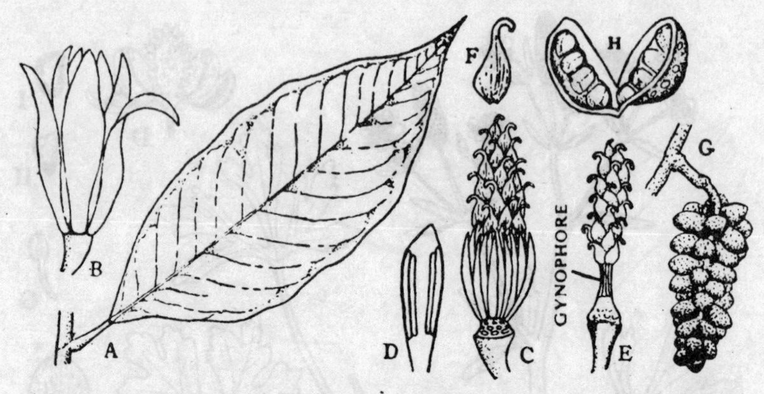

Figure 4. *Magnoliaceae. Michelia champaca. A,* a leaf; *B,* a flower; *C,* stamens and carpels spirally arranged on the thalamus; *D,* a stamen with four anther-lobes; *E,* carpels (free); *F,* a carpel; *G,* aggregate fruits (follicles); and *H,* a follicle dehiscing.

elongated thalamus; ovules 1 or few in each carpel. **Fruit:** an aggregate of berries or follicles. **Seed:** albuminous. Endosperm of the seed non-ruminated. *Floral formula*—$\oplus$ ⚥ $P_{\infty} A_{\infty} \underline{G}_{\infty}$.

Examples. Mostly ornamental evergreen trees and shrubs, e.g. *Magnolia grandiflora, M. fuscata, Michelia champaca, M. alba,* tulip tree (*Liriodendron*), *Talauma,* etc.

Family 3—*Annonaceae* (850 sp.—100 sp. in India)

Habit: shrubs and trees, sometimes climbers. **Leaves:** simple, alternate, distichous (phyllotaxy ½)and exstipulate. **Flowers:** regular, bisexual, and hypogynous; often aromatic. **Perianth:** usually in three whorls of three members each; sepals 3 and petals 6 in two whorls. **Androecium:** stamens numerous, free, arranged spirally round the slightly elongated thalamus; filament short or absent; anther-lobes linear, 4; with prolonged connective. **Gynoecium:**

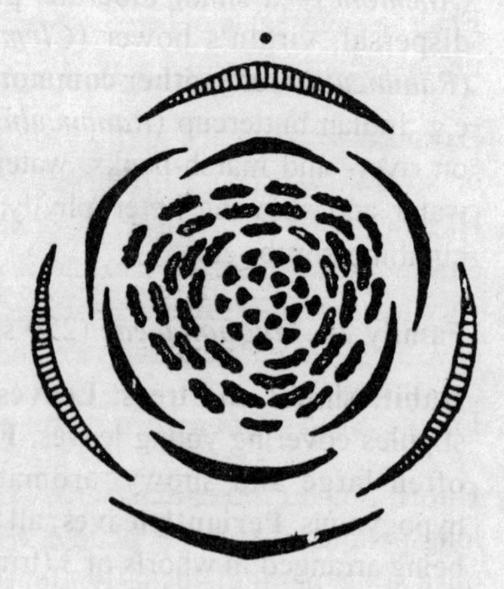

Figure 5. Floral diagram of *Annonaceae* (*Artabotrys*).

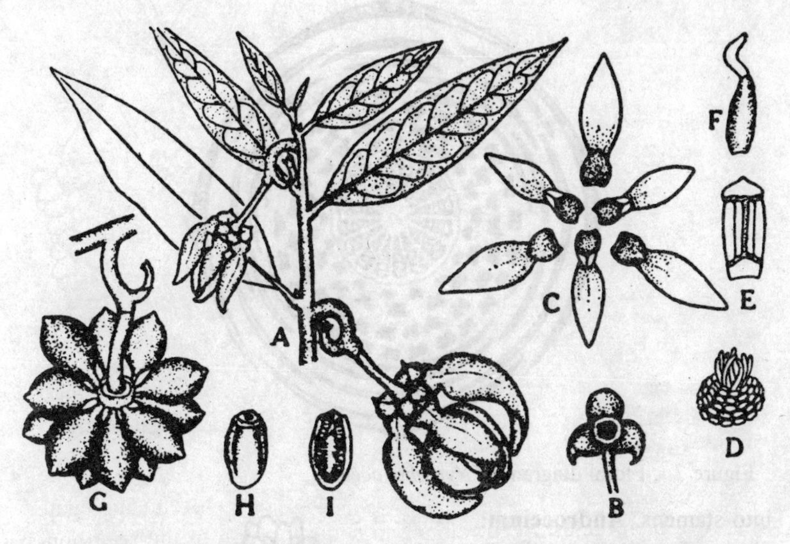

Figure 6. *Annonaceae. Artabotrys. A,* a branch with two flowers; *B,* calyx; *C,* petals spread out; *D,* Stamens and carpels; *E,* a stamen with four anther-lobes; *F,* a carpel; *G,* an aggregate of berries; *H,* a seed; and *I,* the seed cut longitudinally showing the ruminated endosperm.

carpels numerous, free or connate; ovules one to many in each carpel. **Fruit:** an aggregate of berries. **Seed:** the endosperm distinctly ruminated, i.e. marked by irregular wavy lines. *Floral formula*—$\oplus$ ⚥$K_3C_{3+3}A_{\infty}\underline{G}_{\infty}$.

Examples. Custard-apple (*Annona squamosa;* B. ATA; H. SHARIFA or SITAPHAL)—fruit edible, bullock's heart (*A. reticulate;* B. NONA; H. RAMPHAL)—fruit edible, *Artabotrys odoratissimus* (B. & H. KANTALI-CHAMPA)—flowers very fragrant, *Unona discolor* (B. LAVENDAR-CHAMPA) very fragrant, mast tree (*Polyathia longifolia;* B. DEBDARU; H. DEVADARU or ASHOK)—ever-green tall tree, leaves used for decoration, etc.

Family 4—*Nymphaeaceae* (100 sp.—11 sp. in India)

Habit: aquatic perennial herbs. **Leaves:** usually floating, borne on a long petiole, cordate or peltate. **Flowers:** often large, showy, solitary, on a long pedicel, usually floating; bisexual, regular and usually perigynous, sometimes hypogynous or even epigynous; thalamus fleshy and goblet-shaped. **Perianth leaves:** several, free; sepals usually 4, gradually merging into petals; petals numerous, gradually merging

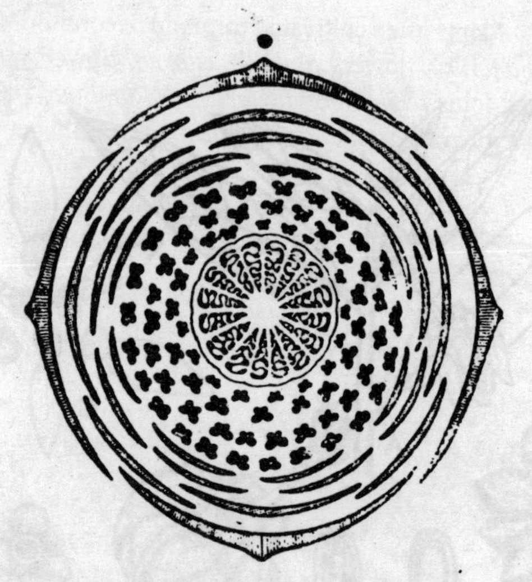

Figure 7. Floral diagram of *Nymphaeaceae*.

into stamens. **Androecium:** stamens numerous, free, usually perigynous, adnate to the fleshy thalamus that envelops the carpels. **Gynoecium:** carpels several, either free on the fleshy thalamus, as in lotus, or syncarpous lying embedded in the thalamus and surrounded by it; ovary unilocular with one ovule or multilocular with many ovules on superficial placentation; stigmas sessile, free or united; radiating, often with horn-like appendages. **Fruit:** a berry. **Seeds:** solitary and exalbuminous, or many with both perisperm and endosperm; spongy aril is often present and helps the seed to float. *Floral formula*—$\oplus$ ⚥ $K_4C_\infty A_\infty G_{(\infty)}$ or $_\infty$.

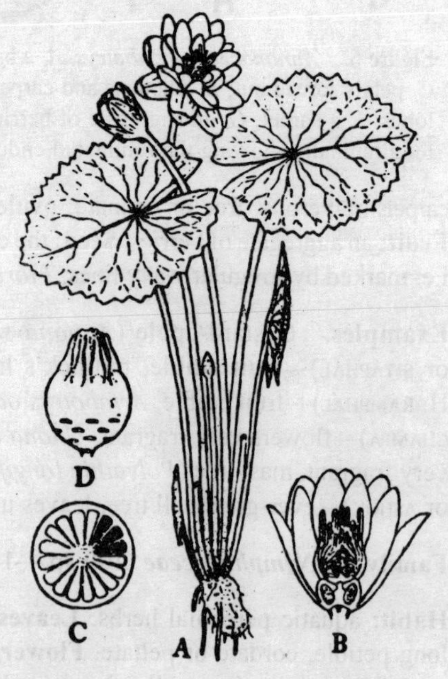

Figure 8. Water lily (*Nymphaea lotus*). *A,* an entire plant; *B,* a flower cut lengthwise; *C,* ovary in transection; *D,* a young fruit. [See also Fig. I/IIIB].

Examples. Plants often cultivated for pond decoration-water lilies such as *Nymphaea lotus*—flowers white, *N. rubra*—flowers red, *N. stellata*—flowers blue, lotus (*Nelumbo nucifera* = *Nelumbium speciosum*), giant water lily (*Victoria amazonica* = *V. regia*)—it bears huge tray-like leaves; see Fig. IV/1), and *Euryale ferox* (B. & H. MAKHNA)—fruits prickly but seeds edible and nutritious.

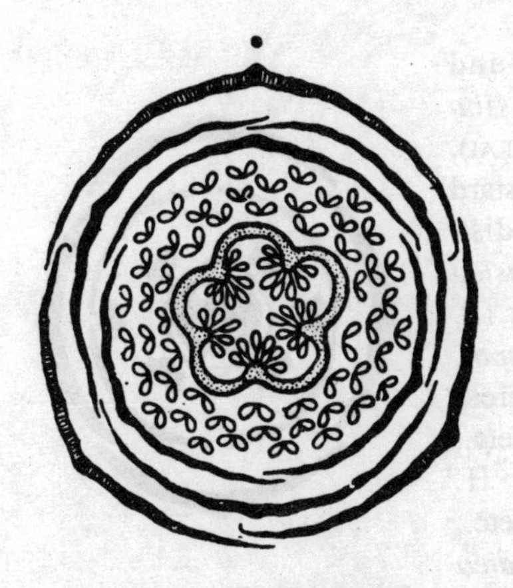

Figure 9. Floral diagram of *Papaveraceae* (*Argemone*).

Family 5—Papaveraceae (250 sp.–40 sp in India)

Habit: mostly herbs with milky or yellowish latex. **Leaves:** radical and cauline, simple and alternate, often lobed. **Flowers:** solitary, often showy, regular, bisexual and hypogynous. **Calyx:** sepals typically 2, rarely 3, free, very caducous. **Corolla:** petals 2 + 2, rarely 3 + 3 , biseriate, large, free, rolled or crumpled in the bud, very caducous; aestivation imbricate. **Androecium:** stamens ∞, free. **Gynoecium:** carpels (2 – ∞), syncarpous; ovary superior, unilocular, placentation parietal; stigmas distinct or sessile and rayed over the ovary; ovules many. **Fruit:** a septicidal capsule dehiscing by valves or a capsule opening by pores. **Seeds:** many with oily enadosperm. *Floral formula*—$\oplus ⚥ K_{2 \text{ or } 3} C_{2+2 \text{ or } 3+3} A_{\infty} \underline{G}_{(2-\infty)}$,

Examples. *Papaver,* e.g. opium poppy (*P. somniferum*)—opium, a narcotic drug, is the latex (containing several alkaloids—morphine nicotine, etc.) obtained from unripe fruits by incision; morphine (morphia) is used as a hypnotic drug; poppy seeds are used as a condiment (POSTO); they also yield an oil, garden poppy (*P. orientale*)—flowers ornamental, prickly or Mexican poppy (*Argemone mexicana*)—a prickly weed bearing yellow flowers, seeds yield an oil, etc.

Family 6—*Cruciferae* (over 3,000 sp.—174 sp. in India)

Habit: annual herbs. **Leaves:** radical and cauline, simple, alternate, often lobed. **Inflorescence:** a raceme (corymbose towards the top). **Flowers:** regular and cruciform, bisexual and complete, hypogynous. **Calyx:** sepals 2 + 2, free, in two whorls, imbricate. **Corolla:** petals 4, free, in one whorl, valvate, cruciform, with distinct limb and claw **Androecium:** stamens 6, in two whorls, 2 outer short and 4 inner long (tetradynamous). **Gynoecium:** carpels (2), syncarpous; ovary superior, ovary, at first 1-chambered, later 2-chambered owing to the development of a *false septum,* with often many ovules in each cell; placentation parietal. **Fruit:** a siliqua or silicula. **Seeds:** exalbuminous; remain attached to a wiry framework called *replum* (see p. 120). *Floral formula*—⊕⚥K_{2+2} C_4 A_{2+4} $\underline{G}_{(2)}$.

Examples. Useful plants: oils and condiments: mustard (*Brassica campestris*), *B. juncea* (B. & H. RAI), white mustard (*B. alba*), black mustard (*B. nigra*), etc.; **vegetables:** radish (*Raphanus sativus*), cabbage (*Brassica oleracea var. capitata*), cauliflower (*B. oleracea* var. *botrytis*), turnip (*B. rapa*), *B. rugosa* (B. & H. LAI), garden cress (*Lepidium sativum;* B. & H. HALIM) etc.; **ornamental:** candytuft (*Iberis;* H. CHANDNI), wallflower (*Cheiranthus*), etc.; **other common plants:** *Rorippa indica* (=*Nasturtium indicum*) and *Eruca sativa*—*common* weeds.

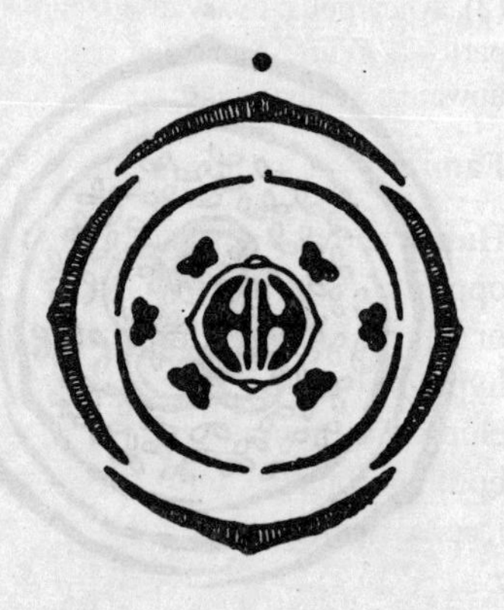

Figure 10. Floral diagram of *Cruciferae*.

Description of Mustard Plant (*Brassica campestris;* Fig. 11). A cultivated

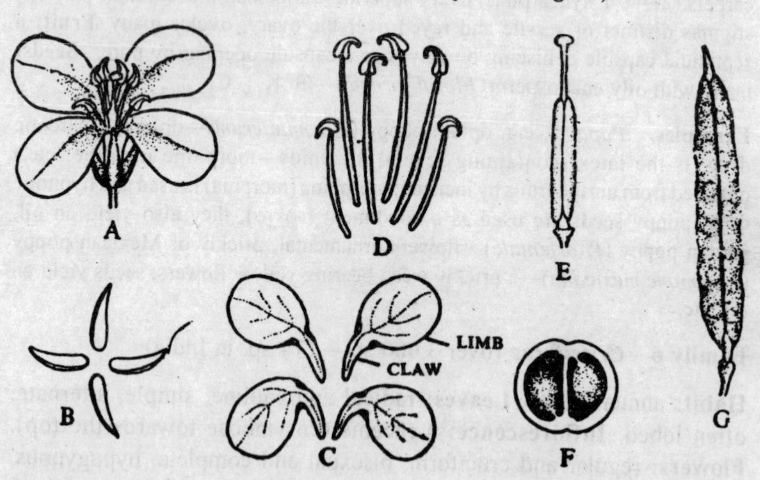

Figure 11. *Cruciferae.* Mustard (*Brassica campestris*) flower. *A,* a flower—cruciform; *B,* calyx; *C,* corolla; *D,* androecium showing tetradynamous stamens; *E,* gynoecium showing two carpels united; *F,* ovary in transection showing parietal placentation and false septum; and *G,* a fruit—siliqua.

winter herb. **Leaves:** simple, alternate, radical and cauline, lyrate. **Inflorescence:** a raceme. **Flowers:** regular, bisexual, hypogynous, cruciform, and bright yellow in colour. **Calyx:** sepals 2 + 2, free, imbricate. **Corolla:** petals 4, free, cruciform, valvate, with distinct claw and limb. **Androecium:** stamens 6, free, 4 inner long and 2 outer short (tetradynamous). **Gynoecium:** carpels (2), syncarpous; ovary divided into 2 chambers by a false septum; placentation parietal. **Fruit:** a narrow, pod-like siliqua opening into 2 valves from base upwards. **Seeds:** many, small, globose and exalbuminous; attached to replum.

Family 7—*Capparidaceae* (650 sp.—53 sp. in India)

Habit: herbs, climbing shrubs, and trees. **Leaves:** mostly alternate, rarely opposite, simple or palmately compound; stipules, if present, minute, or spiny. **Flowers:** regular (actinomorphic), sometimes zygomorphic, hypogynous or perigynous, and bisexual; thalamus in some cases elongated (gynophore) between the stamens and the pistil, sometimes both androphore and gynophore develop. **Calyx:** sepals usually 2 + 2, free. **Corolla:** petals 4, free. **Androecium:** stamens usually many,

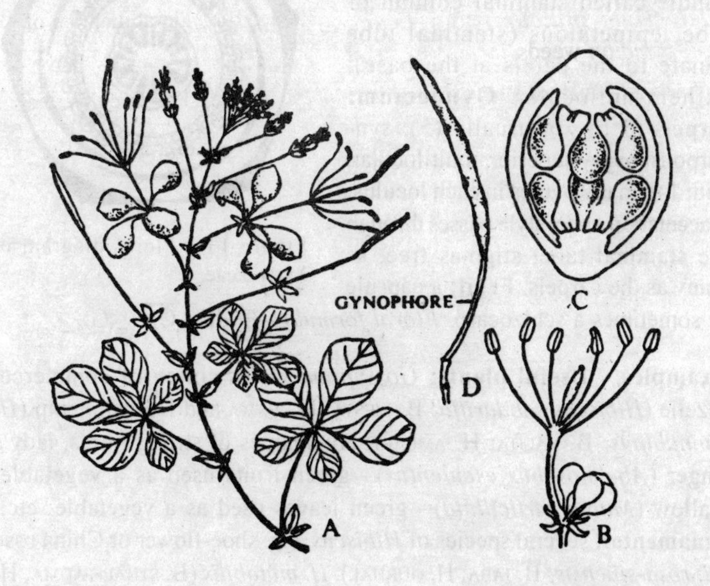

Figure 12. *Capparidaceac. Gynandropsis gynandra. A,* a branch with leaves and flowers; *B,* a flower; *C,* section of ovary showing parietal placentation and *D,* a fruit.

sometimes 6, as in *Gynandropsis,* free. **Gynoecium:** carpels typically (2), syncarpous; ovary superior, 1-celled, or chambered by false partition wall; placentation parietal; ovules many. **Fruit:** an elongated capsule or a berry. *Floral formula*—⊕ ⚥ $K_{2+2}C_4\ A_\infty$ or $_6G_{(2)}$.

Examples. *Gynandropsis gynandra* (B. HURHURE; H. HURHUR), *Polanisia icosandra* (= *Cleome viscose;* B. HALDE-HURHURE; H. HULHUL), *Capparis* (200 sp.), e.g. *C. sepiaria, C. horrida, C. aphylla, C. spinosa,* etc., and *Crataeva nurvala* (=*C. religiosa;* B. BARUN; H. BARNA).

Family 8—*Malvaceae* (1,000 sp.—105 sp. in India)

Habit: herbs, shrubs and trees. **Leaves:** simple, alternate and palmately-veined; stipules 2, free lateral. **Flowers:** regular, polypetalous, bisexual, hypogynous, copiously mucilaginous. **Calyx:** sepals (5), united with *epicalyx* (a whorl of bracteoles). **Corolla:** petals 5, free, aestivation twisted. **Androecium:** stamens ∞ monadelphous, i.e. united into one bundle called staminal column or tube, epipetalous (staminal tube adnate to the petals at the base); anthers unilocular. **Gynoecium:** carpels (5 to ∞), usually (5), syncarpous; ovary superior, multilocular, with 1 to many ovules in each loculus; placentation axile; style passes through the staminal tube; stigmas free, as many as the carpels. **Fruit:** a capsule or sometimes a schizocarp. *Floral formula*—⊕ ⚥ $K_{(5)}\overline{C_5A_{(\infty)}}\underline{G}_{(5-\infty)}$.

Figure 13. Floral diagram of Malvaceae.

Examples. **Useful plants:** *Gossypium* yields cotton of commerce, rozelle (*Hibiscus sabdariffa;* B. MESTA; H. PATWA and Deccan hemp (*H. cannabinus;* B. NALITA; H. AMBARI) are sources of strong fibres, lady's finger (*Abelmoschus esculentus*)—green fruits used as a vegetable, mallow (*Malva verticillata*)—green leaves used as a vegetable, etc.; **ornamental:** several species of *Hibiscus,* e.g. shoe-flower or China rose (*H. rosa-sinensis;* B. JABA; H. GURHAL), *H. mutabilis* (B. STHAL-PADMA; H. GUL-AJAIB), etc., and hollyhock (*Althaea rosea*); **shade tree:** Portia tree (*Thespesia populnea*); **other common plants:** *Sida cordifolia* (B. BERELA; H. BARIARA), *Urena lobata* (B. BAN-OKRA; H. BACHATA), *Hibiscus vitifolius* (B. & H. BAN-KAPAS), *Abutilon indicum* (B. PETARI; H. KANGHI), etc.

Description of China Rose Plant (*Hibiscus rosa-sinensis;* Fig. 14). A much-branched shrub. **Leaves:** simple, alternate, palmately 3-veined at the base, 8–10 cm long, margin serrate, petiole long. **Flowers:** solitary and axillary on a

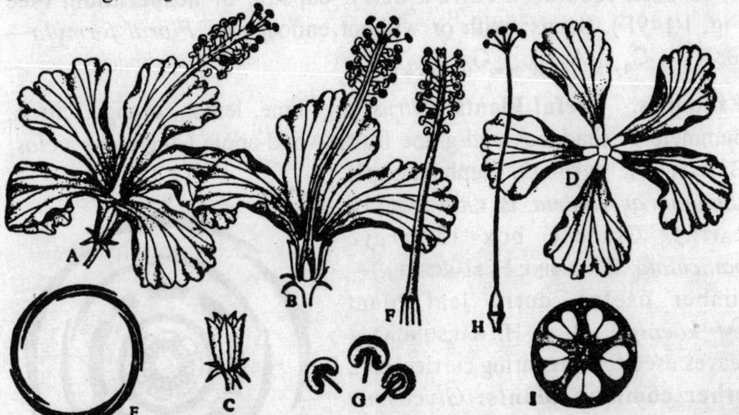

Figure 14. *Malvaceae.* China rose (*Hibiscus rosa-sinensis*) flower. *A,* an entire flower; *B,* the same split open longitudinally showing the four whorls, more particularly the staminal column with the style passing through it; *C,* calyx with epicalyx; *D,* corolla opened out; *E,* twisted aestivation of corolla; *F,* androecium showing monadelphous stamens; *G,* one-celled anthers—young and mature (dehiscing); *H,* gynoecium showing five carpels united; and *I,* ovary in transection showing axile placentation.

long peduncle articulated to the pedicel, large and showy, red in colour, regular, bisexual, hypogynous; bracteoles 5 or more in the form of a whorl known as epicalyx. **Calyx:** 5-lobed. **Corolla:** petals 5, free; aestivation twisted clockwise or anti clockwise. **Androecium:** stamens numerous, united into a bundle (monadelphous), epipetalous. adnate to the petals at the base; anthers free, reniform, I-lobed. **Gynoecium:** carpels (5), connate; style passing through the staminal coloumn; stigmas 5; ovary 5-locular; placentation axile. **Fruit:** not formed; in other species of *Hibiscus* a loculicidal capsule.

Family 9—*Rutaceae* (900 sp.—66 sp. in India)

Habit: shrubs and trees (rarely herbs). **Leaves:** simple or compound, alternate or rarely opposite, *gland-dotted.* **Flowers:** regular, bisexual and hypogynous; *disc* below the ovary prominent. **Calyx:** sepals 4 or 5, free or slightly connate. **Corolla:** petals 4 or 5, free, imbricate. **Audroecium:** stamens variable in number, generally twice as many as petals or numerous in *Citrus* and *Aegle,* free or united in irregular bundles (polyadelphous). **Gynoecium:** carpels generally (4) or (5), or (∞) in *Citrus,* syncarpous, or sometimes free at the base and united above, either sessile or seated on the disc; ovary usually 4– or 5– locular; multilocular in *Citrus,* with axile placentation; ovules 2– ∞ (rarely 1) in each loculus. **Fruit:** a berry, capsule or hesperidium (see Fig. I/149F). **Seeds:** with or without endosperm. *Floral formula*— $\oplus ⚥ K_{4-5}\ C_{4-5}\ A_{8,\ 10\ or\ \infty}\ \underline{G}_{(4,\ 5\ or\ \infty)}$.

Examples. Useful Plants: *Citrus* (e.g. lime, lemon, orange, citron, pummelo or shaddock and grape fruit), wood-apple (*Aegle marmelos;* B. BAEL; H. SIRIPHAL), elephant-apple (*Limonia acidissima;* B. KATH-BAEL; H. KAITH), Chinese box (*Murraya paniculata;* B. KAMINI; H. MARCHULA)—timber useful, curry leaf plant (*M. koenigii;* B. & H. BARSUNGA)—leaves used for flavouring curries, etc.; **other common plants:** *Glycosmis arborea* (B. ASHHOURA; H. BANNIMBU), *Clausena pentaphylla* (B. & H. PANKARPUR), etc.

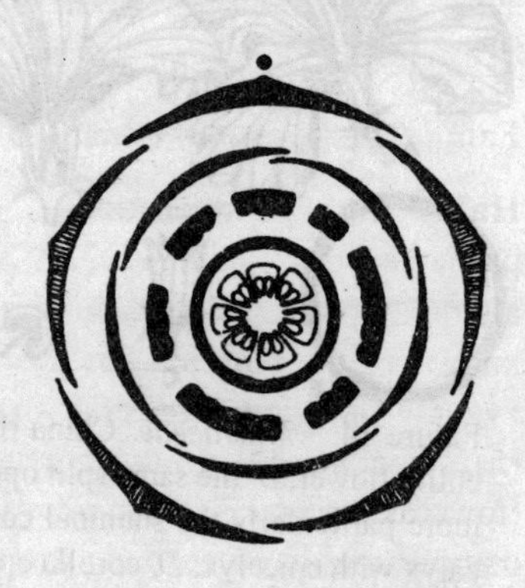

Figure 15. Floral diagram of *Rutaceae.*

Common species of *Citrus:* sour lime (*C. aurantifolia*), sweet lime (*C. limetta*), lemon (*C. limon*), citron (*C. medica*), pummelo or shaddock (*C. grandis*), Mandarin orange (*C. reticulata;* B. KAMALA; H. SANGTRA), sweet orange (*C. sinensis*), grape fruit (*C. paradisi*); etc.

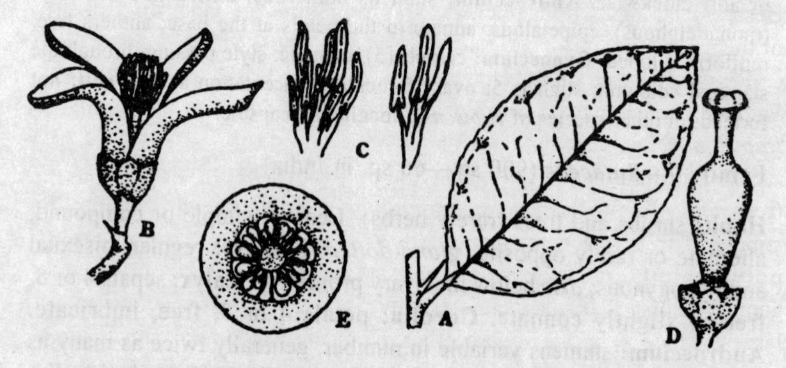

Figure 16. *Rutaceae.* Sour lime (*Citrus aurantifolia*). *A,* a leaf; *B,* a flower; *C,* stamens (polyadelphous); *D,* pistil (on a disc) and calyx; and *E,* section of ovary showing axile placentation.

Description of *Murraya paniculata*. An evergreen shrub or a small tree. **Leaves** pinnately compound (imparipinnate), leaflets alternate, 3–9, oblique in shape. **Flowers** white, fragment, in axillary or terminal corymbs, sometimes solitary, regular, bisexual and hypogynous. **Calyx**—sepals 5, connate, minute, glandular. **Corolla**—petals 5, free, oblong-lanceolate, imbricate. **Androecium**—stamens 10, free, longer and shorter, inserted on an elongated disc. **Gynoecium**—carpels (2), syncarpous; ovary superior, slender, 2-locular (rarely 4- or 5-locular), with a solitary ovule. In each loculus; style linear; stigma capitate. **Fruit** an ovoid berry, 1- to 2-seeded, red or deep, orange when ripe.

Family 10—*Leguminosae* (12,000 sp.—951 sp. in India)

Habit: herbs, shrubs, trees and climbers. **Roots** of many species, particularly of *Papilionaceae,* have nodules (see Fig. III/2). **Leaves:** alternate, pinnately compound, rarely simple, with a swollen leaf-base known as the *pulvinus*. **Flowers:** bisexual and complete, regular or zygomorphic, hypogynous or slightly perigynous. **Calyx:** sepals usually (5), sometimes (4). **Corolla:** petals usually 5, with the odd one posterior (towards the axis), sometimes 4, free or united. **Androecium:** stamens usually 10 or numerous, sometimes less than 10, free or united. **Gynoecium:** carpel 1; ovary 1-celled. with 1 to many ovules, placentation marginal. **Fruit:** a legume or pod. **Seed:** mostly exalbuminous.

This is the second biggest family among the dicotyledons (being only second to *Compositae*), and from the economic standpoint this is probably the second most important family (ranking second to *Gramineae*) because the pulses which are rich in proteins belong to it. Besides, the leguminous plants with root-nodules are natural fertilizers of the soil.

Primarily based on the characters of the corolla and the androedum, *Leguminosae* has been divided into the following three sub-families: *Papilionaceae, Caesalpinieae* and *Mimoseae.*

(1) **Papilionaceae** (754 sp. in India). **Habit:** herbs, shrubs, rarely trees and climbers. **Leaves:** compound, unipinnate, rarely simple; stipules 2, free. **Inflorescence:** usually a raceme. **Flowers:** zygomorphic, polypetalous and papilionaceous. **Calyx:** sepals usually (5), gamosepalous. **Corolla:** petals usually 5, free, the posterior one is the largest called *vexillum;* this partly covers the two lateral ones called *wings* which in their turn cover the two innermost ones united into a boat-shaped *keel;* aestivation vexillary (se Fig. I/115). **Androecium:** stamens ten, diadelphous—(9)+1, rarely free monadelphous. **Gynoecium**

and fruit: as described above. *Floral formula*—% ⚥$K_{(5)}C_5A_{(9)+1}\underline{G}_1$.

Examples. Useful plants: pulses (rich in proteins), e.g. pea (*Pisum sativum;* Fig. 17). pigeon pea or red gram (*Cajanus cajan;* B. ARAHAR; H. RAHAR), Bengal gram (*Cicer arietinum*), green gram (*Phaseolus aureus;* B. & H. MOONG), black gram (*P. mungo;* B. KALAI; H. URID), lentil (*Lens culinaris;* B. & H. MASUR), *Lathyrus sativus* (B. & H. KHESARI), soyabean (*Glycine max = G. soja*), etc.; **vegetables:** country bean (*Dolichos lablab;* B. SHIM; H. SHEM), cow pea (*Vigna sinensis;* B. BARBATI; H. RAUNG), sword bean (*Canavalia ensiformis;* B. MAKHAN-SHIM; M. BARA-SHEM), French bean (*Phaseolus vulgaris*), etc.; **natural fertilizers:** *Sesbania cannabina* (B. & H. DHAINCHA), sesban (*S. sesban;* B. JAINTI; H. JAINT) lucerne or alfalfa (*Medicago sativa*—also an excellent fodder); **timber tree:** Indian redwood (*Dalbergia sissoo*); **ornamental:** sweet pea (*Lathyrus odoratus*), lupin (*Lupinus*), rattlewort (*Crotalaria sericea;* B. ATASH; H. JHUNJHUNIA), butterfly pea (*Clitoria ternatea;* B. APARAJITA; H. GOKARNA), coral-tree (*Erythrina variegata;* B. MANDAR; H. PANJIRA), *Sesbania grandiflora* (B. BAKPHUL; H. AGAST), etc.; **other useful plants:** groundnut (*Arachis hypogaea;* see Fig. III/31—seeds yield 43–46% of edible oil, Indian hemp (*Crotalaria juncea*) yields very strong bark fibres, etc.; **other common plants:** Indian telegraph plant (*Desmodium gyrans*), cowage (*Mucuna pruriens*), wild pea (*Lathyrus aphaca*), *Tephrosia purpurea*—a common weed, etc. (For description of pea flower see p. 117).

Distinguishing Characters of Leguminosae

	Papiliouaceae	*Caesalpinieae*	*Mimoseac*
Leaves	usually 1-pinnate, rarely simple, stipels often present	1- or 2-pinnate, rarely simple, stipels absent	bipinnate, stipels present or absent
Flowers	papilionaceous	zygomorphic	regular, small
Inflor.	racemose	racemose	spherical head
Calyx	gamosepalous	polysepalous, times gamosepalous	gamosepalous
Corolla	potypetalous, posterior petal largest and outermost, aestivation vexillary	polypetalous, posterior petal smallest and innermost, aestivation imbricate	gamopetalous, all petals equal, aestivation valvate
Androecium	stamens ten, (9) + 1, rarely (10) or 10	ten or fewer	often indefinite, sometimes definite

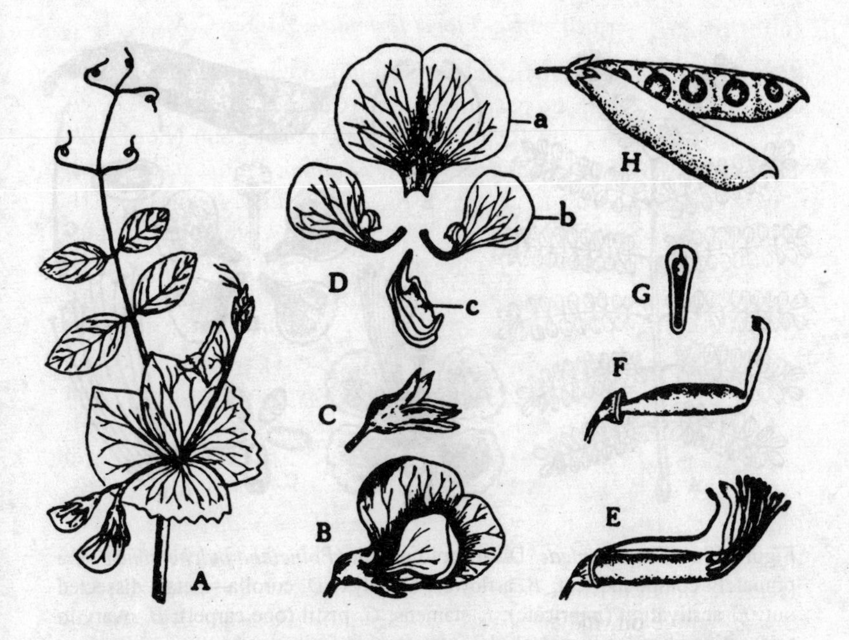

Figure 17. *Papilionaceae*. Pea (*Pisum sativum*). *A*, a branch. *B*, a flower—papilionaceous (w also Fig. I/115); *C*, calyx; *D*, corolla—petals opened out (*a*, vexillum; *b*, wing; *c*, keel); *E*, stamens—(9) + 1, and pistil; *F*, pistil—1 carpel (note the ovary, style and stigma); *G*, ovary in section showing marginal placentation; and *H*, a fruit—legume.

(2) ***Caesalpinieae*** (110 sp. in India). Shrubs and trees, rarely climbers or herbs. **Leaves:** unipinnate or bipinnate, rarely simple, as in camel's foot tree (*Bauhinia*). **Inflorescence:** commonly a raceme, **Flowers:** zygomorphic and polypetalous. **Calyx:** sepals usually 5, polysepalous (sometimes gamosepalous). **Corolla:** petals usually 5, free, imbricate, the upper smallest one always innermost. **Androecium:** stamens ten or fewer, free, Gynoecium: as in Papilionaceae. *Floral formula*— ⚥ $K_5C_5A_{10}\underline{G}_1$.

Examples. Useful plants: tamarind (*Tamarindus indica*)—fruits widely used for sour preparations, Indian laburnum (*Cassia fistula;* B. SHONDAL; H. AMALTASH)—heart-wood very hard and durable, and flowers ornamental, etc.; **medicinal:** *Saraca indica* (B. ASOK; B. SEETA-ASOK)—also ornamental, fever nut (*Caesalpinia crista;* B. NATA; H. KAT-KARANGA), etc.; **dye:** sappan or Brazil wood (*Caesalpinia sappan;* B. & H. BAKAM)—

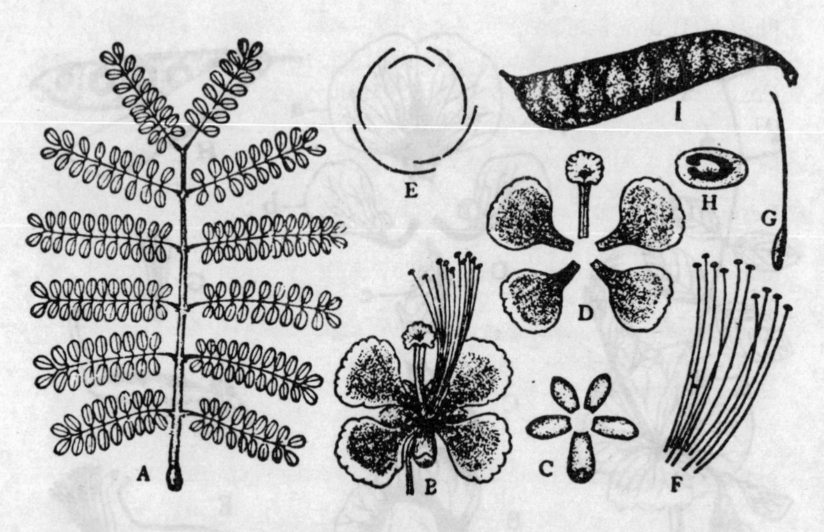

Figure 18. *Caesalpinieae.* Dwarf gold mohur (*Poinciana pulcherrima*). *A,* a pinnately compound leaf; *B,* a flower; *C,* calyx; *D,* corolla—petals dissected out; *E,* aestivation (imbricate); *F,* stamens; *G,* pistil (one carpel); *H,* ovary in transection showing marginal placentation; *I,* a fruit.

wood yields a valuable red dye extensively used for dyeing silk and wool, starch coloured with this dye forms 'ABIR' used in 'HOLY' festival, and pods yield a high percentage of tannin, **ornamental:** camel's foot tree (*Bauhinia purpurea* and *B. variegata;* B. KANCHAN; H. KACHNAR), gold mohur (*Delonix regia;* B. KRISHNACHURA; H. GULMOHR; see

Figure 19. *Caesalpinieae. Cassia sophera. A,* a branch with a pinnately compound leaf (leaflets in 6–10 pairs) and an inflorescence; *B,* stamens and pistil; *C,* pistil (one carpel); and *D,* a fruit (legume) partially opened.

Fig. I/105), dwarf gold mohur or peacock flower (*Poinciana pulcherrima;* B. RADHACHURA; H. guletura—Fig. 18), etc.; **other common plants:** *Cassia sophera* (B. KALKA-SUNDE; H. KASUNDA; Fig. 19), *C. occidentalis* (H. BARA-KASUNDA), ringworm shrub (*C. alata;* B. & H. DAD-MARDAN), *C. tora* (B. & H. CHAKUNDA), *C. auriculata,* etc.

Description of Dwarf Gold Mohur Plant (*Poinciana pulcherrima;* Fig. 18). A much-branched shrub. **Leaves:** bipinnately compound, leaflets many. **Inflorescence:** a raceme. **Flower:** zygomorphic, bisexual and hypogynous. **Calyx:** sepals 5, free, odd one outermost. **Corolla:** petals 5, free, spotted, odd one innermost and smallest, imbricate. **Androecium:** stamens 10, free; filaments slender and long. **Gynoecium:** carpel 1; ovary superior, 1-celled and many-ovuled. **Fruit:** a flat pod, with many seeds.

(3) ***Mimoseae*** (87 sp. in India). Shrubs and trees, sometimes herbs or woody climbers. **Leaves:** commonly bipinnate. **Inflorescence:** a head

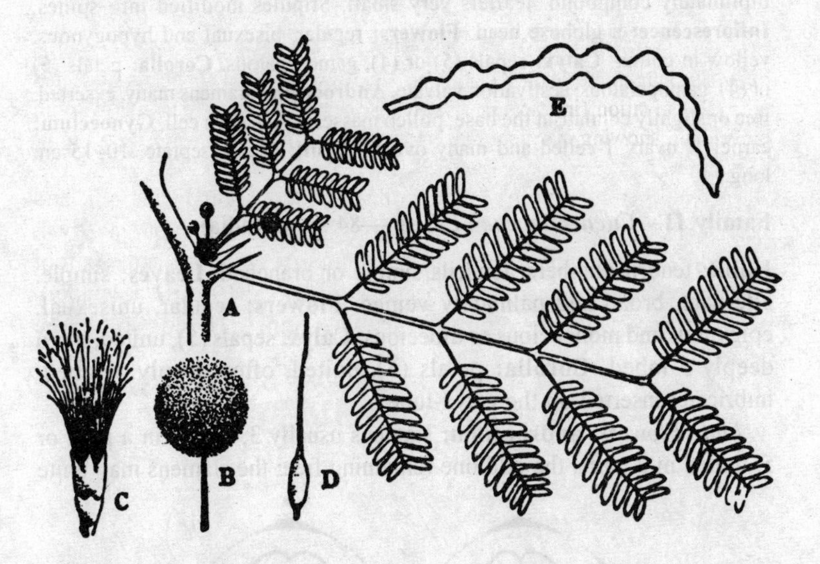

Figure 20. *Mimoseae.* Gum tree (*Acacia nilotica*). *A,* a branch with bipinnate compound leaves; *B,* an inflorescence (head); *C,* a flower; *D,* pistil (one carpel); and *E,* a fruit (lomentum).

or a spike. **Flowers:** regular, often small and aggregated in spherical heads. **Calyx:** sepals (5) or (4), generally gamosepalous, valvate. **Corolla:** petals (5) or (4), mostly gamopetalous; aestivation valvate. **Androecium:** stamens numerous or 10, rarely 8 or 4, free or sometimes united at the base; pollen often united in small masses. Gynoecium: as in Papilionaceae. *Floral formula*—⊕⚥ $K_{(5-4)}$ $C_{(5-4)}$ $A_{\infty \text{ or } 10}$ $\underline{G}_1$.

Examples. Useful plants: catechu (*Acacia catechu;* B. KHAIR; H. KAITH) yields a kind of tannin called catechu which is obtained by boiling chips of heart-wood, gum tree (*A. nilotica* = *A. arabica;* B. BABLA; H. BABUL), Australian *Acacia* (e.g. *A. moniliformis*), many species of *Acacia* are sources of tannin and fuel, *Albizzia lebbek* (A. & H. SIRISH)—a timber tree, *A. procera*—wood suitable for tea chests, many species of *Albizzia* are sources of fuel, rain tree (*Pithecolobium saman*)—planted as a shade tree, etc.; **other common plants:** sensitive plant (*Mimosa pudica;* B. LAJJABATI; H. LAJWANTI or CHHUIMUI), *Pithecolobium dulce* (B. & H. DEKANI-BABUL), nicker bean (*Entada gigas;* B. & H. GILA), Jerusalem thorn (*Parkinsonia aculeata*) and *Prosopis spicigera* (B. & H. SHOMI).

Description of gum tree *Acacia nilotica:* (Fig. 20). A tree, **Leaves:** alternate, bipinnately compound. leaflets very small. Stipules modified into spines. **Inflorescence:** a globose head. **Flowers:** regular, bisexual and hypogynous, yellow in colour. **Calyx:** sepals (5) or (4), gamosepalous. **Corolla:** petals (5) or (4). gamopetalous; aestivation valvate. **Androecium:** stamens many, exserted, free or slightly connate at the base; pollen masses 2–4 in each cell. **Gynoecium:** carpel 1; ovary 1-celled and many ovuled. **Fruit:** a pod, septate, 10–15 cm long.

Family 11—*Cucurbitaceae* (750 sp.—84 sp. in India)

Habit: tendril climbers; tendrils simple or branched. **Leaves:** simple, alternate, broad and palmately veined. **Flowers:** regular, unisexual, epigynous and monoecious or dioecious. **Calyx:** sepals (5), united, often deeply 5-lobed. **Corolla:** petals (5), united, often deeply 5-lobed, imbricate; inserted on the calyx-tube.

Male Flowers: **androecium:** stamens usually 3, united in a pair, or 5, united in 2 pairs, the odd one remaining free; the stamens may unite

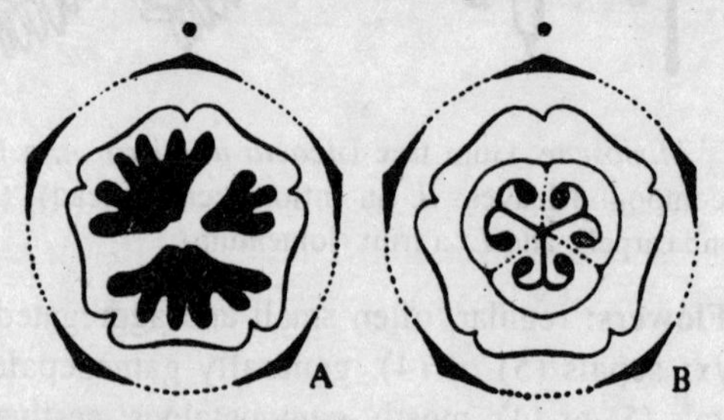

Figure 21. Floral diagrams of *Cucurbitaceae. A,* male flower, *B,* female flower.

by their whole length or by their anthers only; each anther 1- or 2-lobed; paired ones 2- or 4-lobed; anther-lobes sinuous, i.e. twisted like a transverse S. *Floral formula*—⊕ ♂ $K_{(5)}C_{(5)}A_{3 \text{ or } 5}$.

Female Flowers: **gynoecium:** carpels (3), syncarpous; ovary inferior, unilocular; placentation parietal but often the placentae intrude far into the chamber of the ovary making the latter falsely trilocular; ovules many; style 1; stigmas 3 which are often forked. **Fruit:** a pepo. *Floral formula*—⊕ ♀ $K_{(5)}C_{(5)}\overline{G}_{(3)}$.

Plants of this family are mostly used as vegetables, a few yield delicious summer fruits, and a few are medicinal.

Examples. Vegetables: sweet gourd or musk melon (*Cucurbita moschata*), pumpkin or vegetable marrow (*C. pepo*), giant gourd (*C. maxima*), bottle gourd (*Lagenaria siceraria;* B. LAU; H. LAUKI), snake gourd (*Trichosanthes anguina;* B. CHICHINGA; H. CHACHINDA), *T. dioica*

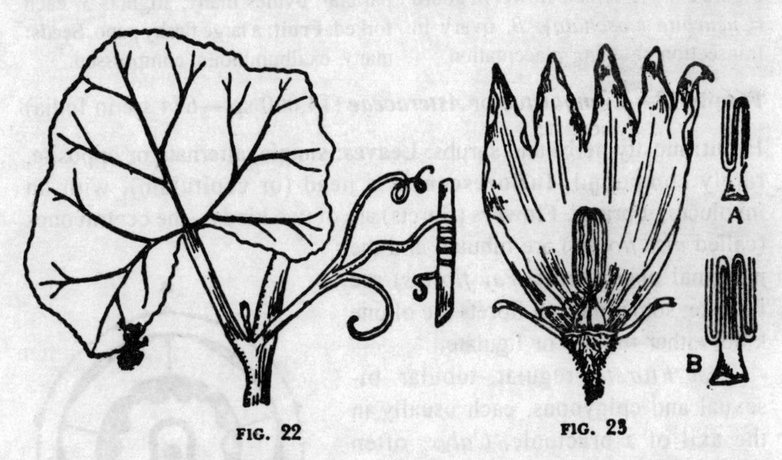

Figure 22. *Cucurbitaceae.* Gourd (*Cucurbita moschata*). Portion of a branch with a leaf and a tendril. Figure 23. Male flower of the same. *A,* one stamen; *B,* two stamens united together.

(B. PATAL; H. PARWAL), bitter gourd (*Momordica charantia;* B. KARALA & UCHCHE; H. KARELA), *M. cochinchinensis* (B. KARALA; H. CHATTHAI), ash or wax gourd (*Benincasa hispida;* B. CHALKUMRA; H. PETHA), ribbed gourd (*Luffa acutangula*), bath sponge or loofah (*L. cylindrica*), etc.; **fruits:** water melon *Citrullus lanatus*), melon (*Cucumis melo*) and cucumber (*C. sativus*); **medicinal:** colocynth (*Citrullus colocynthis;* B. MAKAL; H. INDRAYAN), *Coccinia indica* (B. TELAKUCHA; H. KUNDARU), etc.

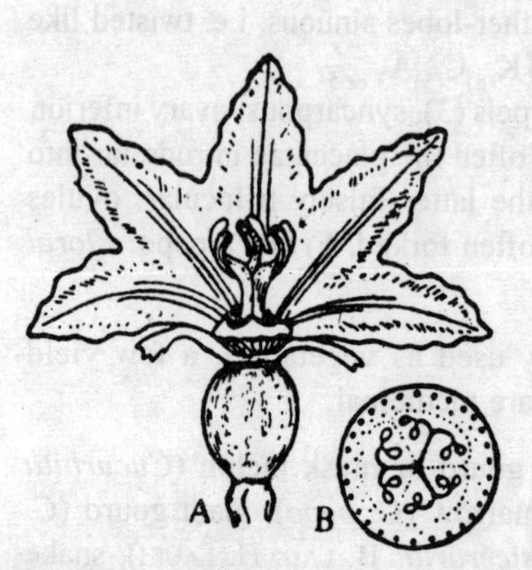

Figure 24. *A,* female flower of gourd (*Cucurbita moschata*); *B,* ovary in transection showing placentation.

Description of Gourd plant (*Cucurbita moschata;* Figs. 22–24). A large climbing herb, hairy all over; tendril opposite the leaf, 2- to 4-fid. **Leaves:** broad, long-petioled, palm--ately veined. **Flowers:** solitary, large, yellow in colour, regular unisexual (monocious). **Calyx:** sepals (5), connate; lobes linear or leafy. **Corolla:** petals (5), connate, campanulate. *In male flowers:* **androecium:** stamens 3, united in a pair, the odd one remaining free; anthers united, one 1-celled and two 2-celled, sinuous, *In female flowers:* **gynoecium:** carpels (3), syncarpous: ovary inferior, 1-celled; placentation parietal; ovules many; stigmas 3, each forked. **Fruit:** a large fleshy pepo. **Seeds:** many, exalbuminous, compressed.

Family 12—*Compositae* or *Asteraceae* (14,000 sp.—674 sp. in India)

Habit: mostly herbs and shrubs. **Leaves:** simple, alternate or opposite, rarely compound. **Inflorescence:** a head (or capitulum), with an involucre of bracts. **Flowers** (florets) are of two kinds—the central ones (called *disc florets*) are tubular, and the marginal ones (called *ray florets*) are ligulate; sometimes all florets are of one kind, either tubular or ligulate.

Disc Florets: regular, tubular, bisexual and epigynous, each usually in the axil of a bracteole. *Calyx:* often modified into pappus, or into scales, or absent **Corolla:** petals (5), gamopetalous, tubular. **Androecium:** stamens 5, epipetalous filaments free but anthers united (syngenesious). **Gynoecium:** carpels (2), syncarpous, ovary inferior, 1-celled with one basal ovule; style 1, stigmas 2. **Fruit:** a cypsela. *Floral formula*—⊕⚥ K *pappus* or o $\overline{C_{(5)}A_{(5)}}\overline{G}_{(2)}$.

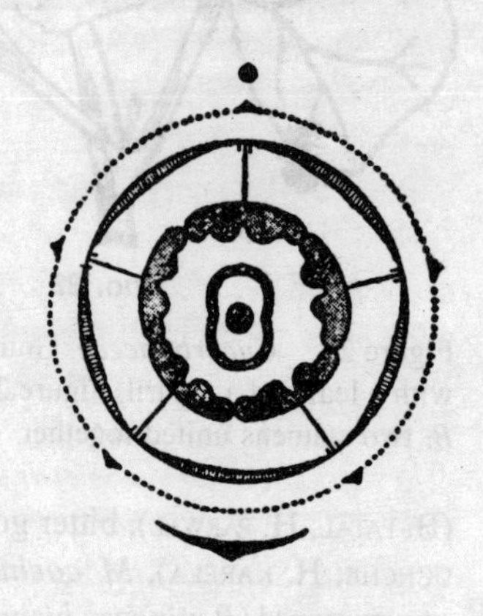

Figure 25. Floral diagram of *Compositae* (disc floret).

Ray Florets: zygomorphic, ligulate, unisexual (female) or sometimes neuter, as in sunflower, and epigynous, each usually in the axil of a

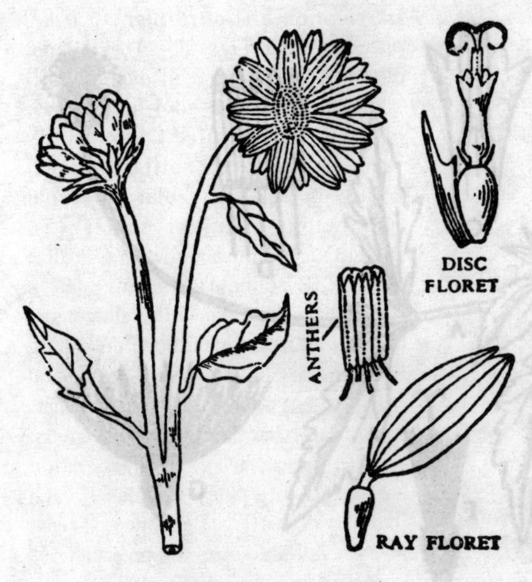

Figure 26. *Compositae*. Sunflower (*Helianthus annuus*). Branch with two heads; disc floret (bisexual) with a bracteole at the base; anthers (syngenesious); ray floret (female or neuter).

bracteole. **Calyx:** as in disc floret. **Corolla:** petals (5), gamopetalous, ligulate (strap-shaped). **Gynoecium:** as in the disc floret. **Fruit:** the same. *Floral formula*—% ♀ K *pappus* or ○ $C_{(5)}A_0\overline{G}_{(2)}$.

Examples. Useful plants: ornamental: sunflower (*Helianthus annuus*), marigold (*Tagetes patula*), *Chrysanthemum, Dahlia, Zinnia, Cosmos,* etc.; **vegetables:** chicory (*Cichorium intybus;* B. & H. KASNI), endive (*C. endivia*), lettuce (*Lactuca sativa*), *Enhydra fluctuans* (B. HALENCHA; H. HARUCH), etc.; **oils:** sunflower (*Helianthus*)—seeds yield a cooking oil, and safflower (*Carthamus tinctorius;* B. KUSUM; H. KUSUM)—seeds yield a cooking oil but its chief use is in the manufacture of paints and soaps; **medicinal:** santonin (*Artemisia cina*), *Eupatorium ayapana, Wedelia calendulacea* (B. BHRINGARAJ; H. BHANGRA), *Eclipta alba* (B. KESHARAJ; H. SAFED BHANGRA), etc.; **insecticides:** a few species of *Chrysanthemum* (*Pyrethrum*); **other common plants:** *Tridax procumbens* (Fig. 27), *Xanthium strumarium* (B. & H. OKRA), *Ageratum, Vernonia, Sonchus, Eupatorium odoratum,* etc.

Description of Sunflower Plant (*Helianthus annuus;* Fig. 26). Sunflower is an annual garden herb. **Leaves:** simple, opposite, often the upper ones alternate. **Inflorescence:** head or capitulum (large in some cases), with an involucre of bracts. **Flowers:** central florets, called disc florets, are tubular and bisexual, and marginal florets, called ray florets, are zygomorphic, ligulate and often

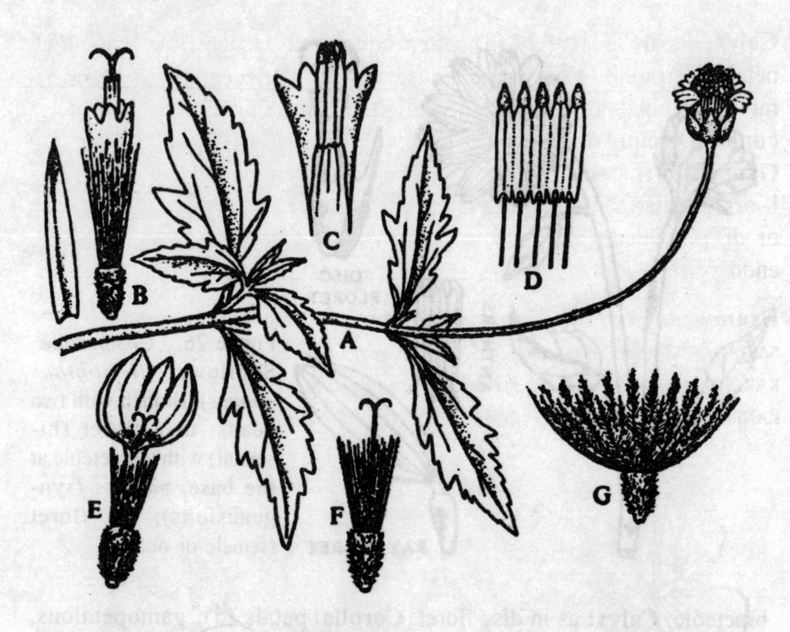

Figure 27. *Compositae. Tridax procumbens. A,* a branch with a head; *B,* a disc floret with a bracteole; *C,* corolla (split open) with epipetalous stamens; *D,* syngenesious stamens (split open); *E,* a ray floret; *F,* pistil and pappus; and *G,* a fruit (cypsela) with pappus (parachute mechanism).

neuter. *Disc florets*—regular, tubular, bisexual and epigynous. **Calyx:** modified into two scales. **Corolla:** gamopetalous, 5-lobed. **Androecium:** stamens 5. epipetalous and syngenesious, forming a tube around the style. **Gynoecium:** carpels (2) syncarpous; ovary inferior. 1-celled, with one basal ovule; style 1, but stigmas 2. **Fruit:** a cypsela. *Ray florets*—zygomorphic. ligulate, female or neuter, and epigynous. **Calyx:** as in disc florets or absent. **Corolla:** ligulate. 5-lobed. **Stamens:** absent. **Gynoecium:** as in disc florets, but style and stigmas sometimes absent making the flower neuter.

Family 13—*Apocynaceae* (1,400 sp.—67 sp. in India).

Habit: herbs, shrubs, trees, twiners and lianes; with latex. **Leaves:** simple, opposite or whorled, rarely alternate. **Flowers:** regular, bisexual and hypogynous, in cymes.

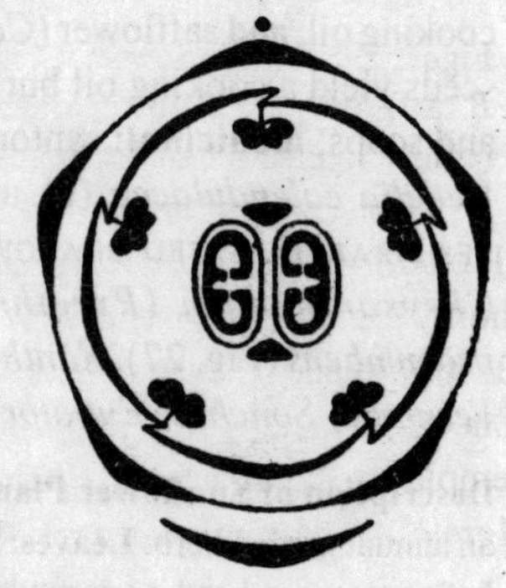

Figure 28. Floral diagram of *Apocynaceae*.

Calyx: sepals (5), rarely (4), often united only at the base. **Corolla:** petals (5), rarely (4), gamopetalous, twisted. **Androecium:** stamens 5, rarely 4, epipetalous, included within the corolla-tube; anthers usually connate around the stigma. Disc present, ring-like or glandular. **Gynoecium:** carpels 2 or (2), apocarpous or syncarpous; ovary superior, l- or 2-locular, with 2 ovules in each. **Fruit:** a pair of follicles, or berry or drupe. **Seeds:** often with a crown of long silky hairs, mostly endospermic. *Floral formula*—$\oplus \text{⚥}\ K_{(5)}\overline{C_{(5)}A_5}\underline{G}_{2 \text{ or } (2)}$.

Examples. Useful plants: medicinal: *Rauwolfia serpentina* (B. SARPAGANDHA; H. SARPGAND), *Holarrhena antidysenterica* (B. KURCHI; H. KARCHI), yellow oleander (*Thevetia peruviana;* B. KALKE-PHUL; H. PILA-KANER)—seeds very poisonous, devil tree (*Alstonia scholaris*; B. CHHATIM;

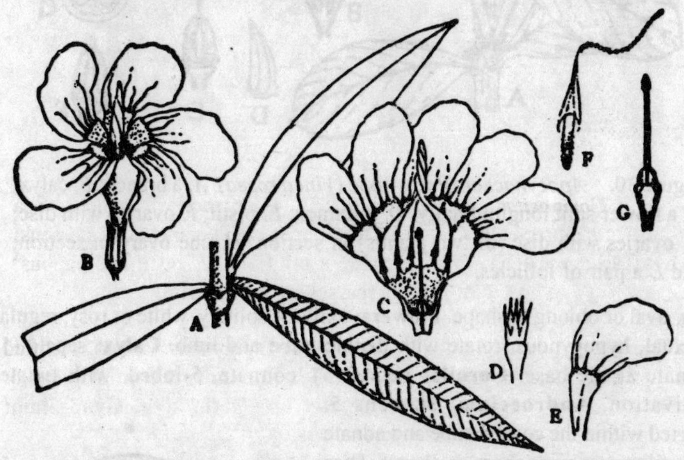

Figure 29. *Apocynaceae.* Oleander (*Nerium indicum*). *A,* a whorl of leaves, *B,* a flower; *C,* a flower opened out; *D,* calyx; *E,* a petal; *F,* a stamen (connective with hairy appendage); and *G,* pistil.

H. CHATIUM), etc.; **fruits:** *Carissa carandas* (B. KARANJA; H. KARONDA)—a thorny shrub; **ornamental:** periwinkle (*Lochnera rosea* = *Vinca rosea;* B. NAYANTARA; H. SADABAHAR—Fig. 30), oleander (*Nerium indicum;* B. KARAVI; H. KANER—Fig. 29), *Ervatamia coronaria* (B. TAGAR; H. CHANDNI), temple or pagoda tree (*Plumeria rubra;* B. KATGOLAP; H. GOLAINCHI), *Aganosma caryophyllata* (B. MALATI; H. MALTI) *Allamanda,* etc.

Description of Periwinkle Plant (*Vinca rosea;* Fig. 30). An erect or procumbent herb or undershrub containing latex. **Leaves:** opposite, 4–5 cm

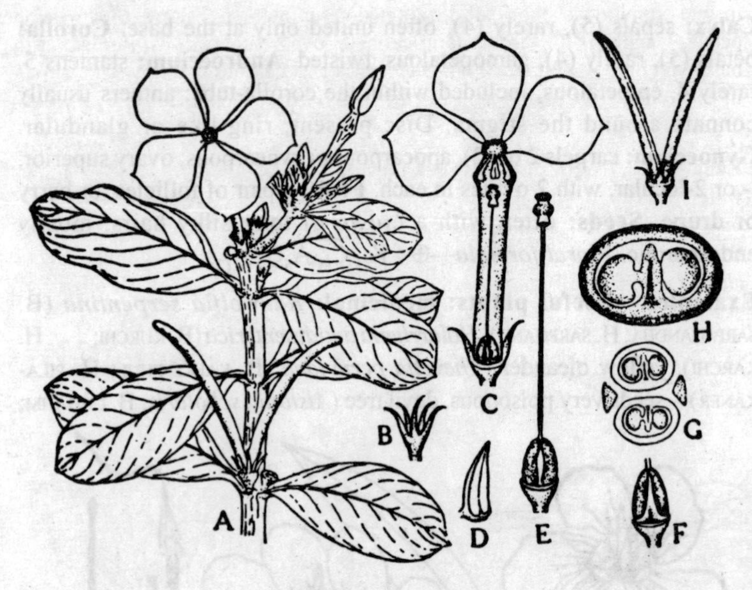

Figure 30. *Apocynaceae.* Periwinkle (*Vinca rosea*). *A,* a branch; *B,* calyx; *C,* a flower split longitudinally; *D,* a stamen; *E,* pistil; *F,* ovaries with disc; *G,* ovaries with disc (of two glands) in section; *H,* one ovary in section; and *I,* a pair of follicles.

long, oval or oblong in shape. **Flowers:** axillary, solitary, white or rosy, regular, bisexual, hypogynous, rotate with distinct tube and limb. **Calyx:** sepals (5), connate at the base. **Corolla:** petals (5). connate, 5-lobed. with twisted aestivation. **Androecium:** stamens 5, inserted within the corolla-tube and adnate to it; filaments very short or absent. Disc of 2 large glands. **Gynoecium:** carpels 2, with 2 free ovaries but 1 style and 1 stigma, annulated. **Fruits:** a pair of slender, erect follicles.

Family 14—*Convolvulaceae* (1,600 sp.—157 sp. in India)

Habit: mostly twiners. **Leaves:** simple, alternate and exstipulate. **Inflorescence:** cymose. **Flowers:** regular, bisexual, hypogyiious, often large and showy. **Sepals:** 5, usually free, imbricate and persistent. **Petals:** (5), united, funnel-shaped, twisted

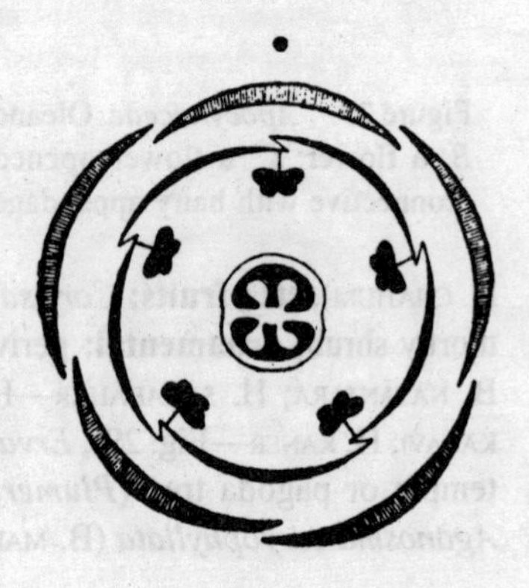

Figure 31. Floral diagram of *Convolvulaceae* (*Ipomoea*).

in bud, sometimes imbricate. **Stamens:** 5, epipetalous, alternating with the petals. **Carpels:** (2), rarely more, connate; ovary superior, with a disc at the base, 2-celled, with 2 ovules in each cell, or sometimes 4-celled with 1 ovule in each cell; placentation axile. **Fruit:** a berry or a capsule. *Floral formula*—⊕⚥ $K_5 \overline{C_{(5)} A_5} \underline{G}_{(2)}$.

Examples. Useful plants: vegetables: sweet potato (*Ipomoea batatas*), and water bindweed (*Ipomoea aquatica = I. reptans*); **medicinal:** *Ipomoea paniculata* (B. & H. BHUL-KUMRA), and Indian jalap (*Operculina turpethum;* B. TEORI; H. PITOHARI); **ornamental:** morning glory (*I. purpurea*), railway creeper (*I. palmata;* Fig. 32), moon flower (*I. grandiflora*), *I. nil* (B. & H. NIL-KALMI),

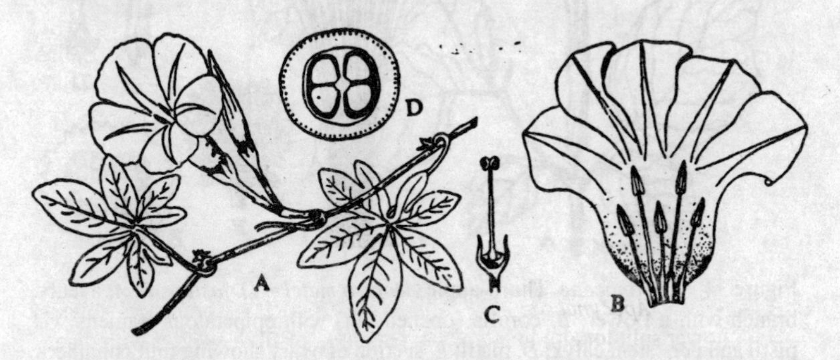

Figure 32. *Convolvulaceae.* Railway creeper (*Ipomoea palmata*). *A,* a branch; *B,* corolla with epipetalous stamens (opened out); *C,* pistil; and *D,* section of ovary showing axile placentation.

I. quamoclit (=*Quamoclit pinnata;* B. KUNJA-LATA; H. KAMLATA), woodrose (*I. tuberosa*)—a liane, *Convolvulus,* etc.; **other common plants:** dodder (*Cuscuta reflexa;* B. SWARNA-LATA; H. AKASH-BEL—see Fig. I/6). *Evolvulus nummularius* and *E. alsinoids*–common weeds, etc.

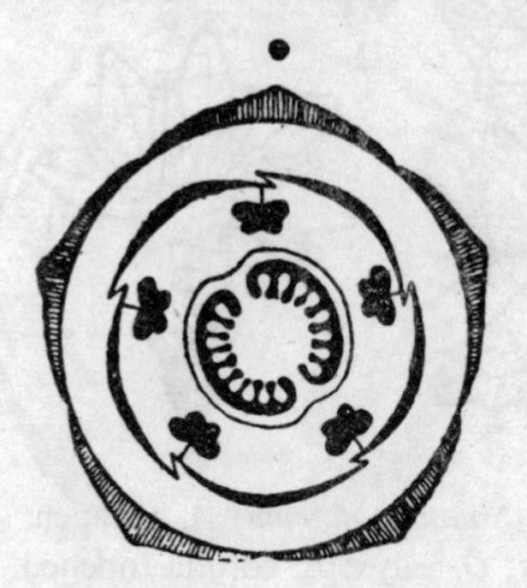

Figure 33. Floral diagrams of *Solanaceae.*

Family 15—*Solanaceae* (2,000 sp.—58 sp. in India)

Habit: herbs and shrubs. **Leaves:** simple, sometimes pinnate as in tomato, alternate. **Flowers:** regular, bisexual, hypogynous. **Calyx:** sepals (5), united, persistent. **Corolla:** petal (5), united, usually funnel- or cup-shaped, valvate or twisted in bud. **Androecium:** stamens 5, epipetalous, alternating with the corolla-lobes; anthers apparently connate. **Gynoecium:** carpels (2), syncarpous; ovary superior, obliquely placed (Fig. 33), 2-celled or sometimes 4-celled owing to the development

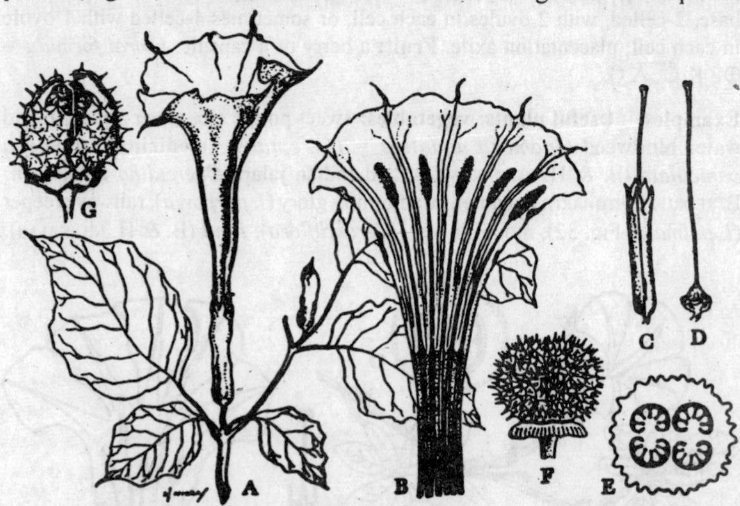

Figure 34. *Solanaceae.* Thorn-apple (*Datura metel* = *D. fastuosa*). *A,* a leafy branch with a flower; *B,* corolla (opened out) with epipetalous stamens; *C,* pistil and persistent calyx; *D,* pistil; *E,* section of ovary showing four chambers (varying from 3 to 5); *F,* a young fruit; and *G,* a mature fruit (capsule).

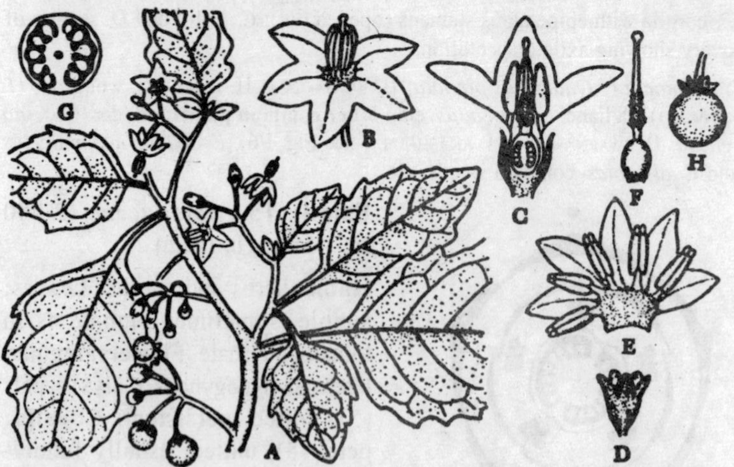

Figure 35. *Solanaceae.* Black nightshade (*Solanum nigrum*). *A,* a branch; *B,* a flower; *C,* a flower cut longitudinally; *D,* calyx; *E,* corolla (opened out) with epipetalous stamens; *F,* pistil; *G,* ovary in transverse section showing axile placentation: and *H,* a fruit (berry).

of a false septum, as in tomato and *Datura;* ovules many; placentation axile. **Fruit:** a berry or capsule with many seeds. *Floral formula*— $\oplus ⚥ K_{(5)} \overline{C_{(5)} A_5} \underline{G}_{(2)}$.

Examples. *Solanum* with 1,500 species is the largest genus of the family. **Useful plant: vegetables:** potato (*Solanum tuberosum*), brinjal (*S. melongena*), tomato (*Lycopersicum esculentum*), etc.; **condiment:** red pepper or chilli (*Capsicum annuum*)—fruits are pungent and stimulant; **medicinal:** deadly nightshade (*Atropa belladonna*), thorn apple (*Datura metel = D. fastuosa;* Fig. 34)—seeds very poisonous, bitter-sweet (*Solanum dulcamara;* B. & H. MITHABISH), *S. surat tense* (=*S. xanthocarpum;* B. KANTIKARI; H. KATITA), *S. indicum* (B. BRIHATI; H. BIRHATTA), *Withania somnifera* (B. ASWAGANDHA; H. ASGAND), etc.; **narcotic:** tobacco (*Nicotiana tabacum*)—tobacco of commerce and also a source of nicotine—an insecticide; **fruits:** gooseberry (*Physalis peruviana;* H. RASBHARI) and tomato (*Lycopersicum esculentum*); **ornamental:** *Petunia,* queen of the night (*Cestrum nocturnum;* B. HASNA-HANA; H. RAT-KI-RANI), etc.; **other common plants:** black nightshade (*Solanum nigrum;* H. GURKAMAI—Fig. 35), wild gooseberry (*Physalis minima*) and wild tobacco (*Nicotiana plumbaginifolia*).

Family 16—***Acanthaceae*** (2,200 sp.—409 sp. in India).

Habit: herbs, shrubs and a few climbers; cystoliths often present in the leaf and the stem. **Leaves:** simple, entire and opposite, usually decussate. **Inflorescence:** a spike or a cyme. **Flowers:** zygomorphic, often bilabiate or oblique, bisexual and hypogynous, often with conspicuous bracts and

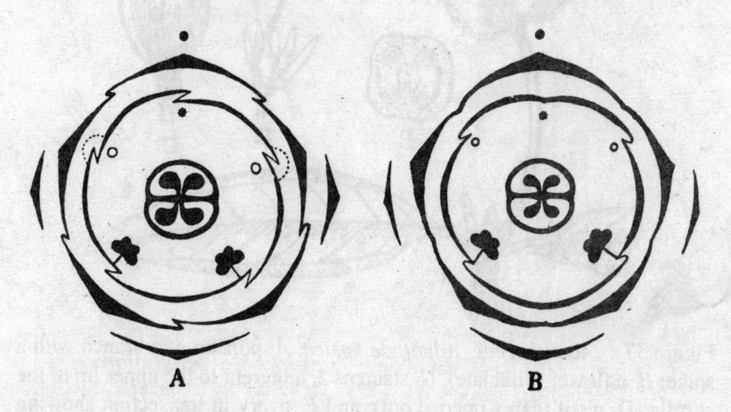

Figure 36. Floral diagrams (two types) of *Acanthaceae*.

bracteoles. **Calyx:** sepals (5), rarely (4), gamosepalous, twisted or imbricate in bud. **Corolla:** petals (5), connate in a 2-lipped or oblique corolla, twisted or imbricate in bud. **Androecium:** stamens 2 or 4, if 4 didynamous, epipetalous; staminodes often present; disc often conspicuous. **Gynoecium:** carpels (2), syncarpous; ovary superior, 2-celled, with 2-∞ ovules in each; placentation axile. **Fruit:** a 2-valved capsule. **Seeds:** usually on curved hooks (jaculators) which press the fruit from inside; the latter then bursts with a sudden jerk and scatters the seeds (see Fig. I/159). *Floral formula*—% ⚥ $K_{(5)}\overline{C_{(5)}A}_{2 \text{ or } 4}\underline{G}_{(2)}$.
Examples. **Medicinal:** *Andrographis paniculata* (B. KALMEGH; H. MAHATITA)—an effective remedy for liver complaints in children, and *Adhatoda vasica* (B. BASAK; H. ADALSA)—an excellent remedy for cough; **ornamental:** *Barleria cristata*—flowers white or rose-coloured, *Ruellia tuberosa* (see Fig. I/159), *Thunbergia grandiflora*—a climber etc.; **other common plants:** *Rungia* and *Dicliptera*—common weeds.

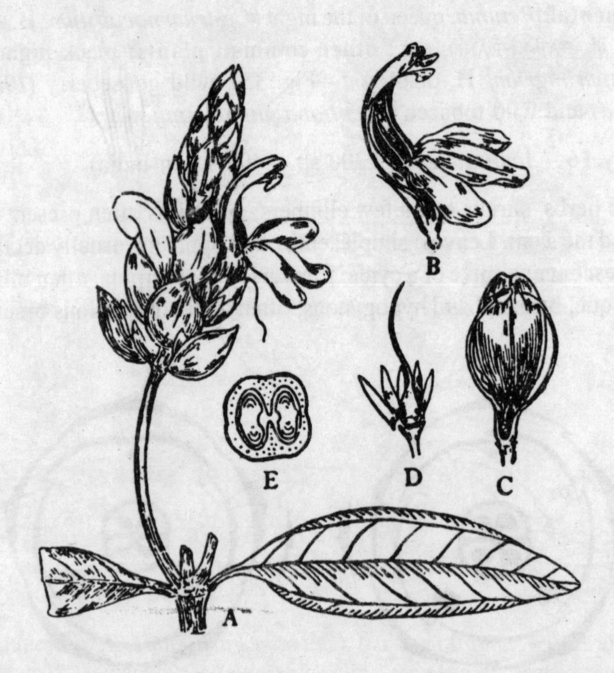

Figure 37. *Acanthaceae. Adhatoda vasica. A,* portion of a branch with a spike; *B,* a flower (bilabiate), *C,* stamens 2, adherent to the upper lip of the corolla; *D,* pistil (calyx opened out); and *E,* ovary in transection showing axile placentation.

Figure 38. Floral diagram of *Labiatae*.

Family 17—*Labiatae* (3,500 sp.—391 sp. in India)

Habit: herbs and undershrubs, with square stem. **Leaves:** simple, opposite or whorled, with oil-glands. **Flowers:** zygomorphic, bilabiate, hypogynous and bisexual. **Inflorescence.** verticillaster (see p. 90), sometimes reduced to true cyme. **Calyx:** sepals (5), gamosepalous, unequally 5-lobed, persistent. **Corolla:** petals (5), gamopetalous, bilabiate, i.e. 2-lipped; aestivation imbricate. **Androecium:** stamens 4, didynamous, sometimes only 2, epipetalous.

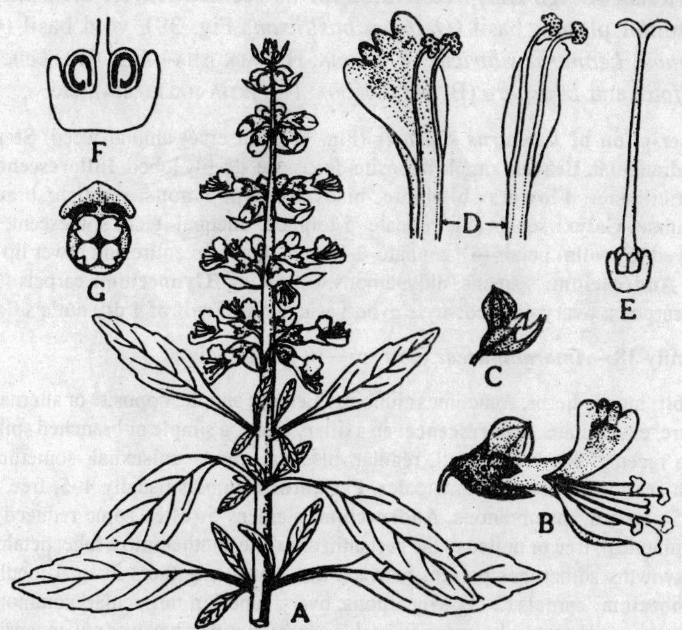

Figure 39. *Labiatae.* Basil (*Ocimum basilicum*). *A,* a branch with inflorescences. *B,* a flower—bilabiate (note the didynamous stamens); *C,* calyx, *D,* corolla split open with epipetalous stamens; *E,* pistil (note the gynobasic style); *F,* ovary with the disc in longitudinal section; and *G,* a fruit of four nutlets enclosed in the persistent calyx.

Gynoecium: carpels (2), syncarpus; disc prominent; ovary 4-lobed and 4-celled, with one ovule in each cell (basal placentation); style gynobasic, i.e. develops from the depressed centre of the lobed ovary; stigma bifid. **Fruit:** a group of four nutlets, each with one seed. *Floral formula*— % ⚥ $K_{(5)}\overline{C_{(5)}A_{(4)}}\underline{G}_{(2)}$.

Labiatae abounds in volatile, aromatic oils which are used in perfumery and also as stimulants.

Examples. Useful plants: medicinal: sacred basil (*Ocimum sanctum;* B. & H. TULSI), garden mint (*Mentha viridis;* B. PUDINA; H. PODINA), peppermint (*M. piperita*)—source of peppermint oil and menthol, thyme (*Thymus*)—source of thyme oil and thymol, rosemary (*Rosmarinus*)—yields oil of rosemary, patchouli (*Pogostemon heyneanus*)—yields patchouli oil, lavender (*Lavandula vera*)—yields lavender oil, etc.; **ornamental:** sage (*Salvia;* see,Fig. I/137), *Coleus* (see Fig. I/32), marjoram (*Origanum*)—cultivated for its scented leaves, etc.; **other common plants:** basil (*Ocimum basilicum;* Fig. 39), wild basil (*O. canum*), *Leonurus sibiricus* (B. DRONA; H. HALKUSHA—Fig. 40), *Leucas linifolia* and *L. aspera* (B. SWET-DRONA; H. CHOTA-HALKUSHA), etc.

Description of *Leonurus sibiricus* (Fig. 40). An erect annual weed. **Stem:** quadrangular. **Leaves:** simple opposite-decussate, deeply lobed. **Inflorescence:** verticillaster. **Flowers:** bilabiate, bisexual, hypogynous, reddish; bracts subulate. **Calyx:** sepals (5), connate, 5-toothed, unequal. teeth spinescent, 5-nerved. **Corolla:** petals (5), connate, 2-lipped, upper lip entire and lower lip 3-fid. **Androecium:** stamens, didynamous; conniving. **Gynoecium:** carpels (2), syncarpous; ovary 4-lobed; style gyno-basic; 2-fid. **Fruit** of 4 dry nutlets.

Family 18—*Amaranthaceae* (850 sp.—46 sp. in India).

Habit: mostly herbs, sometimes climbing. **Leaves:** simple, opposite or alternate, entire, exstipulate. **Inflorescence:** an axillary cyme, a simple or branched spike, or a raceme. **Flowers:** small, regular, bisexual, rarely unisexual, sometimes with scarious bracts and bracteoles. **Perianth:** members usually 4–5, free, or (4–5), united, membranous. **Androecium:** stamens 5 (often some reduced to staminodes), free or united to the perianth, or to one another into a tube; petaloid outgrowths sometimes present between the stamens; anthers 2- or 4-locular. **Gynoecium:** carpels (2–3), syncarpous; ovary superior, unilocular, commonly with one ovule (many, however, in cock's comb). **Fruit:** a berry or nut or capsule (dehiscent). **Seed:** endospermic. *Floral formula*— ⊕ ⚥ $P_{4-5 \text{ or } (4-5)}A_5\underline{G}_{(2-3)}$.

Examples. Leafy vegetables: *Amaranthus caudatus, A. gangeticus, A. blitum* (=*A. oleracea*)—a tall annual widely cultivated, *A. polygamus,* etc.; **ornamental:** *A. tricolor*—leaves variously coloured and showy, *A. paniculatus*—a tall plant bearing long crimson or golden yellow pendulous

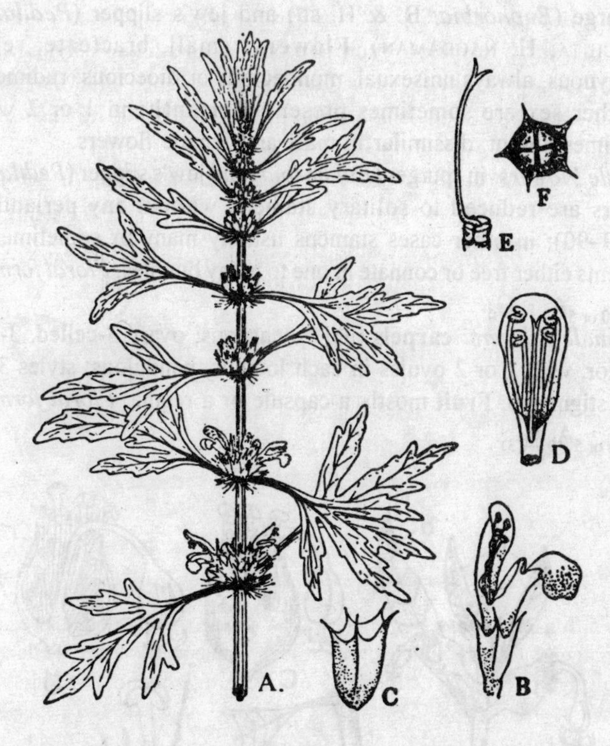

Figure 40. *Labiatae. Leonurus sibiricus. A,* a branch with opposite leaves and inflorescences; *B,* a flower (bilabiate); *C,* calyx; *D,* stamens—didynamous and epipetalous; *E,* pistil (note the gynobasic style and 4-lobed ovary); and *F,* fruit of four nutlets enclosed in persistent calyx.

spikes, cock's comb (*Celosia*), e.g. *C. cristata* bearing red flowers, *C. argentea* bearing white flowers, *C. plumosa* bearing yellow flowers, button flower or globe amaranth (*Gomphrena globosa*), etc.; **weeds:** prickly amaranth (*Amaranthus spinosus*), *A. viridis,* chaff-flower (*Achyranthes aspera*) with long spinous spikes, *Digera arvensis*—a slender weed, *Alternanthera amoena*—a common prostrate weed. *Aerua lanata* with a dense coating of hairs, *A. scandens*—leaves and branches reddish, *Allmania nodiflora* commonly used as a border for garden beds, *Deeringia celosioides*—a rambling climber bearing globose red berries, etc.

Family 19—*Euphorbiaceae* (7,000 sp.—374 sp. in India).

Habit: herbs, shrubs and trees, often with acid milky juice. **Leaves:** simple, usually alternate; stipules usually present. **Inflorescence:** varying—racemose or cymose, or mixed, or a cyathium (see p. 89), as in spurge (*Euphorbia;* B. & H. SIJ) and jew's slipper (*Pedilanthus;* RANGCHITA; H. NAGDAMAN). **Flowers:** small, bracteate, regular, hypogynous, always unisexual, monoecious or dioecious; rudiments of the other sex are sometimes present. **Perianth:** in 1 or 2 whorls, sometimes absent, dissimilar in male and female flowers.

Male Flowers: in spurges (*Euphorbia*) and jew's slipper (*Pedilanthus*) flowers are reduced to solitary stamens without any perianth (see pp. 89–90); in other cases stamens usually many or sometimes few; filaments either free or connate in one to many bundles. *Floral formula*—
⊕♂ $P_{0 \text{ or } 5}A_{1-\infty}G_{0}$.

Female Flowers: **carpels** (3), syncarpous; ovary 3-celled, 3-lobed, superior, with 1 or 2 ovules in each loculus, pendulous; styles 3, each bifid; stigmas 6. **Fruit** mostly a capsule or a regma. *Floral formula*—
⊕♀ $P_{0 \text{ or } 5}A_{0}\,\underline{G}_{(3)}$.

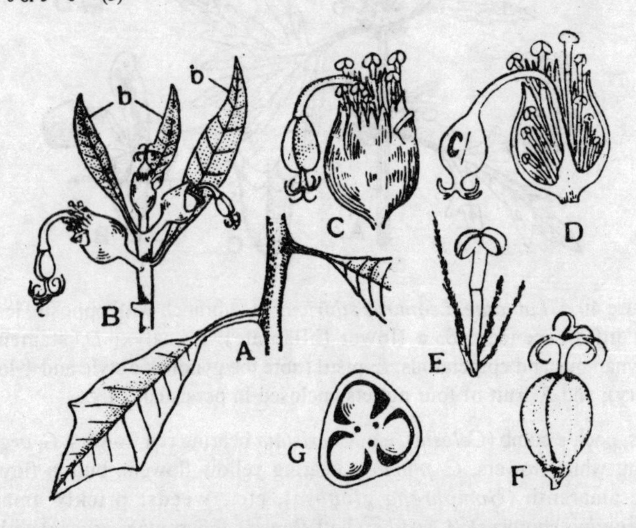

Figure 41. *Euphorbiaceae.* Poinsettia (*Poinsettia pulcherrima*). *A,* a portion of a branch; *B,* a branch with three inflorescences; *b,* petaloid bracts; *C,* an inflorescence (cyathium); *D,* cyathium cut longitudinally showing the centrally placed female flower surrounded by numerous male flowers; *E,* a male flower (reduced to a stamen only) with bract and bracteoles at the base; *F,* female Bower (reduced to a pistil only); *G,* ovary in transverse section.

Examples. Useful plants: castor (*Ricinus communis;* (Fig. 42)—seeds yield castor oil, emblic myrobalan (*Emblica officinalis;* B. AMLA; H. AMLIKA)—fruits rich in vitamins, medicinal and also used for tanning.

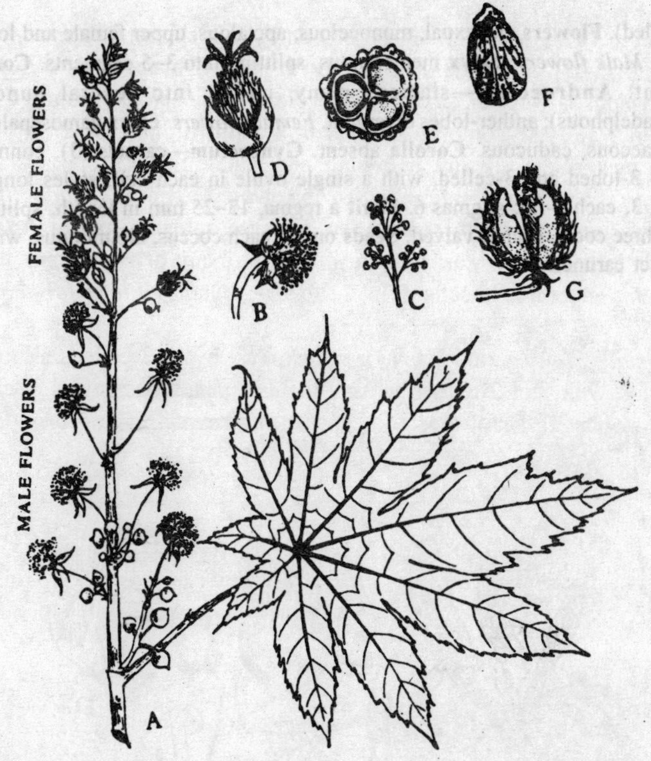

Figure 42. *Euphorbiaceae.* Castor (*Ricinus communis*). *A,* a branch with a leaf and an inflorescence; *B,* a male flower; *C,* branched stamens; *D,* a female flower; *E,* ovary in transverse section; *F,* a seed; and *G,* a fruit (regma) splitting.

Baccaurea (B. LATKAN; H. LUTKO)—aril, pulpy and edible, tapioca (*Manihot esculenta*)—tuberous roots yield a starchy food, garden crotons (*Codiacum variegatum*) with ornamental variegated leaves, poinsettia (*Poinsettia pulcherrima,* Fig. 41)—ornamental *Jatropha curcas* and *J. gossypifolia* grown as hedge plants, spurges (*Euphorbia* with about 2,000 species), e.g. *E. neriifolia* (B. MANSHA-SIJ; H. SIJ), *E. antiquorum, E. splendens,* etc.; **other common plants:** *Croton sparsiflorus*—a weed, *Acalypha indica*—a weed, *Tragia involucrata*—a nettle, *Euphorbia pilulifera*—a weed, etc.

Description of Castor Plant (*Ricinus communis;* Fig. 42). An evergreen, much-branched, tall shrub. **Leaves** simple, alternate, broad, palmately lobed and veined, lobes usually 7 or 9. **Inflorescence** a terminal raceme (subpanicled). **Flowers** unisexual, monoecious, apctalous, upper female and lower male. *Male flowers;* **calyx** membranous, splitting into 3–5 segments. **Corala** absent. **Androecium**—stamens many, united into several bundles (polyadelphous); anther-lobes divergent. *Female flowers:* **calyx** gamosepalous, spathaccous, caducous. **Corolla** absent. **Gynoecium**—carpels (3), connate; ovary 3-lobed and 3-celled, with a single ovule in each cell; styles long or short, 3, each 2-fid; stigmas 6. **Fruit** a regma, 13–25 mm in length, splitting into three cocci, each 2-valved. **Seeds** one in each coccus, albuminous, with a distinct caruncle.

Chapter 3

Selected Families of Monocotyledons

Family 1—*Liliaceae* (over 3,000 sp.)

Habit: herbs and climbers, rarely shrubs, with rhizome or bulb. **Leaves:** radical or cauline or both, simple, alternate. **Flowers:** regular, trimerous, bisexual, hypogynous, solitary or in raceme, panicle or spike, or in cyme (umbellate). **Perianth:** perianth leaves (tepals) 6, in two whorls, usually free (polyphyllous), sometimes united (gamophyllous); aestivation valvate. **Androecium:** stamens 6, in two whorls, rarely 3, free or united with the perianth (epiphyllous). **Gynoecium:** carpels (3), syncarpous; ovary superior, 3-celled; ovules usually ∞; placentation axile. **Fruit:** a berry or capsule. **Seed:** albuminous. *Floral formula*—$\oplus$ ⚥ $P_{3+3}A_{3+3}\underline{G}_{(3)}$ or $\overline{P_{(3+3)}A_{3+3}}\underline{G}_{(3)}$.

Examples. Useful plants: vegetables: onion (*Allium cepa;* Fig. 44), garlic (*A. sativum*), leek (*A. porrum*), shallot (*A. ascalonicum*), etc.; **medicinal:** *Asparagus racemosus* (B. SATAMULI; H. SATWAR), *Smilax zeylanica;* B. KUMARIKA; H. CHOBCHINI—see Fig. I/63), Indian aloe (*Aloe vera;* B. GHRITA-KUMARI; H. GHIKAVAR), etc.; **ornamental:** lily (*Lilium*), glory lily (*Gloriosa superba;* see Fig. I/77C), day lily (*Hemerocallis fulva*), dagger plant or Adam's needle (*Yucca gloriosa;* see Fig. I/92), dragon plart (*Dracaena*), asphodel (*Asphodelus*), butcher's broom (*Ruscus;* see Fig. I/58A), etc.; **fibre-yielding:** bowstring hemp (*Sansevieria roxburghiana;* B. MURVA; H. MARUL), etc.

Figure 43. Floral diagram of *Liliaceae*.

Description of Onion Plant (*Allium cepa;* Fig. 44). A cultivated herb with tunicated bulb. Bulb surrounded by inner fleshy and outer dry scales. **Leaves:** radical, cylindrical, hollow, sheathing. **Inflorescence:** a terminal umbel on the leafless flowering stem or scape. Bracts 2, sometimes 3, membranous, enclosing the young umbel. **Flowers:** small, white, regular, bisexual, hypogynous, sometimes replaced by bulbils. **Perianth:** of 6 lobes, connate below, campanulate. **Androecium:** stamens 6, free; filaments narrow or dilated at the base. **Gynoecium:** carpels (3), syncarpous; ovary 3-lobed and 3-celled; style short, filiform; stigma minute; ovules usually 2 in each cell. **Fruit** a membranous capsule.

Description of Asphodel Plant (*Asphodelus tenuifolius*). An annual herb. **Leaves:** radical, slender, erect, semi-terete, fistular. **Scapes:** several, long, each ending in a laxly racemose inflorescence. Bracts scarious. **Flowers:** small, regular, bisexual, hypogynous. **Perianth** of six segments, white, each with a brownish costa, free or slightly united at the base, marcescent. **Androecium:** stamens 6. shorter than the perianth; flaments flattened at the base enclosing the ovary; anthers linear, versatile. **Gynoecium:** carpels (3), syncarpous, ovary 3-celled, with two ovules in each; stigma 3-lobed. **Fruit:** a globose capsule, broader in diameter, usually with one seed in each cell. **Seed:** albuminous.

Family 2—*Scitamineae* (over 1,200 sp.)

Habit: herbs, rarely woody and tree-like, e.g. traveller's tree (*Ravenala*), underground stem usually in the form of a slender or stout rhizome;

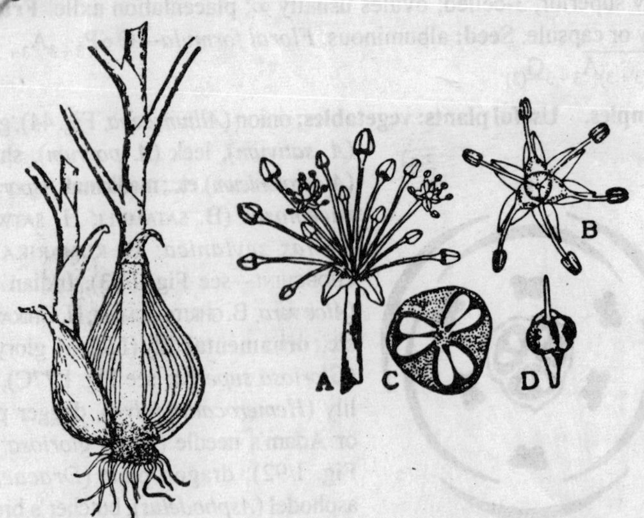

Figure 44. *Liliaceae.* Onion (*Allium cepa*). *Left,* an onion plant; *A,* an inflorescence. *B,* a flower; *C,* ovary in transverse section showing axile placentation; and *D,* pistil.

aerial stem distinct or 'false' made of sheathing leaf-bases, and the flowering stem or scape pushing out through the 'false' stem and ending in an inflorescence. **Leaves:** spiral or distichous, and sheathing. **Inflorescence:** a raceme, spike or spadix with often large spathes, either terminal or axillary. **Flowers:** zygomorphic, mostly bisexual and epigynous; bracts often spathaceous. **Perianth** of six segments in two whorls. **Androecium:** stamens varying (see sub-families). **Gynoecium:** carpels (3), syncarpous; ovary inferior and trilocular; placentation axile; ovules usually many, **Fruit:** a berry or capsule. **Seeds:** with perisperm, sometimes with aril. This family has, been divided into four sub-families, mainly depending on the number of stamens.

(1) ***Musaceae*** (150 sp.). **Leaves:** spiral, rarely distichous, sheathing but with no ligule. **Perianth** petaloid in two series—one with 5 limbs united and another solitary and free. **Stamens** in 2 whorls, 5 perfect and the 6th one sterile or absent. In *Ravenala,* however, all the six stamens are fertile. *Floral formula*—% ⚥ $P_{(5)+1}A_{3+2}\overline{G}_{(3)}$.

Examples. *Musa* (35 sp.), e.g. banana (*M. paradisiaca*)—a dessert fruit, plantain (*M. sapientum*)—green fruit used as a vegetable, *M. superba* and *M. nepalensis*—ornamental, fruits not edible, *M. textilis* yielding commercial Manila hemp or abaca, traveller's tree (*Ravenala madagascariensis;* B. PANTHA-PADA—see Fig. I/85), ornamental, bird of Paradise (*Strelitzia reginae*)—ornamental, etc.

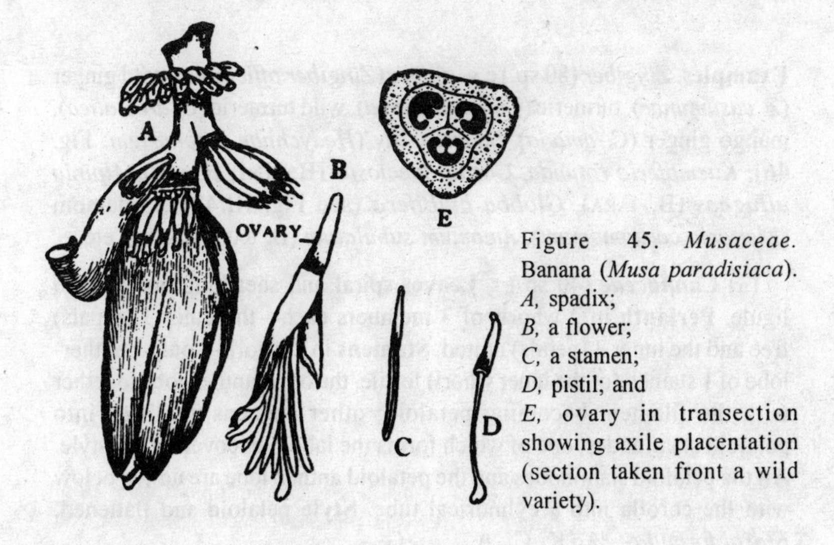

Figure 45. *Musaceae*. Banana (*Musa paradisiaca*). *A,* spadix; *B,* a flower; *C,* a stamen; *D,* pistil; and *E,* ovary in transection showing axile placentation (section taken front a wild variety).

(2) ***Zingiberaceae*** (700 sp.). **Leaves:** distichous, sheathing, with a distinct ligule. **Perianth** of 6 segments in 2 whorls, generally distinguishable into calyx and corolla. **Stamens** in 2 whorls—only 1 perfect, adnate to corolla-throat (this is the posterior one of the inner whorl), the other 2 stamens of this whorl are united to form a 2-lipped labellum; the anterior stamen of the outer whorl is absent and the remaining 2 modified into petaloid staminodes or absent. **Style** slender, passing through the two anther-lobes. *Floral formula*—$\%\ ⚥\ K_{(3)}C_3A_1\overline{G}_{(3)}$.

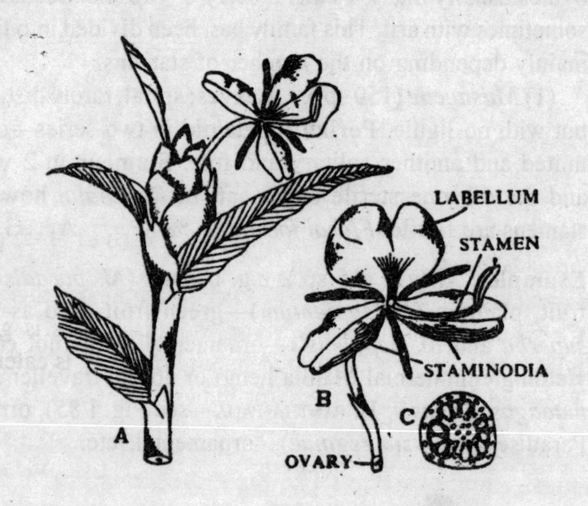

Figure 46. *Zingiberaceae.* Butterfly lily (*Hedychium coronarium*). *A,* a branch with inflorescence; *B,* a flower; and *C,* ovary in transection showing axile placentation.

Examples. *Zingiber* (80 sp.) e.g. ginger (*Zingiber officinale*), wild ginger (*Z. casumunar*), turmeric (*Curcuma longa*), wild turmeric (*C. aromatica*), mango ginger (*C. amada*), butterfly lily (*Hedychium coronarium;* Fig. 46), *Kaempferia rotunda, Costus speciosus* (B. KUST; H. KEU), *Alpinia allughas* (B. TARA), *Globba bulbifera* (see Fig. III/42), cardamom (*Elettaria cardamomum*), *Amomum subulatum* (B. BARA-ELAICH), etc.

(3) ***Cannaceae*** (40 sp.). **Leaves** spiral and sheathing but with no ligule. **Perianth** in 2 whorls of 3 members each—the outer 3 (sepals) free and the inner 3 (petals) united. **Stamens** in 2 whorls—only 1 anther-lobe of 1 stamen (of the inner whorl) fertile, the other anther-lobe together with the filament becoming petaloid; other stamens modified into petaloid staminodes, one of which forms the labellum covering the style. All the petaloid staminodes and the petaloid anther-lobe are united below with the corolla into a cylindrical tube. **Style** petaloid and flattened. *Floral formula*—$\%\ ⚥\ K_3\ C_{(3)}A_{1(\text{perfect})}G_{(3)}$.

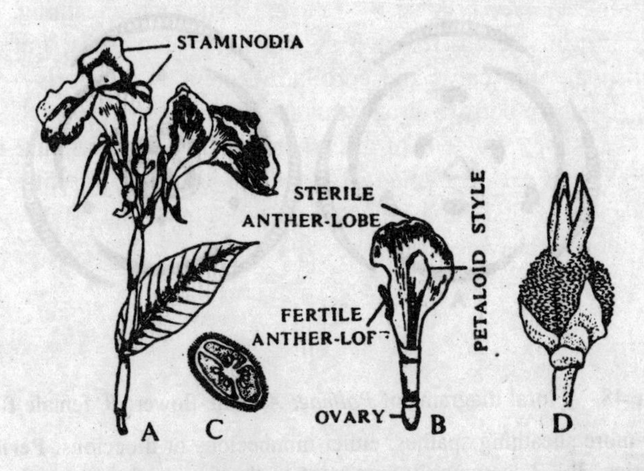

Figure 47. *Cannaceac.* Indian shot (*canna indica*). *A,* a branch; *B,* a flower (perianth and staminodia cut out); *C,* ovary in transection showing axile placentation; and *D,* a fruit.

Examples. Only genus is *Canna* in tropical America. Indian shot (*Canna indica*) with many varieties and hybrids is grown widely in Indian gardens, *C. edulis*—the starchy rhizome is eaten as a vegetable or ground into flour in the West Indies.

(4) ***Marantaceae*** (350 sp.). **Leaves** distichous, sheathing, and with ligule. Floral characters like those of *Cannaceae*. It is distinguished, however, from allied families by the presence of a joint or swollen pulvinus at the junction of the petiole and the leaf-blade. It is chiefly a tropical American family. *Floral formulae*—% ⚥ $P_{3+3}A\frac{1}{2}\overline{G}_{(3)}$.

Examples. *Maranta* (23 sp.), e.g. arrowroot (*Maranta arundinacea*), some ornamental species are *Maranta sanderiana, M. zebrina, M. bicolor, M. viridis,* etc., *Calathea* (150 sp.)—several species ornamental, *C. allouia*—small potato-like tubers are eaten as a vegetable in the West Indies, *Clinogyne dichotoma* (B. SITALPATI), and *Phrynium variegatum*—ornamental.

Family 3—*Palmae* (2,500 sp.)

Habit: shrubs or trees, except cane (*Calamus*) which is a climber. **Stem:** erect, unbranched and woody, rarely branched. **Leaves:** usually forming a crown, plaited in bud, sometimes very large, either palmately cut, or divided **(fan palms)** or pinnately cut or divided **(feather palms):** petiole often with sheathing base. **Flowers:** sessile, often produced in immense numbers, regular, hypogynous, unisexual or bisexual, in simple or compound spadix enclosed in

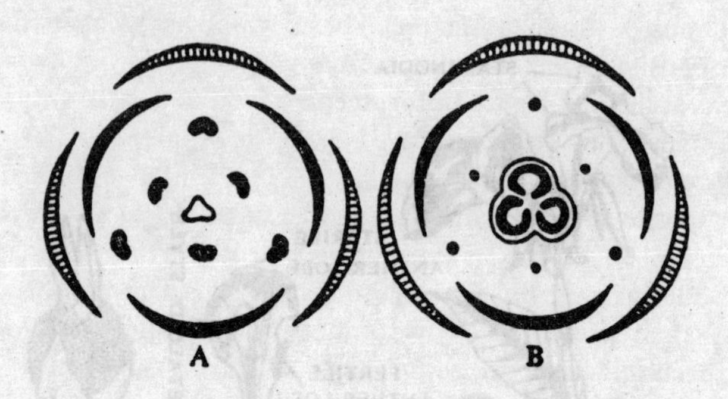

Figure 48. Floral diagrams of *Palmae; A,* male flower; *B,* female flower.

one or more sheathing spathes; either monoecious or dioecious. **Perianth** in two series, 3 + 3, commonly persistent in the female flower. **Androecium:** stamens usually in two series, 3 + 3, filaments free or connate; anthers versatile, 2-celled. **Gynoecium:** carpels (3) or 3. syncarpous or apocarpous; ovary superior, 1- or 3-locular, with 1 or 3 ovules. **Fruit:** a drupe or berry. **Seed:** albuminous. *Floral formulae*—⊕♂ $P_{3+3}A_{3+3 \text{ or } 3}$, or ⊕♀ $P_{3+3}\underline{G}_{(3)}$.

Economically this is one of the most important families. Many palms such as the palmyra palm, toddy palm, date palm, coconut palm, etc., are tapped for toddy (fermented country liquor) or for sweet juice from which jaggery or sugar is made, and they also (except the toddy-palm) yield fruits. Coir fibres of coconut-palms are used for making mats, mattresses and brushes, and also for stuffing cushions. Leaves of many palms are woven into mats, hats and baskets and also used for thatching. Some palms yield oil, e.g. coconut palm, oil palm, etc. Sago palms yield sago, which is obtained by crushing the pith. Betel-nut is used for chewing with betel leaf. Cane is used for making chain, sofas, tables and baskets and also for a variety of other purposes; some canes (rattan canes) grow to a length of 150–180 metres. Many palms are ornamental.

Examples. Fan-palms: palmyra-palm (*Borassus flabellifer*), talipot-palm (*Corypha umbraculifera*) bears a huge inflorescence once only after about 40 years and then dies; oil-palm (*Elaeis guineensis*), doum-palm (*Hyphaene*) shows dichotomous branching; double coconut-palm (*Lodoicea maldivica*), a native of the Seychelles Islands, bears the largest seed (see Fig. I/158), etc.
Feather-palms: cane (*Calamus*—about 375 sp.), Indian sago-palm or toddy-palm (*Caryota urens*), coconut-palm (*Cocos nucifera*), date-palm (*Phoenix sylvestris*), betel-nut-palm (*Areca catechu*), sago-palm (*Metroxylon rumphii*), etc.

Family 4—*Gramineae* or *Poaceae* (10,000 sp.)

Habit: herbs, rarely woody, as bamboos **Stem:** cylindrical, internodes with distinct nodes and (sometimes hollow). **Leaves:** simple, alternate, distichous, with sheathing leaf-base which is split open on one side opposite to the leaf-blade; a hairy structure called the *ligule* is present at the leaf-blade. **Inflorescence:** usually a spike or panicle of spikelets; each spikelet consists of one or few flowers, and bears at the base two *empty glumes,* a little higher up a flowering glume called *lemma,* and opposite to the lemma a somewhat smaller glume known as the *palea* (Fig. 50D). The flower remains enclosed by the lemma and the palea. **Flowers:** usually bisexual, sometimes unisexual and monoecious. **Perianth:** represented by two minute scales called the *lodicules*. **Androecium:** stamens 3, sometimes 6, as in rice and bamboo; anthers versatile and pendulous. **Gynoecium:** carpels (3), often reduced to 1; ovary superior, 1-celled, with 1 ovule; styles usually 2; stigmas feathery. **Fruit:** caryopsis. **Seed:** albuminous. *Floral formula*—$P_{\text{lodicules (2 or 3)}}$ $A_{3 \text{ or } 6}$ $\underline{G}_{(3)}$.

From an economic standpoint *Gramineae* is regarded as the most important family, as cereals and millets, which constitute the chief food of mankind, belong to this family. Most of the fodder crops which are equally important to domestic animals also belong to this family. The importance of bamboo, thatch grass and reed as building materials, and of sugarcane as a source of sugar and jaggery is well known. The importance of sabai grass and bamboo as a source of paper pulp cannot be over-emphasized.

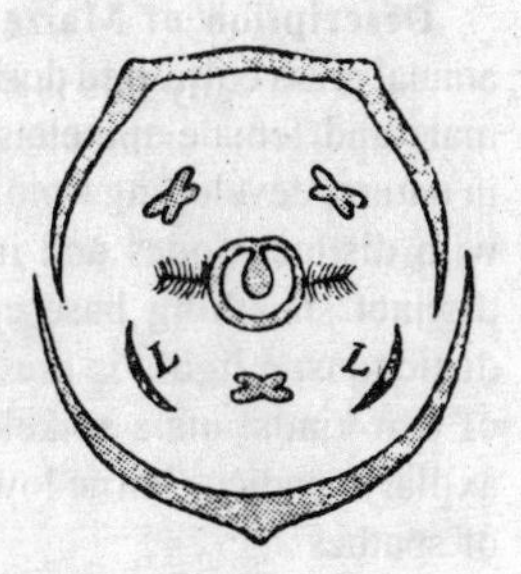

Figure 49. Floral diagram of *Gramineae; L,* lodicule.

Examples. **Useful plants:** cereals such, as rice (*Oryza sativa;* Fig. 50) wheat (*Triticum aestivum*), maize or Indian corn (*Zea mays;* Fig. 51), barley (*Hordeum vulgare*) etc.; millets such as great millet (*Sorghum*) vulgare; (B. JUAR; H. JOWAR), *Pennisetum typhoideum* (B. & H. BAJRA), *Eleusine coracana* (B. & H. MARUA), etc.; sugarcane (*Saccharum officinarum*), thatch grass (*S. spontaneum*), guinea grass (*Panicum maxicum*), bamboo (*Bambusa*), reed (*Phragmites*), giant reed (*Arundo*), lemon grass (*Cymbopogon*), saboi grass (*Ischaemum*), *Vetiveria* (B. & H. KHUS-KHUS), etc.; **other common plants:** various grasses such as dog grass (*Cynodon dactylon;* B. DURBA-GHAS; H. DOOB), love thorn (*Chrysopogon aciculatus*), *Imperata cylindrica,* several species of *Panicum,* etc.

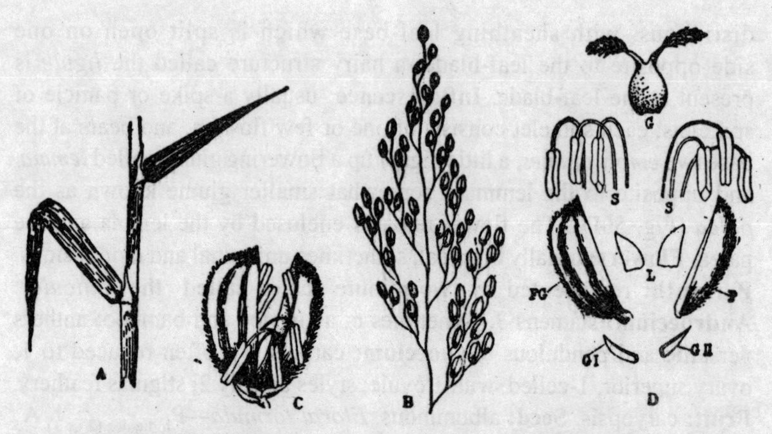

Figure 50. *Gramineae*. Rice (*Oryza sativa*). *A*, portion of a branch with sheathing leaves and ligules; *B*, a panicle of spikelets; *C*, 1-flowered spikelet (note the glumes and stamens); *D*, spikelet dissected out—*G I*, first empty glume; *G II*, second empty glume; *FG*, flowering glume: *P*, palea; *L*, lodicule; *S*, stamens; and *G*, gynoecium.

Description of Maize Plant (*Zea mays;* Fig. 51). This is a tall, stout, annual grass cultivated during the rainy season. The plant is monoecious, bearing male and female spikelets in separate inflorescences. **Roots** are adventitious in nature, developing from the lower nodes; while the **stem** is solid and provided with distinct nodes and internodes. **Leaves** are long, broad and flat, with a distinct sheathing base enclosing the stem; they are simple, alternate and distichous; a ligule is present at the base of the leaf-blade. **Inflorescences** are of two kinds: male spikelets in a terminal panicle, while female spikelets in axillary spadices borne lower down, each spadix remains enclosed by a number of spathes.

Male spikelets occur in pairs—one sessile and the other stalked; each spikelet is 2-flowered. **Glumes** I and II distinctly nerved and empty; glume III, (flowering glume) and glume IV (palea), which are hyaline, enclose a flower **Perianth** is represented by two small fleshy cup-shaped lodicules. **Androecium:** stamens 3; anthers linear and pendulous.

Female Spikelets are densely crowded in several vertical rows on the fleshy rachis and are sessile. Each spikelet with a lower barren, (extremely reduced) floret and an upper fertile (normal) one. **Glumes** I and II membranous, broad and empty; glume III (flowering glume) and glume IV (palea) hyaline and enclose a flower. Lodicules absent or very feebly developed. **Gynoecium:** carpels (3); ovary obliquely ovoid and plano-convex; style 1 (really 2 fused into 1), very long and bi-fid at the tip; styles and stigmas papillose and hang out in a tuft from the top of the spadix. **Fruit** or maize grain is a caryopsis. It is albuminous with

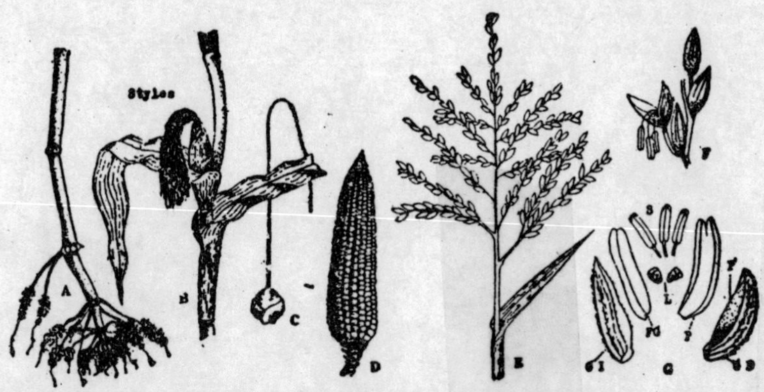

Figure 51. *Gramineae.* Maize or Indian corn (*Zea mays*). *A,* adventitious roots; *B,* female spadix in the axil of a leaf; *C,* female spikelet; *D,* ripe cob; *E,* a panicle of male spikelets; *F,* two pairs of male spikelets; and *G,* a male spikelet dissected out—*G I,* first empty glume; *G II,* second empty glume; P^1, palea of the lower flower; *FG,* flowering glume; *P,* palea of the upper flower; *L,* lodicules and *S,* three stamens of the upper flower.

distinct scutellum (see Fig. I/17B). Flowers are anemophilous, i.e. pollination is brought about by wind and after fertilization the female spadix develops into the maize cob.

Part VIII

EVOLUTION AND GENETICS

Organic Evolution
Genetics

Chapter 1

Organic Evolution

Evolution means the descent of a new form of plant or animal from the pre-existing one. Biologists have been finally led to believe in the doctrine of evolution by the epoch-making deductions of Charles Darwin in 1859, as opposed to that of creation believed earlier, particularly by the theosophists. It is now universally accepted that all higher and more complex forms of life—plants and animals—have evolved from lower and simpler forms. Even the minute organisms like bacteria are not newly formed, as proved by Louis Pasteur, a French scientist and founder of the Science of Bacteriology, in the year 1864. Once life came into existence it became continuous, progressing and changing through successive generations and finally gave rise to the present forms of plants and animals through many millions of years. Evolution is proceeding even now.

According to geologists and astronomers, the earth may be about 4,550 years old. At the initial stage it was only a molten mass at an excessively high temperature. Gradually the earth cooled, and there was spontaneous generation of life—how and exactly when, we are not in a position to say with any amount of certainty. According to fossil records life seems to be already somewhat advanced about 600 million years ago in the form of marine algae and primitive invertebrates. Life must have originated much earlier (possibly more than 2,000 million years ago) in an even simpler form. A very important recent biological discovery is that *nucleic acid* is the basis of life. Therefore, it may be assumed that organic molecules of nucleic acid appeared first under certain physico-chemical conditions prevailing then, and some of the complex molecules acquired the characteristics of life, possibly in the form of aquatic bacteria. Evolution had proceeded from this stage, possibly leading to blue-green algae first and then to green algae as the basis of further descent. Two major events of the early age which had tremendously influenced the course of evolution of plants and animals through the successive ages to the present day may be specially noted: (a) development of the green pigment *chlorophyll* in blue-green algae and green algae at a very early stage, and later in all other green plants, monopolising for all time to come the function of manufacturing food for their own nutrition as well as that of the animals; and (b) *migration* of green plants from water (sea) to land, i.e. from the easy life of the aquatic medium to the difficult life of the land environment under varying external conditions; the establishment of true land plants, evidently adapting themselves to the new conditions, soon opened gateway to the rapid evolution of many new land plants in diversified lines from them (see p. 535); further, the green plants first prepared the ground for habitation and feeding of animals on land; the animals thus invaded land much later.

EVIDENCES OF ORGANIC EVOLUTION

1. **Geological Evidence.** The petrified remains of ancient plants and animals or impressions left by them in rocks are called **fossils**. A fossil (*fossilis,* dug out) is thus a relic of past life—plant or animal—dug out of the earth. The study of plant fossils or animal fossils is called **palaeontology**. Rocks formed in strata in successive geological periods of the earth have been found to bear fossils of particular types of plants and animals (more advanced forms being non-existent then). From fossils, we come to know about the life of the past, i.e. the history of appearance of new groups of plants and animals, and disappearance of old ones in successive stages, their habit and habitat, and their distribution over the earth. It is remarkable that the fossils of any one stratum bear a close resemblance to those of the next or the previous stratum. This leads us to the firm belief that there is a relationship between the plants or the animals which appeared through the successive periods of the earth. In fact, from fossil records it has been possible to trace the origin and the trend of evolution of several groups of plants and animals from the simpler to the more complex forms. Fossils thus bear sound evidence of organic evolution. Fossils are, therefore, of special interest and importance in this respect. They, however, show several wide gaps in the evolutionary history of plants and animals. Because of the missing links the origin of several groups has still remained a mystery.

Possible Origin and Evolution of Plants in Successive Geological Ages.

Cenozoic or Recent Age (beginning 65 million years ago). This is the 'age of angiosperms'. In this age angiosperms rapidly increased in diversified forms and extended all over the earth, forming the dominant vegetation. As the Cenozoic advanced, rapid cooling and drying of the earth's surface resulted, in the destruction of many woody forms and evolution of a great variety of new herbaceous forms (annuals and perennials), representing their dominance at the present time.

Mesozoic or Middle Age (beginning 225 million years ago). This was the 'age of higher gymnosperms' with abundant modern pteridophytes. Several new types of gymnosperms appeared in four distinct groups and formed the dominant vegetation of the earth. Most of the flourishing species of this age soon disappeared, while cycads and conifers have continued up to the present day in reduced numbers. Several new pteridophytes appeared in herbaceous forms, and some of them at least are in existence now. A great event of this age is the first appearance of angiosperms (their ancestry, however, not known yet). They were mostly trees in habit and formed big forests.

Palaeozoic or Ancient Age (beginning 570 million years ago). The lower Palaeozoic was the 'age of algae', while the mid-upper Palaeozoic was the 'age of primitive pteridophytes' with abundant primitive gymnosperms. In the early period of this age marine algae appeared in abundance. A very important event of this period is the first appearance of true land plant—*Psilophyton,* a small upright rootless and leafless plant, together with a few other allied ones. They had then a worldwide distribution. They possibly originated from some green algae, and were probably the ancestors of several groups of pteridophytes in divergent lines. Bryophytes soon appeared for the first time. In the Carboniferous period (345 millions of years ago) of this age giant lycopods (40 m. in height), giant horsetails (30 m. in height), varieties of ferns including tree ferns and a special group of seed-bearing ferns (seed-ferns or pteridosperms) were dominant and produced huge forests. Later they formed extensive coal deposits. In the late Palaeozoic these plants declined and soon disappeared.

Proterozoic (beginning 2,000 million years ago). Doubtful fossils of bacteria, fungi, blue-green algae and green algae.

Archeozoic (beginning 5,000 million years ago). No evidence of living organisms but existence of unicellular life quite probable.

2. **Taxonomic Evidence.** According to resemblances and differences we classify plants and animals into certain well-marked groups, the members of each group resembling one another more closely. It is difficult to conceive of the similarities in forms without having recourse to evolution. Besides, it is seen that between two or more species of a

particular genus there are intermediate forms linking such species. (*intergrading species*). If species were constant the occurrence of such forms could not be accounted for.

3. **Morphological and Anatomical Evidence.** Similarities in morphological and anatomical characters among certain groups of plants, and among certain groups of animals, are very characteristic from the standpoint of evolution. The development of different organs, tissues, advance of sporophyte. reduction of gametophyte, stele, vascular tissue, etc., among plants, and similarly the development of different organs and tissues, nerves, bones, brain, etc., among animals, all in successive stages, show evolutionary tendencies among plants and animals, and lend support to the theory of evolution.

4. **Embryological Evidence.** The study of the nature and development of the embryo shows a great resemblance among certain groups of plants and of animals. For example, the embryos of dicotyledons (in general) look alike; those of mammals also look alike. Similar is the case with other groups of plants and animals. The striking resemblance in structure and development of the embryos in them can only be explained on the basis of evolution, i.e. descent of forms from a common ancestry. Besides, in all cases one fact at least is common, i.e. the embryo develops from the egg-cell or ovum. Sometimes some organs of plants or animals show a striking resemblance to certain forms from which they have possibly been derived. Thus when a fern spore germinates it resembles a filamentous alga; it then assumes a thalloid form resembling a liverwort; and finally it grows into a fern plant. Secdlings sometimes show their resemblance to plants which may be their ancestors. Thus the seedling of Australian *Acacia* shows bipinnate compound leaves like other species of *Acacia,* although adult Australian *Acacia* has only the winged petiole or rachis (phyllode) without the compound leaf. Likewise the frog passes through a tadpole stage resembling a fish which is supposed to be its ancestor.

5. **Evidence from Geographical Distribution.** It has been seen that many allied species of plants in their wild state remain confined to a particular area. The explanation is that they sprang up from a common ancestor in that region and could not migrate owing to some barriers such as high mountains, seas and deserts. Thus we find that double coconut-palm (*Lodoicea*) originated in Seychelles, traveller's tree (*Ravenala*) in Madagascar, *Eucalyptus* in Australia, cacti in the dry regions of tropical America, cactus-like spurges (*Euphorbia*) in the deserts of Africa, etc., often with allied species close together, showing

thereby that all the allied species have evolved from the same ancestral species.

MECHANISM OF ORGANIC EVOLUTION

Variation. Variation is the rule in nature. No two forms, belonging even to the same species, are exactly alike. The differences between them are spoken of as **variations**. Variations are the basis on which evolution works. Variations may take place in one or more organs of plants and animals, and these may be of the following types. (a) *Variation due to change in environment*. This is Lamarck's view (see p. 538). This type of variation influenced by the environment is not, however, believed to be inherited. (b) *Slow but continuous variation from generation to generation*. This is Darwin's view (see p. 540) and, according to him, is the basis of organic evolution. In this type of variation a continuity or gradation is maintained between the individuals of a species and between allied species. (c) *Discontinuous variation or mutation*. This means sudden and sharp variation shown by one or more individuals of a species in any one generation. The individuals show no gradations, as in the previous case. This is De Vries' view (see p. 542). As mutation occurs suddenly and spontaneously there is no knowing when a new form will appear by this process. There are, however, many cases of mutation on record.

Heredity. Heredity means transmission of characteristics and qualities of parent forms to their offspring. This is evident from the fact that a particular species of plant on reproduction gives rise to the same species and to no other. Although no two forms are exactly alike, still offspring bear the closest resemblance to their ancestral forms, and they also resemble one another most closely with, of course, individual differences. Heredity tends to keep the individuals of a species within specific limits, while variation tends to separate them. Variation no doubt is responsible for evolution, while heredity is a check on uncontrolled variation.

Chromosome Mechanism in Heredity. The question arises: What is the physical basis of heredity or, in other words, the mechanism of inheritance of parental characters by the offspring? About the year 1884 it became known, primarily due to the work of Strasburger, that the male gamete of the pollen-tube and the egg-cell of the embryo-sac are directly involved in fertilization and reproduction. About the same year Strasburger and Hertwig revealed the fact that it is through the chromosomes of the two gametes (i.e. reproductive nuclei) that the characters of the parent forms pass to the next generation and so on to the successive generations. The conception of transmission of characters through the media of chromosomes is spoken of as *chromosome mechanism of heredity*. It is obvious that any particular character of the parent (e.g. colour of flower) cannot be found in the chromosome; but it may be safely assumed that something representing that particular character must be present in it. That 'something', obscure though it is, is called the *factor* or *determiner* or **gene** for that particular character. A gene is a material unit of the chromosome (DNA)—in fact, a small length of it—responsible for transmission of a given unit-character or a group of characters as a unit from the parent to the offspring. A gene is thus a hereditary factor or determiner located in the chromosome, and each chromosome has a very large number of genes (several hundreds or thousands) in its body. The genes divide, and all the cells resulting from repeated mitotic divisions have identical sets of them. The chemical basis of inheritance, or in other words, the chemical composition of genes, is the *nucleic acid,* mostly DNA or sometimes RNA as in several viruses. The theory of genes in the chromosomes was first introduced by Morgan in 1926. According to the theory of Morgan and his colleagues the chromosomes are the bearers of hereditary characters, and the genes located in them are responsible for all the characteristics of the parent forms and their transmission to the offspring generation after generation, Genes are thought to be ultra-microscopic particles occurring in pairs (one paternal and one maternal) in linear series in the chromosomes. The behaviour of genes in transmission of characters is, however, very complicated. Another important fact to be noted in this connection is that all gametes always have *haploid* (or n) chromosomes and when they fuse in the process of fertilization the *diploid* (or $2n$) number is restored in the zygote—$n + n = 2n$.

THEORIES OF ORGANIC EVOLUTION

Lamarck's Theory: Inheritance of Acquired Characters. A theory to explain the cause of evolution was put forward by the French biologist Lamarck in 1809. His theory resolves itself into three factors, viz. (a) influence of the environment, (b) use and disuse of parts, and (c) inheritance of acquired characters. Lamarck held the view that environment plays the principal part in the evolution of living organisms. He noted many instances of plants where individuals of the same species grown under different environmental conditions showed marked differences between them. From such observations, Lamarck held the view that plants react to external conditions, and that as a result of cumulative effects produced by the changed conditions through successive generations new species make their appearance. In the case of animals the changes are brought about by the use and disuse of parts. The use or exercise of certain parts results in the development of those parts, while disuse or want of exercise results in the degeneration of the parts. He believed that new characters, however minute, acquired in each generation are preserved and transmitted to the offspring (inheritance of acquired characters). The classic example cited in this connection is that of the giraffe. Lamarck's view was that horse-like ancestors of these animals living in the arid region in the interior of Africa had to feed on the leaves of trees. They had necessarily to stretch their limbs to reach up to the leaves. This use or exercise resulted in the lengthening of the neck and the front legs, and thus a new type of animal made its appearance from a horse-like ancestor. His theory is open to certain objections: one objection is that adaptations due to the influence of the environment are very slight and superficial; another objection is that the inheritance of acquired characters has not yet been proved. In fact, if seeds collected from plants growing away from their original habitat for many years are taken back to the original habitat even after many years the plants lose their acquired characters and return to their original forms.

Biography of Lamarck. Lamarck was a self-taught man. He worked hard throughout his life but always remained poor. After his father's death Lamarck joined the French Army in the Seven Years' War with the Germans. While at Toulon during the war he developed a taste for the study of flowers. After the war he joined a bank in Paris. Now he seriously took to his favourite study of flowers and soon wrote a *French Flora*. He left the bank in pursuit of his study, and travelled in Holland, Germany and Hungary. On his return to Paris, he became known as a famous botanist. He used to write on a variety of science subjects, sometimes erratically, and contributed several articles to the French *Encyclopaedias*, then in preparation. After the French Revolution (1789) when science was officially recognized, Lamarck was appointed Professor of Zoology at the Museum of Natural History in 1794 at the age of 50 although he knew almost nothing of the subject. However, he seriously and assiduously took to the classification of the invertebrates. Soon

Figure 1. Jean Baptiste Lamarck (1744–1829), French biologist and propounder of the first theory of evolution in 1809.

he became a full-fledged biologist. It was Lamarck who first held the view that a species was not constant but changed under the influence of environment and gave rise to new species. Thus he propounded the first theory of evolution by inheritance of acquired characters in 1809. Between 1816 and 1822 Lamarck wrote his important work *Natural History of Invertebrate Animals* in seven volumes. Owing to heavy strain on his eyes he became quite blind for the last ten years of his life. The last two volumes of his work were written out for him by one of his daughters.

Darwin's Theory: Natural Selection. The next theory to solve the problem of evolution was put forward in 1859 by Charles Darwin and published in his *Origin of Species by Means of Natural Selection.* His theory, based on a mass of accurate observations, intensive and extensive studies and prolonged experiments for over 20 years, led the whole scientific world to believe in the doctrine of evolution. His theory, called the *theory of natural selection,* is based on three important factors: (a) over-production of offspring and a consequent struggle for existence, (b) variations and their inheritance, and (c) elimination of unfavourable variations (survival of the fittest).

Struggle for Existence. If all the seeds of any particular plant were to germinate and all seedlings to grow up into full-sized plants, a very wide area would soon be covered by them in course of a few years. If other plants (and also animals) were to increase at this rate, a keen competition, called the struggle for existence, would be set up at once among them because supply of food, water and space would fall far short of the demand. A struggle would soon ensue, resulting in the destruction of huge numbers of individuals.

Variations and their Inheritance. It is known to all that no two individuals, even coming out of the same parent stalk, are exactly alike. There are always some variations, however minute they may be, from one individual to another. Some variations are suited to the conditions of the environment, while others are not. According to Darwin these minute variatiom are preserved and transmitted to the offspring, although no cause for these variations was assigned by him.

Survival of the Fittest. In the struggle for existence the individuals showing variations in the right direction survive, and these variations are transmitted to the offspring; others with unfavourable variations perish. This is what is called by him 'survival of the fittest'. The survivors gradually and steadily change from one generation to another, and ultimately give rise to new forms. These new forms are better adapted to the surrounding conditions.

Natural Selection. Darwin's observations on the variations of domestic animals and cultivated plants served him as a clue to the elucidation of his theory of natural selection. His explanation of natural selection is this: animals and plants are multiplying at an enormous rate. As we know, no two individuals are exactly alike, the new forms naturally show certain variations. Some variations are favourable or advantageous so far as their adaptation to the conditions of the environment is concerned, and others are not so. Owing to an excessive number crowding together a keen struggle for existence ensues. And in this struggle those that have favourable variations and are, therefore, better fitted, naturally survive; the rest perish. Through this survival of the fittest the species change steadily owing to preservation and transmission of minute variations, and gradually give rise to newer forms. Darwin called this process 'natural selection' from analogy to artificial selection. It is the environment that selects and preserves the better types and destroys the unsuitable forms. Although Darwin receives the fullest credit for bringing about the final acceptance of the doctrine of evolution, his theory is open to certain doubts.

Biography of Darwin. As a mere boy Darwin used to take special delight in collecting birds' eggs, insects and rocks and studying habits of birds. He was rather dull in his academic study at school. He was sent to Cambridge to become a minister of religion before entering a church. But during his three years' stay there he used to mix with the Cambridge naturalists. He became a keen beetle-hunter and captured many new species. Now came a definite turning point in Darwin's whole life. He was entertained as a naturalist on board the Admiralty vessel, *H.M.S. Beagle,* which sailed from the shores of England on a long five-year voyage of survey (1831–36) in the South Atlantic and Pacific Oceans. This voyage was of immense value to Darwin and to the whole world. His extensive collections of animals (induding tiny sea-animals) and strange plants, corals and fossils, and his

Figure 2. Charles Darwin (1809–1882), famous English biologist and founder of the theory of evolution in 1859.

observations on strange desert plants, rich tropical forests, different kinds of birds, huge tortoise, large lizards, etc., at St. Jago Island, Brazil, certain parts of South America, New Zealand and Australia, opened a new avenue of study for Darwin. The *Beagle* returned to England late in 1836 via the Cape of Good Hope, with Darwin's shipload of specimens. Darwin now settled down to write a long scientific report in five volumes which took him about twenty years to complete. His intense work for this prolonged period, his keen observations on domestic animals and cultivated plants, his wide study of all connected scientific papers published till then, and his clear thinking finally led him to formulate the theory of evolution. But unfortunately for Darwin, while he was giving final and concrete shape to his ideas of evolution by natural selection he received an essay from a young naturalist, rather an explorer, Alfred Russel Wallace (1823–1913) working independently far away in the Malay Archipélago. Darwin was struck by the ideas expressed in that essay, which tallied almost word for word with his own. Without any of them claiming a priority a joint paper was published on the above subject in 1858 under both the names. Wallace, however, recognized Darwin's superiority as a naturalist and yielded leadership to him. The following year, 1859, Darwin published his epoch-making book *Origin of Species*—the fruit of many years of hard labour and study. The book caused a tremendous stir all over the world. It, however, met with bitter attack from a large section of the people for his daring act against God and religion. But his theory survived, and Darwin came to be recognized as the founder of the theory of evolution.

De Vries' Theory: Mutation. Another theory to explain the cause of evolution was advanced by a Dutch botanist Hugo De Vries in 1901–3. He held that small variations which Darwin regarded as most important from the standpoint of evolution, are only fluctuations around the specific type. These variations are not inheritable. De Vries held that large variations appearing suddenly and spontaneously in the offspring in one generation are the cause of evolution. These variations De Vries called 'mutations'. He observed an evening primrose (*Oenothera lamarckiana*), introduced from America, growing in a field in Holland. Among numerous plants he found two types quite distinct from the rest. These new types had not been described before, and having bred true he regarded them as distinct species. *Oenothera Lamarckian* and the new species were removed to his garden at Amsterdam, and cultivated through many generations. It was found that among thousands of seedlings raised a few appeared that were different from the rest. These when raised, generation after generation, always came true to type. These new forms are known as *mutants*. He concluded that his mutation theory explained the cause of evolution. While De Vries agreed with Darwin's view regarding natural selection weeding out unsuitable forms, he held the view that new species are not formed, as Darwin said, by the slow process of continuous variations.

Figure 3. Hugo De Vries (1848–1935), Dutch botanist and founder of the mutation theory of evolution in 1901–3.

Biography of De Vries. He was educated at Leyden, Heidelberg and Wurzburg. Later he became a professor (1877–1918) at the University of Amsterdam. Once, while on a field trip, he was struck by the appearance of some new forms growing among a mass of evening primrose (*Oenothera lamarckiana*). This attracted him to the study of botany and evolution. His experimental methods of work, specially on *Oenothera,* led him to the rediscovery in 1900 of Mendel's laws of hereditary, and to the elucidation of the theory of evolution by sudden and discontinuous variations which he called mutations. His *mutation theory* (1901–3), as distinct from Darwin's slow and continuous variations by natural selection, explains the cause of evolution, and is regarded as the greatest contribution to the history of evolution. He also went to America to study *Oenothera* in its natural habitat. *Plant Breeding* (1907) is another of his best known works. After his retirement from the University he established an experimental garden at Hiversum and continued his experimental work in producing new forms through many generations of cultures.

GENETIC BASIS OF EVOLUTION

Recent advancement in biological science has revealed that the change in the genetic composition of the population brings about evolution, along with natural selection, among the organisms. The frequency of allelic genes and genotype present in a population are maintained by random interbreeding from generation to generations. But the evolutionary change takes place due to variations introduced by mutations and the recombination of mutant genes. Under a new environment, individuals try to adjust themselves into the new environment. But only the 'fittest' win in the 'Struggle for existence', through adaptations.

During this process, natural selection, in the long run, alters the proportions of allelic genes in favour of the 'fittest'. This theory includes the combination of ideas of Darwin and other biologists like de Vries, Huxley, Dobzhan-sky and others and is also called Neo-Darwinism. This theory explains how 'variation' among the living organisms came into existence what Darwin failed to explain.

Chapter 2

Genetics

Genetics is the modern experimental study of the laws of inheritance (variation and heredity). The first scientific study on genetics was carried out by Gregor Mendel. He entered a monastery in Brunn (Austria), where he carried on his scientific investigations on hybridization of plants. The results of his eight years' breeding experiments were read before the Natural History Society of Brunn in 1865, and in the following year these were published in the transactions of that Society. But his work remained unnoticed until 1900, when three distinguished botanists, Hugo De Vries in Holland, Tschermak in Austria and Correns in Germany, discovered its significance. Since then Mendel's work has formed the basis of the study of genetics. Mendel died in 1884 before he could see his work accepted and appreciated.

Mendel's Experiments. Mendel selected for his work the common garden pea. In the pea he found a number of contrasting characters—flowers purple, red or white; plants tall or dwarf; and seeds yellow or green, smooth or wrinkled. He concentrated his attention on only one pair of characters at a time, and traced them carefully through many successive generations. In one series of experiments he selected tallness and dwarfness of plants. The results he achieved in these experiments were the same in all cases. It did not matter whether he took the dwarf plant as the male and the tall plant as the female, or vice-versa.

Monohybrid Cross. For monohybrid cross only one pair of contrasting characters is taken into consideration at a time. Mendel selected a pea plant, 2 metres in height, and another, 0.5 metre in height. He brought about artificial crossings between the two. The progeny that resulted from these crossings were all tall. This generation, known as the first filial generation or F_1 generation, was inbred. Seeds were collected and sown next year. They gave rise to a mixed generation of talls and dwarfs (but no intermediate) in the ratio of 3:1, i.e. three-fourths talls and one-fourth dwarfs. This generation is known as the second filial generation or F_2 generation. All dwarfs of the F_2 generation bred true, producing dwarfs only, in the third and subsequent generations. Seeds were collected separately from each F_2 tall plant and sown separately. It was seen that one-third of the talls bred true to type, while the other two-thirds again split up in the same ratio of 3:1. The F_2 ratio is, therefore, 1:2:1, i.e. *one-fourth* pure talls, *half* mixed talls, and *one-fourth* dwarfs.

The above scheme of inheritance may be represented as follows. Here T represents the factor for tallness, and t the factor for dwarfness.

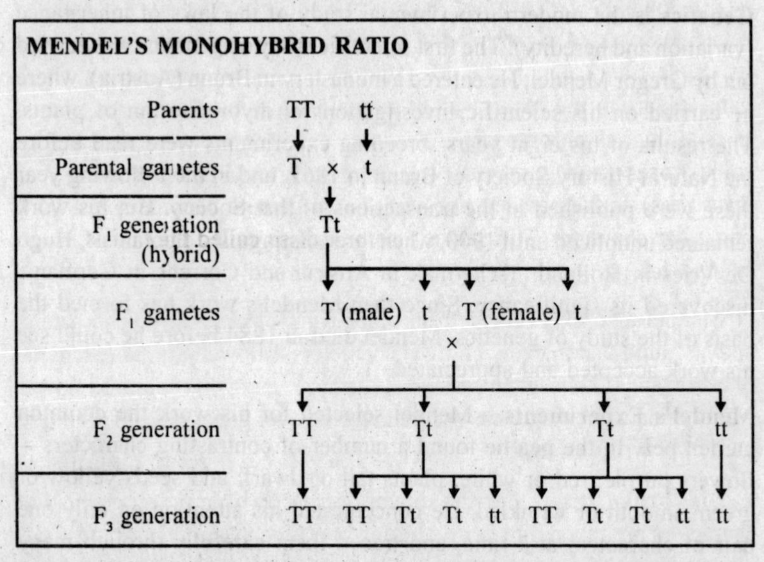

Mendel's laws of Inheritance. From the results of his experiments on carefully selected crossings, Mendel formulated certain laws to explain the inheritance of characters as follows.

1. **Law of Unit Characters.** This means that all characters of the plant are units by themselves, being independent of one another so far as their inheritance is concerned. There are certain factors or determiners (now called *genes*) of unit characters, which control the expression of these characters during the development of the plants.

2. **Law of Dominance.** The characters, as stated above, are controlled by factors or genes. These occur in pairs (arranged in a linear

fashion in the chromosome, as now known) and are responsible for tallness and dwarfness separately. One factor may mask the expression of the other. Thus in the F_1 generation all the individuals are tall, the other character remaining suppressed. The character that expresses itself in the F_1 generation is said to be *dominant*, and the character that does not appear in the F_1 generation is said to be *recessive*. The factor for the recessive character is, however, always present in the F_1 individuals. In the above experiment tallness is the dominant character and the suppressed dwarfness is the recessive character. The contrasting pairs of characters are called **allelomorphs**. Thus tallness and dwarfness are allelomorphs.

3. **Law of Segregation.** The factors for the contrasting characters remain associated in pairs in the somatic cells of each plant throughout its whole life. Later in its life-history when spores (and subsequently gametes) are formed as a result of reduction division, the factors located in homologous chromosomes become separated out, and each of the four spores (and gametes) will have only one factor (tallness or dwarfness) of the pair but not both, i.e., a gamete becomes *pure* for a particular character. This law is also otherwise called the law of *purity of gametes*.

Mendel also experimented on other pairs of alternative characters, and he found that in every case the characters followed the same scheme of inheritance. Thus in the garden pea he discovered that coloured flower was dominant over white flower; yellow seed over green seed; and smooth seed over wrinkled seed.

Dihybrid Cross. For the dihybrid cross two pairs of contrasting characters are taken into consideration at a time. Mendel selected a tall plant with red flowers and a dwarf one with white flowers. Four unit characters are, therefore, concerned in the dihybrid ratio. Factors for tallness or dwarfness and red flowers or white are independently inherited, and may be considered to be located in separate chromosome pairs. Artificial crossing was brought about between these two plants. In the F_1 generation all individuals were tall with red flowers; for tallness is dominant over dwarfness, and coloured flowers dominant over white. When the seeds from the F_1 generation were grown, a segregation of characters showing all possible combinations took place in the following proportions: 9 red talls, 3 white talls, 3 red dwarfs, and 1 white dwarf. Thus 9:3:3:1 is the dihybrid ratio.

Nos. 1, 2, 3, 4, 5, 7, 9, 10, 13 are tall-red = 9
Nos. 6, 8, 14 .... tall-white = 3
Nos. 11, 12, 15 .... dwarf-red = 3
Nos. 16 .... dwarf-white = 1

It will further be noticed that Nos. 1, 6, 11 and 16 are *homozygous* (i.e. they have two similar gametes), breeding true; while the rest are

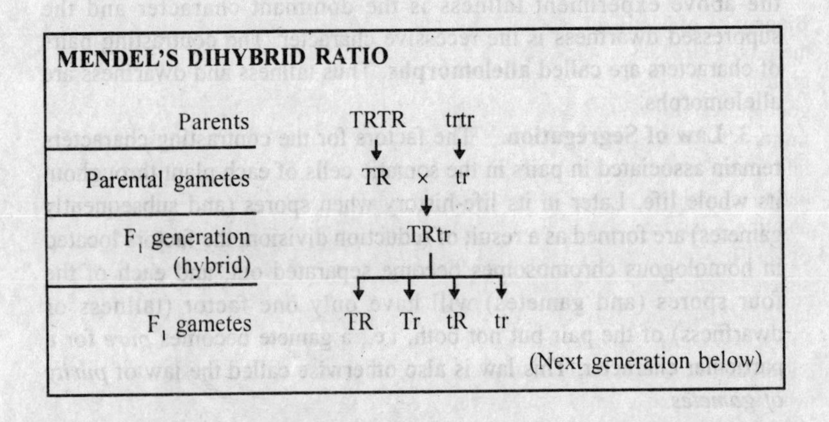

Female gametes of F_1 \ Male gametes of F_1	TR	Tr	tR	tr
TR	TRTR (tall-red) [1]	TRTr (tall-red) [2]	TRtR (tall-red) [3]	TRtr (tall-red) [4]
Tr	TrTR (tall-red) [5]	TrTr (tall-white) [6]	TrtR (tall-red) [7]	Trtr (tall-white) [8]
tR	tRTR (tall-red) [9]	tRTr (tall-red) [10]	tRtR (dwarf-red) [11]	tRtr (dwarf-red) [12]
tr	trTR (tall-red) [13]	trTr (tall-white) [14]	trtR (dwarf-red) [15]	trtr (dwarf-white) [16]

F_2 generation

heterozygous (i.e. they have two dissimilar gametes), segregating in, the next generation.

No. 1 (TRTR) will breed true for tall-red
No. 6 (TrTr) .... tall-white
No. 11 (tRtR) .... dwarf-red
No. 16 (trtr) .... dwarf-white

Biography of Mendel. Mendel developed a love for gardening while still a mere boy. His father, a successful horticulturist, initiated him into the art of grafting of fruit trees. Mendel was a good student at school. Soon after his father met with an accident from a falling tree and became invalid, Mendel had to struggle hard for want of money, and finally left school in 1840 at the age of 18. Through the generous help of his sister, Mendel took a two-year course in Philosophy, and at the advice of Prof. Franz of Brunn, Mendel entered the Monastery at Brunn, and lived there, with a short break, for about 41 years till his death in 1884. While at Brunn he took a course in Theology and became a High School teacher. Without a scientific background he was unsuccessful as a science teacher. Abbe Napp, then in charge of the Monastery, sent Mendel to the University of Vienna to study science for two years. On completion of the course, Mendel returned to the Monastery and joined Brunn High School (1852) as a science teacher. The Monastery had spacious grounds and Mendel carried out most of his experiments on heredity there during the period of 1856–1864. He established the laws of heredity for the first time and put genetics (later also called Mendelism in his honour) on a sound scientific basis (read text). In 1864 Mendel became the Abbe of the Monastery. He could hardly then, owing to heavy official duties, pursue his own scientific work. He became rich and was very generous to others. He never forgot his benevolent sister and was fond of his nephews.

Figure 4. Gregor Johann Mendel (1822–84), Austrian monk, biologist and famous geneticist. He established the laws of inheritance of characters in 1865–66.

Plant Breeding. The subject of plant breeding, although developed in recent times on modern scientific lines after Mendel's discoveries, was known in early times to the Egyptians and Assyrians. Later during the 18th and 19th centuries several artificial crossings were made by many workers and interesting results obtained in the form of new varieties. But it was Mendel who first laid the foundations of plant breeding on a scientific basis and formulated the laws of inheritance of characters. Plant breeding consists in producing new types of offspring by artificial pollination brought about between the flowers of two different species, varieties or even genera. By this process, also called **crossing** or **hybridization,** it has been possible to combine in the offspring certain desired characters of both the parents. In actual practice the stamens of a flower (bisexual) are removed before its anthers mature, while the flower with the gynoecium intact is covered with a paper or muslin bag to prevent natural pollination. When the stigma of this flower matures, pollen from another selected parent is applied to it. The offspring resulting from such a cross are new types, called **hybrids,** which are often more vigorous than the parent forms. This phenomenon is spoken of as **hybrid vigour** or **heterosis**. The **economic importance** of cross-breeding is manifold and almost unlimited, and already much has been achieved in various agricultural and industrial crops regarding their yield, quality and other useful characters.

Part IX

RECENT ADVANCES IN BOTANY

Genetic Engineering
Industrial Microbiology
Biotechnology
Tissue Culture

Chapter 1

Genetic Engineering

The nucleotide sequence in the DNA determines the basic features of all organisms or in other words how organisms should act and behave. A particular nucleotide sequence codes for a specific amino acid chain that is responsible for a particular trait. Therefore, if the nucleotide sequence is wrong, it will code for wrong amino acid chain and bring certain genetic disorder. Such genetic disorders can be corrected by replacing one base pair of nucleotide sequence by another desired base pair or inserting or deleting one base pair. This results in altering the functions of a gene. The science which deals with the correction of defective nucleotide sequence or implementation of desired genes in the chromosomes in order to bring permanent and heritable changes in organisms is called genetic engineering. Genetic engineering has made it easier to detect and cure genetic diseases before the birth of a child. Gene Bank and DNA Clone Bank have been now established so that different types of genes of known functions can be made available. Genetic engineering now is playing important role in the fields of medicine (for the production of insulin, vitamins, hormones, specific antibodies, synthetic DNA, etc.), industry (for the production of organic acids, proteins and other products), agriculture (production of drought resistence, disease resistence, high yielding plants by bringing genetic modification in plants), environmental protection by developing certain microorganism which can quickly breakdown hydrocarbons, plastics, etc.

Chapter 2

Industrial Microbiology

The use of microorganisms in industrial processes and exploitation of their metabolic activities for the production of many useful substances is referred to as industrial microbiology. The beneficial utility of certain microorganisms was known to man since long. The fungus yeast was commercially used by brewers and bakers in the production of alcoholic beverages and bakery products. Only in recent times, with the advancement of knowledge in the field of science and technology, industrial microbiology has advanced to a greater extent. Specially the discovery and production of antibiotics have now established a close relationship between the microbiological and pharmaceutical industries. In industrial microbiology, the microorganisms which can be grown in large quantity and within a short time and which can easily convert raw materials (substrates) into desired products are generally selected. Such important products are various kinds of food such as curd, cheese, etc., different kinds of alcoholic beverages, different chemicals like organic acids, amino acids, etc., various antibiotics, vitamins, steroid drugs, industrial enzymes, etc. Industrial microbiology involves the use of various kinds of microorganisms such as bacteria like *Lactobacillus, Streptococcus, Acetobacter,* etc., fungi like *Aspergiltus, Saccharomyces, Trichoderma, Penicillium, Rhizopus,* etc., cyanobacteria and many others.

Chapter 3

Biotechnology

Biotechnology is a most advanced discipline which involves the use of genetic engineering in altering and manipulating various life forms and the application of such life forms for processing of the various materials to obtain desired, useful industrial products. Biotechnology received due attention in 1970s after the discovery of recombinant DNA technology (otherway called genetic engineering) and the use of biological processes in various industries. Biotechnology has now helped the scientists to undertake the task of mass production of growth hormones, vaccines, insulin, biofertilizers, biopesticides, polypeptide gene therapy, etc. Further biotechnology has brought a new revolution in agriculture, tissue culture, etc. In India, an official agency called National Biotechnology Board (NBTB) was established by the Government of India in 1982, which was later converted to the Department of Biotechnology in 1986. Moreover, there is also a branch of 'International Centre of Genetic Engineering Biotechnology' in Delhi that started functioning in 1998 at Jawaharlal Nehru University (JNU). Other biotechnology-centres are IARI, Delhi University, Indian Institute of Science at Bangalore, Kanpur, Mysore, etc., Regional Research Laboratory at Jammu and also other bodies of like nature.

Chapter 4

Tissue Culture

The inherent property of any cell to divide and develop into the entire organism is known as cellular totipotency. Gottlieb Haberlandt, a German botanist was the first to give the idea about tissue culture in 1902. But the significant progress was made in this line in 1938. The modern tissue culture technique is based on the concept of cellular totipotency. By tissue culture technique young plant or animal cells can be grown in the laboratory-glasswares containing appropriate growth medium (nutrients) in which such cells divide and finally develop into the entire organisms (like the ones to which the cells belonged). These may then be transferred to the natural habitat for further growth and development. This technique helps in propagation and studying morphogenesis of an organisms.

Part X

ECONOMIC BOTANY

General Description
Economic Plants

Chapter 1

General Description

Economic Botany deals with the various uses of plants and plant products as applied to the well-being of mankind. It also includes various practical methods that may be adopted for their improvement in one or more directions as needed by man. The economic uses of plants are varied and, therefore, the scope for improvement is immense to meet man's ever-increasing needs. The primary needs of mankind are, of course, **food**, **clothing** and **shelter**, which, originally the gifts of nature, were subsequently improved upon by man through the application of his scientific knowledge. The gifts of nature are almost unlimited, and thus a variety of useful products are obtained from the plant kingdom.

Methods of Improvement. The methods commonly employed for the improvement of crops in terms of yield, quality, etc., are (1) selection, (2) breeding, (3) improved methods of cultivation, (4) use of adequate amount of chemical fertilizers and manures, (5) selection and use of 'quality' seeds, (6) judicious selection of crops for a particular area, (7) double or even triple croppings, (8) protection against diseases and pests by application of fungicides, insecticides, and pesticides, and (9) proper irrigation by suitable methods.

Selection consists of picking out the best individuals among a field crop in respect of one or more desired economic characters, and collecting the seeds from them for the next sowing. Second and third selections are made in the same way. Finally the promising ones are used for field trials, always keeping the best and rejecting the rest. By this method an advancement of quality and quantity has been achieved in India in a number of crops, e.g. rice, cotton, *Sorghum* (JUAR OR CHOLAM), *Pennisetum* (BAJRA), *Eleusine* (RAGI), etc.

Plant Breeding consists in combining into the offspring certain desirable characters met with in two separate parent plants belonging to two different but allied species, or varieties or sometimes even genera. Some of the important methods of plant breeding are:

1. **Selection.** It is an old method and favours only the survival and propagation of certain plants having suitable characters while rejecting the others. When the selection occurs in nature as a result of evolution (i.e. survival of the fittest), it is called natural selection. Whereas, when certain selected plants are choosen from the field and they are artificially cross-pollinated for obtaining hybrids of superior quality, it is called artificial selection such as **mass selection** (i.e., a group of quality crops are choosen and they are cross pollinated. Their seeds are collected together and sown. The process of selection is repeated in the same way from generation to generation till the uniformity in the desired characters of the hybrids is achieved), **pure line selection** (i.e., a quality individual is obtained from a mixed population through only self pollination), **colonal selection** (i.e., some plants in which fertilization does not take place and, therefore, fruits are seedless, reproduce only vegetatively. Here a single parent plant is involved which procuces carbon copies (new individuals) of its own kind by vegetative reproduction. These progenies (called clones) are very stable. Hence by clonal selection a plant with hybrid vigour can be preserved.)

2. **Hybridization.** Hybridization, a commonly followed method of plant breeding, can easily be applied to both cross-pollinated and self pollinated plants. In this method the characteristics of different parent forms are combined in the offsprings and thereby new crop varieties with certain desired characteristics are obtained. It is also possible, by this method, to produce many new varieties of plants of economic and ornament importance and even to produce disease resistant plants.

3. **Mutation breeding.** The sudden heritable variations appearing suddenly and spontaneously in the offspring in one generation are called mutations. Such variations now can artificially be brought about in the plants and new trails can be obtained. This is the most modern and practical method of crop improvement.

The **economic importance** of this method is immense, and achievements in this direction in various agricultural and industrial crops regarding their yield, quality and other useful characters have already been considerable. Thus, in America, new types of wheat, maize, tomato and potato—all high-yielding and disease-resistant—have been evolved by following the practical method of plant breeding. In Russia new varieties of summer and winter wheat (wheat × couch grass, and wheat × *Elymus*), and of barley (barley × *Elymus*) are some of the outstanding achievements. In India also a considerable amount of work has been done in this direction with desired results in many cases. Thus improved strains of rice, wheat, millets, maize, sugarcane, pulses, oil-seeds, cotton, tobacco, jute, flax, hemp, etc., combining higher yield, better quality and resistance to pests and diseases, have been evolved by cross-breeding selected varieties. The results have been spectacular in some cases. Thus several superior strains of rice have been evolved in different rice-growing States. Further, some new hybrids of rice, e.g. GEB 24, Pusa 33, etc.,—high-yielding, early-maturing, disease-resistant and easy-cooking—have been evolved by the Indian Agricultural Research Institute by cross-breeding selected varieties. New improved varieties of wheat have been similarly produced. A new wheat, New Pusa 4 (or NP 4), evolved by IARI at Pusa (shifted to New Delhi in 1936 after the great Bihar earthquake in 1934) was awarded the first prize many times at International Exhibitions held in America, Australia and Africa. Recently some new strains of wheat have been evolved by this Institute to suit different climatic regions of India. Besides, new millets and maize (see pp. 494–95) have been evolved, which yield 50% more than the common varieties. A recent success of the Institute is the production of a sweet-flavoured tomato with high vitamin content, evolved by crossing a cultivated variety with a wild South American variety. New types of sugarcane evolved at Coimbatore have already become world famous. They are now widely grown, materially contributing to the growth and expansion of the sugar industry in India. There is still ample scope for improvement of several crops for food and industry.

Chapter 2

Economic Plants

Economic plants are numerous and have a variety of uses. Many of them occur in nature, particularly in hills and forests, while a good number of them are cultivated for food and industry. From the economic standpoint such plants may be classified under the following heads: (A) cereals, (B) millets, (C) pulses, (D) vegetables, (E) oil-seeds, (F) fruits, (G) spices, (H) fibres, (I) medicinal plants, (J) timber, and (K) beverages. Plant products to be used as food must contain sufficiently high percentages of carbohydrates, proteins and fats and oils together with vitamins and essential minerals, and these are mainly furnished by cereals and millets, supplemented by pulses, vegetables, vegetable oils and fruits. It may be noted that India is the largest producer of tea, groundnut, sugarcane and spices.

A. **Cereals.** All cereals and millets are rich in starch, and generally contain vitamins A, B and C. They belong to *Gramineae,* and are cultivated as annual crops. Cereals form the main food of mankind and in India they occupy about 60% of the total area under cultivation. Her total annual yield of food grains has now increased to about 125.6 million tonnes on the basis of the steps taken in the following directions: improved method of cultivation and increased use of fertilizers (see p. 252), cultivation of some high-yielding and short-duration varieties (the latter particularly in flood-affected areas), arrangement for proper irrigation, intensive and extensive cultivation, and protection against diseases, pests and rodents (the annual loss on this account being 10–18%). The major cereals are rice, wheat and maize, and the major millets are *Eleusine* (RAGI), *Sorghum* (JUAR or CHOLAM) and *Pennisetum* (BAJRA).

(1) **Rice** (*Oryza sativa*) is the major agricultural crop in India, covering a total area of 37.4 million hectares, which is the world's largest

rice area. Her total yield has now gone up to 52.7 million tonnes. Rice is the staple food in India and in tropical Asia, and feeds over 60% of the world's population. Rice has been in cultivation in India and China from time immemorial. There are several thousand varieties of rice (about 4,000 in India alone). Rice is widely cultivated in India except in northwest India where wheat is the main crop. The plant thrives under conditions of moderately high temperature, plenty of rainfall or irrigation, and heavy manuring. With the introduction of some new varieties of rice (high-yielding, disease-resistant, early-maturing) evolved in India by selection and breeding the average yield of rice has now gone up to 1,720 kg. per hectare per year (of course, much more under the Japanese method of cultivation and also under special conditions) as against 3,117 kg. in Egypt, 3,406 kg. in Japan, 3,472 kg. in Italy and 3,718 kg. in Spain, where fields are heavily manured. In India Tamil Nadu has the highest yield of 1,974 kg. per hectare. There is usually a double or even triple croppings of paddyin India (summer and winter), The latter (AMAN) is far better than the former (AUS) in respect of yield and quality. Some special varieties such as Taichung Native 1, Táinan 3, Kalimpong 1, IR 8, IR 24, Pusa 33, etc.,—all high-yielding and early-maturing—have been recently introduced, with very satisfactory results. The average chemical composition of rice is starch 70–80%, proteins 7%, oils 1.5%, some vitamins (A, B and C). Paddy straw is an important fodder.

(2) **Wheat** (*Triticum aestivum*) is the second staple food of people in India. It does, however, form the principal diet in western countries. There are several varieties of wheat, and these may be broadly classified into *hard* and *soft*. The former varieties are adopted for making SUJI and ATTA, while the latter varieties are used for bread-making. Wheat is cultivated mainly in Uttar Pradesh, Madhya Pradesh, Gujarat, Rajasthan, Haryana and Punjab. Grains are sown in October-December and the crop harvested in March-May. Wheat is a universal crop, i.e. it can be successfully grown in both temperate and tropical countries. The average yield of wheat in India is poor; it is only about 900 kg. per hectare per year, as against 1,396 kg. in Canada, 1,708 kg. in Japan, and 2,468 kg. in Great Britain. Several new varieties of wheat—hardy, disease-resistant, high-yielding, with better milling and bread-making qualities—have been evolved in India (e.g. SONA227 evolved be IARI) with the result that the total annual yield has now gone up to 31.33 million tonnes. Some of the best wheats are grown in Australia, America

and Russia. The average chemical composition is starch 66–70%, proteins 12% and oils 1.5%. Wheat straw is a fodder.

(3) **Maize** (*Zea mays*) is an important cereal food for poorer classes of people. It is cultivated both in the hills and the plains, and it flourishes both in hot and cold climates. The usual sowing season is April-May, and the harvesting season July-August. Each plant commonly bears one cob, sometimes two. The average yield is very low, being about 986 kg. per hectare per year, going up to 1,570 kg. with good varieties grown under favourable conditions. Maize cobs may be 15–25 cm in length, and the grains golden yellow, dull yellow, red, white, etc., in colour. There are several varieties and hybrids. Some new hybrids of maize have been evolved in India, which are high-yielding, disease-resistant and nutritious. A hybrid maize (Texas 26) evolved by the Indian Agricultural Research Institute at New Delhi yields over 2,800 kg. per hectare per year. Maize grains are taken as a substitute for other cereal grains in many rural areas. Commonly these are ground into flour called cornflower. Grains are also either boiled or fried and then eaten. Leaves and stems form a good fodder, and the grains a nutritious food for farm animals. The average chemical composition is starch 68–70%, proteins 10%, and oils 3.6–5%. In addition the grains contain an appreciable quantity of calcium and iron.

B. **Millets.** Smaller grained cereals are commonly called millets, and there are several kinds of them. The more important ones are as follows.

(1) ***Eleusine coracana*** (African millet—RAGI or MARUA) is an important food crop of Karnataka (Mysore), and is extensively cultivated in Karnataka (maximum production in India), Tamil Nadu and Andhra, and to some extent in Maharashtra, Uttar Pradesh, Bihar and Punjab. RAGI being a short duration crop, 2 or 3 croppings are generally practised a year. The plants are dwarf but very hardy, and the spikes are short in a whorl. RAGI is a dryland crop. Commonly grains are sown in June-July and the crop harvested in August-September. The average yield per hectare per year is 785–1,120 kg., sometimes double this quantity in red soil properly irrigated. The grains can be stored for several years without injury. The average chemical composition is starch 73%, proteins 7% and oils 1.5%. The straw is a nutritious fodder for cattle.

(2) ***Sorghum vulgare*** (=*Andropogon sorghum;* great millet—JUAR or CHOLAM) is the best of all millets. It affords nutritious food, nearly as good as wheat. It is widely cultivated in India, particularly in South India, Maharashtra and Gujarat. Two or sometimes three croppings are

generally practised a year. The plants are tall (3–4 metres in height) and stout (the stem often sweetish), and the panicle much-branched. There are several varieties and hybrids of JUAR. Grains are commonly sown in June-July, and the crop harvested in October-November. The average yield per hectare per year is 785–896 kg., often double this quantity in black soil properly irrigated. The average chemical composition is starch 72%, proteins 9% and oils 2%. The straw is a good fodder for cattle.

(3) ***Pennisetum typhoideum*** (pearl millet—BAJRA) is another important millet. It is cultivated practically throughout India. The plants are 1–2 metres in height, and the spikes are long and occur in clusters. BAJRA is grown in regions with low rainfall on both red and black soils. Two croppings are generally practised. Commonly grains are sown in May, and the crop harvested in July-August. The grains often require threshing and husking like paddy, and are cooked like rice, or ground into flour. The average yield per hectare per year is 560–670 kg. or a little more under irrigation. The average chemical composition is starch 71%, proteins 10% and oils 3%. The straw is not commonly used as a fodder for cattle.

C. **Pulses.** As foodgrains pulses stand next to cereals. They are widely cultivated in India as winter crops in rotation with cereals, occupying about 18% of the total area under cultivation. They belong to the family *Papilionaceae*. They are valued as food because of their high protein contents averaging 22–25% (in soya-bean 35% or even more). They contain vitamins A, B and C (particularly when sprouted). Pulses commonly used in India are Bengal gram, black gram, green gram, pea, pigeon pea or red grain and lentil. Pulses are widely used in various culinary preparations, particularly DAL. The plants form good fodder, and having root-nodules for nitrogen fixation (see p. 263), they form excellent green manure. In habit the pigeon pea plant is a shrub; the rest are annual herbs.

(1) **Bengal gram** (*Cicer arietinum*) is extensively cultivated all over India as an important pulse. The crop matures in three months. The gram is nutritious but somewhat difficult to digest. Split seeds are commonly used for preparation of DAL and also some forms of pies. Grams are also boiled, fried or roasted. Soaked seeds fed to horses add to their strength and working capacity. Dried seeds ground into flour are relished by many people. Apart from high protein contents the gram seeds are rich in oil, and are a good source of vitamin A. Because of

such contents the germinating seeds are specially good for the growing children. Average chemical composition is starch 58%, proteins 23.5% and oils 5%. Average yield per hectare per year usually varies from 680–1,130 kg. The straw forms a good fodder.

(2) **Black gram** (*Phaseolus mungo*) is one of the best pulses grown on a large scale throughout India. The seeds have usually a dark-brown skin. The proteins of black gram are easily digestible, and are almost as good as meat. Apart from its use as DAL black gram is used in the preparation of various forms of pies. PAPPAD (wafer) is commonly made of this pulse. Germinating seeds are sometimes taken raw, and are considered nutritious. Average chemical composition is starch 54%, proteins 22% and oils 1%. Apart from the usual protein and vitamin contents the seeds are rich in phosphoric acid. Average yield per hectare per year is 450–550 kg. Husked pods, seeds and straw form a valuable cattle feed.

D. **Vegetables.** (a) Leafy vegetables like cabbage, lettuce, spinach, etc., are rich in vitamins, usually A, B, C and E, and should be included in the daily diet. (b) Several fruits are also used as vegetables, e.g. gourds, beans, lady's finger, brinjal, tomato, etc. (c) Tuber crops are fleshy underground roots or stems laden with a heavy deposit of food material. Of all the vegetables potato (*Solanum tuberosum;* see Fig. I/47), an underground stem-tuber, has possibly the widest use, being available throughout the year in both warm and cold countries. It is in fact a universal article of diet, and is used in a variety of culinary preparations.

E. **Oil Seeds.** There are several species of plants yielding fixed oils, edible and industrial, in high percentages. In oil-seeds India holds a prominent position in the world market with her huge export. Oil-yielding plants occupy about 8% of the total cropped area in India. Some common *edible* oils are groundnut oil, coconut oil, gingelly or sesame oil, mustard oil, sunflower oil, etc., and some common *industrial* oils are castor oil, linseed oil, tung oil, olive oil, etc.

(1) **Groundnut** (*Arachis hypogaea;* see Fig. III/32) is a prostrate or semi-erect annual herb. The seeds yield about 43–46% of edible oil. Groundnut cultivation is now a major agricultural operation in India and occupies the largest area in the world. Tamil Nadu, Andhra, Gujarat and Maharashtra are the principal areas of its cultivation. The average yield per hectare per year is 900–1,100 kg., or more with good varieties. Nuts are nutritious, containing 31% of easily assimilable proteins. They

also contain Ca, P and vitamin B. Nuts are eaten raw, fried or roasted, and used in some confectioneries, but groundnut is mainly cultivated for its edible oil which is used extensively for cooking. It is the principal commercial oil of the Vanaspati industry. In Europe an imitation butter called 'margarine' is manufactured from this oil. Other uses of the oil are: soap-making, illuminant, lubricant, paints, varnishes, tanning, etc. There is a big export of groundnuts, oil and oil-cake to Europe. The oil-cake is a good feed for cattle.

(2) **Castor** (*Ricinus communis;* see Fig. VII/42) is a quick-growing shrub, mainly cultivated in Karnataka and Tamil Nadu and to some extent in Bihar and West Bengal. There are two varieties—one bearing larger seeds and the other smaller seeds. The larger seeds yield an inferior quality of oil (25–40%), while the smaller seeds yield a superior quality of oil (36–40%) or even more. The yield of seeds per hectare per year varies from 450–560 kg. (sometimes up to 1,000 kg.). Castor oil has a variety of uses. The oil extracted from smaller seeds and purified is used as a medicine; it is a safe purgative, as known from time immemorial, possibly because it contains *ricinoleic acid*. The oil extracted from larger seeds is widely used for lubricating machinery, particularly its rolling parts, and for lighting in villages; it gives a bright light without soot. It is the main ingredient of the copal varnish. It is also used for dressing leather and skin in tannery. Its other uses are candle-making and soap-making. Refined castor oil, often perfumed, is a good cooling hair-oil, widely used. Castor cake is a valuable manure but not a cattle feed since it contains *ricin,* a poisonous alkaloid. The oil, however, is free from it. ERI silk-worms are reared on the leaves of the castor plant. India is the largest exporter of castor seeds and castor oil.

F. **Fruits.** India abounds in excellent dessert fruits. Apart from their food value they are rich sources of vitamins. Many of them are made available out of season in the form of various preserves, either as slices or as jam, jelly, pickle, marmalade, chutney (sweet or hot), etc. Some common edible, palatable and nutritious Indian fruits are: mango, papaw, pineapple, banana, orange, litchi, grapes, apple, guava, plum, peach, etc.

G. **Spices.** Spices are certain aromatic and pungent plant products used for seasoning and flavouring various food preparations and fruit and vegetable preserves, and enhancing their taste. They are extensively used in cookery and confectionery; hot or sweet chutneys, beverages, etc., and chewed with or with betel leaf. India is the only country in the world that produces almost all kinds of spices, and she exports about 52 tonnes of annually, particularly pepper, ginger, cardamom and turmeric. Some of the common spices are as follows.

(1) **Nutmeg** is the aromatic kernel of the seed of *Myristica fragrans,* while its aril which is more aromatic and more useful is the mace of commerce. *Myristica* is a big evergreen tree of the Moluccas; in India it grows abundantly in the Western Ghats Range. Each female tree bears a few thousand fruits, oval in shape and more or less 3 cm in length. The fruit has a hard shell which breaks into two pieces, exposing a large seed. The aril (mace) is bright-red in colour and deeply lobed. It is rich in a volatile oil and is very aromatic. The mace forms an important spice, often together with the kernel, and is used all over India as a seasoning and flavouring material for various curries and confectioneries. It is also used in medicine. Usually 100 seeds produce about 85 gms of dried mace. Nutmeg contains a yellowish fat (a fixed oil) called 'nutmeg butter'. Because of its agreeable aroma it is used in perfumery, hair-lotions and ointments.

(2) **Cinnamon** is the dried brown or dark-brown bark peeled off from the twigs of *Cinnamomum zeylanicum,* a small tree of Sri Lanka, cultivated in South India. The bark contains a volatile oil, tannin, sugar and gum. It is aromatic and tastes sweet. It is extensively used for flavouring various food preparations and fruit and vegetable preserves. The oil of cinnamon is used in certain drugs as all intestinal antiseptic.

(3) **Camphor** is obtained front *Cinnamomum camphora,* a tall tree of China, Japan and Formosa (Taiwan); planted in some gardens in India. Camphor is extracted by steam distillation from old wood cut into chips. Sometimes, as in Florida and Sri Lanka (Ceylon), young twigs and leaves are used in the same way for the extraction of camphor. It has a strong but agreeable aroma, and is widely used in very small quantities in various food preparations, chutneys, perfumery and medicines. Camphor industry is practically a monopoly of Japan. Camphor is very slightly soluble in water but readily so in alcohol and ether. It volatilizes very slowly. Synthetic camphor is also in wide use now.

(4) **Bay Leaf** is the dried leaf of *Cinnamomum tamala,* a medium-sized tree, growing in many parts of India, abundantly in the Khasi Hills. Dried leaves are widely used as a flavouring spice in a wide range of food preparations, chutneys and fruit and vegetable preserves.

A few other plant products commonly used as important spices are: cardamoms, cloves, ginger, garlic, red pepper or chilli, black pepper, black cumin, cumin, ajowan, coriander, etc.

H. **Fibres.** Commercial vegetable fibres may be classified as (a) floss fibres or lint, e.g. cotton (*Gossypium*), silk cotton (*Bombax*) and kapok (*Ceiba*); (b) bast fibres, e.g. jute (*Corchorus*), sunn hemp (*Crotalaria*) and rhea or ramie (*Boehmeria*); (c) coir fibres, e.g. coconut fibres; and (d) leaf fibres, e.g. bowstring hemp (*Sansevieria*) and American aloe (*Agave*).

(1) **Cotton** is the hairy covering of the seed of *Gossypium,* and is the most important textile fibre of commerce. The fibres are removed from the seed by a process called 'ginning', and spun into yarn and woven into various kinds of garments, screens, sheets, canopies, sails and a variety of other things. The qualities of cotton fibres are length, strength, fineness, silkiness, etc. Indian cottons are in these respects poor in quality, having short staples (12.7 to 25.4 mm), while Egyptian cottons have staple lengths of 31.7 to 38 mm and American cottons 38 to 50.8 mm. Upland American cotton, naturalized in India, is extensively cultivated here. The cultivation of long staple foreign cottons in Indian soil has not proved to be a success yet. Of all the Indian cottons the Broach cotton of Gujarat is the finest. Although the area under cotton in India is the largest in the world, her total output is much less than that of other cotton-producing countries. The average annual yield of cotton lint in India is only about 97 kg. per hectare; while in the U.S.A. it is 177 kg., in Japan 203 kg., in U.S.S.R. over 272 kg., and in Egypt about 416 kg. Maharashtra, Gujarat, Madhya Pradesh, Punjab and Rajasthan are the important cotton-growing areas in India. The indigenous cottons of India like *Gossypium herbaceum* yield 3–10 quintals of seed-cotton per hectare per year; while Upland American cotton (*G. hirsutum*) and hybrid cottons yield 25–30 quintals, sometimes up to 40 quintals.

(2) **Jute** is the best fibre of *Corchorus capsularis* and *C. olitorius*. Jute is widely cultivated in the low-lying areas of West Bengal (mainly), Assâm, Bihar and Orissa, and to some extent only in Uttar Pradesh, Meghalaya and Tripura. It is extensively cultivated in Bangladesh. Jute is essentially a rainy season crop thriving under conditions of flooding at a later stage. Fibres mature with the formation of the fruits. After harvesting, the jute plants are retted in water for about 10 to 15 days, sometimes more. The fibres are then stripped off the stalks by hand. They are then washed, dried in the sun and finally baled. The annual yield usually varies from 823 to 1,646 kg. per hectare, with an average

of 1,300 kg. With the extension of cultivation the annual jute production in India has gone up to 64.54 lakh bales of 180 kg. each. Jute fibres are extensively used for making gunny bags, cheap rugs, carpets, cordage, hessian (coarse cloth), etc. There are now 69 jute mills in and around Calcutta, and 1 in Assam.

I. **Medicinal Plants.** Our forests abound in medicinal herbs, shrubs and, trees, yielding many valuable drugs. It is estimated that they number over 4,000 species. Of them about 2,500 to 3,000 species are in general use in some form or other. The Eastern and Western Himalayas and the Nilgirl Hills are known to be the natural abodes of many such plants. A good number of them are now being cultivated in different States on experimental and commercial basis. The Central Drug Research Institute at Lucknow is carrying on research on indigenous medicinal plants.

(1) **Sandalwood** (*Santalum album*) is the monopoly of Karnataka and Kerala. The heartwood yields 5 to 7% of a yellow aromatic volatile oil—the sandalwood oil. The oil has many medicinal properties: it is cooling, astringent and useful in biliousness, vomiting, fever, thirst and heat of the body. The seed-oil is used in skin diseases.

(2) **Garlic** (*Allium sativum*) is a strong-smelling, whitish scaly bulb, the smell being due to the presence of a sulphur-containing volatile oil present in all parts of the plant. The plant is a small perennial bulbous herb cultivated throughout India. Garlic is used as a condiment, particularly in fish and meat preparations, and in various fruit and vegetable preserves. It has some important medicinal properties. It is an effective remedy for high blood pressure, rheumatic and muscular pain, and for giddiness and sore eyes. It is a good tonic for the lungs. Besides, it is digestive and carminative (relieves flatulence and pain in the bowels). It heals intestinal and stomach ulcers, and is in fact regarded as Nature's best antiseptic for the alimentary canal. It is highly efficacious in torpid liver and dyspepsia.

(3) **Thorn-apple** (*Datüra fastuosa*) is a common fleshy poisonous annual herb, containing a few alkaloids. Seeds are extremely poisonous. Fresh leaves and juice are narcotic and anodyne, sometimes applied to relieve pain and inflammation in rheumatism and gout. They are also used in the treatment of epilepsy and obstinate headache. Smoking leaves in the form of cigarettes often gives immediate relief in the case of asthma, and possibly also whooping cough. Daturine of seeds is often used as a substitute for belladonna. A narcotic drug *stramonium* is prepared from dried leaves and flowering tops.

(4) **Sacred basil** (*Ocimum sanctum*) is a common rigid herb. Leaves have expectorant property, i.e. ejecting phlegm—the thick slimy matter—from the throat. The juice is a household remedy for cough, cold and bronchitis in children. It is also applied to the skin in ringworm and other skin diseases. It is also a good remedy for ear-ache. An infusion of leaves is an effective stomachic in gastric disorders in children.

(5) **Penicillin** is the secretary product of a soil fungus called green mould (*Penicillium notatum*). It is a wonder drug of modern times. It is most effective on a wide range of infectious diseases caused by bacteria.

J. **Timber.** Timber is the wood (heart-wood) used for various building purposes: houses, boats, bridges, ships, etc., for making furniture, packing boxes, tea-chests, cabinets, shelves, matchsticks and boxes, plywood, etc., and for railway sleepers. Timber and firewood (fuel) together with many other important forest products constitute the forest wealth of a country. To be self-sufficient in them a country should normally have about one-third of the total land area under forest. In this respect India lags behind, having only 22.2%. In India, Madhya Pradesh has now the largest forest area, while Assam occupies second place. The useful timber trees of Indian forests number over 75 species. The quality of timber depends on its hardness, strength, weight, presence of natural preservatives, such as tannin, resin, etc., durability against heat, moisture, and insect attack, work-ability, grains, colour, porosity, and capacity for taking polish and varnish. Some of the best timber trees are: teak (*Tectona*), sal tree (*Shorea*), Indian redwood (*Dalbergia*), jarul tree (*Lagertroemia*), mahogany (*Swietenia*), pine (*Pinus*), deodar (*Cedrus*), etc.

K. **Beverages.** They are agreeable liquors meant for drinking. Tea, coffee and cocoa are common such beverages.

(1) **Tea** is the dried and prepared leaves of *Thea* (=*Camellia*) *assamica* and *T. sinensis,* and many hybrids. Tea is now a universal drink. From the tea garden to the tea cup there is, however, a long history. There are different grades of tea. The terminal bud with *two* leaves forms *fine* tea; the same with *three* leaves forms *medium* tea; and the same with *four* leaves forms *coarse* tea. The average yield of manufactured tea in India has now gone up to 1,360 kg. per hectare per year, and 1.8 kg. of green leaves usually make 0.45 kg. of cured tea. The yield of green leaves per plant is more or less 0.9 kg. India is the largest producer and exporter of tea in the world. She produces more than half of the world's total

output. Over 50% of India's total output is produced in Assam with an yield of 1,680 kg. (rising up to 2,800 kg.) per hectare. There are 750 tea gardens (excluding many small ones) in Assam (mostly in Upper Assam) and 350 in North Bengal. Tea is also grown on a large scale in Tamil Nadu, Kerala and Karnataka. India now produces about 560 million kg. of tea and exports about 220 million kg., earning a foreign exchange of about Rs. 270 crores next only to jute. Tea grows in the plains and in the hills. It may be noted that Darjeeling tea is famous for its very agreeable flavour and taste, and thus sells at much higher rates. Manufactured tea contains 4–5% of tannins (catechins), which are responsible for colour and strength of the infusion, 3.5–5% of caffeine which is a stimulant for the heart, and a little volatile oil to which the aroma of the tea is due. Tea bushes require pruning every year for adequate 'flushing'. Proper irrigation and manuring lead to increased yield.

(2) **Coffee** is a favourite drink in South India. Seeds of *Coffea arabica* and *C. robusta,* particularly the former, are the sources of coffee. The aroma of the coffee powder develops on proper and skilful roasting. Certain chemicals are also added for this purpose. A coffee bush usually yields 0.45–0.9 kg. of cured coffee. Coffee contains several vitamins and also caffeine (an alkaloid) to the extent of 0.75–1.5%. The main coffee plantations are in the low hills of South India—Karnataka, Tamil Nadu and Kerala. India's coffee production has now gone up to about 1 lakh tonnes, and her export to about 57,000 tonnes. Brazil and Kenya are the world's largest suppliers of coffee. India ranks third in the world in her coffee production.

(3) **Cocoa** is prepared from the seeds of *Theobroma cacao,* a small tree. Each tree commonly bears 70–80 fruits, each measuring 15–22 × 7–10 cm. Each fruit contains numerous seeds. The fruits are cut or broken open, and the seeds dried, roasted and powdered. Cocoa powder makes a refreshing and nourishing drink. With the addition of certain ingredients chocolate is made out of this powder. Cocoa has also a food value. It contains theobromine and caffeine (1% or less), proteins (15%), starch (15%) and fatty oil (30–50%). On an average 50 pods, each having about 30 good seeds, yield over 1 kg. of cured cocoa. Cocoa-butter is used in medicine. Cocoa is extensively cultivated in tropical America, the West Indies, Brazil, Ghana and Kenya. It is also cultivated in Java and Ceylon. The world's supply comes mainly from Brazil, Kenya and Ghana (Ghana supplying the largest quantity).

Appendix I

Questions

INTRODUCTION

1. What is protoplasm? Give an account of its physical and chemical nature. **2.** Enumerate the important differences between the living and the non-living. **3.** What are the main characteristics of plants by which they can be distinguished from animals? **4.** Classify plants into their main groups. Give the main characteristics of each group. Illustrate your answers with sketches and examples. **5.** Give the main characteristics of living organisms.

PART I. MORPHOLOGY

Chapters 1–2. **1.** Classify plants according to their habit, and give a detailed account of various modes of climbing. **2.** What are autotrophic and heterotrophic plants? Describe the types of heterotrophic plants that you have studied. **3.** Draw a labelled sketch showing the parts of a 'flowering' plant, and briefly mention the functions of those parts.

Chapter 3. **1.** Describe the parts of an exalbuminous seed, and the mode of its germination. Give necessary sketches. **2.** Describe with the aid of sketches the parts of a castor seed or a maize grain, and the mode of its germination. **3.** What do you understand by epigeal germination? Describe the process with reference to a typical seed. **4.** What are the essential conditions necessary for the germination of a seed? Devise an experiment to prove them. **5.** Write, short notes on endosperm, scutellum, coleoptile, epicotyl and vivipary.

Chapter 4. **1.** What are the characteristics of the root by which it can be distinguished from the stem? **2.** Give an account of the modified forms of roots. Describe them with sketches and examples. **3.** What are adventitious roots? Describe with sketches and examples at least five types of such roots. **4.** What are the normal functions of the root? Point out how this organ adapts itself to meet certain specialized functions. **5.** Write short notes on endogenous, fusiform root, haustoria, epiphytic root, root-hair and fasciculated roots. **6.** What is a root? Describe the different regions of a root. State the functions of roots.

Chapter 5. **1.** Describe a vegetative bud as seen in a longi-section. Of what use is it to the plant? Describe the various types of buds. **2.** Name and describe the various kinds of underground stems. Why are they not regarded as roots? What are their functions? **3.** Describe a potato tuber, and give reasons in support of its morphological nature. How do you distinguish between a stem-tuber and a root-tuber? **4.** Describe, with sketches and examples, the various modifications of stems for vegetative propagation. **5.** Write short notes on decumbent, caudex, rhizome, bulb, bulbil, tendril, thorn, phylloclade and cladode. **6.** Describe the various types of aerial modified stems.

Chapters 6–7. **1.** Describe the parts of a typical leaf, and give an account of the various modifications it undergoes. **2.** What is venation? Give the principal types of veins. Mention the functions of the system of veins. **3.** Distinguish between a simple leaf and a compound leaf, a compound leaf and a short branch. What are the main types of compound leaves? **4.** Give a short account of phyllotaxy, and explain what is meant by orthostichy and genetic spiral. **5.** What are the normal functions of the leaf? For what other purposes may it be utilized? **6.** How do plants protect themselves against injury by animals? **7.** Note any points of morphological interest connected with the following plants: pea, *Smilax,* glory lily; *Polygonum,* rose, *Naravelia,* pitcher plant, *Cardanthera,* Australian *Acacia,* nettle, *Hemiphragma* and snake plant. **8.** Describe the different types of modified leaves with diagrams and examples in each case.

Chapters 8–9. **1.** What is an inflorescence? Describe the simple racemose types. Give sketches and examples. **2.** What kind of inflorescence do you find in gold mohur, aroid, banana, sunflower, coriander, *Ocimum* and grass? Describe any four of them with sketches. **3.** Describe the parts of a typical flower, and indicate the functions of these parts. **4.** Describe, with sketches and examples, the structure of hypogynous, perigynous and epigynous flowers. Describe the thalamus of *Gynandropsis* and passion-flower. **5.** Discuss: 'A flower is a modified shoot'. **6.** Explain and cite instances of adhesion and cohesion in flowers. **7.** What is a pollen grain? Describe a mature pollen grain and a mature embryo-sac. What happens when a pollen grain germinates? **8.** What is a placenta, and where do you find it? Describe the different types with sketches and examples. **9.** Draw a neat diagram of an anatropous or orthotropous ovule, and label the parts. **10.** Describe, with sketches, the common forms of ovules. **11.** Write notes on any five of the following: panicle, helicoid, verticillaster, hypanthodium, hermaphrodite, gynophore, spathe, involucre, bilabiate, corona, epicalyx, pollinium, didynamous and apocarpous.

Chapters 10–11. **1.** What is cross-pollination? How is it effected? **2.** What are the characteristics of entomophilous and anemophilous flowers? **3.** Describe the mode of pollination in sunflower, *Salvia,* maize and *Vallisneria.* **4.** What are the contrivances met with in flowers to prevent self-pollination? **5.** Give a detailed description of the process of fertilization in an angiosperm. What is double fertilization? **6.** Write notes on cleistogamy, spur, anemophily, monoecious, dichogamy, protandry, egg-cell and synergids.

Chapters 12–14. **1.** Describe the changes that take place in the ovule leading to its conversion into the seed. **2.** How do you classify fruits? Describe the principal types. **3.** Describe the fruits of pea, mustard, cotton, mango, tomato, apple, gourd and pineapple. **4.** Describe in botanical terms the edible parts of the following fruits: apple, cashew-nut, cucumber, litchi, fig, pineapple, mango, coconut, jack and orange. **5.** Give a concise account of how seeds and fruits are dispersed by wind. Of what importance is the distribution to the species? **6.** Write notes on aril, perisperm, mesocarp, legume, capsule, berry, siliqua, pepo, sorosis and censor mechanism.

PART II. HISTOLOGY

Chapter 1. **1.** What is protoplasm? Where is it found in plants? What are the different kinds of movements exhibited by it? **2.** Describe the structure of the nucleus, and state the main functions performed by it. **3.** Describe the parts of a typical plant cell, and give a short account of their functions. **4.** What is the modern idea of cell-structure as revealed by the 'electron' microscope? **5.** Give a clear account of mitochondria, ribosomes, Golgi body, DNA and RNA. **6.** Describe the microscopic structure of starch grains and aleurone grains. How would you demonstrate their presence in the plant tissue? What kind of food do they represent? **7.** What is cellulose? What modifications may it undergo? How would you distinguish between cellulose and lignin? **8.** Enumerate and briefly describe the most important reserve materials in plants. **9.** Give an account of the occurrence of mineral crystals in plants. How do you recognize them? **10.** Write short notes on middle lamella, vacuole, bordered pits, cystolith, cytoplasm, chromosome and raphides. **11.** Give an account of 'somatic cell-division', and indicate the importance of the process. **12.** State the main differences between mitosis and meiosis.

Chapters 2–3. **1.** What is a tissue? What are the principal kinds of tissues found in plants? Describe their microscopic structure and state their functions. **2.** Where do you find a sieve-tube? Draw and describe its structure. **3.** Describe the apical meristem of a stem or a root, as seen in a longi-section. **4.** Describe the internal structures as seen in transverse section of (a) sunflower stem and (b) maize stem. **5.** What are stomata? Describe their structure and functions. How do they behave when the atmosphere is dry? **6.** Describe the tissue-elements of a typical vascular bundle. What are the different types of vascular bundles? **7.** Write notes on tracheid, trachea, latex cell, sclerenchyma, collenchyma, endodermis and pericycle.

Chapters 4–6. **1.** Describe the anatomical structure of a dicotyledonous stem. **2.** Give a detailed description of the internal structure of a monocotyledonous stem. **3.** Describe the structure of a monocotyledonous root, as seen in a transverse section, and compare it with that of a dicotyledonous root. **4.** Describe the anatomical structure of a dorsiventral leaf, and state the functions of the different tissues met with in it. **5.** Write notes on hard bast, bicollateral bundle, conjunctive tissue, protoxylem, epiblema, mesophyll and palisade parenchyma.

Chapter 7. **1.** How does a dicotyledonous stem grow in thickness? **2.** Describe with neat sketches the origin and activity of cambium in a dicotyledonous root. **3.** What are annual rings? How are they formed? Describe their anatomical structure. **4.** How are cork and lenticel formed? Describe, with neat diagrams, their anatomical structure, and state their functions.

PART III. PHYSIOLOGY

Chapters 1–4. **1.** What are the common types of soils? How are they formed? Give an account of their physical and chemical properties. **2.** What is humus? What is its utility in plant growth? How do you estimate the humus content of a soil? **3.** Enumerate the essential chemical elements that enter into the composition of a green plant. What are the methods usually followed to determine them? **4.** How does an ordinary green plant get its supply of carbon and nitrogen? What special means of nitrogen supply are found in *Leguminosae?* **5.** What is osmosis? What role does it play in plant physiology? Devise an experiment to demonstrate it. **6.** Write notes on turgidity, plasmolysis, capillarity, nitrification and nodule bacteria. **7.** Describe nitrogen cycle.

Chapter 5. **1.** What is root pressure? How do you demonstrate and measure it? Explain the significance of the process. **2.** What is transpiration? Devise an experiment to show the rate of transpiration from a twig. **3.** Devise an experiment to prove unequal transpiration from the two surfaces of a leaf. What is the significance of this difference? **4.** Devise a simple experiment to show that a transpiring twig produces suction. Comment on the processes concerned. **5.** What do you understand by 'ascent of sap'? Discuss the main forces concerned in the process. **6.** What is transpiration? Describe the different factors which affect the rate of transpiration. State the significance of transpiration. **7.** What is the importance of transpiration in plant life? **8.** Write notes on manometer, bleeding, leaf-clasp, potometer, and exudation.

Chapters 6–9. **1.** Give a concise account of carbon-assimilation by green plants. **2.** How would you prove experimentally that oxygen is evolved during photosynthesis? **3.** How would you prove experimentally that light and carbon dioxide are indispensable for the formation of starch in photosynthesis? **4.** What are the external conditions necessary for photosynthesis? Clearly explain the influence of these conditions on the process. **5.** How would you prove experimentally that non-green parts of plants cannot photosynthesize? **6.** What are proteins? How are they synthesized in plants? **7.** What are autotrophic and heterotrophic plants? Give a short account of various modes of nutrition of the latter. **8.** What special modes of nutrition are found in carnivorous plants? Describe a few common types. **9.** Describe the various organs and tissues which lay up stores of reserve food. What are the common forms of reserve food stored up in the seed?

How are they utilized? **10.** Give a clear account of digestion and assimilation of food in green plants.

Chapters 10–11. **1.** What is respiration? Discuss the mechanism of respiration in plants. **2.** How would you prove experimentally that plants respire? **3.** Describe the behaviour of plants deprived of oxygen, and demonstrate it by an experiment. **4.** Describe the nature of exchange of gases between the green plant and the atmosphere. **5.** Distinguish between respiration and photosynthesis. **6.** Write notes on the following: anaerobic respiration, fermentation, metabolism, respiroscope, glycolysis, enzymes and Krebs cycle. **7.** What is fermentation?

Chapters 12–14. **1.** What is growth? What is the influence of external conditions on growth? **2.** How would you measure growth in length of the stem? **3.** What is meant by irritability in plants? Give instances? **4.** Describe heliotropism, geotropism and hydrotropism, and demonstrate experimentally any one of these processes. **5.** Give a short account of hormones and vitamins occurring in plants. **6.** Write notes on heliotropic chamber, clinostat, auxanometer, and grand period of growth. **7.** Describe the methods by which the 'flowering' plants reproduce themselves vegetatively.

PART IV. ECOLOGY

Chapters 1–2. **1.** Define ecology. Enumerate the factors which influence ecological grouping of plants. Give examples. **2.** What are the characteristic features of hydrophytes? Cite familiar examples. **3.** What are halophytes? State what you know of their special characteristics. Give examples. **4.** Describe the characteristic features of xerophytes. Give examples. **5.** What is mangrove vegetation? Where do you find it in India? Describe, with examples, the main features of such a vegetation. **6.** What is ecosystem? Describe the following in an ecosystem: abiotic factors, biotic factors and energy flow.

PART V. CRYPTOGAMS

Chapters 1–2. **1.** How do you classify cryptogams? State the differences between (a) Algae and Fungi, and (b) Bryophyta and Pteridophyta. **2.** Give a brief account of the structure and mode of reproduction in *Oscillatoria* and *Chlamydomonas*. **3.** Describe the life-history of *Ulothrix* or *Spirogyra*. **4.** Describe the life-history of

Vaucheria. **5.** Write notes on zygospore, antherozoid, oogonium, coenocyte, isogamy, hormogonia, pyrenoid, palmella stage and gonidia.

Chapters 3–5. **1.** Give a brief account of bacteria, and state what you know of their harmful and beneficial effects. **2.** Give the life-history of any saprophytic fungus that you have studied. **3.** Compare the life-history of *Spirogyra* with that of *Mucor*. **4.** Some yeast cells are put into sugar solution. State and describe the changes that you may expect in (a) yeast cells, and (b) sugar solution. **5.** Give a brief account of the life-history of *Agaricus* or *Penicillium*. **6.** Give a general account of lichens.

Chapters 6–7. **1.** Describe the life-history of *Riccia* or *Marchantia,* and trace the alternation of generations in it. **2.** What do you understand by alternation of generations? Illustrate your answer by reference to a moss plant or a fern plant. Indicate the limits of the two generations. **3.** Describe the gametophytic generation of moss, or the sporophytic generation of fern. **4.** Describe the prothallus of fern. What phase does it represent in the life-history of the plant? **5.** Describe the sporangiferous spike of *Equisetum* and the nature of its spores, or describe the gametophytic generation of this plant. **6.** Briefly describe the life-history of *Marsilea,* tracing clearly the alternation of generations in it. **7.** Describe the sporophytic generation of a heterosporous plant. **8.** Write notes on: Gemma cup, calyptra, rhizophore and prothallus.

PART VI. GYMNOSPERMS

Chapters 1–2. **1.** Describe the life-history of *Cycas*. **2.** Describe the megasporophyll and microsporophyll of *Cycas*. **3.** Describe the ovule of *Cycas,* as seen in a longitudinal section. **4.** Describe the mode of pollination and fertilization in *Cycas*. **5.** Brielly describe the life-history of *Pinus*. **6.** Describe how pollination and fertilization take place in *Pinus*. **7.** Describe young male and female cones of *Pinus*. **8.** Describe the ovule of pine as seen in longitudinal section. **9.** Give the differences between gymnosperms and angiosperms.

PART VII. ANGIOSPERMS

Chapters 1–2. **1.** Define species, genus, family, variety and nomenclature and illustrate them by suitable examples. **2.** Give the outline of any modern system of classification of 'flowering' plants. **3.** Enumerate and explain the differences between dicotyledons and monocotyledons. **4.** Give diagnostic characters of the following families: *Cruciferae, Malvaceae, Papilionaceae* and *Cucurbitaceae.* Mention at least three economic plants belonging to each family. **5.** Give the main characteristics of the family *Nymphaeaceae.* Illustrate your answer and cite at least two examples. **6.** Describe (a) the inflorescence and (b) the androecium of the following families: *Cruciferae, Compositae* and *Labiatae.* **7.** Mention the respective families where you find the following morphological characters: monadelphous, syngenesious, epipetalous, didynamous, apocarpous and inferior ovary. **8.** Describe the family *Cucurbitaceac* or *Solanaceae* with necessary sketches and mention at least three examples of economic importance. **9.** Refer any five of the following plants to their respective families: *Ranunculus, Brassica, Hibiscus, Pisum, Cucurbita, Solanum* and *Ocimum,* and give the characteristics of any one of these families.

Chapter 3. **1.** Give the main characteristics of the family *Palmae,* and mention at least three plants of economic importance belonging to it. **2.** Describe *Liliaceae* with necessary sketches and examples. **3.** What is the economic importance of *Gramincae?* Cite at least five examples. Give the important morphological characteristics of the family.

PART VIII. EVOLUTION AND GENETICS

Chapters 1–2. **1.** What evidences can you cite in support of the theory of evolution? Clearly explain any three of them. **2.** State briefly Darwin's contribution towards the idea of organic evolution. **3.** Give a concise idea of some of the important theories advanced from time to time to explain organic evolution. **4.** Explain Mendel's monohybrid cross. Tabulate the results up to F_2 generation. **5.** Briefly explain Mendel's laws of inheritance. **6.** Give in a tabulated form the results of Mendel's dihybrid cross. **7.** What is the practical importance of Mendel's experiments? **8.** How would you proceed to determine whether a plant possessing a particular dominant character is homozygous or heterozygous?

PART IX. RECENT ADVANCES IN BOTANY

Chapters 1–4. **1.** What is biotechnology? What is the significance of this branch? **2.** Write notes on tissue culture and industrial microbiology.

PART X. ECONOMIC BOTANY

Chapters 1–2. **1.** Describe the methods commonly employed for improvement of field crops. **2.** What are cereals? Of what importance are they to human beings? Give a short account of at least two important cereals widely cultivated in India. **3.** Give a short account of wheat, maize and *Sorghum*. Where are they cultivated in India? **4.** Enumerate the important vegetable oils of India. What are the sources of these oils? Give a brief account of their uses. **5.** Mention some important dessert fruits of India, and state the edible parts in them. **6.** What are the common timber trees of India? State their uses. **7.** Write notes on the uses of the following plants: nutmeg, cardamom, cloves, ginger, garlic, pulses, potato, sugarcane and sandalwood. **8.** What are the common beverages used in India? Where are the plants yielding such products cultivated in India? Give a brief account of their manufacture and use. **9.** Describe three medicinal plants and their medicinal value.

Appendix II

Glossary of Names of Plants

Botanical Names in *italics;* English names in Roman; Indian names in CAPITALS: A. for Assamese, B. for Bengali, G. for Gujarati, H. for Hindi, K. for Kannada, K'. for Konknni, M. for Malayalam, M'. for Marathi. O. for Oriya, P. for Punjabi, T. for Tamil, and T'. for Telugu.

Abelmoschus esculentus (lady's finger) = A., O. & M'. BHENDI; B., H. & P. BHINDI; G. BHINDA; K. BHENDE KAYI; K'. BHENDDI; M. & T. VENDAKKA; T'. BENDA

Abrus precatorius (crab's eye or Indian liquorice) = A. LATUM-MONI; B. KUNCH; H. & P. RATTI; K. GULAGANJI; K'. GUNJ, DEVCHARA-DOLLE; M. KUNNI; M'. KUNI; O. KAINCHA, GUNJA; T. KUNDOOMONY; T'. GURUGINJA

Abutilon indicum = A. JAPA-PETARI; B. PETARI; G. DABALI; H. KANGHI; K. THURUBI GIDA, SHREE-MUDRE GIDA, KISANGI; K'. MUDRO; M. & T. PERINTHOTTY; M', MUDRA; O. PEDIPEDIKA; P. PILI-BUTI; T'. THUTIRIBENDA

Acacia nilotica = *A. Arabica* (gum tree) = A. TORUAKADAM; B. BABLA; G. KALO-ABAVAL; H. BABUL, KIKAR; K. KARI JALI; K'. BABULZHADD; M. & T. KARU-VELAM; M'. BABHUL; O. BABURI; P. KIKAR; T'. NALLATUMMA

Acacia catechu (catechu) = A., B. & M'. KHAIR; G. KHER; H. & P. KATHA, KHAIR; K. KAGGALI, KACHU; K'. KAIRAZHADD, KAIRAKUKH; M. KADARAM; O. KHAIRA; T. KADIRAM; T'. KHADIRAMU

Acalypha indica = B. MUKTAJHURI; G. VANCHI KANTO; H. KUPPI; K. KUPPI GIDA, TUPPAKEERE; K'. MAKDDA-XHEMPDDI; KOLEA-XHEMPDDI; M. & T. KUPPA-MANI; M'. KHOKALI; O. INDRAMARISHA; P. KOKALI KUPPAMANI; T'. KUPPIN-TAKU

Achryanthes aspera (chaff-flower) = A. UBTISATH; B. APANG; G. SAFED AGHEDO; H. LATJIRA; K. UTTARANI; K'. DOVO-AGHADO; M. KATALADY; M. AGHADA; O. APAMARANGA; P. PUTH KANDA, KUTRI; T. NAHIROORVY; T'. ATTARENI

Acorus calamus (sweet flag) = A., B. & H. BOCH; G. GODAVAJ; K. BAJE; K'. WAIKHAND; M. VAYAYMBU; M'. WEKHAND; O. BACHA; P. WARCH, BOJ, BARI; T. VASAMBOO; T'. VASA

Adhatoda vasica = A. BANHAKA; B. BASAK; G. ADULSO; H. ADALSA; K. ADU-SOGE, KURCHI GIDA, ADDALASA; K'. ODDUSO; M. ADALODAKAM; M. ADULSA; O. BASANGA; P. BANSA SUBJ, BASUTI; T. ADATODAY; T'. ADDASARAMU

Aegle marmelos (wood-apple) = A. & B. BAEL; G. BILIVA-PHAL; H. SIRIPHAL; K. BILVA PATRE; K'. BELFOLI; M. KOOVALAM; M'. BEL; O. BELA; P. BIL; T. VILVAMARAM; T'. BILVAMU

Agave americana (American aloe or century plant) = B. & H. KANTALA; G. JANGLI-KANVAR; K. KATTALE; K'. GAIPAT; M. NATTUKAITA; M'. GHAYPAT; O. BARA-BARASIA; P. WILAYATI KANTALA; T. ANAKUTTILAI; T'. BONTHARAKASI

Albizzia lebbek (SIRIS tree) = A., B., H., M'. & P. SIRISH; G. PITOSARSHIO; K. SHIRISHA BAGE, HOMBAGE; K'. XRIXI; M. KARINTHAKARA; O. SIRISA; T, VAGAI; T'. DIRISANA

Allium cepa (onion) = A. PONORU; B., H. & P. PIYAZ; G. DUNGARI; K. NEERULLI, ULLAGADDI; K'. PIAO, KANDEAPAT, M. ULLI; M'. KANDA; O. PIAZO; T. VENGAYAM; T'. NEERULLI

Allium sativum (garlic) = A. NAHARU; B. RASUN; G. LASAN; H. & P. LASHUN; K. BELLULLI; K'. LOSSUN; M. VELUTHULLI; M'. LASUN; O. RASUNA; T. VENGAYAM; T'. VELLULLI

Alocasia indica = A. & B. MANKACHU; G. ALAVU; H. MANKANDA; K. MANAKA; K'. ALLU, AUUM; M'. ALU; O. MANASARU; P. ARVI

Aloe vera (Indian aloe) = A. CHALKUNWARI; B. GHRITAKUMARI; G. KUNVAR; H. GHIKAVAR; K. LOLESARA; K'. KANTTIKOR, KATKUNVOR; M. KATTARVAZHA; M'. KORPHAD; O. GHEEKUANRI; P. KAWARGANDAL, GHIKUAR; T. KUTTILAI; T'. KALABANDA

Alstonia scholaris (devil tree) = A. CHATIAN; B. CHHATIM; H. CHATIUM; K. SAPTA PARNA, MADDALE, KODALE; K'. SANTNARUKH; M. EZHILAMPALA, M'. SATVIN; O. CHHATIANA, CHHANCHANIA; P. SATONA; T. ELILAIPILLAI; EDAKULAPALA

Amaranthus spinosus (amaranth) = A. KATAKHUTURA; B. KANTANATE; G. TAN-JALJO; H. & P. CHULAI; K. MULLU KEERE (or HARIVE) SOPPU; K'. DOVI-BHAJI; M. MULLANCHEERA; M'. KATE MATH; O. KANTANEUTIA, KANTAMARISHA; T. MULLUKKERAI; T'. MUNDLA THOTAKURA

Amorphophallus campanulatus = A. & B. OL; G. & M'. SURAN; H. ZAMIKAND; K. SUVARNA (or CHURNA) GEDDE; K'. SURAN, LUTTIEZHADD, SOTRI; M. CHAENA; O. OLUA; P. ZAMIN KANDA; T. KARUNAKILANGU; T'. THIYA KANDHA

Anacardium occidentale (cashew-nut) = A. KAJU-BADAM; B. HIJLI-BADAM; G., H., M'. & P. KAJU; K. GODAMBI, GERUPAPPU; K'. KAZ; M. KASHUMAVU, KAPPALUMAVU; O. LANKA BADAM; T. MUNDIRI; T'. JIDIMAMIDI

Ananas comosus (pineapple) = A. MATI-KOTHAL; B. ANARAS; G., H. & M'. ANANAS; K'. ANOS;. M. ANNANAS; T. ANASSAPPALAM; T'. ANASAPANDU

Andrographis paniculata = A. KALPATITA, B. & H. KALMEGH, MAHATITA; G. KIRYATO; K. NELA BEVU, KALA MEGHA; K., KIRYATEM; M. KIRIYATHTHU; M'. PALEKIRAIET; O. BHUINIMBA; P. CHARAITA; T. NELAVEMBU; T'. NELAVEMU

Annona reticulata (bullock's heart) = A. ATLAS; B. NONA; G., H., M'. & P. RAMPHAL; K. RAMA PHALA; K'. ANON; M. ATHA; O. NEUA BADHIALA; T. RAMSITA; T'. RAMAPHALAMU

Annona squamosa (custard-apple) = A. ATLAS; B. ATA; G. & M. SITAPHAL; H. & P. SHARIFA, SITAPHAL; K. SEETHA PHALA; K'. ATER; M. SEEMA-ATHA; T. SEETHA; T'. SEETAPHALAMU

Anthocephalus indicus = A., B., H. & P. KADAM; G. & O. KADAMBA; K. KADAMBBA MARA, KADAVALA; K'. KODOMB. M. KADAMBU; M'. KADAMB; T'. KADAMBA, KADIMI

Arachis hypogaea (peanut or groundnut) = A., B. & O. CHINA-BADAM; G. MAFFALI; H. & P. MUNGPHALI; K. NELAGADALE, SHENGA, KALLEKAI; K'. MUSBIBIKNAMZHADD; M. & T. NILAKKADALAI; M'. BHUI-MUG; T'. VERU SANAGA

Areca catechu (areca- or betel-nut) = A. TAMBUL; B., G., M'. & P. SUPARI; H. KASAILI; K. ADIKE; K'. MADDI; M. ADAKKA; O. GUA; T. PAKKU; T'. POKA

Argemone mexicana (prickly or Mexican poppy) = A. KUHUMKATA; B. SHEAL-KANTA; G. DARUDI; H. PILADHUTURA; K. DATTURADA GIDA ARISINA UMMATI; K'. MARANDDI, HOLDUVO-DUTRO; M. SWARNAKHYEERI; M. PIWALA DHOTRA; O. KONTAKUSUMO; P. KANDIARI; T. BRAHMADANDU; T'. BRHMADANDI

Aristolochia gigas (pelican fower) = A., B. & O. HANSHA-LATA; K. KURI GIDA; K'. POPOT-VAL; M. GARUDAKKODI; M'. POPAT VEL; P. BATKH PHUL; T. ADATHINA-PALAI

Artabotrys odoratissimus = A. KOTHALI-CHAMPA; B., G. & H. KANTALI-CHAMPA; K. MANORANJINI, KANDALA SAMPIGE; K'. HIRVO-CHAMPO; M. & T. MANO-RANJINL; M'. HIRWA CHAPHA; O. CHINI-CHAMPA; P. CHAMPA; T. MANORAN-JITHAM

Artocarpus heterophyllus (jack tree) = A. KOTHAL; B. KANTHAL; G. MANPHANASA; H, KATAHAR; K. HALASU; K'. PONOS; M. & T. PILA; M'. PHANAS; O. PANASA; P. KATAR

Asparagus racemosus = A. SHATMUL; B. SATAMULI; H. & P. SATAWAR: K. SHATAVARI; K'. SATAVIR, SAN-JUANU-VAL; O. CHHATUARI; M., M'. & T. SATHAVARI; T'. PILLITEGA

Averrhoa carambola (carambola) = A. KORDOI-TENGA; B. KAMRANGA; G. KAMARAKHA; H. & P. KAMRAKH; K. KAMARAKI, KAMARAK; K'. KORMONGARUKH; M. IRIMPANPULI; M'. CAMARANGA; O. KARMANGA; T. KAMARANKAI; T'. TAMARTA

Azadirachia indica (margosa) = A. MOHA-NIM; B., H. & P. NIM, NIMBA; G. LIMBA, K. OLLE BEVU; K'. KODDU-NIM, KADU-LIMBU; M. VEEPU; M'. KADU LIMB; O. NIMBA; T. VEMBU; T' VEPA

Baccaurea sapida = A. LETEKU; B. LATKAN; H. LUTKO; K. KOLI KUKKE; K'. LUTKO; P. KALA BOGATI

Bambusa tulda (bamboo) = A. BANTH; B., H. & P. BANS; G. KAPURA; K. HEBBIDIRU, UNDE BIDRU; K'. KONDO, MAN; M. MULAH; M'. VELBONDI; T. SAMBARKEERAI; T', VEDURU

Basella rubra (Indian spinach) = A. PURAI; B. PUIN; H. O. & P. POI; K. KEMPU BAYI BASALE; K'. BONDI-VAL; M. SAMPARCHEERA; M'. VELBONDI; T. SAMBARKEERAI; T'. KURUBACHHALI

Batatas edulis, see *Ipomoea batatas*

Bauhinia variegata (camel's foot tree) = A. B. & M'. KANCHAN; G. KOVIDARA; H. & P. KACHNAR, K. ULIPE, BILI MANDARA; K'. TEMBRI, KUDO; M. MANDARUM; O. KANCHANA; T. TIRUVATTI; T'. ADAVIMANDARA

Benincasa hispida (ash gourd) = A. KOMORA; B. CHALKUMRA; G. KOHWLA; H. & P. PETHA; K. BOODU GUMBALA; K'. BHELLPOTRI; M. KUMPALAM; M'. KOHAIA; O. PANI-KAKHARU; T. KUMPALY, T'. PULLA GUMMUDI

Beta vulgaris (beet) = A. BEET-PALENG; B. BEET-PALANG; G. & M'. BEET; H. & P. CHUKANDAR; K. BEET ROOT; K'. BETARAB; O. PALANGA SAGA, BEET; T'. BEETUDUMPA

Biophytum sensitivum (sensitive wood-sorrel) = A. & B. BAN-NARANGA, G. JAHARERA; H. LAJALU; K. HORA MUNI; K'. LOJERI; M. MUKKUTTI, THINDA NAZHI; M'. LAJARI

Blumea lacera = A. KUKUR-SHUTA; B. KUKUR-SONGA; G. KALARA; H. & P. KOKRONDA; K. GANDHARI GIDA; K'. BURANDO; M'. BURANDO; O. POKA-SUNGA; T. KATUMULLANGI; T' KARUPOGAKU

Boerhavia diffusa (hogweed) = A. PONONUA; B. & M'. PUNARNAVA; G. GHETULI; H. SANTH; K. BALAVADIKE, GONAJALI, RAKTA PUNARNAVA; K'. PUNARNAVO; M. THAZHUTHAMA; O. GHODAPURUNI; P. BISKHAPRA, ITSIT; T. MUKKARATAI; T'. PUNARNABA

Bombax ceiba (silk cotton tree) = A. SIMUL; B. SIMUL, G. RATOSHEMALO; H. & P. SIMAR, SIMBAL, K. BOORUGA. KEMPUBOORUGA; K'. KAPXIN; M.&

T. ELAVU, MULLILAVU; M'. KATE SAVAR; O. SIMULI; T'. KONDABURAGA, SALMALI

Borassus flabellifer (palmyra-palm) = A. & B. TAL; G. & M'. TAD; H. & P. TAR; K. TALE MARA, TATI NUNGU; K'. TAD-MADD; M. KARIMPANA; O. TALA; T. PANAI; T'. TEDI

Brassica campestris (mustard) = A. SARIAH; B. SARISHA, G. SAFED-RAI H. & P. SARSON; K. SASIVE; K'. SANSUAM; M. KATUKU; M'. MOHORI; O. SOROSHA; T. KARUPPUKKADUGU; T'. AAVA

Bryophyllum pinnatum (sprout-leaf plant) = A. PATEGAZA, DUPORTENGA; B. PATHURKUCHI; H. ZAKHMI-I-HAYAT; K. KADU BASALE, K'. PANFUTI; M. PUNNILA; M'. PANPHUTI; O. AMARPOI; P. PATHUR-CHAT; T. RANAKALLIS, T'. RANAPALA

Butea monosperma (flame of the forest) = A., B. & M'. PALAS; G. KHAKARA; H. & P. DHAK; K. MUTTUGA; M. CHAMATHA; O. PALASA; T. SAMITHU, PALASAM; T'. MODUGA

Caesalpinia pulchrrima, see *Poinciana pulcherrima*

Cajanus cajan (pigcon pea or red gram) = A. RAHAR-MAH; B. ARAHAR; G. TUVARE; H. RAHAR; K. THOGARI KALU; K'. TURDAL; M. THUVARA; M'. TUR, O. HARADA; T. THOVARAY; T'. KANDULU

Calotropis gigantea (madir) = A. AKON-GOCH; B. AKANDA; G. AKADO; H. & P. AK; K. EKKADA GIDA; K'. RUIEZHADD; M. & T. ERUKKU; M'. RUI; O. ARKA; T'. JILLEDU

Cannabis sativa (hemp) = A., B., H. & P. BHANG; GANJA; G., P. & T. GANJA; K. GANJA GIDA, BHANGI; K'. BHANG; M. KANCHAVU; M'. BHANG; O. BHANGA, GANJEI; T'. GANJA CHETTU

Capsicum annuum (chilli) = A. JOLOKIA; B. LANKA, MARICH; G. LALMIRCHI; H. & P. LAL-MIRCH; O. LANKAMARICHA; K'. MIRSANG, SANG, MIROLLI; M. MULAGU; M'. MIRCHI; O. LANKAMARICHA; T. MILAGU; T'. MIRAPAKAYA

Cardiospermum halicacabum (balloon vine) = A. KOPALPHOTA; B. KAPALPHUTKI, SHIBJHUL; G. KARODIO; K. BEKKINA BUDDE GIDA, ERUMBALLI, K'. KOPOLFODI; M. VALLIYUZHINJA; M'. KAPALPHDI; O. PHUTPHUTKIN; P. HAB-UL-KULKUL; T. MODAKATHAN; T'. BUDDAKAKKIRA, KASARITAGE

Carica papaya (papaw) = A. AMITA; B. PAYPAY; H. & P. PAPITA; M. OMAKKA, KAPPILANGA; M'. POPAI; O. AMRUTA RUTA BHANDA; T. PAPALI; T'. BOPPAYI

Carissa carandas = A. KORJA-TENGA; B. KARANJA; H. & P. KARAUNDA; K. KAVALI GIDA, KARANDA; K'. KANDDAM, KARVONDDAM; M. ELIMULLU; M'. KARVANDA; O. KHIRAKOLI; T. KALAKKAI; T'. KALIVI VAKA

Carthamus tinctorius (safflower) = A. &. B. KUSUMPHUL; G. KUSUMBO; H. & P. KUSAM; K. KUSUBI, KUSUME; K'. KUSUMPHUL, KUSUMBO; M. SINDOORAM; M'. KARDAI; O. KUSUMA; T. KUSUMBA; T'. AGNISIKHA

Carum copticum, see *Trachyspermum*

Cassia fistula (Indian laburnum) = A. SONARU; B. SHONDAL; G. GARMALA; H. & P. AMALTASH; K. KAKKE GIDA, HONNAVARIKE; K'. IZABFI; M. & T. KONNAI; M'. BAHAWA; O. SUNARI; T'. RELA, KOLAPONNA

Cassia sophera = A. MEDELUA; B. KALKASUNDA; G. KASUNDARI; H. & P. KASUNDA, K. KASAMARDA; K'. AKASSIA; M. PONNARAN or PONNAMTHAKARA; M'. KALA-KASBINDA; O. KUSUNDA; T. PONNAVEERAN; T'. TAGARA

Cassytha filiformis = B. AKASHBEL; H. AMARBELI; K. AKASHA BALLI, MANGANA UDIDARA; K'. SORGVAL; M. AKASAVALL; M'. AKASHVALLI; O. AKASHA BELA; P. AMIL, AMARBELI

Casuarina equisetifolia (beef-wood trec) = A., B., H. & P. JHAU; G. VILAYATI SARU; K. SARVE MARA, GALI MARA; K'. SURICHEM-ZHADD; M. CHOOLA-MARUM, KATTADIMARUM; M', KHADSHERANI; O. JHAUN; T. SAVUKKU; T'. SARAVU

Celosia cristata (cook's comb) = A. KUKURA-JOA-PHUL; B. MORAG-PHUL; G. LAPADI; H. JATADHARI; K. MAYURA SHIKHI; K'. VILUD; M. KOZHIPULLU; M'. KOMBADA; O. GANJACHULIA; P. KUKURPHUL; T'. KODI JUTTU TOTAKURA

Centella asiatica (Indian pennywort) = A. MANIMUNI; B. THULKURI; G. KAR-BRAHMI; H. & P. BRAHMIBOOTI; K. ONDELAGA, BRAHMI SOPPU; K'. BRAHMI; M. KODANGAL, KOTAKAN; M'. BRAHMI; O. THALKUDI; T. VULLARAI

Cestrum nocturnum (queen of the night) = A. & B. HAS-NA-HANA; H. RAT-KI-RANI; K. RATRI RANI HOOVU; K'. RATRICHI-RANI

Chrysopogon aciculatus (love thorn) = A. BONGUTI; B, CHORKANTA; K. GANJI-GARIKE HULLU; K'. CHOR-KANTO; O. GUGUCHIA; P. CHORKANDA

Cicer arietenum (Bengal gram) = A. BOOT-MAH, B. CHHOLA; G. H. & P. CHANA; K. KADALE, CHANNA; K'. CHONE; M. & T. KADALAI; M. HARABHARA; O. BUTA; T'. SENAGALU

Cinnamomum camphora (camphor) = A. & B. KARPUR; G., H. & M'. KAPUR; K. KARPURADA GIDA; K'. KAMFUR; M. KAPPURAVRIKSHAM; O. KARPURA; P. KAPUR; T. KARUPPURAM; T'. KAPPURAMU

Cinnamomum tamla (bay leaf) = A. TEJPAT, MAHPAT; B. TEJPATA; G. & H. TEJPAT; K. KADU DALCHINNII; K'. TEZPAT; M'. TAMAL; O. & P. TEJPATRA; T. TALISHAPPATTIRI; T', TALLISHAPATRI

Cinnamomum zeylanicum (cinnamon) = A., B., G., M'., O. & P. DALICHINI: H. DARCHINI; K. DALCHINNI, LAVANGA CHAKKE; K'. TIKHI; M. & T. ILLAVANGAM; T'. DALCHINA CHEKKA

Cissus quadrangularis = A., B. & H. HARHJORA; K. MANGARA VALLI, SANDU BALLI; K'. KAND-VAL; M. NILAM PARANTA; M'. KANDAWEL; O. HADAVANGA; P. GIDARDAE, DRUKRI; T. PIRANDAI; T'. NALLERU

Citrullus lanatus (*water melon*) = A. KHORMUJA; B. TARMUJ; G. KARIGU; H. & P. TARBUZA; K. KALLANGADI BALLI; K'. KALLING; M. & T. KUMMATTIKKALI; M'. KALINGAD; O. TARABHUJA

Citrus aurantifolia (sour lime) = A. NEMU-TENGA; B. KAGJI-NEBU; G. LIMBU; H. NIMBOO; K. NIMBE; K'. LIMBIN; M. CHERUNARAKAM; M'. KAGADI LIMBU; O. NARANGO; P. GALGAL; T. ELIMICHCHAM; T'. NARINJA

Citrus grandis (pummelo or shaddock) = A. REBAB-TENGA; B. BATABINEBU; G. OBAKOTRU; H. & P. CHAKOTRA; K. CHAKKOTHA; K'. TORANJ; M. BAMBLEE-NARAKAM; M'. PAPANAS; O. BATAPI; T. BAMBALMAS

Citrus reticulata (orange) = A. KAMALA-TENGA; B. KAMALA; G. SUNTRA; H. NARANGI; K. KITTALE; K'. SONTRA, TANJERIN; M. NARAKAM; M'. SANTRA; O. KOMOLA; P. SANGTRA; T. NARANGAM; T'. KAMALAPHALAM

Clitoria ternatea (butterfly pea) = A., B. & O. APARAJITA; G. GARANI; H. APARAJIT, GOKARNA; K. GIRI KARNIKE, SATUGADA GIDA; K'. GOKARNI, SHANKA-PUSHPA; M. SANKHUPUSHPAM; M'. GOKARNA; P. APARIJIT, NILU LOFL; T. KAKKATAN; T'. SANKHAM PUVULU

Coccinia indica = A. BELIPOKA; B. TELAKUCHA; H. KUNDARU; K. THONDE KAYI, KAGE DONDE; K'. TENDLI; M. KOVEL; M'. TONDALE; O. KUNDURI, KAINCHIKA-KUDI; P. GHOL; T. KOVARAI; T'. KAKIDONDA

Colocasia esculenta (taro) = A. & B. KACHU; H. KACHALU, ARVI; K. KESAVINA GEDDE, SAVE GEDDE; K'. TEREM; M. CHEMPU; M'. KASALU; O. SARU; P. KACHALU; T. SAMAKILANGOO; T'. CHEMA

Coriandrum sativum (coriander) = A., B., O. & P. DHANIA; G. DHANE; K. KOTHAMBARI, HAVEEJA; K'. KOTHIMBIR, DANIA; M. & T. KOTTAMALLI; M'. KOTHIMBIR; T'. DHANIYALU

Crotalaria juncea (Indian or sunn hemp) = A. SHON; B. SHONE; G., H. & P. SAN; K. APSENABU, SANNA SENABU; K'. KHULKHULO; M. THANTHALAKOITI; M'. KHULKHULA; O. CHHANAPATA; T. SANAPPAI; T'. JANNAMU

Crotalaria sericea (rattlewort) = A. GHANTA-KORNA; B. ATASHI; H. JHUNJHUNIA; K. GIJIGIJI GIDA; K'. GHAGRI: M. THANTHALAKOTTI; M'. GHAGRI; O. JUNKA; P. JHANJHANIAN

Cucumis melo (melon) = A. BANGI; B. PHUTI; G. TARBUCH; H. & P. KHARBUZA, PHUTI & KAKRI; K. KARABUJA, KEKKARIKE; K'. CHIBUD, CHIBOLL, ME LANV; M. & T. THANNIMATHAI; M'. KHARBUJ; O. KHARBUJA

Cucumis sativus (cucumber) = A. TIANH; B. SASHA; G. KAKRI; H., M. & P. KHIRA; K. SOUTHE KAYI; K'. TOUXIN; M. MULLENVELLARI, O. KARUUDI; T. MULLUVELLARI; T'. DOSAKAYA

Cucurbita moschata (sweet gourd) = A. RONGA-LAU; B. MITHA-KUMRA; H. MITHAKADDU; K. SEEGUMBALA; K'. DUDIN; M. MATHANGAI; M'. KALA BHOPALA; O. MITHA KOKHARU; P. HALWA-KADDU; T. POOSANIKAI

Curcuma longa (turmeric) = A. HOLODHI; B. HALOOD; B. & M'. HALAD; H. & P. HALDI; K. ARISINA; K'. HOLLDI, GHOR HOLLOD; M. MANGAL; O. HALADI; T. MANJAL; T'. PASUPU

Cuscuta reflexa (dodder) = A. AKASHI-LOTA, RAVANAR-NARI; B. SWARNALATA; G. AKASWEL; H. AKASHBEL; K. BADANIKE, BANDALIKE, MUDITALE; K. AMARVAL; M'. AMAR VEL; O. NIRMULI; P. AMARBEL

Cynodon dactylon (dog grass) = A. DUBORI-BON; B. DURABAGHAS; G. DURVA; H. & P. DOOB; K. GARIKE HULLU, KUDIGARIKE; K'. HARIALI, DHURVA; M. & T. ARUGAMPULLU; M'. HARALI; O. DUBA GHASA; T', GERICHA GADDI

Dalbergia sissoo (Indian redwood) = A. SHSHOO; B. SISSOO; G. SHISHAM; H. & P. SHISHAM, TAHLI; K. BIRADI, BINDI, SHISSU; K'. BIRONDI; M. VEETI; M'. SHISAVI. O. & T. SISU

Datura fastuosa (thorn-apple) = A. SHISHOO; B. SISSOO; G. SHISHAM; H.& P. DHUTURA; K. DATTURA, UMMATTI; K'. DUTRO; M. UMMAM M'. DHOTRA; O. DUDURA; T. OOMMATHAI; T'. UMMATHA

Daucus carota (carrot) = A., B., G., H. & M. GAJAR; K. GAJJARI; O. GAJOR; T. GAJJARA KELANGU; T'. GAJARAGEDDA

Delonix regia (gold mohur) = A. RADHACHURA; B. KRISHNACHURA; G., H., M'. & P. GULMOHR; K. SFEME SANKESWARA, KEMPU TURAI; K'. GULMOHR; M. MARAMANDARAM; O. RADHACHUDA; T. MAAYILKONNAI

Dillenia indica = A. OUTENGA; B., H.& P. CHALTA; G. CARAMBAL; K. MUCHHILU, KATLEGA; K'. KARAMAL; M. VALLAPUNNA; M'. KARAMAL; O. OU; T. UVATTEKU; T'. UVVA

Dioscorea bulbifera (wild yam) = A. GOCH-ALOO; B. GACHH-ALOO; G. SAURIYA; H. & P. ZAMINKHAND; K. HEGGENASU, KANTA GENASU; K'. KARANDEM, CHIRPUTTAM-VAL; M. KATTUKACHIL; M'. KADU KARANDA; O. DESHI-ALOO, PITA-ALOO; T. KATTUKKILANGU; T'. PENDALAMU.

Dolichos lablab (country bean) = A. UROHI; B. SHIM; G. AVRI; H. & P. SEM; K. AVARE BALLI; K'. VATTANNO; M. SIMAPAYARU; M'. PAVATA, VAL, O. SIMA; T. AVARAI; T'. CHIKKUDI

Duranta plumieri = A. JEORA-GOCH; B. DURANTA -KANTM; H. & P. NILKANTA; K. DURANTHA KANTI; K'. RAMI; M'. DURANTA; O. BILATI KANTA, BEMJUATI

Eclipta alba = A. KEHORAJI: B. KESARAJ; G., H. & P. BHANGRA; K. GARUGADA GIDA, GARUGALU; K'. BHONGRO; M. & T. KAYYANYAM, KAITHONNI; M'. MAKA; O. KESHDURA; T'. GUNTAGALIJERU, GALAGARA

Eleusine coracana = B. MARUA; G. NAVTO; H. & K. RAGI; K'. NACHNI; M. MUTHARI; M'. NACHANI; O. MANDIA; P. KODRA. MANDWA; T. KELVARAGU; T' RAGI, CHOLLU

Emblica officinalis, see *phyllanthus*

Enhydra fluctuans (water cress) = A. HELACHI-SAK, MONA-SAK; B. & P. HALENCHA; H. HARUCH; K'. HARUCHZHADD; M'. HARKUCH; O. HIDIMICHI, PANI SAGA

Entada gigas (nicker bean) = A. GHILA; B., H., O. & P. GILA; G. SUVALI-AMLI; K. GARDALA, HALLEKAYI BALLI; K'. GORBI, M. KAKKUVALLY; M' GARBI; T. CHILLU; T'. GILLATIGAI

Enterolobium, see *Pithecolobium*

Ervatamia coronaria = A. KOTHONA-PHUL; B. & M'. TAGAR; H. & P. CHANDNI; K. NANDI BATLU, NANJA BATLU; K'. NINTIZHADD, ANAND, M. & T. NANTHIAR, VATTAM; O. TAGARA; T'. NANDIVARDHANAMU

Erythrina variegata (coral tree) = A. MODAR; B. MANDAR; G. PANARAWAS; H. PANJIRA; K. HARIVANA, VARJIPE; K'. PONGRA; M. & T. MURUKKU; M'. PANGARA; O. PALDHUA; P. DARAKHT FARID, PANGRA; T'. BADITA CHETTU

Eugenia, see *Syzygium*

Euphorbia antiquorum = B. BAJBARAN; G. TANDHARI; K. BONTE GALLI, CHADARA GALLI; K'. NILKANTTI, NIVALKANI; M. CHATHHIRAKKALLI; M'. CHAUDHARI NIWDUNG; O. DOKANA SIJU; P. DANDA THOR, TIDHARA SEHUD; T. SHADRAIKALLI; T'. BONTHAKALI

Euphorbia neriifolia = A. SIJU; B. MANSHASIJ; G. THOR; H. SIJ; K. ELE GALLI; M. & T. ILAKKALLI; K'. MONXA-SIZ; M'. CHAUDHARI NIWDUNG; O. PATARA SIJU; P. GANGICHU; T. AKUJEMUDU

Euphorbia pulcherrima (poinsettia) = A. LALPAT; B., M'. & P. LALPATA; K. POINSETTIA GIDA; K'. TAMBDDEM-PAT; O. PANCHUTIA; P. LAL-PATTI; T. MAYILKUNNI

Euryale ferox = A. NIKORI; B. H. & P. MAKHNA; M'. PADMA KANT, MAKHANU; O. KANTA PADMA; T'. MELLUNIPADAMANU

Feronia limonia, see *limonia*

Ferula assa-foetida (asafoetida) = A., B., G. H:, M'. HING; K'. INGU, HINGU; K'. ING; M. KAYAM; O. HENGU; T'. INGUVA

Ficus bengalensis (banyan) = A. BOR-GOCH; B. BOT; H. & P. BARH; G. & M. WAD; K. AALADA MARA; K'. VODD, M. PEERALU; O. BARA; T. AADUMARAM; T'. MARRI

Ficus carica (fig) = A. DIMORU; B. DUMUR; G. UMBARO; B. & P. ANJIR; K. ATHI; K' ANJIR; M & T. ATHTHYMARAM; M'. UMBAR; O. DIMURI; T'. BODDA

Ficus religiosa (peepul) = A. ANHOT; B. ASWATTHA; G. JARI; H. & P. PIPAL; K. ARALI, ASWATHA; K'. PIMPOLL; M. ARAYALU; M'. PIMPAL; O. ASWATHA; T. ARASU; T. ASWATHAM

Foeniculum vulgare (ansi or fennel) = A. GUAMOORI; B. PANMOURI; G. WARIARI; H. & P. SAUNF; K. DODDA JEERIGE, DODDA SOMPU;

K'. BODDIXEP, FUNCH VERDOXE; M'. BADISHEP; O. PAN MOHURI; T', PEDDA JEELAKARRA

Gardenia jasminoides (cape jasmine) = A. TOGOR; B., H. & P. GANDHARAJ; G. DIKAMALI; K. SUVASANE MALLE; K'. JASMIN; M'. GANDHRAJ; O. SUGANDHARAJ

Gloriosa superba (glory lily) = A. & B. ULATCHANDAL; G. & M'. KHADYANAG; H. KALIARI, KULHARI; K. SHIVASHAKTI, LANGULIKE; K'. VAGABOSKHE, VAGA-CHEODAULEO; M. MANTHONNI, PARAYANPOOVA; O. PANCHAANGULIA; P. GURH-PATNI, KULHARI; T. KALAPAL-KILANGU; T'. AGNISIKA

Gossypium sp. (cotton) = A. KOPAH; B., H. & P. KAPAS; G. RUI; K. KATHI; M. KURUPARATHY; M'. KAPUS; O. KOPA; T. PARATHY; T'. PRATTI

Gyanandropsis gynandra = A. BHUTMULA; B. HURHURE; G. ADIYAKHARAM; H. HURHUR; K. NARAMBELE SOPPU; K'. HURHURO; M. KATTUKATUKU; M'. TIL-VAN; O. ANASORISIA, SADA HURHURIA; P. HULHUL; T. NAIKA-DUGU; T'. VAMINTA

Helianthus annuus (sunflower) = A. BELI-PHUL; B. & O. SURJYAMUKHI; G. SURYAMUKHI; H. & P. SURAJMUKHI; K. SURYAKANTHI; K'. SURIAFUL; M. & T. SURIYAKANTI; M'. SURYAPHUL; T'. SURYAPHUL; T'. SURYAKANTAN

Hibiscus esculentus, see *Abelmoschus*

Hibiscus mutabilis = A. & B. STHAL-PADMA; G. UPALASARI; H. GULAJAIB; K. BETTA DAVARE, KEMPU SURYAKANTHI; K'. TAMBODI-BHENDDI; M. CHINAPPARATTI; M'. GULABI BHENDI; O. THALA-PADMA; P. GUL-AJAIB; T. SEMBARATTAI

Hibiscus rosa-sinensis (china rose or shoe-flower) = A. JOBA; B. JABA; G. JASUNT; H. GURHAL, JASUM; K. KEMPU DASAVALA; K'. DOXIN, LAMPIANY; M. CHEMPARATHY; M'. JASWAND; O. MANDARA; P. GURHAL, JIA PUSHPA; T. SAMBATHOOCHEDI; T'. DASANI

Hibiscus sabdariffa (rozelle) = A. MESEKA-TENGA; B. KURCHI; H. & P. PATWA; K. KEMPU PUNDRIKE; K'. TAMBDDI-AMBADI; M. PULICHI; M'. LAL-AMBADI; O. KHATA KAUNRIA; T'. SEEMA GONGURA

Hiptage bengalensis (=*H. madablota*) = A. MADHOILOTA; B. & O. MADHABILATA; G. MADHAVI; H. MADHULATA; K. MADHAVI LATHE; K'. HALADA-VAL, DIS-DUSTI; M. SITAPU; M'. MADHUMALATI; P. MADHULATA, BANKAR; T. KURUK-KATN MADAVI

Holarrhena antidysenterica = A. DUDKHORI; B. KURCHI; G. INDRA-JAVANU; H. KUTAJ, KARCHI; K. KODACHAGA, KODAMURUKA, KORJU; K'. KUDO; M. KODAKAPPALA; M'. KUDA; O. PITA KORUA; P. INDER JAU, KAWAR, T'. AN-KUDU, KODAGA

Hydrocotyle, see *Centella*

Impatiens balsamina (balsam) = A. DAMDEUKA; B. DOPATI; H. GULMENDI; K. GOURI HOOVA, BASAVANA PADA; K'. CHIDDO, BALSOM; M. & T., BALSAM; M'. TERADA; O. HARAGOURA; P. MAJITI, BANTIL, PALLU; T'. GORINTAPOOLA-CHETTU

Ipomoea aquatica = *I. reptans* (water bindweed) = A. KALMAU; B. & H. KALMISAK (-G); G. NALINIBHAJI; K. BILI HAMBU; K'. NALINI-BHAJI; M. KALAMBI, NAL; O. KALAMA SAGA; P. NALI, KALMI SAG; T'.TUTICURA

Ipomoea batatas (sweet potato) = A. & B. MITHA-ALOO; G. SHAKKARIA; H. & P. SHAKARKAND; K. GENASU; K'. KONGAM-ZHADD; M. MADHURAKI ZHANGU; M. BATALA; O. CHINI-ALOO, KANDA-MULA; T'. KANDAMOOLA

Ixora coccinea = A. & B. RANGAN; H. GOTAGANDHAL, RANJAN; K. MALE HOO GIDA, KEPALE; K'. PILKOLI, PITKOLI; M. & T. CHETHTHY; THETTY; M'. MAKADI; O. KHADIKA PHULA, RANGANI; P. RUNGAN; T'. BANDHU, KORANI, TOGARU

Jasminum sambac (jasmine) = A. JUTI-PHUL; B. BELA; H. MUGRA; G. BAT-MOGRI; K. GUNDU MALLIGE; K'. MOGRIN, MOGRA; M. MULLA, M'. MOGARA; O. NALI BAIGABA, VERENDA; T. ADALAI; T'. NEPALEMU

Jatropha gossypifolia = A. BHOTERA; B. H. & P. LALBHARENDA; K. CHIKKA KADU HARALU; M'. VALAVATI ERAND; O, NALI BAIGABA; VERENDA; T. ADALAI; T'. NEPALEMU.

Jussiaea repens = A. TALJURIA; B. KESSRA; K. NEERU DANTU, KAVAKULA, K'. KALAFUR-PAN; M. NIRGRAMPU; M'. PAN LAWANG; T. NIRKIRAMPU; T'. NIRU-YAGNIVENDRAMU

Lagenaria siceraria (bottle gourd) = A. JATI-LAU; B. & O. LAU; H. LAUKI; K. EESUGAYI BALLI, HALU GUMBALA; K'. TAMBDDO-DUDHI, BOBRO; M. & T. CHORAKKAI; M'. DUDHYA BHOPALA; P. GHIYA; T'. ANAPA

Lagerstroemia speciosa = A. AJAR; B., H. & P. JARUL; K. HOLE DASAVALA, CHELLA, BENDEKA; K'. TAMAN; M. NIRVENTEKKU; M'. TAMAN; O. PATOLI; T. PUMARUTHU

Lantana indica (lantana) = G. GHANIDALIA; K. LANTAVANA GIDA, HESIGE HOOVA; K'. GHANNIARI; M. PUCHEDI; M'. GHANERI; O. NAGA-AIRI; P. DESI LANTANA; T. ARIPPU; T'. LANTANA

Lathyrus sativus = A. KOLA-MAH; B., H. & O. KHESARI; G. MATER; K. CHIKKA TOGARI, VISHA TOGARI, KESARI BELE; K'. FEIJANV; M'. LAKH; P. KISARI DAL

Lens culinaris (lentil) = A. MOSOOR-MAH; B., H., M'. & P. MASUR; G. MASURIDAL; K. MASURU BELE, LENTEL GIDA; K'. MUSURIDAL; O. MASURA; T. MUSURLU

Leonurus sibiricus = A. RONGA-DORON; B. DRONA; H. HALKUSHA, GUMA; K'. DRON; O. KOILEKHIA; T'. ENUGUTUMNI

Lepidium sativum (garden cress) = A. & B. HALIM-SAK; G. ASALIYA; H. HALIM; K. KURTHIKE, KURATHIRUGI; M'. ALIV; O. HIDAMBA SAGA; P. HALON; T'. ADIYALU, ADELI

Leucas linifolia = A. DORON, DURIJM-PHUL; B. SWET-DRONA; G. JHINA-PANNI KUBO, H. CHOTA-HALKUSA; K. GANTU THUMBE, KARJAL GIDA; K'. TUMBO; M. THUMPA; M'. DRONAPUSHPI, GUMA; O. GAISA; P. GULDODA; T. THUMBAI; T'. TAMMA CHETTU

Limonia acidissima (elephant-apple) = A. & B. KATH-BAEL; G. KOTHA; H. & P. KAITHA; K. KADU BILVA PATRE, NAYI BELA; K'. HOTI-BEL; M. BLANKA; M'. KAWATH; O. KAINTHA; T. VELAMARUM; T'. VELAGA

Linum usitatissium (linseed) = A. TICHI; B. TISHI; G. JAVA; H. & P. ALSHI; K. SEEME AGASE BEEJA; K'. SONBIA; M'. JAWAS; O. PEPSI; T. AALI-VIRAI

Loranthus longiflorus = A. ROGHU-MALA; B. MANDA; G. VANDO; H. BANDA; K. SIGARE BADANIKE; K'. DAKTTI-BENDOLL; M. ITHTHIL; M'. BANDFUL; O. MALANGA, MADANGA; P. PAND; T. PULLURUVI; T'. BAJINNIKI, BADANIKA

Luffa acutangula (ribbed gourd) = A. JIKA; B. JHINGA; G. SIROLA; H. & P. KALITORI; K. HEERE BALLI; K'. XIRGONSALLEM; M. PEECHIL, PEECHINGAI; M'. DODAKA; O. JAHNI; T. PEECHANKA; T'. BEERAKAYA

Luffa cylindrica (bath sponge or loofah) = A BHOL; B. DHUNDULI H. & P. GHIYATORI; K'. GONSALLEM; M'. GHOSALE;' O. PITA TARADA; T'. GUTTI BEERAKAYA, NETI BEERAKAYA

Lycopersicum esculentum (tomato) = A. BELAHI-BENGENA; B. BILATI-BEGOON; H. & P. TAMBETA; G. TAMETA, TOMATO; K. TOMATO; M. & T. THAKKALIK-KAI; M'. TAMBETA; O. BILATI BAIGANA; T'. THAKKAILI, TAMATA

Mangifera indica (mango) = A., B. & H. AAM; G. & O. AMBO; M. MAVU; T. MAMARAM. MANGA; T'. MAMIDI, AMRAMU

Martynia annua = *M. diandra* (tiger's nail) = A. & B. BAGH-NAKHI; G. VICH-CHIDA; H. SHERNUI; K. HULI NAKHA, GARUDA MOOGU; K'. VAGA-NAKXEO; M. & T. KAKKACHUNDU, PULINAGAM; M'. WINCHAURI; O. BAGHA NAKHI; P. HATRAJORI; T. GARUDA MUKKU

Mentha viridis (spearmint or garden mint) = A. PODINA; B., G., H., M. & O. PUDINA; M. PUTIYINA; T'. PODINA, PUDINA

Michelia champaca = A. & P. CHAMPA-PHUL; B. SWARNACHAMPA; G. RAE CHAMPAC; H. CHAMPAK; K. SAMPIGE; K'. CHAMPO, CHAMPEACHEMZHADD; M. & T. CHEMPAKAM; M'. SONCHAPHA; O. CHAMPA; T'. SAMPAKA

Mimosa pudica (sensitive plant) = A. LAJUKI-LOTA; B. LAJJABATI-LATA; G. LAJJAWANTI; H. & P. LAJWANTI; K. MUTTIDARE MUNI, MUDUGU DAVARE, K'. LOJEHOKOL; M. THOTTALVADI; M'. LAJALU; O. LAJAKULI, LAJKURI; T. THOTTA-SINIGI; T'. PEDDA NIDRAKANTHA

Mirabilis jalapa (four o'clock plant or marvel of Peru) = A. GODHULI-GOPAL; B. KRISHNAKOLI; H. GULABBAS; K. SANJE MALLIGE, GULBAKSHI; BHADRAKSHI; K'. MERENDAM, SHABDULI; M. NALUMANICHEDI; M'. GULBAKSH; O. RANGANI, BADHULI; P. GUL-E-ABSASI; T. ANDIMANDARAJ; T'. CHANDRAKANTHA

Momordica charantia (bitter gourd) = A. TITA-KERALA; B. KARALA, UCHCHE; G., H. &. P. KARELA; K. HAGALALAYI; K'. KARATEM; M. & T. PAVAL, PAVAKKAI; M'. KARLE; O. KALARA; T'. KAKARA

Moringa oleifera (drumstick or horse radish) = A. & O. SAJANA; B. SAJINA; G. SARAGAMA; H. SAINJNA; K. NUGGE MARA, MOCHAKA MARA; K'. MOXING; M. MURINGA; M'. SHEVAGA; P. SAONJNA; T. MURUNGAI; T. MUNAGA

Morus alba and *M. nigra* (mulberry) = A. NOONI; B. TOONT; G. TUTRI; H. & P. SHAH-TOOT; K. KAMBALI GIDA, RESHME HIPPALI GIDA; K'. AMOR; M. MALBERRY; M'. TUTI; O. TUTAKOLI; T'.. POOTIKAPALLU

Mucuna pruriens (cowage) = A. BANDAR-KEKOA; B. ALKUSHI; G. KIVANCA; H. & P. KAWANCH; K. NASAGUNNI, NAYI SONKU BALLI; K'. KATKUTLLI, KASKULLI; M. NAI-KORUNA; M'. KHAJ KUIRA; O. BAIDANKA

Murraya paniculata (Chinese box) = A. KAMINI-PHUL; B., H. & O. KAMINI; K. KADU KARI BEVU, ANGARAKMA GIDI; K'. KARBEL, KARIAPAT; M. MARA-MULLA; M'. PANDHARI KUNTI; P. MARUA; T. KATTUKARUVEPPILAI; T'. NAGA-GOLUGI

Musa paradisiaca (barnana) = A. KOL; B. KALA; G. & H. KELA; K. BALE GIDA, BALE HANNU; K'. KEL; M. VAZHA; M'. KADALI, KEL; O. KODOLI, ROMBHA; T. VAZHAI; T'. ARATI KADALI

Nelumbo nucifera = *Nelumbium speciosum* (lotus) = A. PODUM; B. & O. PADMA; G. & M'. KAMAL; H. & P. KANWAL; K. KAMALA, TAVARE; K'. DO-VEM-SALLOK; M. THAMARA; T. THAMAṚAI; T'. TAMARA

Nerium indicum (oleander) = A. KORBI-PHUL; B. KARAVI; G. & M'. KANHER; H. & P. KANER; K. KANIGALU; K'. GUL-BAKAULI, TAMBDDO-KANER; M. & T. ARALY; O. KARABI; T'. GANNERU

Nyctanthes arbortristis (night jasmine) = A. SEWALI-PHUL; B. SHEWLI, SHEPHALI; G. RATRANE; H. HARSHINGAR; K. PARIJATA; K'. PARDAK; M. PAVIZHA MULLA; M'. PARIJATAK; O. SINGADHARA; P HARSANGHAR; T. PAVELAM; T'. PARIJATHAM

Nymphaea lotus (water lily) = A. BHET; B. SHALOOK; G. NILOPAI; H. & P. NILOFAR; K. KENDAVARE, KANNAIDILE; K'. XALLOK, KAMAL; M. & T. AMPAL; M'. LALKAMAL; O. KAIN, KUMADA; T'. KALTTVA

Ocimum sanctum (sacred basil) = A. TULASHI; B., G., H. & P. TULSI; K. SREE TULSI, VISHNU TULSI; K'. TULSI; M. & T. THULASI; M'. TULAS; O. TULASI; T'. LAXMI TULASI, ODDHI

Oldenlandia corymbosa = B. & P. KHETPAPRA; G. PARPAT; H. DAMAN-PAPPAR; K. HUCHHU NELA BEVU, KALLU SABBASIGE; K', PARIPAT; M'. PITPAPADA; O. GHARPODIA; T. VERINELLAVEMU

Opuntia dillenii (prickly pear) = A. SAGORPHENA; B. PHANIMANSHA; G. NAG-NEVAL; H. NAGPHANI; K. PAPAS KALLI, CHAPPATE KALLI; K'. TAPTAM, PATYO; M. ELAKKALLI; M'. PHADYA NIWDUNG; O. NAGAPHENI; P. CHITARTHOR; T. SAPPATHTHIKKALLI; T'. BRAHMA JEMUDU

Oroxylon (=*Oroxylum*) *indicum* = A. BHAT-GHILA; B. SONA; G. PODVAL; H. ARLU; K. PATAGANI, SONEPATTA, TIGUDU; K'. TETU; M. PATHIRI; M. TETU; O. PHANPHANIA; PHAPANI; P. SANNA; T. PAYYALANTHA; T'. PAMPINI

Oryza sativa (paddy) = A., B. & H. DHAN; G. CHOKHA; K. BHATHA, NELLU; K'. BHAT; M. NELLU; M'. BHAT; O. DHANA; P. CHAWAL; T. ARISHI; T'. VARI, DHANYAMU

Oxalis repens = *O. corniculata* (wood-sorrel) = A. SENGAI-TENGA, TENGECHI; B. AMRULSAK; H. CHUKA-TRIPATI, KHATRI-PATTI; K. PUTTAM PURALE; K'. AMBOXI; M. PULIYARILA; M'. AMBOSHI; O. AMBILITI, AMLITI; P. KHATTI-BUTI; T'. SAVIRELA

Pandanus odoratissimus (screwpine) = A. KETAKIPHUL; B. & G. KETAKY; H. & P. KEORA; K. TALE HOOVU, KEDIGE; K'. ATTO, KEVDDO, KERKI; M. KAITHA; M'. KEWADA; O. KIA; T. THAZHAI; T'. MOGIL

Papaver somniferum (opium poppy) = A. AFUGOCH; B. AFING; G., H. & P. APHIM; K. GASA GASE, APPEMU GIDA; M. & T. GASHAGASHA; M'. APHU; O. APHIMA; T'. NALLAMANDU, GASAGASALU

Passiflora foetida (passion flower) = A. JUNUKA; B., H. & P. JHUMKALATA; K. KUKKI BALLI; K'. SAIBACHE-GAVE, CHAGAM-FUL, KRIXNA-KAMAL; M. KRISTHU-PAZHAM; M'. KRISHNA KAMAL; O. JHUMUKALATA; T. SIRUPPUNAIKKALI; T'. JUKAMALLE, TELLAJUMIKI

Pennisetum typhoideum (pearl millet) = B., H., O. & P. BAJRA; K. SAJJE, KAMBU; K'. BAZRI; M. KAMPAM, M'. BAJARI; T. KAMBU, BAJRA; T'. SAJJA, SAJJALU

Phaseolus aureus (green gram) = A. MOGU-MAH; B. & H. MOONG; G. MUGA; K. HESARU; K'. MUG, HIRVE-MUG; M. CHERUPAYARU; M'. HIRAVE MUG; O. JHALN-MUG; P. MUNG; T. PACHAPAYARU; T'. PESALU

Phaseolus mungo (black gram) = A. MATI-MAH-, B. MASH KALAI; G. URAD; H. URID; K. UDDU; K'. UDID; M. UZHUNNU; M'. UDID; O. MUGA; P. MASH; T. ULUNNU; T'. MINUGULU

Phyllanthus emblica = *Emblica officinalis* (emblic myrobalan) = A. AMLOKI; B. AMLA, AMLAKI; G. AMBALA; H. AMLIKA; K. NELLI-BETTADA NELLI, NELLI-ISNELLI; K'. ANVALLO; M. & T. NELLIKKAI; M'. AWALA; O. ANLA; P. AMLA; T'. USIRI

Piper betle (betel) = A., B., G., H. & P. PAN; K. VEELE DELE, YELE BALLI; K'. KHAVCHE-PANN, PATTI-PANN; M. & T. VETHILA; M'. NAGWELI; O. PANA; T'. TAMALAPAKU

Piper longum (long pepper) = A. PIPOLI; B. PIPOOL; G. PIPARA; H. PIPLI; K. HIPPALI; K'. PIMPOLI; M. THIPPALI; M', PIMPALI; O. PIPALI; P. DARFILFIL, MAGHAN; T'. PIPPALU

Piper nigrum (black pepper) = A. JALUK; B. GOLMARICH; G. GOLMIRCH; K. KARI MENASU; K'. MIRIAM, MIRIAKON, KALE-MIRIAM; M. KURU-MULAGU; M'. KALI MIRI; O. GOLA MARICHA; P. KALI MARCH; T. MILAGOO; T'. SAVYAMU

Pistia stratiotes (water lettuce) = A. BORPUNI; B. PANA; G. JALAKUMBHI; H. & P. JALKHUMBI; K. ANTARA GANGE; K'. GANGAVAJI; M, MUTTAPPAYAL, M'. GANGAVATI; O. BORA JHANJI; T. AGASATHAMARAI; T'. AKASATAMARA

Pisum sativum (pea) = A. MOTOR; B., H., O. & P. MATAR; G. VATANA; K. BATANI, VATANI; K'. VATTANO; M. PAYARU; M'. WATANE; T. PATTANI; T'. GUNDUSANIGHELU

Pithecolobium saman (rain tree) = A. SIRISH GOCH; K. MALE MARA; K'. SIRIX; M. URAKKAM-THOONGIMARAM; M'. SAMAN; O. BADA GACHA CHAKUNDA, BANA SIRISHA

Plumbago zeylanica = A. AGYACHIT; B. CHITA; G. CHITRAMULA; H., M'. & P. CHITRAK; K. BILI CHITRA MOOLA; K'. KODU-VAL. M. & T. KODUVELI; O. DHALACHITA

Plumeria rubra (temple tree) = A. GULANCHI; B. KATGOLAP; G. RHAD-CHAMPO; H. & P. GOLAINCHI; K. HALU SAMPIGE; K'. KHUR CHAMPO; M. EEZHAVA-CHEMPAKAM; M'. KHUR CHAPHA; O. KATHA CHAMPA; P. GULCHIN; T . ERRA NOOROO, VARAHALU

Poinciana pulcherrima (dwarf gold mohur or peacock flower) = A. KRISHNA- CHURA; B. RADHACHURA; O. SANDHESHRO; H. GULETURA; K. KENJIGE GIDA, RATNA GANDHI; K'. KOMBEA-XENKAREM, SHANKARACHEM-ZHADD; M. RAJMALI; M'. SHANKASUR; O. KRUSHNACHUDA, GODIBANA; P. KRISHANACHURA; T. MAYIRKONRAI; T'. TURAYI

Polianthes tuberosa (tuberose) = A., B. & O. RAJANIGANDHA; H. & P. GUL-SHABO; K. SUGANDHA RAJA, NELA SAMPIGE, SANDHYA RAGA; K'. DOVI-LILI, GULCHHADI; M'. GULCHHADI; T. NILASAMANGI; T'. SUKANDARAJI

Polyathia longifolia (mast tree) = A. & O. DABADARU; B. DEBDARU; G. ASHO PALO; H. & M'. ASHOK; K. PUTRAJEEVI, KAMBADA MARA; K'. AXOK; M. ARANMARAM; P. DEVIDARI; T. NETTILINGAM; T'. ASOKAMU, DEVADARU, NARAMAMIDI

Polygonum sp. = A. BIHLONGONI; B. PANIMARICH; K'. UDKA-MARIX; M. MOTHALA MOOKA; O. MUTHI SAGA; P. NARRI, T. AATALARIE

Portulaca oleracea (purslane) = A. HANHTHENGIA; B. NUNIA-SAK; G. LONI; H. & P. KULFA-SAG; K. DODDA GONI SOPPU; K'. GHOLCHI-BHAJI, GHOL-BHAJI; M. KARICHEERA; M'. GHOL, O. BALBALUA; T. KARIKEERAI; T'. PEDDAPAVILI KURA

Pothos scandens = A. HATI-LOTA; G. MOTO PIPAR; K. ADKE BEELU BALLI, AGACHOPPU; K'. ANJAN-VAL; M. ANAPPARUVA; M'. ANJAN VEL; O. GAJA PIPALI; P. GAZPIPAL

Prosopis spicigera = A. SOMIDH; B., H., M'. & O. SHOMI; G. KANDO; K. VUNNE, PERUMBE; K'. XOMBI; M. PARAMPU; P. JAND; T. PERUMBAI; T'. JAMBI

Psidium guayava (guava) = A. MODHURI-AM; B. PAYARA; G. JAMFAL; H. & P. AMRUD; K. SEEBE, CHEPE, PERALA; K'. PER; M. PERAKKA; M'. PERU; O. PIJULI; T. KOYYA; T'. JAMA

Pterospermum acerifolium = A. KONOKCHAMPA; B. MOOCHKANDA; H. KANAKCHAMPA; K. MUCHUKUNDA GIDA; K'. MOOCHKUND; M'. MUCHKUND; O. MOOCH-KUNDA; T. VENNANGU; T'. MAISAKANDA, MUSHKANDA

Punica granatum (pomegranate) = A. & B. DALIM; G. DADAM; H. & P. ANAR; K. DALIMBE; M. MATALAM; M'. DALIMB; O. DALIMB; T. MADULAM; T'. DANIMMA

Quamoclit pinnata (=*Ipomoea quamoclit*) = A. KUNJALOTA; B. KUNJALATA, TORULATA; H. & P. KAMLATA; K. KAMALATHE; K'. GONESPUSHPA, CHITRANJI; M'. GANESH PUSHPA; O. KUNJALATA; T'. KASIRATNAM

Quisqualis indica (Rangoon creeper) = A. MADHABI-LOTA; B. SANDHYA-MAIATI; G. BARMA SINIVEL; H. & P. LAL-MALTI; K. RANGOON KEMPUMALLE; K'. FIRNGUICHAMELI; M'. LAL CHAMELI; O. MODHUMALATI; T. RANGOON MALLI; T'. RANGOONMALLE

Raphanus sativus (radish) = A., B., M'. & O. MULA; H. & P. MULI; K. MOO-LANGI; K'. MULLO; M. MULLANKI; T. & T'. MULLANGI

Rauwolfia serpentina = A. CHANDO; B. CHANDRA, SARPAGANDHA; G. SARPA-GANDHA; H. & P. SARPGAND, D. CHOTA CHAND; K. SARPAGANDHI, SHIVANABHI BALLI, SUTRANABHI; K'. SORPA-GANDH; M. AMALPORIYAN; M'. SARPAGANDH; O. PATALA GARUDA; T'. DUMPARASANA

Ricinus communis (castor) = A. ERI-GOCH; B. & P. ARANDA; G. ERANDI; H. RENDI; K. HARALU; K'. ENDELL; M. & T. AVANAKKU; M'. ERAND; O. JADA; P. RENDI, ARANDA; T'. AMUDAMU

Rumex vesicarius (dock) = A. CHUKA-SAK; B. CHUKA-PALANG; H. CHUKA, KHATTA-PALAK; K. CHUKKI SOPPU, SUKKE SOPPU; M'. CHUKA; O. PALANGA; P. KHATTA-MITHA; T'. CHUKKAKURA

Saccharum officinarum (sugarcane) = A. KUNHIAR; B. & H. AKH; G. SNERDE; K. KABBU; K'. US; M. & T. KARIMPU; M'. USA; O. AKHU; P. GUNNA; T'. CHERUKU

Salmalia malabarica, see *Bombax ceiba*

Sansevieria roxburghiana (bowstring hemp) = A. GUMUNI; B. MURGA, MURVA; H. MARUL; K. MANJINA NARU, GODDTIMANJI; K'. MURVO, SORPAZHADD; M. PAMPINPOLA; O. MURUGA; T. MARUL

Saraca indica (asoka tree) = A., B. & P. ASOK; G. ASUPALA; H. SEETA-ASOK; K. ASHOKADA MARA, KENKALI, ACHANGE; K'. SITACHO-AXOK; M. & T. ASOKAM; M'. SITECHA ASHOK; O. ASOKA; T'. ASOKAMU

Sesamum indicum (gingelly) = A. TISI; B., H., M'. & P. TIL: G. MITHO-TEL; K. YELLU; K'. TIL, M. & T. ELLU; O. KHASA, RASHI; T'. NUVVULU

Sesbania grandiflora = A. & B. BAKPHUL; G. AGATHIO; H. & P. AGAST; K. AGASE, CHOGACHI; K'. BAKFUL, AGASTO; M. AGATHI; M'. AGASTA; O. AGASTI; T. AGATHYKKEERAI; T'. AVIS

Sesbania sesban = A. JOYANTI; B. JAINTI; G. RAYSANGANI; H. & P. JAINT; K'. JOYONTI, SEVORI; M. SHEMPA; M'. SEVARI; O. JAYANJI; T. SITHAGATHI

Shorea robusta (sal tree) = A., B., H. & P. SAL; G. RAL; K. BILE BHOGE, AASINA MARA, ASCHA, KARNA; K'. XALA; M. MARAMARAM; M'. SHALA, RALVRIKSHA; O. SALA; T. SHALAM; T'. GUGGILAMU

Sida cordifolia = A. BARIALA; B. BERELA; G. JANGLI-METH; H. BARIARA; K. HETHUTHI; K'. JONGLI-METHI; M. KURUMTHOTTI; M'. CHIKANA; O. BISIRIPI; P. KHARENTI; T. KARUMTHOTTEE; T'. CHIRUBENDA

Smilax zeylanica (Indian sarsaparilla) = A. HASTIKARNA-LOTA; B. KUMARIKA; H. CHOBCHINI; K'. UPRATTI-KANTTI, VOIRFATTI-KANTTI; M'. GHOT VEL; O. KUMBHATUA, KUMARIKA; P. USHBA; T'. KONDATAMARA, PHIRANGI

Solanum nigrum (black nightshade) = A. POKMOU; B. GURKI; G . PILUDU; H. GURKAMAL MAKOI; K. KARI KACHI GIDA, KEMPU KACHI, KAKA MUNCHI; K'. KANGANI; M. MULAGUTHAKKALI; M'. KANGANI; O. NUNNUNIA; P. MAKO; T. MANATHAKKALI; T'. KAMANCHICHETTU

Solanum tuberosium (potato) = A., B., H., O. & P. ALOO; G. PAPETA; K. ALUGEDDE; M. & T. URULAKKIZHANGU; M'. BATATA; T'. URULAGDDA, BAGALADUMPA

Sorghum vulgare (great millet) = A. JOU-DHAN; B. & G. JUAR; H.& P. JOWAR; K. BILL JOLA; K'. ZOWAR; M. & T. CHOLAM; M'. JAWAR; O. BAJARA; T'. JONNALU

Tagetes patula (marigold) = A. NARJIPHUL; B. & H. GENDA; K. CHENDU HOOVA, SEEME SHAVANTIGE; M'. GULJAPHIRI; O. GENDU; P. GENDA, GUTTA; T'. BANTI

Tamarindus indica (tamarind) = A. TETELI; B. TENTUL; G. AMIL; H. & P. IMLI; K. HUNISE MARA; K'. & M'. CHINCH; M. & T. PULI; O. KAINYA, TENTULI; T'. CHINTHA

Tamarix dioica = A. JHAU-BON; B. & H. BON-JHAU; K. SEERA GIDA; K'. ZAO; M'. JAO; O. DISHI-JHAUN, THARTHARI; P. PILCHI

Tectona grandis (teak) = A. & B. SHEGOON; G. & H. SHAGWAN; K. TEGADA MARA, SAGUVANI; K'. SAYLO, SAYEL; M. & T. THEKKU; M'. SAG; O. SAGUAN; P. SAGWAN; T'. TEKU

Thespesia populnea (portia tree) = B. PARAS; G. PARUSA-PIPALO; H. & P. PARASPIPAL; K. BUGURI, HOOVARISI, JOGIYARALE; K'. BHENDI-ZHADD; M. & T. POOVARASU; M'. BENDICHA JHAR; O. HABALI; T'. GANGARAVI

Thevetia peruviana (yellow oleander) = A. KARABI; B. KALKE-PHUL; G., H. & P. PILA-KANER; K. KADUKASI KANAGALU; K'. ARKAFUL, KANER; M. & T. SIVANARALI; M'. PIWALA KANHER; O. KANIARA, KONYAR; P. ULA; T. PACHHA-GANNERU

Tinospora cordifolia = A. AMOR-LOTA, AMOI-LOTA; B. GULANCHA; G. GADO; H. GURCHA; K. AMRUTA BALLI, MADHU PARNI; K'. GUL-VAL; M. AMRITHU; M'. GULVEL; O. GULUCHI; P. GALE; T. SINDHILKODI; T'. TIPPATIGE

Trachyspermum ammi = *Carum copticum* (ajowan or *ajwan*) = A. JONI-GUTI; B. JOWAN; G. AJAMO; H. & P. AJOWAN; K. OMU, AJAWANA; K'. AZMO; M. AYAMODAKAM, M'. OWA; O. JUANI; T. OMAN; T'. OMAMU

Tragia involucrata (nettle) = A. CHORAT; B. BICHUTI; H. & P. BARHANTA; K. TURACHI BALLI, CHELURA GIDA; K'. KHAJAKOLTI; M. CHORIYANAM; M'. KHAJA-KOLTI; O. BICHHUATI; T. KANJURI; T'. DULAGONDI

Trapa natans (water chestnut) = A. SHINGORI; B. PANIPHAL; G. SHENGODA; H. & P. SINGARHA; K. MULLU KOMBU BEEJA; K'. UDKAFOL, XINGDO; M. KARIMPOLA; M'. SHINGADA; O. SINGADA; T. SINGARAKOTTAI; T'. KUBYAKAM

Tribulus terrestris = B. GOKHRIKANTA; G. GOKHARU; H. GOKHRU; K. SANNA NEGGILU; K'. GOKHRU; M. NERUNJIL; M'. KATE GOKHRU; O. GOKHARA; P. BHAKHRA; T. NERINJI; T'. PALLERU

Trichosanthes anguina (snake gourd) = A. DHUNDULI; B. CHICHINGA; G. PADA-VALI; H. CHACHINDA; K. PADAVALA; K'. PODIUM, PODOLEM; M. PADAVALAM; M'. PADVAL; O. CHHACHINDRA; P. PAROL; T. PUDALAI; T'. POTLA

Trichosanthes dioica = A. & B. PATAL; G. & M'. PARWAR; H. & P. PARWAL; K. KADU PADAVALA; K'. PARVAR; M. PATOLAM; O. PATALA; T. KOUMBUPPUDA-LAI; T'. KOMMUPOTLA

Triticum aestivum (wheat) = A. GHENHU; B. GOM; G. GAHUN; H. & P. GEHUN; K. GODHI; K'. GOUM, M. KOTHAMPU; M'. GAHU; O. GAHAMA; T. GHODUMAI; T'. GOTHI, GODUMULU

Typha elephantina (elephant grass) = B. HOGLA; G. GHABAJARIN; H. PATER; K. ANEJONDA, APU, NARI-BALA; M'. PAN KANIS; O. HAUDAGHASA, HOGOLA; P. PATIRA; T. CHAMBU

Typhonium trilobatum = A. SAMA-KACHU; B. GHET-KACHU; K. KANDA GEDDE; K'. KACHZHADD; M. CHENA; T. KARUN-KARUNAI, ANAIKKORAI; T'. JAMMUGADDI

Urena lobata = A. BON-AGARA; B. BAN-OKRA; H. & P. BACHATA; K. DODDA BENDE, KADU THUTHI; K'. THERVARE; M. OORPUM; M'. VAN-BHENDI; O. JAT-JATIA; T. OTTATTI

Utricularia sp. = B. JHANJI; K. NEERU GULLE GIDA, SEETHASRU BEEJA; K'. VONOSPATI; M. MULLANPAYAL, KALAKKANNAM; M'. GELYACHIVANASPATI; O. BHATUDIA DALA

Vanda roxburghi (orchid) = A. KOPOU-PHUL; B., H. & P. RASNA; G. RASNANAI; X. VANDAKA GIDA; K'. MOTTE-BENDOLL, ISPIRIT-SANT; M. MARAVAZHA; M'. BANDE; O. RASHNA, MADANGA

Vangueria spinosa = A. KOTKORA, MOYEN-TENGA; B. & H. MOYNA; K. CHEGU GADDE ACHHURA MULLU; K'. TERO, ALLU; M'. ALU; O. GURBELI; T. MANAKKARAI; T'. SEGAGADDA

Vigna sinensis = A. NESERA-MAH; B. BARBATI; H. BORA; K. ALASANDI; TADAGANI; K'. CHOVLI-VAL, OLSANDI; M' CHAVLI; O. BOR-GADA; P. PAUNG; T. THATTAPAYERU; T'. ALACHANDALU

Vinca rosea (periwinkle) = A. & B. NAYANTARA; H. SADABAHAR; K. KEMPUKASIKANIGALU, TURUKU MALLIGE; K'. PERPETIM, NITIPUSHPA, SODDAMFUL; M. KASITHUMPA; M'. SADAPHULI; O. SADABIHARI; P. RATTAN JOT

Viscum monoicum (mistletoe) = A. ROGHUMALA; B. BANDA; H. & P. BHANGRA, BANDA; K. HASARU BADANIKE; K'. BENDOLI; M. ITHTHIL; M'. JALUNDAR; O. MALANGA; T. OTTU

Vitis trifolia (wild vine) = B. AMAL-LATA; G. KHAT-KHATUMBO; H. & P. AMAL-BEL; K. NEERGUNDI, NOCHHI, NEERLAKKI; K'. DAKUVAL, NINGTTO. AMBOTT-VAL; M. SORIVALLI; M'. AMBATVEL; O. AMAR-LATA

Vitis vinifera (grape vine) = A., B., H. & P. ANGOOR; G. MUDRAKA; K. DRAK-SHI-BALLI; M. & T. MUNTHIRYVALLY; M'. DRAKSHA-VEL; O. ANGURA; T'. DRAKSHA

Wedelia calendulacea = A. BHIMRAJ; B. BHIMRAJ, BHRINGARAJ; G., H. & P. BHANGRA; K. KESHARAJA, GARGARI; K'. HOLDUVO-BHONGRO; M. PEEKAYYAN-NYAM; M'. PIVALA-BHANGRA; O. BHRUNGARAJA

Withania somnifera = A. LAKHANA; B. ASWAGANDHA; G. ASUNDHA; H. & P. ASGANDH; K. ASWAGANDHI, PENNERU, HIRE MADDINA GIDA; K'. ASKANDH; M. & T. AMUKKIRAM; M'. ASKANDH; O. AJAGANDHA; T'. ASVAGANDHI

Xanthium strumarium (cockle-bur)=A. AGARA; B. & H. OKRA; G. GADIYAN; K. MARALU UMMATHI; K'. XANKESHVOR; M'. SHANKESHVAR; O. CHOTA GOGHURU; P. GOKHRU KALAN; T. MARLUMUTTA; T'. MARULAMATHANGI

Zea mays (Indian corn or maize) = A. MAKOI-JOHA; B. BHUTTA; G. & P. MAKAI; H. MAKKA, BHUTTA; K. MUSUKINA JOLA, GOVINA JOLA; K'. BHUTTO, ZONNVE, MOKHO; M., M'. & T. MAKKACHOLAM; O. MAKKA; T'. MOKKAJONNA

Zingiber officinale (ginger) = A., B. & O. ADA; G. ADHU; H. ADRAK; K. SHUNTI, ALLA; K'. ALEM; M. INCHI; M'. ALE; P. ADARAK; T. INJI; T'. ALLAM

Zizyphus mauritiana = *Z jujuba* (Indian plum) = A. BAGARI; B. KUL, G. BORADI; H. & P. BER; K. ELACHI, BORE HANNU; K'. & M. BOR; M. & T. ELANTHAI; O. BARKOLI; T'. REGU

Index